Elastic Waves in Solids

Applications to Signal Processing

Elastic Waves in Solids

Applications to Signal Processing

E. Dieulesaint
Professeur à l'université Pierre et Marie Curie (Paris VI)

D. Royer
Maitre-Assistant à l'École Supérieure de Physique et de Chimie de Paris

translated by
A. BASTIN and M. MOTZ

A Wiley–Interscience Publication

JOHN WILEY & SONS
Chichester · New York · Brisbane · Toronto

Originally published under the title *Ondes elastiques dans les solides,*

British Library Cataloguing in Publication Data:

Dieulesaint, E.
Elastic waves in solids.
1. Elastic waves
2. Solids
I. Title II. Royer, D.
531'.33 QC176.8.W3 80-49980

ISBN 0 471 27836 X

Printed in Great Britain

Foreword

Ultrasonics is the phenomenon of sound waves propagating in solids and liquids at frequencies much higher than the audible range, but at less than about 100 MHz. This practical limit is due to the fragility of the emitter - the classical quarter wave quartz plate - which then becomes too thin, about one mil (25 μm). Ultrasonics has given rise to numerous and valuable applications, which were fully developed during the nineteen sixties: quartz filters for long distance telephony, sonar for marine exploration, memories for the half-frame of colour television, and so on.

Nowadays, scientists and engineers have used a new wide range of frequencies, from 100 MHz to a few GHz, thus implementing new sophisticated, but also useful, functions with "hypersonics": squeezing the width of radar pulses, modulating laser beams, or deflecting them rapidly.

All these recent achievements of science and technology require the use of single crystal chips of complicated atomic structure, and the use of unusual waves propagating in the bulk or along the surface of these anisotropic solids.

This book offers a clear and accurate introduction to the physics of these media and of their electromechanical waves. It starts with a few chapters on basic processes: signal processing, crystallography, and related mathematical topics, such as symmetry groups for crystals and tensor calculus. It deals in detail with the most important of the various waves, and goes on to the main aspects of the associated technology and to their application to devices.

The authors are well known for their applied research and achievements in this field. They have perfected this book while teaching the students in engineering science at the "Ecole de Physique et de Chimie de Paris" where the Curie brothers discovered piezoelectricity some one hundred years ago.

Paul GRIVET
Fellow IEEE

Acknowledgements

The authors would like to express their gratitude to Dr. A. Bastin and to Dr M. Motz who between them have carried the not inconsiderable burden of translating a text of this nature. They are grateful to P. Ravinet, who made a series of useful suggestions, and to Professor Eric Ash for his help and encouragement in implementing the English language edition.

We are also much indebted to Professeur P. Grivet for honouring us by contributing the preface to this volume.

Contents

Preface

This book is intended for senior students majoring in physics, and for graduate students, engineers and professors interested in new developments. It may be useful to explain at this stage the philosophy adopted by its authors.

The study of acoustic* wave propagation in anisotropic solids requires a basic knowledge of crystal symmetry and of the tensors which describe the response of the crystal to external excitations. A description of applications to signal processing often involves the use of Fourier transforms. Our experience with students and engineers from differing backgrounds has convinced us of the need to clarify these concepts. This accounts for the material covered in the first three chapters.

Chapter 4 contains the definition of the stress and strain fields, the calculation of the elastic potential energy and the enumeration of the components of the elastic tensors in crystals. It is not necessary, on a first reading, to follow every detail of the proofs relating to the reduction of these tensors.

Chapter 5 which deals with elastic wave propagation is fundamental. The analysis of propagation in unbounded media ends with the plotting of slowness surfaces of crystals such as silicon, rutile and sapphire. Starting from these examples the reader can find the solution in other practical cases. As regards bounded media, our

* In this book, the terms "elastic wave" and "acoustic wave" are used interchangeably.

study is concerned only with propagation on a plane surface (Rayleigh waves) and in a medium consisting of a layer on a substrate (Love waves). Consequently, the reflection or refraction of elastic waves is discussed under only a few simple conditions, subsequently useful for treating these two types of waves.

Waveguides are excluded from the scope of the book. In our opinion, the contents of Chapter 1 enable the reader to understand the waveguide applications of strip guided waves (Lamb waves), which are briefly mentioned in the last chapter.

Piezoelectricity modifies the propagation conditions and, under specific symmetry conditions, allows the propagation of a particular surface wave (Bleustein-Gulyaev waves). This is the subject of the second part of Chapter 6 whose first part is devoted to the definition of piezoelectric effects and their tensor description. Although the reader may leave out a few of the proofs, especially those establishing the restrictions imposed on the piezoelectric tensors by the crystal symmetry, it is important that he knows how to interpret the results summarized in the various tables.

Chapter 7 discusses elastic wave generation. Only the most commonly used piezoelectric transducers are described, namely the loaded resonator for bulk waves, and the interdigital transducer for surface waves. The algebraic formulation of the behaviour of the latter is complicated and, despite several simplifying assumptions, some of the calculations are tedious. Once the principle of Rayleigh wave generation is understood, a reader with only limited time may move directly on to the method of discrete sources.

Elastic waves interact with light waves. This property has been exploited, especially since the development of lasers, not only for the deflection and modulation of optical beams but also for the measurement of the characteristics of acoustic beams. This interaction is discussed in Chapter 8.

Applications of elastic waves to signal processing are described in the last chapter. Various functions are considered such as, time delay, pulse compression and bandpass filtering. The frequency domain ranges from a few megahertz to several gigahertz. This field is constantly evolving. Therefore our aim has been to illustrate principles and not to quote performances. Accordingly we have chosen photographs of delay lines operating at frequencies sufficiently low for details of the transducers to be visible.

Although the applications discussed refer to electronics, this book should be useful to engineers working with acoustic waves in other areas, such as metallurgy where they can be used for the detection of cracks and flaws (non-destructive testing), or in medicine where they can reveal foreign bodies or tumours.

We have had regretfully to omit a few subjects, such as the coupling between charge carriers and acoustic waves. However, the amplification phenomena which result from this coupling have not to date found any practical application.

As a complement to the text, exercises with detailed solutions are to be found at the end of each chapter.

Chapter 1

General Concepts. Waves. Signals. Linear Systems

This first chapter contains three main sections.

The first section recalls the characteristics of progressive plane waves, stationary waves, guided waves, and the definition of the group velocity of a wave packet.

The second section deals with elastic wave propagation and the derivation of the propagation equation for a continuous medium (i.e. a fluid) and for a discrete medium (i.e. a solid at the atomic scale). The aim of Section II is to explain the viewpoint which has been adopted throughout this book, namely that the propagation medium of elastic waves is a solid, usually anisotropic (crystalline); however it is always considered to be continuous and intrinsically non dispersive since the frequency range we are studying, below 10^{10} Hz, is far below the cut-off frequency of crystals.

The third section is concerned with the spectral decomposition of a signal and explains the link between the frequency response and the impulse response of a linear time-invariant system. These concepts, related to Fourier transform techniques, are prerequisites for Chapter IX which describes the applications. The treatment of the properties of the Dirac function and of Fourier transform is not mathematically rigorous. Using distribution theory and detailing convergence criteria would have distracted the reader from the subject, which is the connection between time analysis and spectral analysis. In any event all physical signals satisfy the existence criteria of the relevant integrals.

1.1. GENERAL PROPERTIES OF WAVES

A local departure from equilibrium conditions usually gives rise to a propagating perturbation, this is a progressive wave. The description of a wave involves some parameters namely velocity, frequency, and wave vector, the definition of which does not depend on the nature of the disturbance. If the medium is finite, the progressive wave is reflected by the edges. There appears a stationary wave or a guided wave. We shall now look at this in more detail.

1.1.1. Expression for a Progressive Plane Wave

One way of observing the propagation of a phenomenon is to record, at some fixed point, the time variation of a characteristic quantity.

For the observer located at x_o, the phenomenon is determined by the value u of this quantity at time t_o (fig. 1.1). Let us suppose that the perturbation is not modified when it propagates, suffers no attenuation, and travels at constant velocity V; there the value u is the same at point x and time t such that

$$x = x_o + V(t - t_o)$$

or equivalently:

$$t - \frac{x}{V} = t_o - \frac{x_o}{V}$$

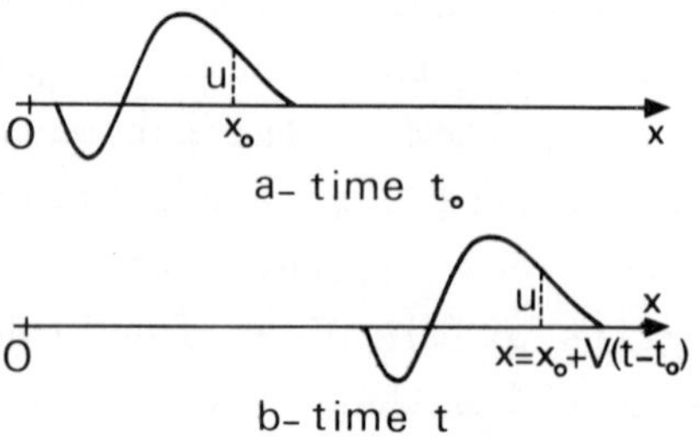

Fig. 1.1. Propagation of a perturbation. The perturbation is defined by the value u at position x_o and time t_o. It propagates at velocity V, reaching point x at time t with

$$x = x_o + V(t - t_o)$$

Any function u(x,t) describing the phenomenon depends solely on the expression $(t - \frac{x}{V})$ which has the same value for all x and all t, provided $t - \frac{x}{V}$ remains constant:

$$\boxed{u(x,t) = F(t - \frac{x}{V})} \qquad (1.1)$$

This function $F(t - \frac{x}{V})$ represents a wave travelling the positive x direction. If the propagation is in the negative x direction then:

$$u(x,t) = F(t + \frac{x}{V})$$

The displacement which results from the propagation in the same medium of two waves in opposite directions is:

$$u(x,t) = F_1(t - \frac{x}{V}) + F_2(t + \frac{x}{V}).$$

Among all possible disturbances, harmonic vibrations about a mean position are very important since other perturbations result from the superposition of these vibrations.

Let us consider the acoustic wave emitted by a plane membrane vibrating with displacement

$$u_M = A \cos \omega t$$

At a distance x, this wave is expressed by

$$u = A \cos \omega(t - \frac{x}{V})$$

u is the displacement of a "slice" of air, but could also represent the pressure or density variation in a slice parallel to the membrane.

If T denotes the period of the vibration, the frequency f = 1/T is related to the angular frequency ω by

$$\omega = \frac{2\pi}{T} = 2\pi f.$$

Then u can be expressed as

$$u = A \cos 2\pi(\frac{t}{T} - \frac{x}{\lambda})$$

where the wavelength λ = VT is the distance travelled by the perturbation in a period. Since the phenomenon is not changed after time T, λ represents the space interval which lies, at a given time, between two identical states of the fluid (air), for instance two successive maxima of u.

It is also useful to write

$$u = A \cos (\omega t - \frac{\omega x}{V}) = A \cos (\omega t - kx)$$

where $k = \omega/V = 2\pi/\lambda$ is known as the wave number. The quantity kx measures the difference in phase at a given time of the phenomenon at x and the origin, in our case between the vibration of the slice located at x and the vibration of the membrane.

Let us call ϕ the overall phase of the wave:

$$\phi = \omega t - kx$$

the wave number k is related to the variation of ϕ with x by:

$$k = - (\frac{\partial \phi}{\partial x})_t \qquad (1.2)$$

whereas the angular frequency ω represents the time variation of ϕ at a given point:

$$\omega = (\frac{\partial \phi}{\partial t})_x \qquad (1.3)$$

So there exists a time-space correspondence.

Time		Space	
Period T Angular frequency ω	$\omega = \frac{2\pi}{T}$	Wave length λ Wave number k	$k = \frac{2\pi}{\lambda}$

The propagation velocity $V = \omega/k$ is called the phase velocity. It is the velocity of an observer who always sees the vibration in the same phase:

$$\omega(t - \frac{x}{V}) = \text{constant}$$

so that the wave seems motionless to him.

In the above example, the particles located in a plane parallel to the membrane move in phase and the movement of every point in this wave plane is known when the distance x between the plane and the origin is specified. The expression for the vibration of a given point P (defined by vector $\vec{x}$) involves the unit vector $\vec{n}$ perpendicular to the wave plane (fig. 1.2). The coordinate x of P is

$$x = \vec{n} \cdot \vec{x}.$$

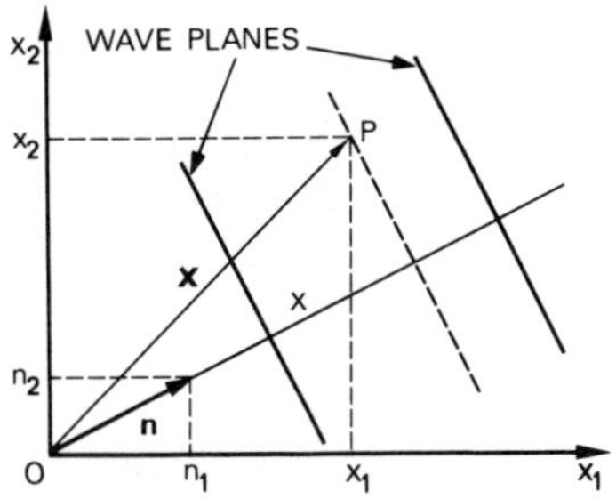

Figure 1.2. Plane wave. At any point P such that $\vec{x} = \overrightarrow{OP}$, the vibration, created by the wave propagating along unit vector $\vec{n}$, is

$$u(\vec{x},t) = F(t - \frac{\vec{n}.\vec{x}}{V})$$

Inserting this into (1.1), one obtains

$$\boxed{u = F(t - \frac{\vec{n}.\vec{x}}{V})} \qquad (1.4)$$

Developing (1.4) in an orthonormal coordinate system involves the components x_1, x_2, x_3 and the direction cosines n_1, n_2, n_3 of the propagation direction (they are the three components of $\vec{n}$):

$$u = F(t - \frac{n_1x_1 + n_2x_2 + n_3x_3}{V})$$

For the harmonic plane wave:

$$u = A \cos \omega(t - \frac{\vec{n}.\vec{x}}{V})$$

or

$$u = A \cos (\omega t - \vec{k}.\vec{x})$$

where the wave vector

$$\vec{k} = \frac{\omega}{V}\vec{n} = \frac{2\pi}{\lambda}\vec{n} = k\vec{n}$$

has been introduced.

The amplitude A of the wave has been taken as constant. It is clear that if the propagating wave undergoes attenuation or amplification, the amplitude depends on the point P, i.e. on $\vec{x}$:

$$u = A(\vec{x}) \cos (\omega t - \vec{k}.\vec{x})$$

Complex representation of sine functions - Although they are real, many quantities encountered in physics can be represented by complex numbers. This use of complex numbers, which is common for harmonic phenomena, relies on the fact that every linear relationship between complex numbers, provided the coefficients are real, is valid for the real and imaginary parts separately. In particular the equations of physics are often linear with real coefficients.

For a harmonic quantity

$$u = A \cos (\omega t + \Phi)$$

the relevance of the complex representation

$$u_c = A \exp i(\omega t + \Phi), \text{ so that } u = \mathrm{Re}(u_c)$$

($\mathrm{Re}(u_c)$ denotes the real part of u_c) arises from the specific properties of the exponential function

- in a sum of vibrations at the same frequency, the time dependent expression $\exp(i\omega t)$ can be factorised out:

$$\Sigma u_c = \exp(i\omega t)\ \Sigma A \exp i\Phi$$

- the time derivative is the product with $i\omega$:

$$\frac{du_c}{dt} = i\omega u_c$$

- the square of the amplitude is given by

$$A^2 = u_c u_c^*$$

where u_c^* is the complex conjugate of u_c.

A plane harmonic wave, in this notation is written as:

$$u_c = A \exp i(\omega t - \vec{k}.\vec{x})$$

1.1.2. Reflexion of a Plane Progressive Wave

When the propagation medium is no longer unbounded and homogeneous especially when its properties vary suddenly within a wavelength, the wave may be entirely or partly reflected. This is also true at a boundary between two media with different properties. The boundary conditions depend on the nature of the wave. They are simpler if the perturbation cannot propagate in one medium and if one characteristic quantity, vanishing on one side of the boundary, also vanishes at the boundary because of continuity requirements.

1.1.2.1. Normal incidence. Stationary waves

When the incident wave

$$u_I(x,t) = F_I(t - \frac{x}{V})$$

is normal to the boundary surface, the reflected wave propagates in the opposite direction:

$$u_R(x,t) = F_R(t + \frac{x}{V})$$

Let us assume that the vibration vanishes at the boundary for all time (perfectly reflecting boundary).[1] If the origin of the x-axis is taken at the boundary:

$$u_I(0,t) + u_R(0,t) = 0 \quad \text{for all } t$$

so:

$$F_R(t) = - F_I(t) \quad \text{for all } t$$

Hence the reflected wave

$$u_R = - F_I(t + \frac{x}{V})$$

is always symmetric to the incident wave with respect to the origin.

The resulting wave is then

$$u = u_I + u_R = F_I(t - \frac{x}{V}) - F_I(t + \frac{x}{V})$$

In the harmonic case:

$$u = A\{\cos \omega(t - \frac{x}{V}) - \cos \omega(t + \frac{x}{V})\}$$

$$u = 2A \sin \omega t \sin \frac{\omega x}{V}$$

The vibration appears as the product of two distinct functions, one of which depends on time and the other on

---

1. It is unnecessary to specify the nature of the wave or of the boundary, provided there is one characteristic quantity that vanishes at the boundary. For example the particle displacement of an elastic wave vanishes at a fixed wall, and its pressure variation vanishes at a free surface. Other examples show that at the end of an open line there is no electric current and that at the end of a short circuited line the voltage vanishes. At a perfectly conducting boundary, an electromagnetic wave has no tangential electric field. It is also implicitly assumed that the quantity is conserved on a reflexion. The more complex problem of elastic wave reflexion at a boundary between two crystals is dealt with in §5.2.

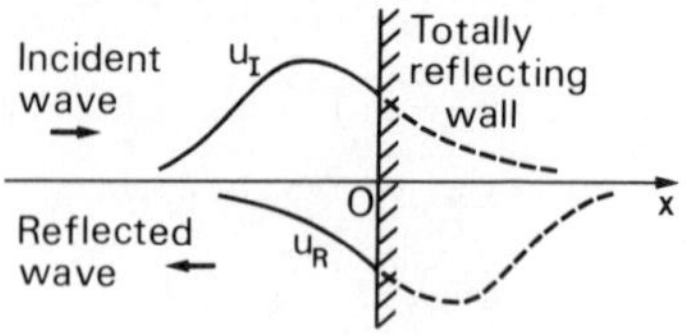

Figure 1.3. Total reflexion of a progressive wave. The reflected perturbation is symmetric with respect to the incident perturbation about an origin O, taken at the boundary.

space. The concept of propagation vanishes, and we observe a stationary wave, characterized by the fact that all points vibrate in phase and with different amplitudes, as shown in Fig. (1.4) where a few successive states of vibration are depicted.

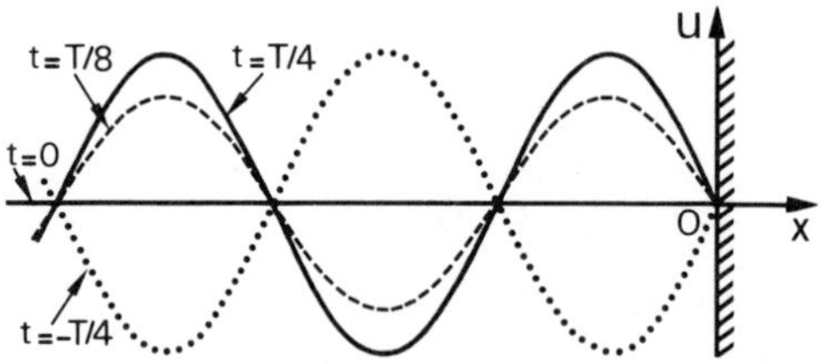

Figure 1.4. Stationary wave. The total reflexion of a progressive sine wave generates a stationary wave system.

The points of co-ordinate x such that

$$\frac{\omega x}{V} = -n\pi \qquad n = 0, 1, 2, 3, \ldots$$

where u is always zero, are called nodes. Half-way between the nodes are the maxima. The interval between two successive nodes is half a wavelength.

If the reflexion is not total, the amplitude A_R of the reflected wave is smaller than that of the incident wave

A_I, part of which is transmitted into the second medium. The resulting wave

$$u = A_I \cos \omega(t - \frac{x}{V}) - A_R \cos \omega(t + \frac{x}{V})$$

$$u = (A_I - A_R) \cos \omega(t - \frac{x}{V}) + A_R \{\cos \omega(t - \frac{x}{V}) - \cos \omega(t + \frac{x}{V})\}$$

is the sum of a progressive wave and a stationary wave:

$$u = (A_I - A_R) \cos \omega(t - \frac{x}{V}) + 2A_R \sin \omega t \sin \frac{\omega x}{V}$$

It is customary to define the standing wave ratio (SWR) by the ratio of the extreme values of the vibration

$$SWR = \frac{A_I + A_R}{A_I - A_R}$$

The SWR, which is infinite for pure stationary waves, and is equal to 1 for a progressive wave, is directly related to the reflexion coefficient $r = A_R/A_I$:

$$SWR = \frac{1 + r}{1 - r}$$

Let us now consider the case of a wave which is forced to travel between two parallel, perfectly reflecting walls, located at $x = 0$ and $x = -L_1$. In order that the stationary wave

$$u = 2A \sin \omega t \sin \frac{\omega x}{V}$$

vanish also at the boundary $x = -L_1$, it is necessary that

$$\frac{\omega L_1}{V} = n\pi, \qquad n \text{ integer}$$

or

$$f = n\frac{V}{2L_1}$$

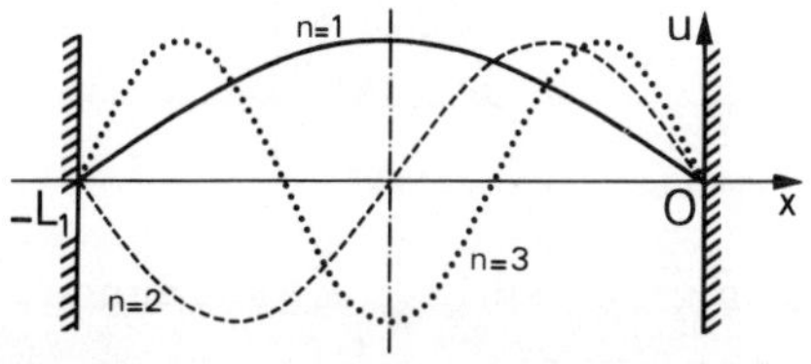

Figure 1.5. Resonator. The first vibrational modes of a resonator consisting of two perfectly reflecting boundaries.

The only stationary waves that can exist have frequencies which are integer multiples of $V/2L_1$. The first eigenmodes of this resonator are shown in Fig. 1.5. The even (odd) modes are antisymmetric (symmetric) with respect to the median plane.

1.1.2.2. Oblique incidence. Guided waves

Let us now suppose that the incident wave vector $\vec{k}$ is no longer normal to the boundary. In our co-ordinate system (Fig. 1.6), the incident wave is expressed by

$$u_I = A \cos (\omega t - k_1 x_1 - k_2 x_2)$$

Let $\vec{k}'$ be the wave vector of the wave wholly reflected by the plane $x_1 = 0$

$$u_R = - A \cos (\omega t - k_1' x_1 - k_2' x_2)$$

Since the resulting vibration

$$u = 2A \sin (\omega t - \frac{k_1 + k_1'}{2} x_1 - \frac{k_2 + k_2'}{2} x_2) \sin (\frac{k_1 - k_1'}{2} x_1 + \frac{k_2 - k_2'}{2} x_2)$$

has to vanish everywhere on the plane $x_1 = 0$, k_2' and k_2 are equal; and, allowing for the conservation of wave number ($k = k' = \omega/V$), we get $k_1' = \pm k_1$. Only $k_1' = - k_1$ provides a non vanishing value of u:

$$u = 2A \sin (\omega t - k_2 x_2) \sin k_1 x_1$$

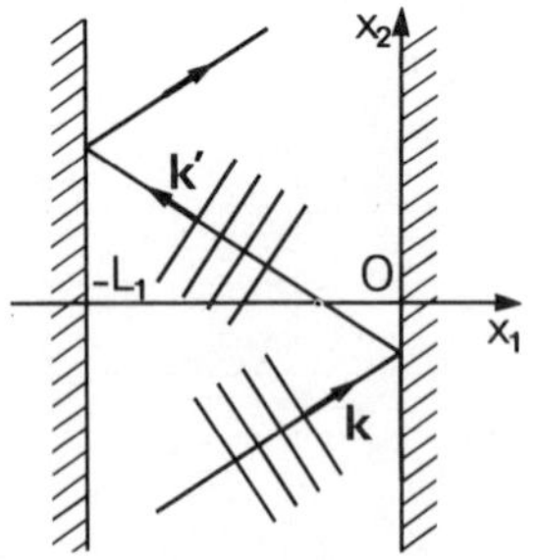

Fig. 1.6. Reflexion with oblique incidence. The wave propagates in the x_2 direction through successive reflexions at parallel boundaries.

The incident and reflected wave vectors are symmetric with respect to the reflecting boundary.

For u to vanish also on the second boundary $x_1 = -L_1$, it is required that

$$k_1 = n\frac{\pi}{L_1} \qquad n = 1, 2, 3, \ldots$$

The final expression is

$$u = 2A \sin(\omega t - k_2 x_2) \sin\frac{n\pi}{L_1} x_1$$

which represents a wave propagating along the x_2-axis between the nodal boundary planes $x_1 = 0$ and $x_1 = -L_1$. This wave is guided by the reflecting walls, and its phase velocity $V_G = \omega/k_2$ is higher than $V = \omega/k$ obtained for the free wave; in fact, along the guide axis, the wave planes pass faster since they propagate at an angle. To every value of the integer n, there corresponds a mode. The nth mode has (n-1) equidistant nodal planes, parallel to and between the walls.

The wave vectors $k_G = k_2$ of the guided wave and k of the free wave are related by

$$k_G^2 = k^2 - k_1^2 = k^2 - \left(\frac{n\pi}{L_1}\right)^2$$

The smallest possible value of k: $k_c = \pi/L_1$ corresponds to the maximum wavelength $\lambda_c = 2L_1$ which is called the cutoff wavelength. Only waves with frequencies above the cutoff frequency $f_c = V/2L_1$ can propagate in the guide. For $f < f_c$, the wave vector k_G is imaginary and the wave is absorbed on entering the guide.

A wave guide is an example of a dispersive structure, where the propagation velocity depends on the frequency; the effect of dispersion is linked to the geometry of the guide.[1] The angular frequency ω is not proportional to the wave number k_G. There is now a more complicated relation between them, known as the dispersion equation.[2]

---

1. In a thin plane, there is, however, a non dispersive elastic mode, called the zero-order mode (see § 9.3.3.), without any equivalent in electromagnetism.

2. If several waves can propagate with different velocities in the material and if, on reflexion at a boundary, one wave is converted into another, the dispersion relation is more complicated. This is generally true for elastic waves.

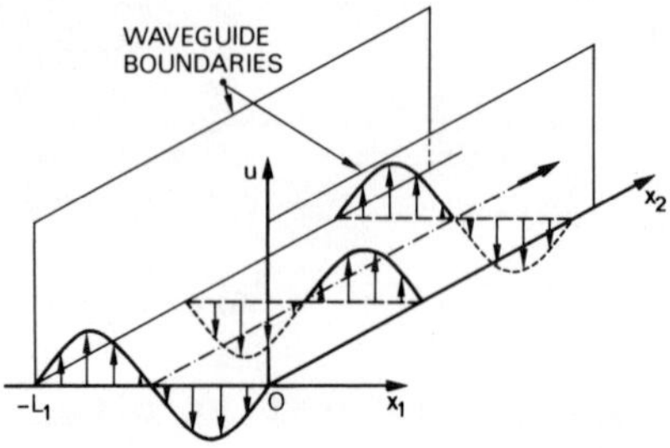

Figure 1.7. Guided wave propagation. A guided wave has nodal planes at the boundaries, and also in between if the mode order is greater than 1. Here, n = 2 (first antisymmetric mode).

$$\omega = Vk = V\{k_G^2 + (\frac{n\pi}{L_1})^2\}^{1/2} \qquad (1.6)$$

It is illustrated, in reduced co-ordinates, in Fig. 1.8. for the first few modes. The guide is very dispersive in the vicinity of the cutoff frequency and its overtones.

If we have two additional reflecting walls at $x_2 = 0$ and $x_2 = -L_2$, a new situation results from the superposition of the incident wave

$$u_I = 2A \sin(\omega t - k_2 x_2) \sin(\frac{n_1 \pi}{L_1} x_1)$$

and the wave reflected by the wall at $x_2 = 0$

$$u_R = -2A \sin(\omega t + k_2 x_2) \sin(\frac{n_1 \pi x_1}{L_1})$$

The stationary wave

$$u = u_I + u_R = -4A \sin k_2 x_2 \sin(\frac{n_1 \pi x_1}{L_1}) \cos \omega t$$

has to vanish at $x_2 = -L_2$. Hence

$$k_2 = \frac{n_2 \pi}{L_2} \qquad n_2 = 1, 2, 3, \ldots$$

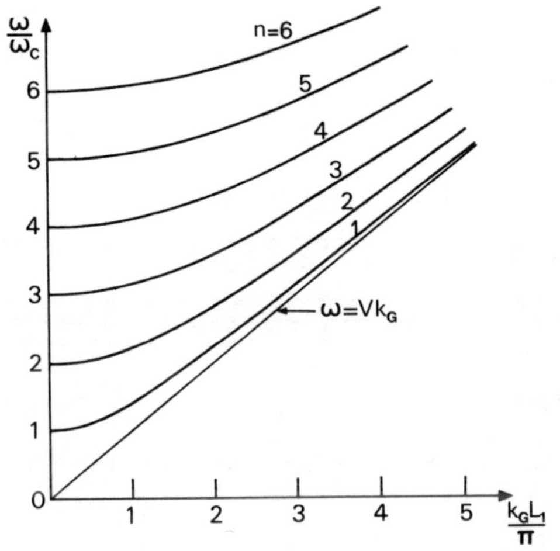

Figure 1.8. Dispersion curves of a waveguide. At angular frequency ω the reduced phase velocity V_G/V of the nth mode is the slope of the line joining the origin and the corresponding point on the curve of this mode.

The allowed frequencies for this wave

$$u = -\ 4A \sin\left(\frac{n_1 \pi x_1}{L_1}\right) \sin\left(\frac{n_2 \pi x_2}{L_2}\right) \cos \omega t$$

are given by

$$\frac{\omega^2}{V^2} = k^2 = k_1^2 + k_2^2$$

or, equivalently,

$$f = \frac{V}{2} \left\{\frac{n_1^2}{L_1^2} + \frac{n_2^2}{L_2^2}\right\}^{1/2} \qquad (1.7)$$

In this two-dimensional cavity, the frequencies of the only modes that can exist are given by (1.7). In order to describe the three dimensional case, it is sufficient to allow for propagation along the x_3-direction perpendicular to the x_1x_2 plane. This guided wave, with wave vector k_3 is

$$u_I = 4A \sin\left(\frac{n_1 \pi x_1}{L_1}\right) \sin\left(\frac{n_2 \pi x_2}{L_2}\right) \cos(\omega t - k_3 x_3).$$

Let the guide be closed at $x_3 = 0$ by a reflecting wall. The interference between u_I and the reflected wave

$$u_R = -\,4A \sin\left(\frac{n_1 \pi x_1}{L_1}\right) \sin\left(\frac{n_2 \pi x_2}{L_2}\right) \cos(\omega t + k_3 x_3)$$

gives rise to a stationary wave of arbitrary frequency

$$u = 8A \sin\left(\frac{n_1 \pi x_1}{L_1}\right) \sin\left(\frac{n_2 \pi x_2}{L_2}\right) \sin k_3 x_3 \cos \omega t$$

The parallelopiped obtained by closing the last end at $x_3 = -\,L_3$ will only allow vibrational modes of the form

$$u = 8A \sin\left(\frac{n_1 \pi x_1}{L_1}\right) \sin\left(\frac{n_2 \pi x_2}{L_2}\right) \sin\left(\frac{n_3 \pi x_3}{L_3}\right) \sin \omega t$$

and of frequency

$$f = \frac{V}{2} \left\{ \left(\frac{n_1}{L_1}\right)^2 + \left(\frac{n_2}{L_2}\right)^2 + \left(\frac{n_3}{L_3}\right)^2 \right\}^{1/2}$$

Note. These simple formulae for guides and resonators are valid only insofar as the reflexion does not affect the wave. They are valid for an elastic wave propagating between walls in a fluid but it will become clear especially from § 5.2, that they do not hold for an elastic wave propagating inside a crystal, because the wave, when it is reflected, gives rise to several waves with various polarizations and velocities. There is, however, an exception: the wave which creates a displacement of particles parallel to the boundaries and normal to the wave vector. For this particular wave, which is conserved by reflexion, formulae (1.5, 1.7), and the curves in Fig. 1.8 are valid. We shall refer to this when discussing in 9.3.3. strip dispersive guides with "transverse" waves.

1.1.3. Velocity of a Wave Group

So far we have considered only monochromatic waves, i.e. sine waves of infinite duration, and fixed amplitude and frequency. By themselves, these pure waves are of little practical interest; an observer looking at the propagation of a monochromatic wave gains no more information from this monotonous stream than from a steady flow of a homogeneous fluid. The transmission of information requires some kind of anomaly, a change in some characteristic quantity. A river may carry information through a floating tree trunk or a variation in flow. Similarly, transport of information via a wave requires the variation of at least one of the parameters, the amplitude or phase. This complex wave, now bearing some information, is no longer monochromatic since the amplitude or the phase is modulated by the signal to be transmitted (see § 1.3.2), but it can be considered as the sum of an infinity of monochromatic waves of different amplitudes and frequencies. This is a wave packet or a wave group. Taking the wave number k as variable:

$$u(x,t) = \int_{-\infty}^{\infty} A(k) \exp i(\omega t - kx)\, dk \qquad (1.8)$$

where A(k) is the amplitude density of the wave packet.

Two cases should be considered depending on whether the propagation medium is dispersive or not. For a path length x_o, in a non dispersive medium where the velocity V does not depend on ω, each component of the group undergoes the same delay $t_o = x_o/V$. After this time t_o, the group comes out unchanged. Therefore, in a non dispersive medium, a complex wave propagates without any distortion. This property is stated in different ways:

- the propagation velocity does not depend on frequency
- the phase shift $\Phi = -\omega x_o/V$ caused by a path length x_o is proportional to the frequency
- the group delay, defined by

$$\tau_g = -\frac{d\Phi}{d\omega} = \frac{x_o}{V}$$

  does not depend on frequency.

Let us now consider the propagation of a wave group in a dispersive medium. In practice the frequency f_o of the carrier is much greater than the signal frequency, so that the amplitude density A(k) has significant values only in a small region in the vicinity of $k_o = \omega_o/V(\omega_o)$. It is thus sufficient to have a first order expansion about k_0 of the dispersion equation:

$$\omega(k) \simeq \omega(k_0) + \left(\frac{d\omega}{dk}\right)_{k_0} (k - k_0).$$

$V_g = (d\omega/dk)_{k_0}$ has the dimensions of velocity. Inserting this quantity into (1.8), leads to

$$u(x,t) = \exp i\{\omega(k_0) - k_0 V_g\}t \int_{-\infty}^{\infty} A(k) \exp - i\{k(x-V_g t)\}dk$$

which can be written in the form

$$u(x,t) = e^{i\Omega_0 t} F(x - V_g t) \qquad (1.9)$$

where $u(x,0) = F(x)$ and $\Omega_0 = \omega(k_0) - k_0 V_g$.

Equation (1.9) shows that during a time t, the wave group has travelled $x = V_g t$. So, in a dispersive medium, a wave packet centered at wave number k_0 propagates with velocity

$$\boxed{V_g = \left(\frac{d\omega}{dk}\right)_{k_0}} \qquad (1.10)$$

known as the group velocity.

Although it is not obvious from the above result, because of the first order expansion of the dispersion relation (which is equivalent to equating the $\omega(k)$ curve at a given point to its tangent) the wave packet is distorted during the propagation. For example, some elementary waves in the front of the group having started early and travelled slowly, may be passed by other waves which started later but travelled faster. The peak of this group, where the relative position of waves is always changing, is located at time t and position x such that the high amplitude components interfere constructively. It is thus necessary that the components about k_0 be in phase, i.e.

$$\phi(k) = \omega(k)t - kx = \text{constant for } k = k_0$$

or

$$\left(\frac{d\phi}{dk}\right)_{k_0} = \left(\frac{d\omega}{dk}\right)_{k_0} t - x = 0$$

By this method of stationary phase we obtain once more the result that the wave packet has a group velocity V_g, as given by (1.10).

If there is no attenuation V_g is also the transport velocity of the energy localized in the wave packet.

For example, let us refer again to the wave guide of § 1.1.2.2. Equation (1.6) immediately provides the velocity of a guided wave group:

$$V_g = \frac{d\omega}{dk_G} = Vk_G\{k_G^2 + \frac{n^2\pi^2}{L_1^2}\}^{-1/2}$$

Since the phase velocity is

$$V_G = \frac{\omega}{k_G} = \frac{V}{k_G}\{k_G^2 + \frac{n^2\pi^2}{L_1^2}\}^{1/2}$$

the product

$$V_gV_G = V^2$$

is a constant, equal to the square of the velocity of the free wave. The frequency dependence of these velocities is illustrated in Fig. (1.9).

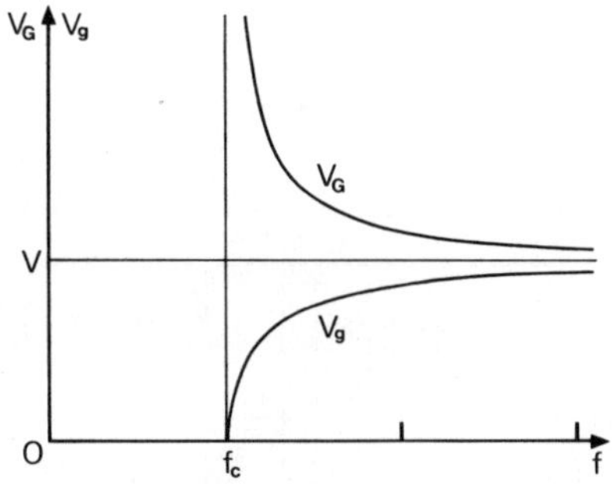

Fig. 1.9. In a waveguide, the phase velocity V_G is larger than the free wave velocity V, whereas the group velocity V_g is smaller than V.

1.2. ELASTIC WAVE PROPAGATION

The foregoing concepts refer to waves in general, whatever the nature of the vibration and the propagation mechanism. Elastic waves are matter waves, they arise from particle movements and they exist in fluid or solid media only, whereas electromagnetic waves can also propagate in a vacuum. Their propagation is governed by

Newtonian mechanics. We consider it in detail first for a free particle fluid and then for a crystal whose atoms are bonded together. The former case is dealt with at a macroscopic scale, the latter at the atomic scale.

1.2.1. Continuous Medium. Macroscopic Viewpoint

Let us return to the plane wave emitted, in a fluid (air), by a plane vibrating membrane. When it is at rest, there is a uniform pressure p_o in the fluid. When moving, let us say in the positive x-direction, the membrane compresses the air slice OA (Fig. 1.10). This is an unstable situation: the fluid expands, compressing in its turn the neighbouring slice AB. The wave thus travels by means of a succession of compressions and expansions; the pressure p(x,t) is therefore a function of space and time. The forces exerted on sides x and x + dx of any slice MN do not reach equilibrium, and the slice moves to M'N'.

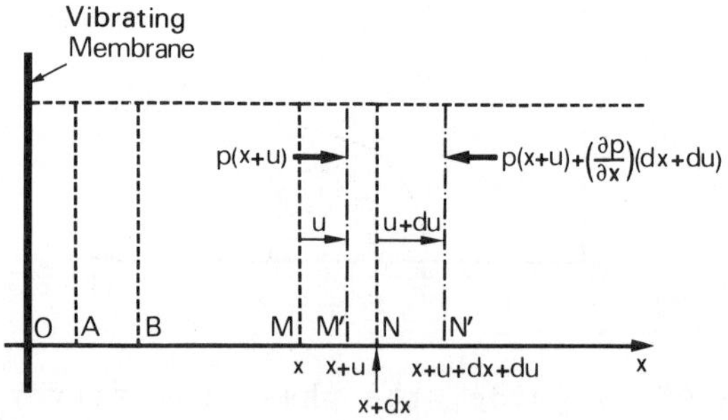

Fig. 1.10. The vibration of a membrane in a fluid generates a longitudinal wave which, when propagating compresses or expands the fluid slices OA, AB, ... MN.

Let u(x,t) be the displacement of the slice x at time t. The displacement of the slice x + dx is then u + du such that

$$du = \left(\frac{\partial u}{\partial x}\right) dx$$

If the cross sectional area of the fluid is s, the force exerted on the slice by the fluid on the left hand side of the wave plane passing at M' is

$$F_{M'} = sp(x + u,t)$$

while the force exerted by the fluid on the right hand side is

$$F_{N'} = -sp(x + u + dx + du,t) = - s[p(x + u,t) + \frac{\partial p}{\partial x}(dx+du)]$$

The resultant force acting on section M'N' of the fluid is

$$dF = F_{M'} + F_{N'} = - s \frac{\partial p}{\partial x} (dx + du)$$

$$dF = - s \frac{\partial p}{\partial x} (1 + \frac{\partial u}{\partial x}) dx$$

and induces an acceleration $\gamma = \dfrac{\partial^2 u}{\partial t^2}$ to the mass $dm = \rho_0 s dx$ of the slice MN (ρ_0 denotes the mass per unit volume of the fluid at rest, i.e. under pressure p_o). The fundamental law of dynamics now gives $dF = \gamma dm$, or equivalently

$$\rho_0 s \frac{\partial^2 u}{\partial t^2} dx = - s \frac{\partial p}{\partial x} (1 + \frac{\partial u}{\partial x}) dx$$

or

$$\frac{\partial^2 u}{\partial t^2} = - \frac{1}{\rho_0} \frac{\partial (\Delta p)}{\partial x} (1 + \frac{\partial u}{\partial x}) \tag{1.11}$$

where the pressure $p(x + u,t)$ at M'N' has been written

$$p(x + u,t) = p_o + \Delta p.$$

The pressure change Δp is related to the volume change Δv of the slice MN through the compressibility coefficient χ:

$$\chi = - \frac{1}{V}(\frac{\Delta v}{\Delta p}).$$

This coefficient, always positive, is sufficient to describe the elastic properties of a non viscous fluid. The volume has changed from $dv_o = s\,dx$ under pressure p_o (position MN) to $dv = s(1 + \partial u/\partial x)dx$ under pressure $p_o + \Delta p$ (position M'N'). The dilatation is

$$\frac{\Delta v}{v} = \frac{dv - dv_o}{dv_o} = \frac{\partial u}{\partial x}$$

Inserting $\Delta p = - \dfrac{1}{\chi} \dfrac{\partial u}{\partial x}$ into (1.11), yields

$$\frac{\partial^2 u}{\partial t^2} = \frac{1}{\rho_0 \chi} \frac{\partial^2 u}{\partial x^2} (1 + \frac{\partial u}{\partial x}) \tag{1.12}$$

The proportionality between the pressure variation and the volume variation is exact only if the latter is much less than 1

$$\frac{\partial u}{\partial x} << 1.$$

With this assumption, equation (1.12) yields, with $V^2 = \frac{1}{\rho_0 \chi}$

$$\boxed{\frac{\partial^2 u}{\partial t^2} = V^2 \frac{\partial^2 u}{\partial x^2}} \tag{1.13}$$

The extrapressure $\Delta p = -\frac{1}{\chi}\frac{\partial u}{\partial x}$ obeys the same differential equation as the mass per unit volume. This partial differential equation, which governs the behaviour in space and time of the fluid, has been obtained by looking at the effect of a fluid slice on its neighbours. Any function of the form

$$u(x,t) = F(t \pm \frac{x}{V})$$

obeys equation (1.13); the disturbance propagates with velocity V.

Equation (1.13), which is known as the wave propagation equation, is quite general and applies to many different waves (mechanical and electrical).

Important Comments

1. Polarization. In the case of a fluid which was chosen for the sake of simplicity, the displacement $\vec{u}$ is perpendicular to the wave planes i.e. parallel to the wave vector. The wave polarization is longitudinal. This may not be the case in a solid: the vibration may be perpendicular to the wave vector so that the wave has transverse polarization (Chap. 5). However, the medium must allow for such shear strain; thus non viscous fluids can transmit only longitudinal waves.

2. Limiting Frequency. Is there a limit to the frequency of elastic waves which can propagate in a gas in which molecules are always moving? The membrane vibration creates a succession of maxima and minima of the fluid pressure (or density). The wave can propagate only if the distance between a maximum and the adjacent minimum is much greater than the mean free path l_m of the molecules

(l_m is the mean distance travelled by a molecule between two collisions). Otherwise, a single free path is enough for the molecules to fill up the low density region, and the perturbation carried by the wave disappears. This condition requires that the wavelength be much greater than the mean free path:

$$\lambda >> l_m \rightarrow f << \frac{V}{l_m}$$

Typical values are for oxygen V = 315 m/s at 0°C; l_m = 4,9 cm under pressure 10^{-3} mm Hg: the limiting frequency is 6400 Hz.

1.2.2. Atomic Chains. Microscopic Viewpoint

In a crystal, the equilibrium position of atoms is fixed. For very short wavelengths, i.e. of the order of magnitude of the interatomic distance, the medium is no longer continuous at the scale of the spatial variations, and this implies the existence of a limiting frequency for elastic waves. Our aim, in this section, is to point out this limit in a monatomic crystal. The atomic pattern is a three-dimensional regular lattice (§ 2.2.1) defined by the intersections of three sets of straight lines, parallel to three principal directions $\vec{a}$, $\vec{b}$, $\vec{c}$, assumed to be mutually orthogonal (fig. 1.11). A longitudinal elastic plane wave propagating along one of these directions (say $\vec{a}$) makes all atoms of planes perpendicular to the wave vector move simultaneously, so that rows parallel to $\vec{a}$ all vibrate in the same way and it is sufficient to consider a single chain of identical equidistant particles (fig. 1.12).

If one atom of this chain is moved, because of the binding forces its two neighbours will also move and so the process continues. The chain is distorted while the disturbance propagates. We take only the nearest neighbour interaction into account, assuming that the strain is small enough for the elastic restoring forces F_1 and F_2 exerted on an atom to be proportional to the deviation from the equilibrium interatomic distance.

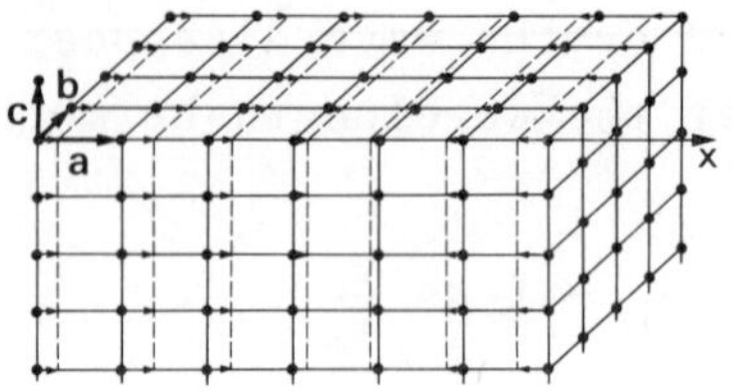

Figure 1.11. Crystal lattice, distorted by the occurrence of a plane longitudinal wave propagating in the $\vec{a}$ direction. We need only look for the atomic displacements in a row parallel to $\vec{a}$.

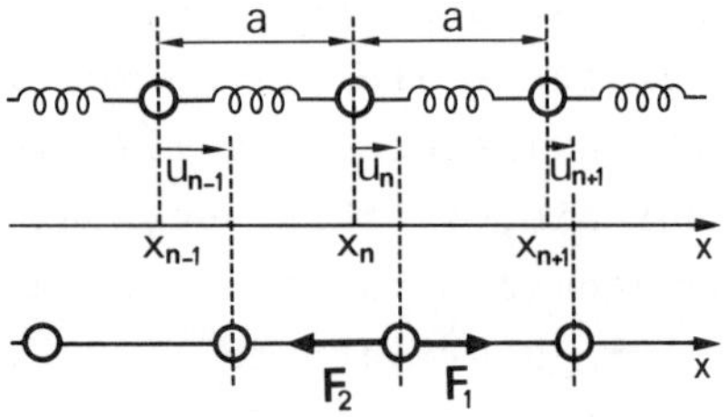

Fig. 1.12. Propagation of a perturbation along an atomic chain. The springs represent the atomic bonds.

Let u_n be the displacement of the nth atom, located at x_n at equilibrium. The forces exerted by the (n-1)th and (n+1)th on the nth atom are:

$$F_1 = K(u_{n-1} - u_n), \quad \text{and} \quad F_2 = K(u_{n+1} - u_n)$$

The equation of motion for the nth atom, given its mass M, is then

$$M \frac{\partial^2 u_n}{\partial t^2} = K(u_{n+1} + u_{n-1} - 2u_n) \qquad (1.14)$$

Let us force one atom, taken as the origin, to vibrate as:

$$u_o = A\, e^{i\omega t}$$

In the steady state, in the absence of any attenuation the motion of every atom is identical to that of the source, but with a phase shift

$$u_n = A\ e^{i(\omega t+\Phi_n)}$$

Substituting the expression for u_n in (1.14), leads to

$$(-\ M\omega^2 + 2K)A\ e^{i(\omega t+\Phi n)} = KA\ e^{i\omega t}(e^{i\Phi n+1} + e^{i\Phi n-1})$$

or

$$-\ M\omega^2 + 2K = K\{e^{i(\Phi n+1-\Phi n)} + e^{i(\Phi n-1-\Phi n)}\} \quad (1.15)$$

The right hand side has to be real. The phase shift $\Delta\Phi$ between successive particles has thus to be constant throughout the chain

$$\Phi_n - \Phi_{n-1} = \Phi_{n+1} - \Phi_n = \Delta\Phi$$

Since the atoms are equidistant at rest, the phase shift Φ_n is proportional to the distance x_n

$$\Phi_n = -\ kx_n \rightarrow \Delta\Phi = -\ ka$$

where k is, by definition, the wave number (see equation 1.2). The displacement is then expressed by

$$u_n = A\ e^{i(\omega t-kx_n)}$$

which corresponds to a longitudinal wave of phase velocity $V = \omega/k$.

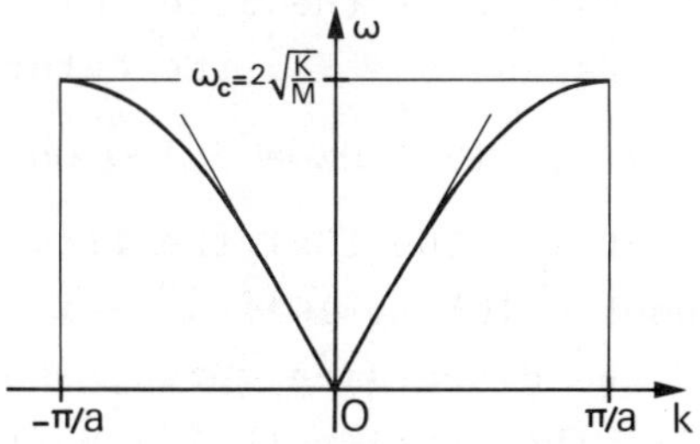

Fig. 1.13. Monatomic chain. Dispersion curves $\omega(k)$ for the longitudinal mode. At any frequency below the cutoff frequency ω_c, two waves can propagate, in both directions, with opposite wave numbers.

Inserting $\Delta\Phi = -ka$ into (1.15) we obtain the dispersion relation

$$M\omega^2 = 2K(1 - \cos ka) = 4K \sin^2 \frac{ka}{2}$$

or

$$\boxed{\omega = 2\sqrt{\frac{K}{M}}\ \left|\sin \frac{ka}{2}\right|}$$

This periodic dispersion curve is plotted in figure 1.13 in the interval $[-\pi/a, +\pi/a]$ known as the first Brillouin zone. For small wave numbers ($ka \ll 1$) it is a straight line:

$$\omega \simeq \sqrt{\frac{K}{M}}\ ak$$

the slope $V_o = a\sqrt{K/M}$ of which yields the propagation velocity of low frequency elastic waves. When the wavelength is of the order of magnitude of the interatomic distance a (ka comparable to π), the medium is dispersive. Since $|\sin ka/2|$ is bounded by 1, the frequency of elastic waves in a crystal is bounded by the cut-off frequency

$$f_c = \frac{1}{\pi}\sqrt{\frac{K}{M}} = \frac{V_o}{\pi a}$$

At this frequency, the group velocity $V_g = d\omega/dk$ vanishes, since the dispersion curve has a horizontal tangent. The ratio $u_{n+1}/u_n = e^{-ika}$ is equal to (-1), showing that neighbouring atoms vibrate out of phase with each other.

Order of magnitude. The velocity V_o of low frequency elastic waves in solids lies between 10^3 and 10^4m/s. Interatomic distances are of the order of a few angströms. Taking $V_o = 5.10^3$ m/s and a = 5Å, the cutoff frequency is

$$f_c = 3{,}2\ 10^{12}\ \text{Hz} = 3{,}200\ \text{GHz}.$$

This is so high a value that the frequency range considered in this book - 10^6 to 10^{10} Hz - is located at the very beginning of the dispersion curve, where ω and k are linearly related. The wavelength $\lambda = V_o/f$, ranging from a few microns to a few millimeters - is much greater than interatomic distances, and, as far as the wave is concerned, the medium looks continuous.

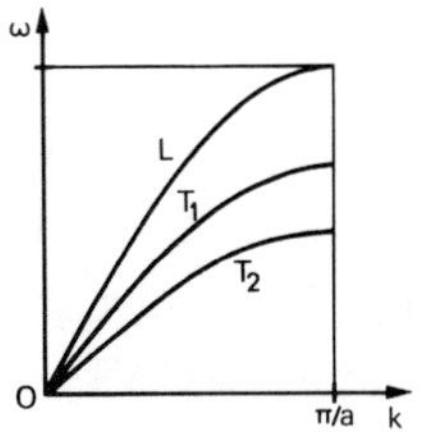

Fig. 1.14. Monatomic chain. Dispersion curves (k > 0) for the three modes.

Comments

1. Transverse modes. The atomic displacement may be transverse, e.g. parallel to the basis vectors $\vec{b}$ or $\vec{c}$ of the lattice (Fig. 1.11). The restoring forces are then different and there will be three dispersion curves for the single propagation direction $\vec{a}$ corresponding to the following waves

- the longitudinal wave L: particle displacement (or polarization) along $\vec{a}$
- the transverse wave T_1: polarization along $\vec{b}$
- the transverse wave T_2: polarization along $\vec{c}$.

For another direction, these curves will differ since the phase velocity depends on the wave vector direction.

For frequencies which are low, compared with the cut-off frequency, the crystal is not dispersive but it is anisotropic.

2. Optical branches. If there are two kinds of atoms in the chain (exercise 1.6) the results are still valid for the centre of mass of the molecule but, in addition to a curve quite similar to that of figure (1.13) and called the acoustic branch, there is another one of high frequency called the optical branch (Fig. 1.26). These two branches differ mainly at high wavelengths ($ka \ll 1$); the overall motion of the molecule takes place at low frequencies (acoustic branch) while the motion of atoms around the fixed center of mass is at a high

frequency (optical branch), equal to the longitudinal vibrational frequency of the molecule. Including transverse modes, the dispersion curve consists of six branches. This is a general result: if the unit cell has p atoms, there are 3p branches, (3p - 3) of which are optical.

3. Thermoelastic waves. The elastic waves considered so far are coherent waves linked to the collective motion of atoms, but in every material there are natural or thermoelastic waves due to the random oscillations of atoms, which account for its thermal behaviour (temperature dependent specific heat).

1.3. SIGNALS, LINEAR TIME INVARIANT SYSTEMS

All devices described in the last chapter transform electric signals by means of elastic waves, i.e. transform the input signal $s_1(t)$ into an output signal $s_2(t)$. Whatever the system, electric filter, mechanical filter, or any other quadripole, it is possible to establish general relations between $s_1(t)$ and $s_2(t)$, provided that the system is linear and time invariant.

The linearity ensures that, if $s_2'(t)$ and $s_2''(t)$ are the responses to the inputs $s_1'(t)$ and $s_1''(t)$, then the input $\lambda s_1'(t) + \mu s_1''(t)$ causes the output $\lambda s_2'(t) + \mu s_2''(t)$, for any scalar λ and μ. The time invariance means that, whatever τ and $s_1(t)$, the response to the input $s_1(t+\tau)$ is $s_2(t+\tau)$.

1.3.1. Real and Fictitious Signals. Dirac Pulse

The function $s(t)$, denoting the time variation of a physical quantity, is real, continuous, and of finite duration. It is, however, useful to introduce nonphysical functions that have remarkable mathematical properties. Among these, the major ones are:

- the complex harmonic function $e^{i2\pi ft}$, of frequency $f = \omega/2\pi$, which extends from $-\infty$ to $+\infty$;
- the Dirac pulse $\delta(t)$, of infinitesimal duration, infinite value at $t = 0$, and unit integral;

- the Heaviside function Y(t), representing a unit step at time t = 0.

In discussing the response to the first two functions, we shall introduce two characteristic functions of the system and establish a relationship between them.

The correct definition of the Dirac pulse involves distribution theory. A distribution T is a linear function, continuous within the space $\mathscr{D}$ of infinitely differentiable functions with closed support, and which assigns a scalar $T(\psi)$ to every function ψ of $\mathscr{D}$.

Thus, a locally summable function g may define a distribution T_g by

$$T_g(\psi) = \int_{-\infty}^{+\infty} g(x)\psi(x)\, dx \qquad (1.16)$$

There are other means of generating distributions. The Dirac distribution δ makes its value at point x_0 correspond to any function $\psi(x)$ - continuous at x_0 - at that point

$$\delta_{x_0}(\psi) = \psi(x_0)$$

Instead of this distribution δ_{x_0}, physicists use a so called Dirac "function" $\delta(x - x_0)$ which defines δ_{x_0} through an equation similar to (1.16)

$$\delta_{x_0}(\psi) = \int_{-\infty}^{+\infty} \delta(x - x_0)\ \psi(x)\ dx = \psi(x_0) \qquad \forall\ \psi(x) \qquad (1.17)$$

which implies that $\delta(x - x_0)$ is zero except for $x = x_0$

$$\delta(x) = 0 \qquad \forall\, x \neq 0$$

Accordingly (1.17) can be written

$$\psi(x_0) \int_{-\infty}^{+\infty} \delta(x - x_0)dx = \psi(x_0)$$

Hence

$$\int_{-\infty}^{+\infty} \delta(x)dx = 1 \ \rightarrow\ \delta(0) = +\infty$$

Such conditions are contradictory: the integral of a function which vanishes everywhere except for one point is null and δ_{x_0} cannot be generated by a well behaved function. However, physicists take $\delta(x)$ as the limit of a function which has significant values only in a small interval around 0, and a very high positive peak value, with unit integral (fig. 1.15). For example, when ε tends to zero (remaining positive)

$$\delta(x) = \frac{1}{\pi} \lim_{\varepsilon \to 0_+} \frac{\varepsilon}{\varepsilon^2 + x^2}$$

for the maximum $1/\pi\varepsilon$ grows to $+\infty$ and the width 2ε goes to 0, whereas the integral

$$\mathscr{A} = \frac{1}{\pi} \int_{-\infty}^{\infty} \frac{\varepsilon dx}{\varepsilon^2 + x^2}$$

remains constant because, defining $y = x/\varepsilon$,

$$\mathscr{A} = \frac{1}{\pi} \int_{-\infty}^{+\infty} \frac{dy}{1+y^2} = \frac{1}{\pi}[\tan^{-1} y]_{-\infty}^{+\infty} = 1$$

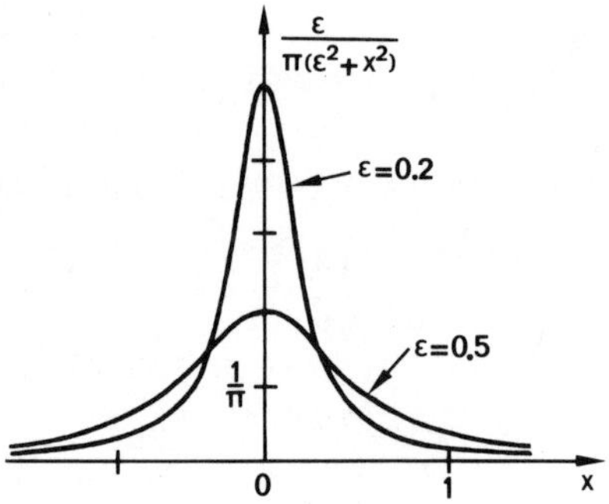

Fig. 1.15. Dirac pulse. The limit of $\frac{\varepsilon}{\pi(\varepsilon^2+x^2)}$ when ε goes to 0_+, satisfies the requirements for the Dirac "function".

The connexion between harmonic functions and the "Dirac function" can be stated in the following way: the Dirac function is the superposition of an infinity of unit amplitude harmonic signals

$$\boxed{\delta(t) = \int_{-\infty}^{\infty} e^{i2\pi ft} df} \qquad (1.18)$$

This equation makes sense within the framework of distribution theory: integrating (formula 1.16) the product of any function $\psi(x)$ with either side of (1.18) yields the same result. In order to prove this, let us note that

$$s(t) = \int_{-\infty}^{\infty} e^{i2\pi ft}\, df = \lim_{\varepsilon \to 0_+} \int_{-\infty}^{\infty} e^{2\pi(ift - \varepsilon|f|)}\, df$$

or

$$s(t) = \lim_{\varepsilon \to 0_+} \left[\int_{-\infty}^{0} e^{2\pi(it+\varepsilon)f}\, df + \int_{0}^{\infty} e^{2\pi(it-\varepsilon)f}\, df\right]$$

$$s(t) = \lim_{\varepsilon \to 0_+} \frac{1}{2\pi}\left(\frac{1}{\varepsilon+it} + \frac{1}{\varepsilon-it}\right)$$

and finally

$$s(t) = \lim_{\varepsilon \to 0_+} \frac{1}{\pi}\, \frac{\varepsilon}{\varepsilon^2+t^2} = \delta(t)$$

Using the Dirac pulse as an input signal is equivalent to applying the set of all harmonic functions simultaneously, from frequencies $-\infty$ to $+\infty$.

Physically, the Dirac pulse is a very brief pulse - compared to the time constants of the system - that, when applied to the system, provides as much information as a point by point analysis for every frequency.

1.3.2. The Spectrum of a Signal. Fourier Transform

Harmonic functions have a fundamental property. They form a complete series in the space of functions of summable modulus and every physical signal s(t) can be decomposed into an infinity of waves

$$\boxed{s(t) = \int_{-\infty}^{\infty} S(f) e^{i2\pi ft}\, df} \qquad (1.19)$$

The function S(f) is the signal spectrum. In order to compute S(f), let us integrate:

$$I(\nu) = \int_{-\infty}^{\infty} e^{-i2\pi\nu t}\, s(t)dt$$

Substituting for s(t) from (1.19)

$$I(\nu) = \int_{-\infty}^{\infty} S(f)\left[\int_{-\infty}^{\infty} dt\, e^{i2\pi(f-\nu)t}\right]df$$

According to (1.18):

$$\int_{-\infty}^{\infty} e^{i2\pi(f-\upsilon)t}\, dt = \delta(f - \upsilon)$$

and

$$I(\upsilon) = \int_{-\infty}^{\infty} S(f)\, \delta(f - \nu) df = S(\nu).$$

Going back to the variable f:

$$\boxed{S(f) = \int_{-\infty}^{\infty} e^{-i2\pi ft}\, s(t) dt} \qquad (1.20)$$

This equation shows that the frequency spectrum of a signal is identical to its Fourier transform. In fact, the Fourier transform of a function g(x) (the absolute value of which is summable) is

$$G(s) = \int_{-\infty}^{\infty} g(x) e^{-i2\pi sx}\, dx$$

This is currently denoted by

$$g(x) \supset G(s)$$

this symbol can also be read from right to left:

$$G(s) \subset g(x).$$

The sign $\subset$ denotes the inverse Fourier transform, which we have shown to exist (1.19):

$$g(x) = \int_{-\infty}^{\infty} G(s) e^{i2\pi sx}\, ds$$

Since a physical signal s(t) is a real function of time:

$$S(-f) = \int_{-\infty}^{\infty} e^{i2\pi ft}\, s(t) dt = S^*(f)$$

The real part of the spectrum is even:

$$Re[S(-f)] = Re[S(f)]$$

while the imaginary part is odd (fig. 1.16):

$$Im[S(-f)] = -Im[S(f)].$$

It is thus sufficient to know one half of the spectrum (positive frequencies) to restore a real signal:

$$s(t) = 2Re \int_{0}^{\infty} S(f) e^{i2\pi ft}\, df$$

If we now consider the signal parity, by introducing the even and odd parts $s_+(t)$, and $s_-(t)$ such that

$$s(t) = s_+(t) + s_-(t)$$

the spectrum is now decomposed into

$$S(f) = \int_{-\infty}^{\infty} s_+(t)e^{-i2\pi ft}\,dt + \int_{-\infty}^{\infty} s_-(t)e^{-i2\pi ft}\,dt$$

or

$$S(f) = 2\int_0^{\infty} s_+(t)\cos 2\pi ft\,dt - 2i\int_0^{\infty} s_-(t)\sin 2\pi ft\,dt$$

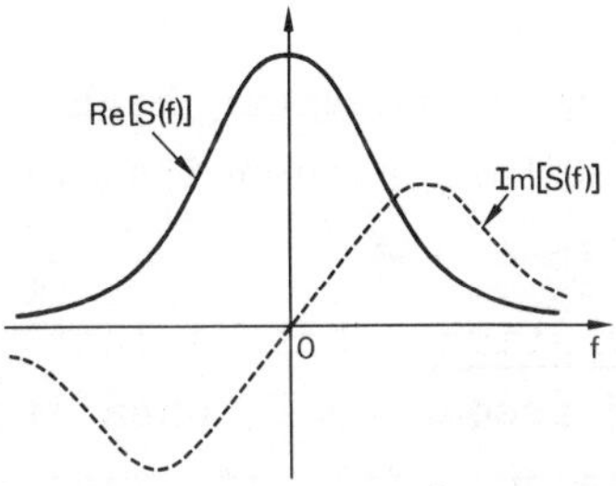

Fig. 1.16. Spectrum of a real signal. The real part is even, the imaginary part is odd.

It is real and even for even signals ($s_- = 0$); imaginary and odd for odd signals ($s_+ = 0$).

Let us now deal with the effect of some signal changes on the spectrum.

Similarity. If S(f) is the spectrum of s(t), then s(at) has spectrum $\frac{1}{|a|}S(\frac{f}{a})$:

$$\int_{-\infty}^{\infty} s(at)e^{-i2\pi ft}\,dt = \frac{1}{|a|}\int_{-\infty}^{\infty} s(at)e^{-i2\pi\frac{f}{a}(at)}\,d(at) = \frac{1}{|a|}S(\frac{f}{a})$$

A time compression of the signal induces a spectrum dilatation and vice versa.

Shift. The spectrum of $s(t+\tau)$ is $e^{i2\pi f\tau}S(f)$:

$$\int_{-\infty}^{\infty} s(t+\tau)e^{-i2\pi ft}\,dt = \int_{-\infty}^{\infty} s(t+\tau)e^{-i2\pi f(t+\tau)}\,e^{i2\pi f\tau}\,d(t+\tau)$$

$$= e^{i2\pi f\tau}\,S(f).$$

A time shift of the signal (e.g. a constant delay) causes a phase shift of the spectrum, proportional to the frequency.

Differentiation. If S(f) is the spectrum of s(t), the spectrum of ds/dt is $i2\pi fS(f)$:

$$\int_{-\infty}^{\infty} \frac{ds}{dt} e^{-i2\pi ft} dt = \int_{-\infty}^{\infty} \lim_{\Delta t\to 0} \frac{s(t+\Delta t) - s(t)}{\Delta t} e^{-i2\pi ft} dt$$

Making use of the above theorem:

$$\frac{ds}{dt} \supset \lim_{\Delta t\to 0} \left[\frac{e^{i2\pi f\Delta t} S(f) - S(f)}{\Delta t}\right]$$

$$\frac{ds}{dt} \supset i2\pi fS(f)$$

The time derivative enhances high frequencies of the spectrum, attenuates low frequencies, and suppresses the zero frequency term.

Amplitude modulation. Let us look for the spectrum of a sine signal of frequency f_o when its amplitude is modulated by an envelope e(t) with spectrum E(f)

$$s(t) = e(t) \cos 2\pi f_o t$$

The Fourier transform of this signal is

$$S(f) = \int_{-\infty}^{\infty} e^{-i2\pi ft} e(t) \cos 2\pi f_o t \, dt$$

$$S(f) = \frac{1}{2} \int_{-\infty}^{\infty} e(t)[e^{-i2\pi(f-f_o)t} + e^{-i2\pi(f+f_o)t}]dt$$

$$S(f) = \frac{1}{2} E(f-f_o) + \frac{1}{2} E(f+f_o)$$

The spectrum of a signal with a carrier is obtained by translating the spectrum of the envelope divided by two in each direction through a distance equal to the carrier frequency (fig. 1.17).

Frequency modulation. The above result can be generalized. Let s(t) be a frequency modulated signal: $s(t) = e(t) \cos \phi(t)$. This is equivalent to saying that the phase is not a linear function of time

$$\phi(t) = \omega_o t + \psi(t)$$

The spectrum of this signal is ($\omega = 2\pi f$)

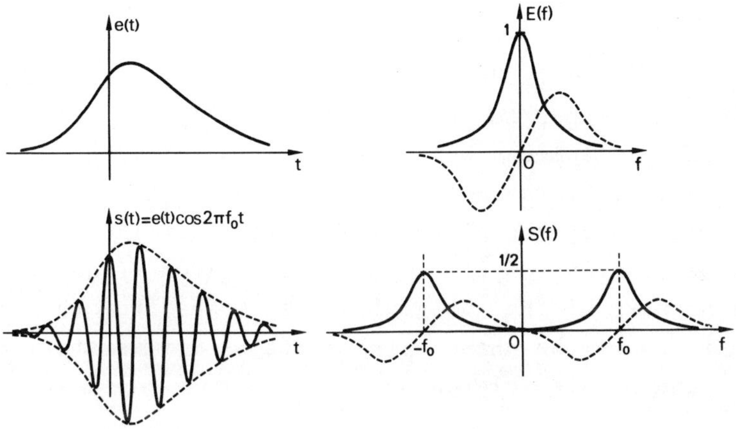

Fig. 1.17. Spectrum of an amplitude modulated sine signal. The spectrum is obtained by translating the spectrum of the envelope divided by two in each direction through a distance equal to the carrier frequency.

$$S(\omega) = \frac{1}{2} \int_{-\infty}^{\infty} e(t) \exp i[(\omega_o - \omega)t + \psi(t)]\, dt$$

$$+ \frac{1}{2} \int_{-\infty}^{\infty} e(t) \exp -i[(\omega_o + \omega)t + \psi(t)]dt$$

Denoting by $E(\omega)$ the spectrum[1] of $e(t)e^{i\psi(t)}$

$$E(\omega) = \int_{-\infty}^{\infty} e(t) \exp i[\psi(t) - \omega t]dt \qquad (1.21)$$

$S(\omega)$ is expressed by

$$S(\omega) = \frac{1}{2} E(\omega - \omega_o) + \frac{1}{2} E^*(-\omega - \omega_o)$$

When the envelope is even $(e(-t) = e(t))$ and when the frequency modulation

$$\omega = \frac{d\phi}{dt} = \omega_o + \frac{d\psi}{dt}$$

1. We now use ω for the sake of simplicity. Strictly speaking, we should have two notations for S and E.

is symmetric about ω_o : $\psi(-t) = \psi(t)$, we have

$$E(-\omega) = E(\omega)$$

and

$$S(\omega) = \frac{1}{2} E(\omega - \omega_o) + \frac{1}{2} E^*(\omega + \omega_o) \qquad (1.22)$$

The spectrum is decomposed into two parts, of identical amplitude and opposite phase, centred on ω_o and $-\omega_o$.

This spectrum can be obtained, to a fairly good approximation, by the method of stationary phase. The time variation of the phase $\alpha(t) = \psi(t) - \omega t$ induces very rapid oscillations of $e^{i\alpha(t)}$. At any given frequency ω_p, the major part of the integral (1.21) comes from the interval around t_p where the phase $\alpha_p(t) = \psi(t) - \omega_p t$ is stationary. Thus t_p and ω_p are related by

$$\left(\frac{d\alpha_p}{dt}\right)_{t_p} = 0 \rightarrow \psi'(t_p) = \omega_p = \omega(t_p) - \omega_o \qquad (1.23)$$

and in the vicinity of t_p

$$\alpha_p(t) \simeq \alpha_p(t_p) + \frac{1}{2} \alpha''_p(t_p)(t - t_p)^2$$

Inserting this expansion into (1.21) leads to

$$E(\omega_p) \simeq e^{i\alpha_p(t_p)} \int_{-\infty}^{\infty} e(t) \exp[\frac{i}{2} \alpha''_p(t_p)(t-t_p)^2]dt$$

Assuming the variations of e(t) are slow, compared to the exponential

$$E(\omega_p) \simeq e(t_p)e^{i\alpha_p(t_p)} \int_{-\infty}^{\infty} \exp[\frac{i}{2} \alpha''_p(t_p)(t - t_p)^2]dt.$$

If we change the variable:

$$\alpha''_p(t_p)(t-t_p)^2 = 2\pi y^2 \rightarrow dt = \sqrt{\frac{2\pi}{|\alpha''_p(t_p)|}}\, dy$$

$$E(\omega_p) \simeq \sqrt{\frac{2\pi}{|\alpha''_p(t_p)|}}\, e(t_p)e^{i\alpha_p(t_p)} \int_{-\infty}^{+\infty} e^{i\pi y^2}\, dy$$

and if we remember that

$$\int_{-\infty}^{\infty} e^{i\pi y^2}\, dy = e^{i\pi/4}$$

we are left with

$$E(\omega_p) \simeq \sqrt{\frac{2\pi}{|\alpha''_p(t_p)|}}\, e(t_p) \exp i[\frac{\pi}{4} + \alpha_p(t_p)] \qquad (1.24)$$

or, in terms of the phase $\psi(t)$:

$$E(\omega_p) \simeq \sqrt{\frac{2\pi}{|\psi''(t_p)|}}\, e(t_p) \exp i[\frac{\pi}{4} + \psi(t_p) - \omega_p t_p] \quad (1.25)$$

The right hand side, which implicitly depends on ω_p through t_p and equation (1.23), usually gives a good approximation to the spectrum $E(\omega)$. It is left as an exercise for the reader (exercise 1.7) to apply this result to a linear frequency modulated signal. The absolute value of the spectrum is given simply by

$$|E(f - f_0)|^2 \simeq |t'(f)| e^2[t(f)] \quad (1.26)$$

since (1.23) implies

$$f_p = f(t_p) - f_0 \quad \text{and} \quad \psi''(t_p) = 2\pi f'(t_p) = \frac{2\pi}{t'[f(t_p)]}$$

where $t'(f) = dt/df$ is the time derivative of the instantaneous signal frequency.

To ensure that these concepts do not remain too abstract, we give (Fig. 1.18) the direct and inverse Fourier transforms of some functions of the dimensionless and conjugate variables x and s.

This list immediately provides the spectrum of some common signals. For example, for a pulse of amplitude A, duration Θ, with a carrier at frequency f_0 (Fig. 1.19), we only need to take (third line) the Fourier transform of the pulse $\Pi(x)$ of unit amplitude and unit width

$$\Pi(x) \supset \frac{\sin \pi s}{\pi s}$$

The signal is of the form

$$s(t) = A\Pi(\frac{t}{\Theta}) \cos 2\pi f_0 t$$

The similarity theorem gives the spectrum of the envelope

$$E(f) = \Theta A \frac{\sin \pi\Theta f}{\pi\Theta f}$$

and the modulation theorem yields

$$S(f) = A \frac{\sin [\pi(f+f_0)\Theta]}{2\pi(f+f_0)} + A \frac{\sin [\pi(f-f_0)\Theta]}{2\pi(f-f_0)} \quad (1.27)$$

The distance between the two zeroes about f_0 is $2/\theta$ at a level -3dB wrt the peak value, the width of the spectrum is $.885/\Theta$.

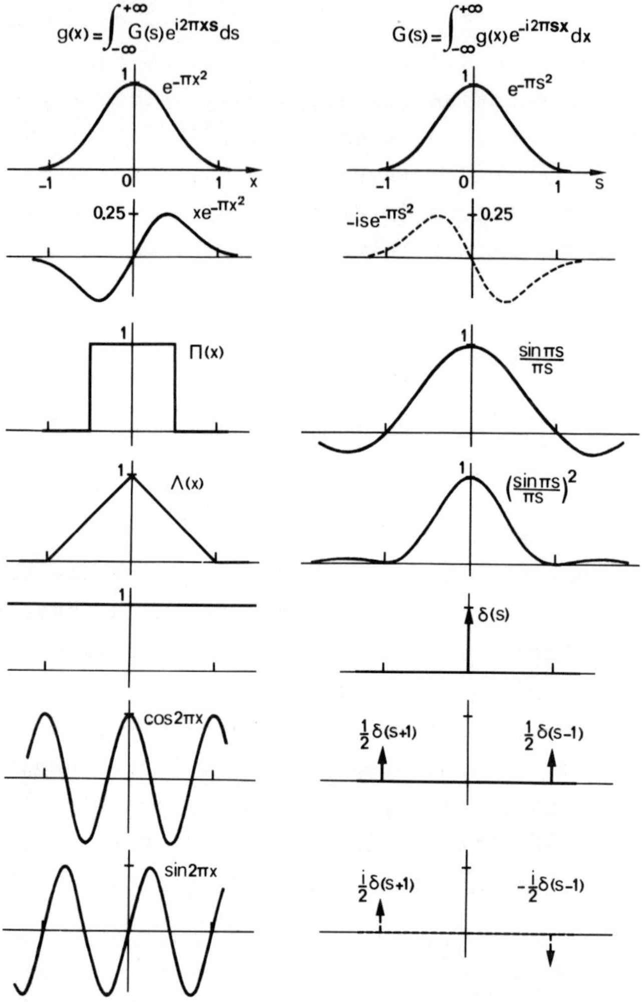

Fig. 1.18. Examples of Fourier transforms. Signals and spectra are always limited in practice; this table should be read from a physical point of view. For example a (sin x)/x function has its spectrum closer to a square signal if it is defined by a larger number of oscillations. Fourier transform theory is more tricky than one would expect at first sight (see the referenced book by E. Roubine). From a mathematical point of view, the correspondence on line 3 (read from the right to the left) is not rigorous since the modulus of (sin x)/x is not summable. The function (sin x)/x has no Fourier transform.

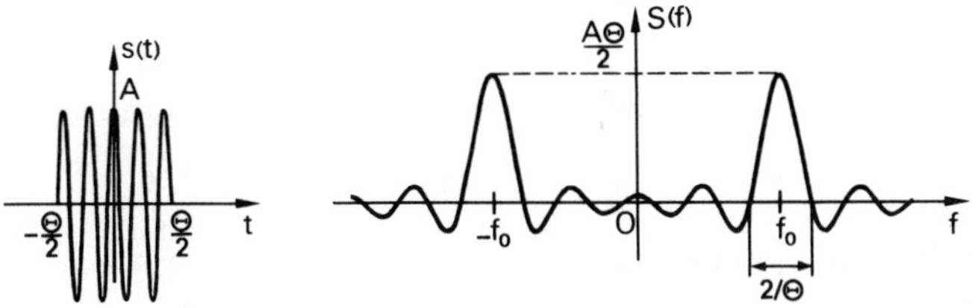

Fig. 1.19. Spectrum of a square pulse with a carrier.

The first line of the table shows that the Fourier transform of a Gaussian function is a Gaussian. The second line is deduced from the first line through the differentiation theorem.

We should recall that the Fourier transform of the Dirac pulse is equal to 1, according to 1.18.

The last two lines show that the spectrum of a sine wave at frequency f_o consists of two lines at f_o and $-f_o$.

We leave it as an exercise to check the other relationships.

We are now able to deal with the response of a linear time invariant system.

1.3.3. Frequency Response of a Linear Time Invariant System

Let $s_1(t) = e^{i\omega t}$ be a harmonic signal of angular frequency ω, and $s_2(t)$ the response to $s_1(t)$. Because of the time invariance and linearity of the system, the response to the time-shifted input signal

$$s_1(t + \tau) = e^{i\omega\tau}\, s_1(t)$$

is

$$s_2(t + \tau) = e^{i\omega\tau}\, s_2(t)$$

Let us define $t_o = t + \tau$; we obtain

$$s_2(t_o)e^{-i\omega t_o} = s_2(t)e^{-i\omega t}$$

This result being true for all t and t_o, it follows that the product

$$s_2(t)e^{-i\omega t} = H(\omega)$$

is independent of time; its value $H(\omega)$ depends on the angular frequency and it constitutes the frequency response of the system; it is also called the transfer function.

To sum up, the response to a harmonic signal $s_1(t) = e^{i\omega t}$ is a harmonic function at the same frequency

$$s_2(t) = H(\omega)e^{i\omega t} \tag{1.28}$$

The action of the system is described by a linear transform, of which every exponential $e^{i\omega t}$ is an eigenfunction with eigenvalue $H(\omega)$.

Let us apply a real cosine input signal:

$$s_1(t) = \cos \omega t = \frac{e^{i\omega t} + e^{-i\omega t}}{2}$$

The output

$$s_2(t) = \frac{1}{2}[H(\omega)e^{i\omega t} + H(-\omega)e^{-i\omega t}]$$

has to be real, so that

$$[H(\omega)e^{i\omega t}]^* = H(-\omega)e^{-i\omega t}$$

or equivalently

$$H^*(\omega) = H(-\omega)$$

Writing this in terms of modulus and phase angle $\Phi(\omega)$:

$$H(\omega) = |H(\omega)|e^{i\Phi(\omega)}$$

provides

$$s_2(t) = |H(\omega)| \frac{e^{i(\omega t+\Phi)} + e^{-i(\omega t+\Phi)}}{2}$$

or

$$s_2(t) = |H(\omega)| \cos(\omega t + \Phi)$$

For a unit amplitude cosine input, the output is a synchronous cosine wave, the amplitude and phase of which are equal to the modulus and phase angle of the frequency response.

1.3.4. Impulse Response

The impulse response $h(t)$ is the output for a Dirac pulse input $\delta(t)$.

From equation (1.17) the expression for the input signal is:

$$s_1(t) = \int_{-\infty}^{\infty} s_1(\tau)\delta(t - \tau)\, d\tau.$$

Because of both the linearity and time invariance of the system the output is

$$\boxed{s_2(t) = \int_{-\infty}^{\infty} s_1(\tau)h(t - \tau)\, d\tau} \qquad (1.29)$$

This operation of s and h is called convolution. By definition, the convolution of two functions f(x) and g(x) is the function

$$C(x) = \int_{-\infty}^{\infty} f(y)\, g\,(x - y)\, dy \qquad (1.30)$$

which is briefly denoted by

$$C(x) = f(x) * g(x)$$

Calculating the result of convolution can be split into three steps, as shown in fig. 1.20.

1) reversal and translation of g(x), to obtain g(x-y)
2) calculating the product g(x-y)f(y)
3) integration (equation 1.30).

Convolution is commutative (it is also associative): changing the variable by substituting u = x-y, yields

$$f_* g = \int_{-\infty}^{\infty} f\,(x - u)\, g\,(u)\, du = g_* f$$

We recall, at this point, that the correlation function is defined by

$$\Gamma(x) = \int_{-\infty}^{\infty} f(y)\, g^*(y - x)\, dy \qquad (1.31)$$

If g is a real function, calculating $\Gamma(x)$ involves the same procedure as for C(x), except for the reversal of g. If g is identical to f, $\Gamma(x)$ is the autocorrelation function of f.

According to (1.29), the response to a harmonic input $s_1(t) = e^{i\omega t}$ is:

$$s_2(t) = \int_{-\infty}^{\infty} e^{i\omega\tau}\, h(t - \tau)\, d\tau$$

which on substituting $u = t - \tau$, becomes

$$s_2(t) = e^{i\omega t} \int_{-\infty}^{\infty} e^{-i\omega u}\, h(u)\, du$$

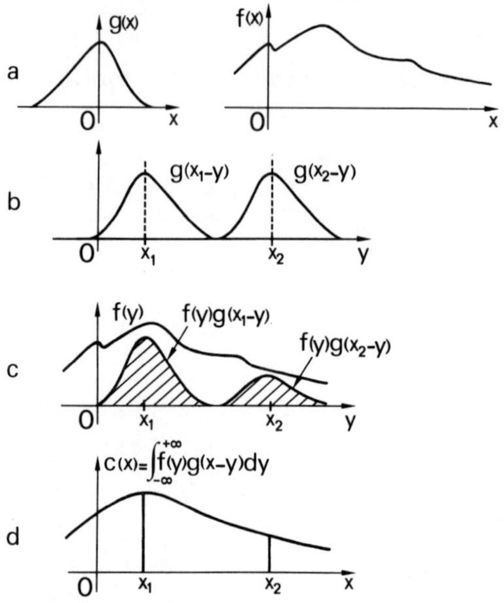

Fig. 1.20. Convolution between two functions.

a) functions f(x) and g(x)

b) inversion and translation of g(x)

c) product of f(y) and g(x - y)

d) integral of this product at every point.

The output/input ratio is simply the transfer function (equation 1.28), so that

$$H(\omega) = \int_{-\infty}^{\infty} e^{-i\omega t}\, h(t)\, dt \tag{1.32}$$

Thus, the transfer function and the impulse response are Fourier transforms of each other.

- the frequency response of a linear time invariant system is the spectrum of the impulse response
- or, inverting (1.32), its impulse response is the inverse Fourier transform of the transfer function.

$$h(t) = \int_{-\infty}^{\infty} e^{i2\pi ft}\, H(f)\, df \tag{1.33}$$

1.3.5. Response to an Arbitrary Signal

We have two methods of calculating the response to any signal of a linear time invariant system.

1) The time analysis provides $s_2(t)$, through the convolution of the impulse response and the input signal:

$$s_2(t) = h(t)_* s_1(t) \tag{1.34}$$

2) The frequency analysis relies on the expansion of the input as a sum of harmonic signals:

$$s_1(t) = \int_{-\infty}^{\infty} S_1(f) e^{i2\pi ft}\, df$$

and, since the system is linear, leads to

$$s_2(t) = \int_{-\infty}^{\infty} S_1(f)\; H\,(f) e^{i2\pi ft}\, df$$

Input →	Invariant Linear System	→ Output
Harmonic: $e^{i2\pi ft}$	$H(f)$	Harmonic: $H(f)e^{i2\pi ft}$
Sinusoidal: $\cos 2\pi ft$	$H(f)=\lvert H(f)\rvert e^{i\Phi(f)}$	Dephased sinusoidal: $\lvert H(f)\rvert \cos(2\pi ft + \Phi)$
Dirac impulse: $\delta(t)$		Impulse response: $h(t)$
Signal: $s_1(t)$ ↕ of spectrum: $S_1(f)$	convolution with $h(t)$ ↕ multiplication by $H(f)$	signal: $s_2(t)=h(t)*s_1(t)$ ↕ of spectrum: $S_2(f)=H(f)S_1(f)$

Fig. 1.21. Response of a time invariant linear system.

The output signal spectrum is the product of the input signal and the transfer function of the system

$$S_2(f) = H(f)\; S_1(f) \tag{1.35}$$

This property also comes out from (1.34), for the Fourier transform of the convolution product of two functions is equal to the scalar product of the Fourier transforms of both functions.

Going through a linear invariant system also implies a phase shift $\Phi(\omega)$, equal to the phase angle of the frequency response. When this phase shift is not proportional to the frequency, the system is dispersive. The time delay of a group of wave components of the signal whose frequencies lie about ω_0 is defined, as in § 1.1.3., by

$$\tau_g = -\left(\frac{d\Phi}{d\omega}\right)_{\omega_0} \qquad (1.36)$$

Table 1.21 is a summary of these results, the arrows indicate the pairs of functions related by a Fourier transform.

1.3.6. Connexion Between a Signal Spectrum and the Radiation Pattern of a Source

Fourier transform techniques are encountered in several fields of physics. They link some pairs of conjugate variables:

- frequency f and time t in signal theory
- momentum $\vec{p}$ and position $\vec{r}$ in quantum mechanics
- "direction" $\sin\alpha$ and aperture of the source x/λ as indicated below.

The radiation from a source involves two problems

- the emission by an antenna, which itself provides the radiated energy (active element)
- the diffraction of a parallel beam by an aperture which re-emits the energy of an incident beam (passive element).

Let us consider a linear source along the x-axis (fig. 1.22) and let

$$u(x) = A(x)\cos[\omega t + \Phi(x)]$$

be the vibration at point P of the source, located at x. This can be written as

$$u(x) = \mathrm{Re}[g_0(x)e^{i\omega t}]$$

where $g_0(x) = A(x)e^{i\Phi(x)}$ denotes the complex amplitude distribution of the source. At distance r, the element [x, x + dx] creates a vibration

$$g_o(x)\ e^{-i2\pi r/\lambda}\ dx$$

$2\pi(r/\lambda)$ is the phase shift due to the path length r. The overall vibration at the point M defined by the vector $\vec{R} = \overrightarrow{OM}$ is the integral

$$G_o(\vec{R}) = \int_{-\infty}^{\infty} g_o(x)\ e^{-i2\pi\frac{r}{\lambda}}\ dx \qquad (1.37)$$

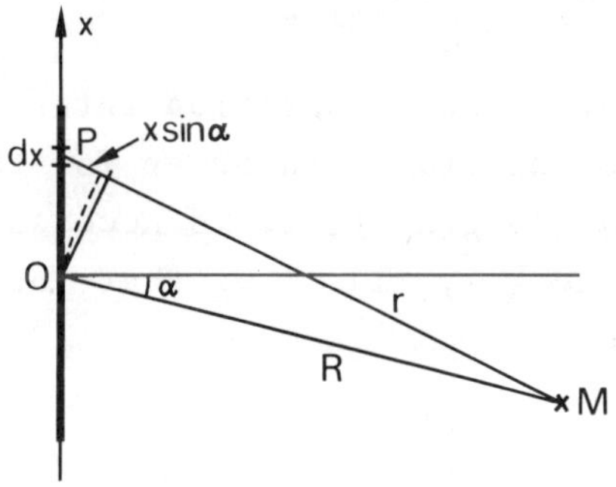

Fig. 1.22. Radiation of a linear source. The vibration originated at P reaches M with a phase delay $2\pi\frac{x}{\lambda}\sin\alpha$ with respect to the vibration originating at the source centre.

Far from the source, i.e. at distances such that R >> x, a good approximation is

$$r = R + x\sin\alpha$$

And equation (1.37) becomes

$$G_o(\vec{R}) = e^{\frac{-i2\pi R}{\lambda}} \int_{-\infty}^{\infty} g_o(x)\ e^{-i2\pi\frac{x}{\lambda}\sin\alpha}\ dx$$

Two factors appear, the former $e^{-i(2\pi R/\lambda)}$ is the phase shift at the average distance R, the latter is the "angular" part

$$G_o\ (\sin\alpha) = \int_{-\infty}^{\infty} g_o(x) e^{-i2\pi\frac{x}{\lambda}\sin\alpha}\ dx$$

At large distances, the angular distribution $G_o(\sin\alpha)$ of the radiation is related to the source amplitude $g_o(x)$ through Fourier transform. In order to put this in the canonical form, we denote $s = \sin\alpha$ and take the wave length λ as the unit length.

$$G(s) = \int_{-\infty}^{\infty} g(\tfrac{x}{\lambda}) e^{-i2\pi(\frac{x}{\lambda})s} d(\tfrac{x}{\lambda}) = \frac{G_o(s)}{\lambda} \text{ with } g(\tfrac{x}{\lambda}) = g_o(x)$$

The function G(s), which is proportional to $G_o(s)$, is called the **radiation** pattern or the angular spectrum of the source. Inverting the Fourier transform, yields

$$g(\tfrac{x}{\lambda}) = \int_{-\infty}^{\infty} G(s)\, e^{i2\pi\frac{x}{\lambda}s} ds.$$

To be accurate, the integration interval is [-1, +1] since s is a sine function; in order for this integration to be nearly equal to $g(x/\lambda)$, it is necessary that G(s) be nearly zero outside [-1, +1], i.e. that the source be sharply directive.

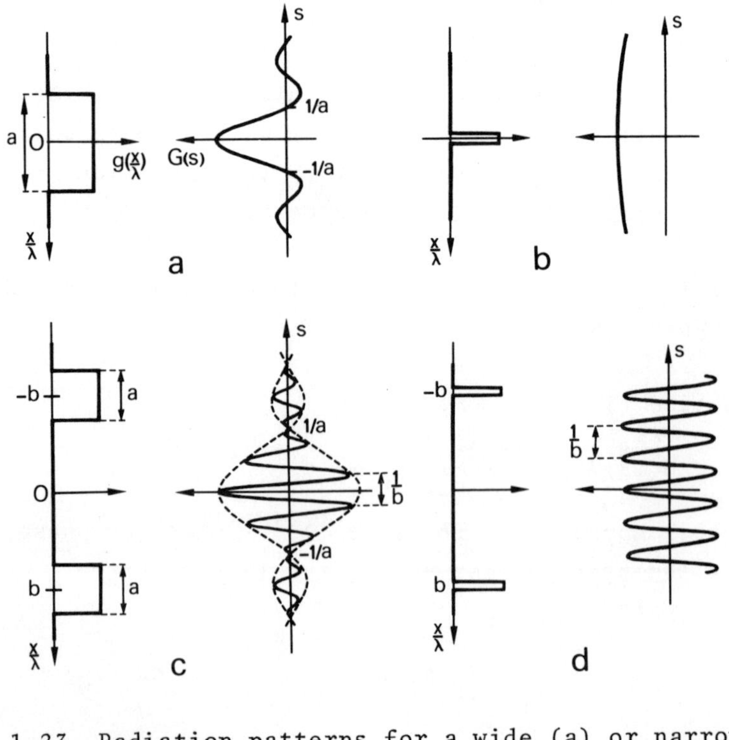

Fig. 1.23. Radiation patterns for a wide (a) or narrow (b) source, and for two synchronous wide (c) or narrow (d) sources.

The following correspondences will stress the analogy between signals and spectra:

Signals	Sources
time t	position (in wave length units) x/λ
frequency f	direction $s = \sin\alpha$
signal s(t)	amplitude distribution of the source $g(x/\lambda)$
spectrum S(f)	radiation diagram G(s).

The previous examples are easily dealt with by making use of the Fourier transform table (fig. 1.18).

- The radiation pattern of an antenna of length $a\lambda$ is

$$\frac{\sin \pi sa}{\pi s}$$

- An aperture which is small compared to λ has a flat angular spectrum;
- The radiation pattern of a pair of antennae, with length a and spacing $2b\lambda$ is

$$2\,\frac{\sin \pi sa}{\pi s}\cos 2\pi bs;$$

- Two apertures small with respect to the wavelength yield a sine spectrum (Young's slits experiment): $\cos 2\pi bs$.

These examples, depicted in fig. 1.23, show that the longer the antenna, the more directive it is.

We have assumed the vibration to be a scalar quantity. The results are still valid for vector vibrations since the radiation is supposed to be observed far from the source. The medium is considered as isotropic.

BIBLIOGRAPHY

General properties of waves

J. Bok and P. Morel. 'Cours de physique. Mécanique-Ondes', Paris: Hermann (1968).

W.C. Elmore and M.A. Heald. 'The physics of waves', New York: McGraw-Hill (1969).

Wave packets. Group velocity

P. Grivet. 'Physique des lignes de haute fréquence et d'ultra haute fréquence', tome I, chap. 5, p. 195. Paris: Masson et Cie (1969).

Atomic chains

L. Brillouin. 'Wave propagation in periodic structures'. New York: Dover (1953).

Distribution theory

L. Schwartz. 'Méthodes mathématiques pour les sciences physiques', p. 76. Paris: Hermann (1965).

Signals. Fourier transforms

E. Roubine. 'Introduction à la théorie de la communication', tome I. Paris: Masson et Cie (1970).
R. Bracewell. 'The Fourier transform and its applications'. New York: McGraw Hill (1965).
A. Papoulis. 'The Fourier Integral and its applications.' New York: McGraw Hill (1965).

EXERCISES

1.1. Prove the equation

$$\langle \mathrm{Re}[u_1] \,.\, \mathrm{Re}[u_2] \rangle = \frac{1}{2}\,\mathrm{Re}[u_1 u^*_2]$$

where $\langle\ \rangle$ denotes the time-average value and $u_k = A_k \exp i(\omega t + \Phi_k)$, $k = 1, 2$.

Solution

$$\langle \mathrm{Re}[u_1].\mathrm{Re}[u_2]\rangle = A_1A_2 \langle \cos(\omega t + \Phi_1)\cos(\omega t+\Phi_2)\rangle$$

$$= \frac{A_1A_2}{2} \langle \cos(2\omega t + \Phi_1 + \Phi_2) + \cos(\Phi_1 - \Phi_2)\rangle$$

Since the mean value of a cosine is zero:

$$\langle \mathrm{Re}[u_1].\mathrm{Re}[u_2]\rangle = \frac{A_1A_2}{2}\cos(\Phi_1-\Phi_2) = \frac{1}{2}\,\mathrm{Re}[u_1u_2^*].$$

1.2. Assuming an adiabatic compression (pv^γ = constant), state the expression for sound velocity in a perfect gas of molecular mass M at temperature θ. Apply the result to air at 0°C ($M = 29$ g, $\gamma = 1.4$).

Solution. The adiabatic compressibility coefficient

$$\chi_\sigma = -\frac{1}{v}\left(\frac{\partial v}{\partial p}\right)_\sigma = \frac{1}{\gamma p_o}$$

is derived from the equation for adiabatic transformations: pv^γ = constant. Inserting this into formula $V = 1/\sqrt{\rho_o\chi}$, we obtain (for a perfect gas) $V = \sqrt{\gamma R\theta/M}$. In the case of air V = 330 m/s.

1.3. Show that, for a progressive wave in a given fluid, the ratio $\Delta p/\dot{u}$ (of extra pressure Δp to the vibration velocity $\dot{u} = \partial u/\partial t$) is a constant.

Solution. Since $u = F(t - \frac{x}{V})$: $\Delta p = -\frac{1}{\chi}\frac{\partial u}{\partial x}$ $= \frac{1}{\chi V} F'$ and $\dot{u} = F'$; therefore $\frac{\Delta p}{\dot{u}} = \frac{1}{\chi V} = \rho_o V$.

This constant $Z = \rho_o V$ is the acoustic impedance per unit area of the medium (in mechanics the impedance is defined as the ratio of force to velocity).

1.4. By expressing the continuity of the displacement u and the pressure p at the boundary between two media of elastic impedances Z and Z' (see exercise 1.3), calculate the reflexion and transmission coefficient (A_R/A_I and A_T/A_I) (fig. 1.24).

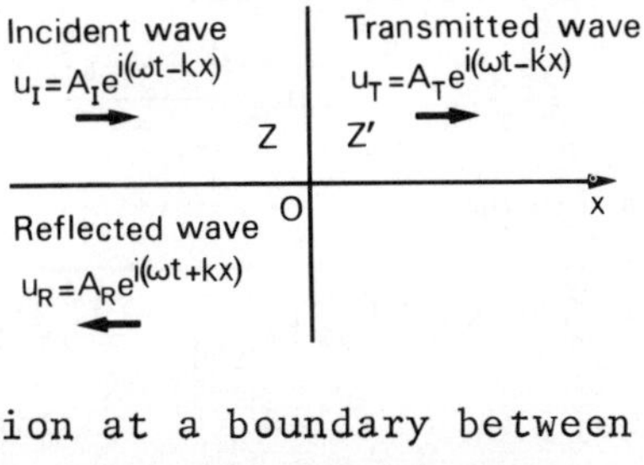

Fig. 1.24. Reflexion at a boundary between two media with different elastic impedances.

Solution. The continuity of displacement at the boundary x = 0 leads to $A_I + A_R = A_T$. Making use of the results of exercise 1.3: $(\Delta p)_I = Z\dot{u}_I$, $(\Delta p)_R = -Z\dot{u}_R$, $(\Delta p)_T = Z'\dot{u}_T$ and the continuity of pressure: $(\Delta p)_I + (\Delta p)_R = (\Delta p)_T$, leads to $Z(A_I - A_R) = Z'A_T$

Hence, the reflexion and transmission coefficients

$$\frac{A_R}{A_I} = \frac{Z-Z'}{Z+Z'} \qquad \frac{A_T}{A_I} = \frac{2Z}{Z+Z'}$$

depend only on the ratio of the elastic impedances. Limiting cases,

1) $\frac{Z'}{Z} = \infty$, $\frac{A_R}{A_I} = -1$, $\frac{A_T}{A_I} = 0$. The reflexion is total and the displacement vanishes at a rigid surface.

2) $\frac{Z'}{Z} = 0$, $\frac{A_R}{A_I} = 1$; the reflexion is total and the pressure differential Δp vanishes at a free surface. For x > 0, there is no matter ($Z' = 0 \rightarrow \rho' = 0$), and consequently no displacement.

1.5. Show that the equation of motion (1.14) of a particle reduces to the propagation equation (1.13) in a continuous medium, when $\lambda >> a$.

Solution. If $\lambda >> a$, the difference $u_{n+1} - u_n$ is infinitesimal, equal to du:

$$\frac{u_{n+1}-u_n}{a} = \left(\frac{\partial u}{\partial x}\right)_{x_{n+1}} \quad \text{and} \quad \frac{u_n-u_{n+1}}{a} = \left(\frac{\partial u}{\partial x}\right)_{x_n}$$

Equation 1.14 becomes

$$M\frac{\partial^2 u}{\partial t^2} = Ka\left[\left(\frac{\partial u}{\partial x}\right)_{x_{n+1}} - \left(\frac{\partial u}{\partial x}\right)_{x_n}\right].$$

In the same way, we are led to equation (1.13)

$$\frac{\partial^2 u}{\partial t^2} = \frac{Ka^2}{M}\frac{\partial^2 u}{\partial x^2} = V_o^2\frac{\partial^2 u}{\partial x^2}$$

since $V_o = a\sqrt{K/M}$ is the low frequency elastic wave velocity.

1.6. Let us consider a chain where all atoms are equidistant, their mass being M_1 for even rank atoms and $M_2 > M_1$ for odd rank atoms (fig. 1.25). Write down the equation of motion for the (2n)th and the (2n+1)th atoms. By setting $u_{2n} = A_1 \exp i(\omega t - 2nka)$ and $u_{2n+1} = A_2 \exp i(\omega t - (2n+1)ka)$, establish the relationship between ω and k, and draw the dispersion curves. Consider the cases $ka \ll 1$ and $ka = \pi/2$.

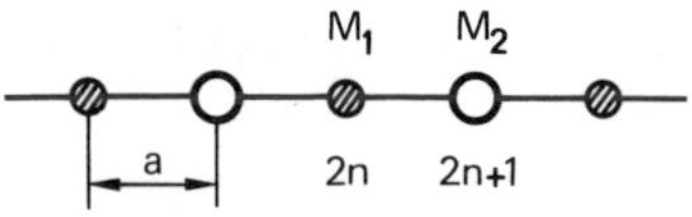

Fig. 1.25. Diatomic chain.

Solution. The equations of motion for the (2n)th and (2n+1)th atoms are analogous to (1.14):

$$M_1 \frac{d^2u_{2n}}{dt^2} = K(u_{2n+1} + u_{2n-1} - 2u_{2n})$$

$$M_2 \frac{d^2u_{2n+1}}{dt^2} = K(u_{2n+2} + u_{2n} - 2u_{2n+1}).$$

Inserting the expressions for the displacements, yields

$$\begin{cases} (2K - M_1\omega^2)A_1 - (2K \cos ka)\, A_2 = 0 \\ -(2K \cos ka)\, A_1 + (2K - M_2\omega^2)\, A_2 = 0. \end{cases}$$

The compatibility of these two homogeneous equations requires

$$\begin{vmatrix} 2K - M_1\omega^2 & -2K \cos ka \\ -2K \cos ka & 2K - M_2\omega^2 \end{vmatrix} = 0$$

and provides the dispersion equation

$$(\omega_{\frac{1}{2}})^2 = K\left(\frac{1}{M_1} + \frac{1}{M_2}\right) \mp K\left[\left(\frac{1}{M_1} + \frac{1}{M_2}\right)^2 - 4\frac{\sin^2 ka}{M_1M_2}\right]^{1/2}$$

The dispersion curve $\omega_1(k)$ is very much like that for the monoatomic chain (fig. 1.26). If $ka \ll 1$, the solutions $\omega_1 \simeq (2K/M_1+M_2)^{1/2}ak$ and $A_1 = A_2$ correspond to the propagation of an elastic wave at velocity $V_o = a\sqrt{2K/M_1+M_2}$, hence the name "acoustic branch" given to the curve $\omega_1(k)$. The other solution, for $ka \ll 1$, gives $\omega_2 = \sqrt{2K(1/M_1+1/M_2)}$ and $M_1A_1 + M_2A_2 = 0$: two neighbouring atoms vibrate with opposite phases and amplitudes A_1 and A_2 such that the centre of mass of the molecule is motionless. If the two particles carry electric charges of opposite signs (ions), one can excite this vibration by means of an electric field of frequency $\omega_2(0)/2\pi$ ($\simeq 10^{12}$ to 10^{13} Hz), in the infrared region, hence the name "optical branch" given to the curve $\omega_2(k)$.

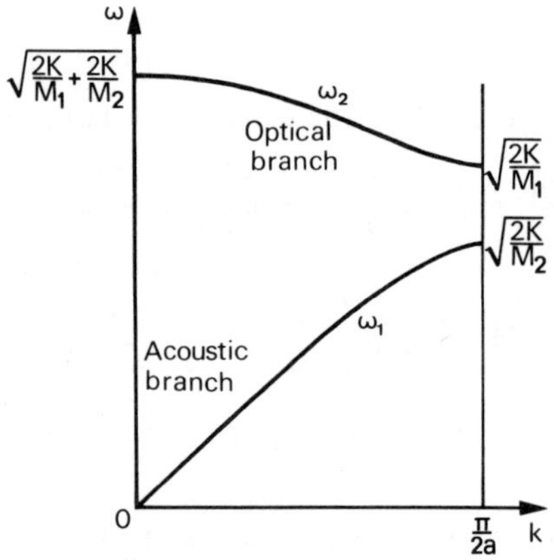

Fig. 1.26. Dispersion curve for the longitudinal mode of a diatomic chain.

For $ka = \frac{\pi}{2}$ $\begin{cases} \omega_1 = \sqrt{2K/M_2} \text{ and } A_1 = 0\text{: lighter atoms are motionless} \\ \omega_2 = \sqrt{2K/M_1} \text{ and } A_2 = 0\text{: heavier atoms are motionless.} \end{cases}$

1.7. By the method of stationary phase, calculate the spectrum of a pulse of duration Θ, the carrier being linearly frequency modulated over a bandwidth B (fig. 9.12):

$$s(t) = \Pi(\frac{t}{\Theta}) \cos \phi (t) \quad \text{with } f = \frac{1}{2\pi}\frac{d\phi}{dt} = f_o + \frac{B}{\Theta} t$$

Solution. For this signal $\psi(t) = \pi\frac{B}{\Theta} t^2$ and according to equation 1.23.

$$\omega_p = 2\pi \frac{B}{\Theta} t_p \rightarrow t_p = \frac{\Theta}{B} f_p$$

Inserting this into equation (1.25):

$$E(f) = \sqrt{\frac{\Theta}{B}} \Pi (\frac{f}{B}) \exp i [\frac{\pi}{4} - \frac{\pi\Theta}{B} f^2].$$

The spectrum $S(f) = 1/2\ E(f-f_o) + 1/2\ E^*(f+f_o)$ is enveloped by two square functions, of width B centered at f_o and $-f_o$. The phase has a parabolic frequency variation. The complete calculation is performed in section 9.3.2. and shows that the method of stationary phase is all the more valid when $B\Theta$ is greater than 1..

1.8. Calculate the Fourier transform of $e^{-\pi x^2}$, given that

$$\int_{-\infty}^{\infty} e^{-\pi x^2} dx = 1.$$

Solution.

$$G(s) = \int_{-\infty}^{+\infty} e^{-\pi x^2} e^{-i2\pi sx} dx = e^{-\pi s^2} \int_{-\infty}^{+\infty} e^{-\pi(x+is)^2} dx.$$

Since the complex function $e^{-\pi z^2}$ has no pole, the integral over the contour Γ (fig. 1.27) vanishes:

$$\int_\Gamma e^{-\pi z^2} dz = \int_{-a}^{+a} e^{-\pi x^2} dx + i\int_0^s e^{-\pi(a+iy)^2} dy$$

$$+\int_a^{-a} e^{-\pi(x+is)^2} dx + i\int_s^0 e^{-\pi(-a+iy)^2} dy = 0$$

When $a \rightarrow \infty$, the only remaining term

$$\int_{-\infty}^{+\infty} e^{-\pi(x+is)^2} dx = \int_{-\infty}^{+\infty} e^{-\pi x^2} dx = 1 \quad \forall s,$$

hence $G(s) = e^{-\pi x^2}$.

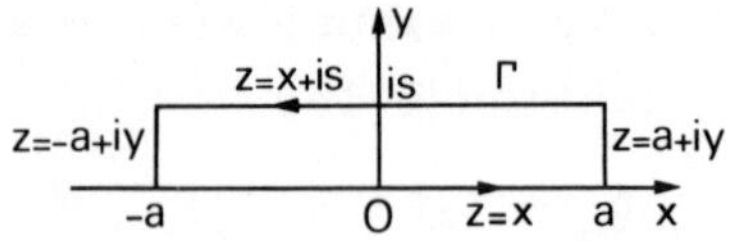

Fig. 1.27. Integration path.

1.9. Prove that $\Lambda(x) = \Pi(x)*\Pi(x)$. From the theorem for the Fourier transform of a convolution, deduce that

$$\Lambda(x) \supset \left(\frac{\sin \pi s}{\pi s}\right)^2.$$

Solution. The convolution integral

$\int_{-\infty}^{\infty} \Pi(y)\,\Pi(x-y)\,dy$, equal to the area shared by the rectangles centered at 0 and at $y = x$, is $\Lambda(x)$ (fig. 1.28). Since the Fourier transform of a convolution product is equal to the product of both Fourier transforms:

$$\Lambda(x) = \Pi(x)*\Pi(x) \supset \left(\frac{\sin \pi s}{\pi s}\right)^2.$$

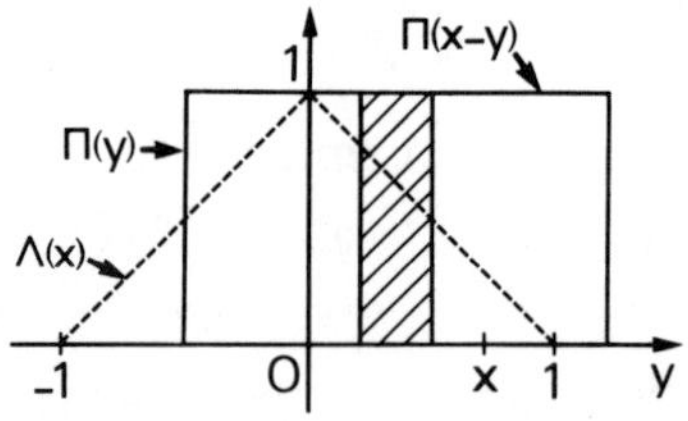

Fig. 1.28. Convolution of two square functions.

1.10. Show that:

$$\sum_{m=-\infty}^{+\infty} e^{-i2\pi mx} = \sum_{n=-\infty}^{+\infty} \delta(x - n).$$

Solution. Let $T(x)$ be $\sum_{m=-\infty}^{+\infty} e^{-i2\pi mx}$. We notice that

$$e^{-i2\pi x}\,T(x) = \sum_{m=-\infty}^{+\infty} e^{-i2\pi(m+1)x} = T(x)$$

$$\Rightarrow (e^{-i2\pi x} - 1)T(x) = 0.$$

The last equation shows that T(x) vanishes for any non integer x. We now split the integral of T(x) by an arbitrary function g(x):

$$I = \int_{-\infty}^{\infty} T(x)g(x)dx = \sum_{n=-\infty}^{+\infty} I_n \qquad \text{with}$$

$$I_n = \int_{n-\frac{1}{2}}^{n+\frac{1}{2}} T(x)g(x)dx \qquad n \text{ integer.}$$

Since, in the interval $[n-\frac{1}{2}, n+\frac{1}{2}]$, T(x) is zero but for x = n:

$$I_n = Cg(n) \quad \text{where} \quad C = \int_{n-\frac{1}{2}}^{n+\frac{1}{2}} T(x)dx$$

The constant C does not depend on n, since T(x + 1) = T(x). Finally, the equation

$$\int_{-\infty}^{+\infty} T(x)g(x)dx = \sum_{m=-\infty}^{+\infty} \int_{-\infty}^{+\infty} e^{-i2\pi mx}g(x)dx$$

$$= C \sum_{n=-\infty}^{+\infty} g(n) \qquad (1.38)$$

becomes with the aid of the Fourier transform G(s):

$$\sum_{m=-\infty}^{+\infty} G(m) = C \sum_{n=-\infty}^{+\infty} g(n) \qquad (1.39).$$

In order to calculate C, we take a particular function as g.

$$g(x) = \Pi(x) \rightarrow G(s) = \frac{\sin \pi s}{\pi s}$$

C = 1 for g(0) = G(0) = 1 and g(n) = G(n) = 0 if $n \neq 0$. From (1.39), we obtain the Poisson relation

$$\sum_{m=-\infty}^{+\infty} G(m) = \sum_{n=-\infty}^{+\infty} g(n)$$

and from (1.38) (which is true for any function

g(x)) the required result is derived

$$T(x) = \sum_{n=-\infty}^{+\infty} \delta(x - n) = \sum_{m=-\infty}^{+\infty} e^{-i2\pi mx} \quad (1.40).$$

1.11. Calculate $\delta[g(x)]$.

Solution. We first calculate the integral $I = \int_{-\infty}^{\infty} f(x)\, \delta\, [g(x)]\, dx$ by performing the change of variable $y = g(x) \rightarrow dy = |g'(x)|\, dx$:

$$I = \int_{-\infty}^{\infty} f[g^{-1}(y)] \frac{\delta(y)\, dy}{|g'[g^{-1}(y)]|} = \frac{f(x_0)}{|g'(x_0)|}$$

where x_0 is a zero of $g(x)$: $x_0 = g^{-1}(0)$. If $g(x)$ vanishes for several values $x_n[g'(x_n) \neq 0]$:

$$\delta[g(x)] = \sum_n \frac{\delta(x - x_n)}{|g'(x_n)|}.$$

Chapter 2
Fundamentals of Crystallography

Elastic waves can propagate in every material, but their amplitude decreases during propagation because the atomic collisions are not purely elastic. The better the medium is ordered then the weaker is the attenuation; losses are greater in a liquid than in a solid, and greater in amorphous or polycrystalline solids than in single crystals. On the other hand, these losses grow very rapidly with increasing frequency, so that liquids are scarcely used above 50 MHz and only single crystals can be used at frequencies of the order of a GHz. Due to the anisotropy of crystals, some directions are more suitable than others for propagating a given type of wave; the wave vector and energy vector are not colinear except for some specific directions. Moreover, the generation of high frequency elastic waves requires piezoelectric crystals. A good understanding of propagation phenomena and elastic wave generation requires a knowledge of crystal properties, and especially of their symmetry.

In this chapter, we analyse crystalline structure starting from the lattice and the primitive cell. Crystals are classified according to their point group, which plays a major role as for the symmetry of macroscopic physical properties. The spatial symmetry, linked to microscopic aspects of the crystals, gains only a mention. Finally we describe the structures of some crystals commonly used in devices.

2.1. DEFINITION OF THE CRYSTALLINE STATE

It is necessary to make a distinction between amorphous solids and crystals. The former, without any specific geometrical form, are extremely viscous liquids[1]. When they melt their fluidity continuously increases with temperature (resins, glass). On cooling, the only visible change is an increase in viscosity; if the temperature is plotted against time, there is no plateau.

On the other hand, for crystal solids, during the cooling of the liquid, the temperature remains constant as soon as polyhedral solid seeds appear; these are crystals. All these polyhedra are convex and similar, i.e. the angles between natural sides of crystals of the same species are always less than 180° and constant, whatever the appearance of the sample.

Quartz polyhedra, for example, are hexagonal prisms with a pyramid at each end. There are exactly 120 degrees between two sides of the prism, and $141^o47'$ between a side of the prism and the adjacent side of a pyramid; $133^o44'$ between two successive sides of a pyramid (fig. 2.1).

The crystal sides may extend in quite different ways, as shown by any cross section perpendicular to the prism axis (fig. 2.2), but the angles are fixed. If we draw, from any point, the perpendicular to the sides of a crystal, we always obtain the same geometric figure. This property, which is known as the law of constant angle, was discovered in the eighteenth century by Romé de l'Isle.

A study of crystals of the same type shows that the crystalline medium is anisotropic and homogeneous.

The appearance of sides of well defined orientation is one consequence of anisotropy. Other consequences include impact shapes (e.g. star shaped crack produced by the action of a point on mica) the effect of corrosion

---

1. Between these two forms (liquid, single crystal), there exist "liquid crystals" characterized by a structure ordered in one or two directions, and polycrystalline solids consisting of an aggregate of microscopic single crystals, the macroscopic behaviour of which is isotropic.

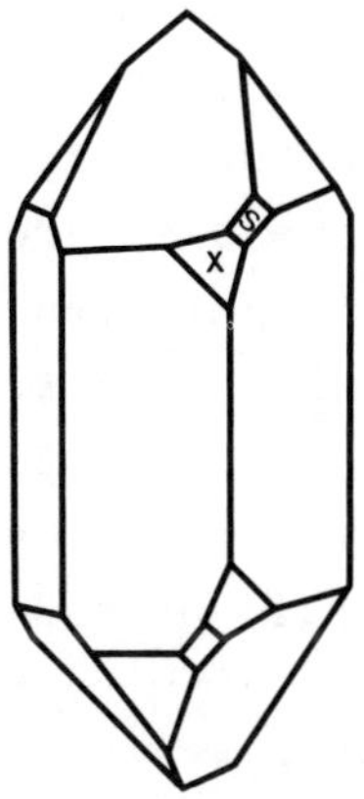

Fig. 2.1. Perfectly grown quartz crystal. In addition to the prism and pyramid sides, there are three pairs of facets s and x which reduce the symmetry about the prism axis to a triad symmetry.

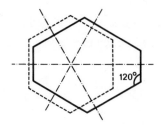

Fig. 2.2. Cross section of the prism in a real quartz crystal. Although the cross section is not a regular hexagon, the angles between successive sides are all 120°.

on different faces of a crystal (e.g. cadmium sulphide), and cleavage along specific planes (mica, calcite). On observing all these effects, R.J. Haüy proposed an ordered, periodic structure of matter at the atomic scale.

This was checked in 1912 in an experiment performed by Friedrich, following a suggestion of Von Laue, on X-ray diffraction. The discontinuous direction dependence of several physical properties such as X-ray diffraction, cleavage, natural orientation of faces, is characteristic of crystals, but most properties are continuous. From our point of view, it is significant that, because of this anisotropy, mechanical properties of crystals such as elastic wave velocity are direction-dependent.

The crystalline medium is homogeneous because different samples of a crystal, provided they are of the same size and orientation, display the same behaviour. On the atomic scale, the crystal is actually discontinuous, but the homogeneity remains in the sense that any point in the crystal has an infinity of equivalent points, along three distinct directions, that is to say, points with the same atomic environment.

2.2. CRYSTALLINE STRUCTURE

The outer appearance of crystals as well as their macroscopic properties strongly encourage the classification of a crystal by the symmetry of the natural orientation of their faces, known as the point group. Counting these symmetry classes requires some knowledge of the atomic structure, which accounts for the macroscopic properties, and that is why first we describe the crystal structure.

2.2.1. Periodicity in crystals. Lattice. Rows. Lattice planes. Cells

A crystalline medium is characterized by the existence of an infinity of geometrical points, all equivalent to a given point O in the crystal. They all have the same atomic environment and can be obtained from each other by integer multiples of three elementary translations defined by $\vec{a}$, $\vec{b}$, $\vec{c}$. A point M equivalent to the origin O is defined by

$$\overrightarrow{OM} = m\vec{a} + n\vec{b} + p\vec{c}$$

where m, n, p are positive or negative integers. The set of all these points, called <u>sites,</u> is a three-dimensional lattice which reflects the crystalline periodicity in any direction (fig. 2.3).

The lattice can be decomposed into one-dimensional simple units (rows) or two-dimensional (lattice planes) or three-dimensional ones (cells).

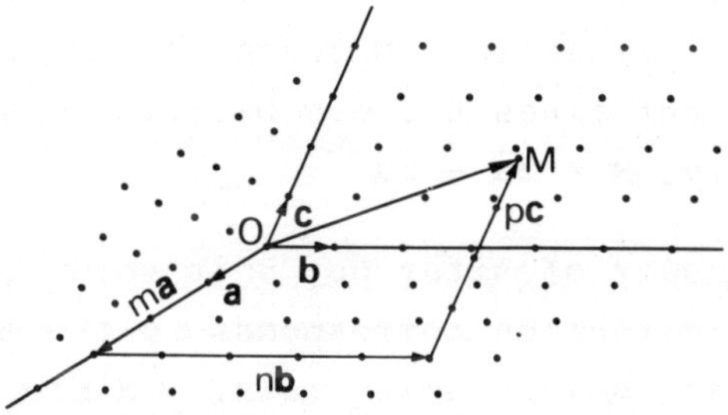

Fig. 2.3. Crystal lattice. All lattice sites are equivalent. They have the same atomic environment as the origin O. They are denoted by $\overrightarrow{OM} = m\vec{a} + n\vec{b} + p\vec{c}$.

<u>Rows</u>. The lattice sites are located at the intersections of three families of parallel straight lines. Fig. 2.4 shows the decomposition into two families of parallel straight lines for a two dimensional lattice, which is easier to sketch on a sheet of paper. On any of the lines, called the rows, the lattice sites are equidistant. Three integers, with no common divisor, u, v, w, define the row [u, v, w] by the vector

$$\vec{R}_{u,v,w} = u\vec{a} + v\vec{b} + w\vec{c}$$

which joins one site to its immediate neighbour. For example, in fig. 2.4 (w = 0), the rows of the D_3-family are denoted $[2,\bar{1},0]$, where $\bar{1}$ means (-1).

A study of fig. 2.4 will also show that the linear density of sites depends on direction; it is high for low index rows, such as D_1 [1,0,0] or D_2 [0,1,0] and low for high index rows, such as D_3 $[2,\bar{1},0]$. For a given

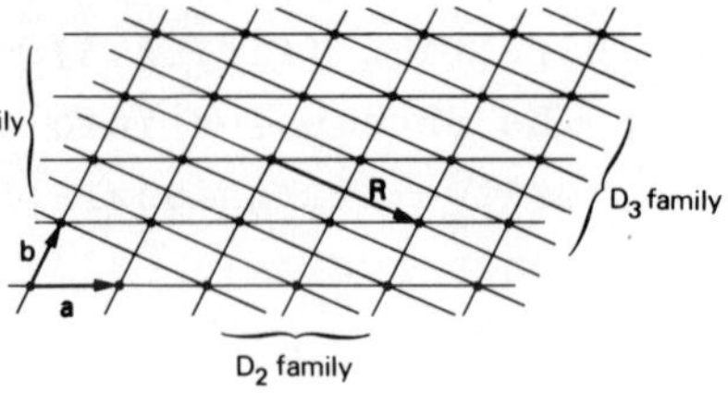

Fig. 2.4. Rows. The sites of a plane lattice lie at the intersections between any two families of straight lines D_1, D_2, D_3, ... For the D_3 family, $\vec{R} = 2\vec{a} - \vec{b}$.

crystal, and density of sites per unit volume, the highest density direction corresponds to the maximum separation between neighbouring rows. Since the atoms are closer to each other, the binding forces are greater for small index rows. In some crystals, this difference in cohesion between neighbouring rows may be so important that the material is divided into fibres (e.g. amianthus).

Lattice planes. It is also possible to separate the lattice into a family of equidistant parallel planes called lattice planes, which can be obtained from each other by the elementary translation of any row that is not included in the planes of the family (Fig. 2.5).

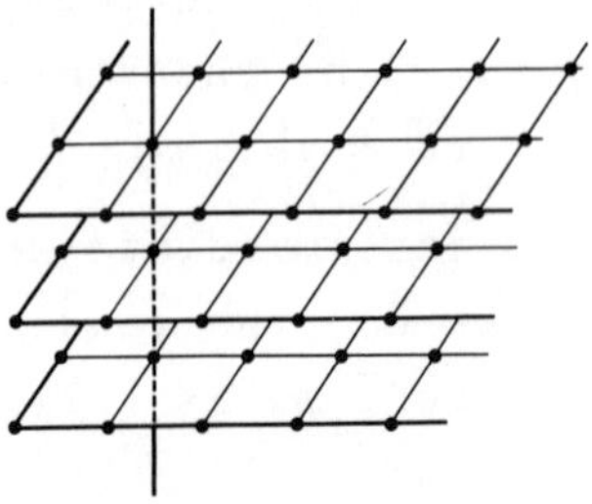

Fig. 2.5. Family of lattice planes.

The notation usually employed to distinguish between families of lattice planes is due to Miller. Since there is a site at the origin and at the end of the three vectors $\vec{a}$, $\vec{b}$, $\vec{c}$, one plane of the family must contain these four points. Generally there are other planes in between so that any basis vectors are divided into equal segments. If every separation between neighbouring planes is a/h along $\vec{a}$, b/k along $\vec{b}$, and c/l along $\vec{c}$, the integers h, k, l are, by definition, the Miller indices of this family of lattice planes. The notation (h, k, l) represents the whole set of planes, not simply a single plane (fig. 2.6). A negative index has a bar above it, e.g. $\bar{k}$. An index is zero for planes parallel to the corresponding axis.

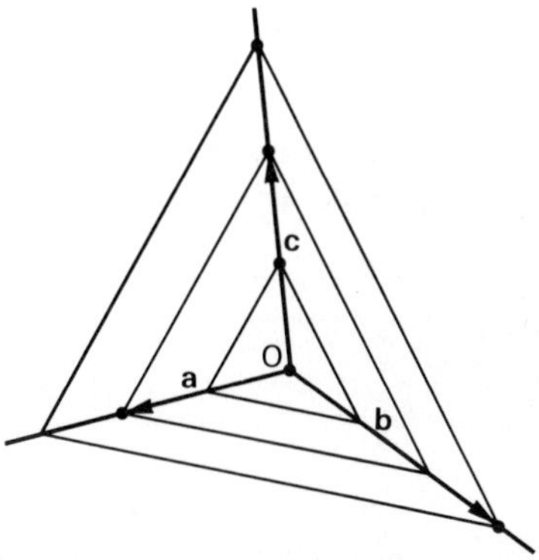

Fig. 2.6. Miller indices. The distance between two neighbouring planes is a/2 along $\vec{a}$, b/3 along $\vec{b}$, c/2 along $\vec{c}$. The Miller indices for this family are 2, 3, 2.

For a given density of sites per unit volume the low Miller index planes (100), (010), (001), are most widely separated and have the highest density of sites per unit area. These planes of high cohesion play a major role: they are eventually cleavage planes since the bonds between them are weak; they also reflect X-rays more than other planes.

Cells. The lattice can also be regarded as a stack of parallelepipeds, all identical to that constructed on the basis vectors $\vec{a}$, $\vec{b}$, $\vec{c}$ (fig. 2.7). The sites are located at the corners of these parallelepipeds, known as cells.

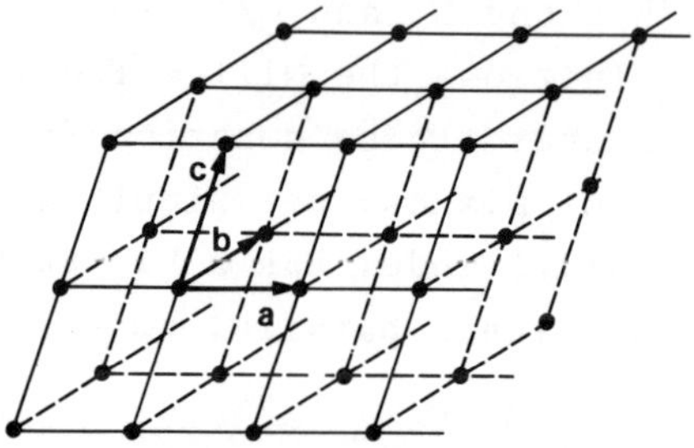

Fig. 2.7. Cells. The lattice can be regarded as consisting of a stack of identical parallelepipeds, known as cells.

The lattice of a given crystal is unique. However, one can generate it by other basis vectors $\vec{a}_1$, $\vec{b}_1$, $\vec{c}_1$, and define other cells, as shown in fig. 2.8 for a two dimensional lattice.

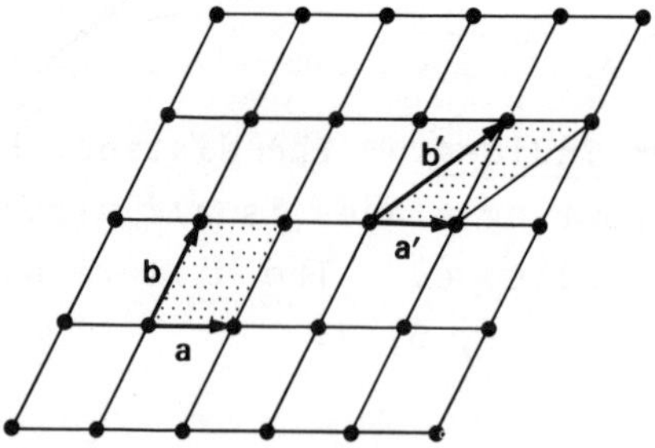

Fig. 2.8. A lattice can be generated from different systems of basis vectors, and therefore described by different cells.

The cells constructed on three basis vectors are unit cells, which means that they contain no sites other than those at the eight corners. All belong to seven other cells, and should be weighted as 1/8 cell site, so that a unit cell contains one site. A multiple (double, triple, ...) cell contains more than one site. Let a cell be defined by the vectors

$$\vec{a}' = u_1\vec{a} + v_1\vec{b} + w_1\vec{c}$$

$$\vec{b}' = u_2\vec{a} + v_2\vec{b} + w_2\vec{c}$$

$$\vec{c}' = u_3\vec{a} + v_3\vec{b} + w_3\vec{c}$$

its volume V' is equal to the absolute value of the scalar triple product $(\vec{a}', \vec{b}', \vec{c}')$.

Since

$$(\vec{a}',\vec{b}',\vec{c}') = \begin{vmatrix} u_1 & v_1 & w_1 \\ u_2 & v_2 & w_2 \\ u_3 & v_3 & w_3 \end{vmatrix} (\vec{a}, \vec{b}, \vec{c})$$

$$V' = pV$$

where $V = |(\vec{a}, \vec{b}, \vec{c})|$ is the unit cell volume and p the absolute value of the determinant. This cell is a multiple cell of order p because it contains p sites (one site occupies a volume V).

A lattice is often defined by a multiple cell which is more symmetric than unit cells. This is so for the face centred cubic lattice (f.c.c.) obtained by putting a site at the centres of the six faces of an ordinary cubic lattice unit cell (Fig. 2.9a). We leave it to the reader to check it is a lattice: the environment of any site is always the same, whether it is at a corner or at the centre of a face of the cube. This cubic cell is quadruple because the six sites in the faces, belonging to two cells, should have a weighting of 1/2 and 8 x 1/8 + 6 x 1/2 = 4. The unit cell which is constructed on the three vectors defined by the centres of the three faces intersecting at O is a rhombohedron (fig. 2.9b).

We shall encounter other examples of multiple cells in §2.3.2.3. devoted to lattice classification.

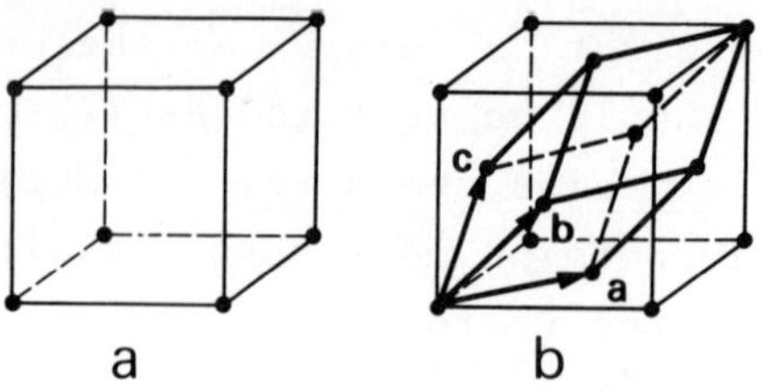

Fig. 2.9. Face centred cubic lattice (f.c.c.). The f.c.c. lattice is obtained (b) by inserting a site at every face centre of the primitive cubic lattice (a). The cubic cell is then a 4-fold cell. The unit cell is the rhombohedron derived from $\vec{a}$, $\vec{b}$, $\vec{c}$.

2.2.2. Atomic Structure

The atomic structure of a crystal is determined without any ambiguity once the lattice and the group of atoms, to be located at each site, are known (fig. 2.10).

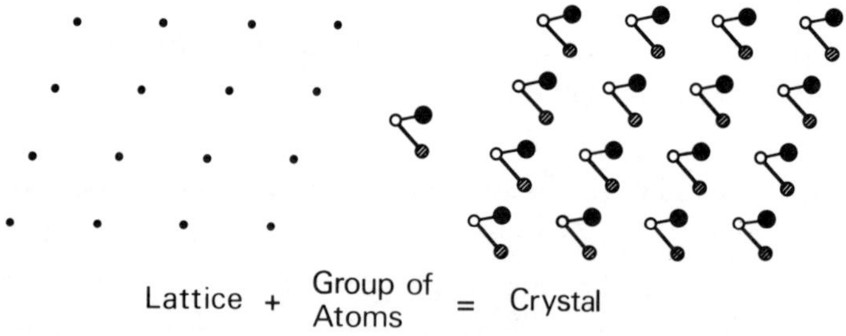

Fig. 2.10. Crystal atomic structure. The crystal consists of a lattice with an atomic group at every lattice site.

The simplest group is a single atom; the corresponding crystal is obtained by placing one atom at each site of the lattice. This is the case for many metals with a face centred cubic lattice, such as copper (3.61 Å) silver (4.071), gold (4.070), aluminium (4.041), nickel and platinum. The numbers in brackets show the size of a cubic cell (in angstroms). This cell contains four atoms, e.g. with co-ordinates (000, $\frac{1}{2}\frac{1}{2}0$, $\frac{1}{2}0\frac{1}{2}$, $0\frac{1}{2}\frac{1}{2}$) in cubic cell units, the other atoms being considered as belonging to other cells. Fig. 2.11a. shows this structure when seen from above. The number of lines within the circles indicates the location of the atoms, each line standing for one quarter of a cube edge. Other metals (lithium, sodium, potassium, chromium ...) have a body centred cubic lattice in which there is one atom at each site and the cubic cell contains two atoms located at 000, and $\frac{1}{2}\frac{1}{2}\frac{1}{2}$ (fig. 2.11b).

Other monatomic crystals such as silicon and germamium have a more complicated diamond type structure. For these materials, the lattice is face centred cubic but the quadruple cell will contain 8 atoms instead of 4, located at 000, $\frac{1}{2}\frac{1}{2}0$, $\frac{1}{2}0\frac{1}{2}$, $0\frac{1}{2}\frac{1}{2}$, $\frac{1}{4}\frac{1}{4}\frac{1}{4}$, $\frac{3}{4}\frac{3}{4}\frac{1}{4}$, $\frac{3}{4}\frac{1}{4}\frac{3}{4}$, $\frac{1}{4}\frac{3}{4}\frac{3}{4}$ (fig. 2.11c). The last four points are obtained from the first four by a translation equal to one quarter of the main diagonal. The basic group consists of two atoms, at 000 and $\frac{1}{4}\frac{1}{4}\frac{1}{4}$. This result could also be derived by starting from the rhombohedral unit cell, which contains two atoms (fig. 2.11c).

If the basic group is made of two different atoms, we have the structure of zinc blende (ZnS) and of gallium arsenide (GaAs).

Another example of a two-atom material is caesium chloride, which has a simple cubic lattice, the basic group consisting of one Cl^- ion at 000 and one Cs^+ ion at the cube centre ($\frac{1}{2}\frac{1}{2}\frac{1}{2}$).

2.3. POINT GROUPS OF CRYSTALS

The periodicity of crystals can be demonstrated only by experiments carried out at the scale of a unit cell, i.e. a few angstroms; they can be performed using X-ray diffraction. From a macroscopic point of view, the elementary crystal translations are infinitesimal so that the periodicity has no direct bearing on physical properties

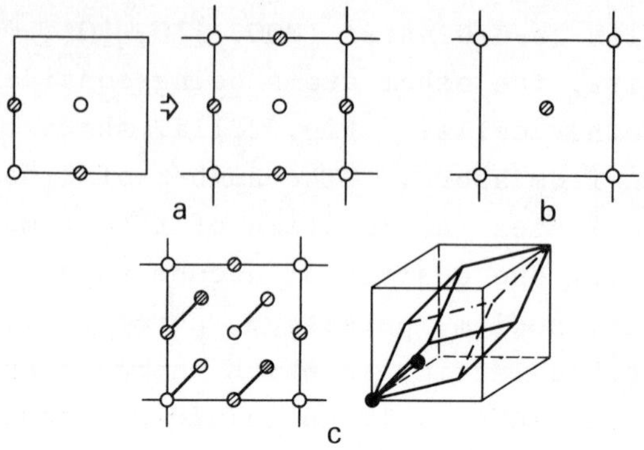

Fig. 2.11. Examples of cubic structures.
a) face centred cubic;
b) body centred cubic;
c) face centred cubic with a diatomic elementary group (diamond type): 4-fold cubic cell and rhombohedral unit cell.

of the crystal, which looks continuous; only the anisotropy is involved at this stage. It is not total, since for a given direction, there may be other equivalent directions along which macroscopic properties are the same. The set of all symmetry operations that put the crystal into a position which cannot be distinguished from the original position by macroscopic measurements is called the point group of the crystal. One says "point" because translations being excluded, there is at least one point which is invariant through all these symmetry operations.

The point group of a crystal can be determined by the symmetry of the set of perpediculars to natural sides, the orientation of which is only one of the physical properties. This group of lines is invariant for a given species; we could therefore use this criterion, linked to macroscopic properties to classify the various types of point group.

We are in fact going to use another procedure, relying on the above results for crystal structure: once we have defined the point groups and set up some useful equations, we shall discuss first the lattice point groups, and then the crystal point groups. Accounting for the lattice periodicity, we select the few allowed symmetry elements and derive necessary relationships between them. We are then left with seven symmetry classes, each of which defines a crystal system. It may happen that different lattices belong to the same crystal system. Bravais has shown that there are only fourteen types of lattice. The overall crystal symmetry, at most equal to the lattice symmetry, is obtained by subtracting some elements from the lattice symmetry. The 32 crystal point groups are then assigned in a natural way to the seven crystal systems.

2.3.1. Point Groups

2.3.1.1. Definitions

Two types of point symmetry elements may be distinguished.

1) Direct symmetry elements, which are the rotation axes. A crystal has an n-fold direct rotation axis, which will be denoted by A_n, if the crystal is unchanged by a $2\pi/n$ rotation about A_n.

2) Inverse symmetry elements can be decomposed into a symmetry centre and inverse rotation axes. A centre of symmetry C corresponds to the symmetry with respect to a point, which is also called inversion. A crystal has an n-fold inverse axis if the crystal remains unchanged after a $2\pi/n$ rotation and after a symmetry operation with respect to a point on the axis. The product (written as a dot product) of both is commutative:

$$\bar{A}_n = A_n . C = C . A_n .$$

The symmetry with respect to a plane is a particular case of an inverse axis: Fig. 2.12 shows that a diad (2-fold) inverse axis is identical to a plane of symmetry, or a mirror M, perpendicular to the axis at the centre of the inversion:

$$\bar{A}_2 \equiv M.$$

Similarly, a centre of symmetry can be regarded as a 1-fold inverse axis: $\bar{A}_1 = C$.

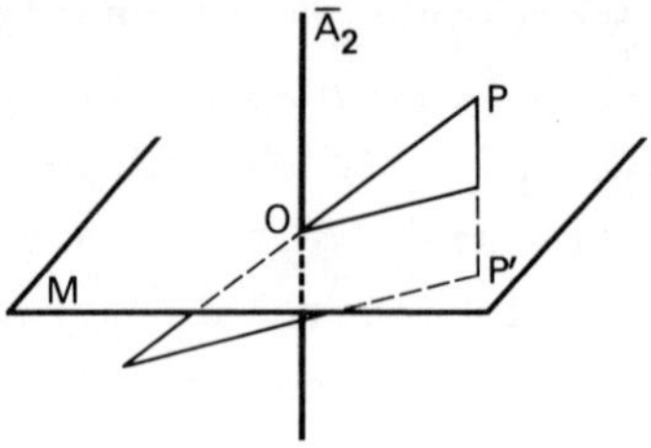

Fig. 2.12. An inverse diad axis ($\bar{A}_2$) is equivalent to a perpendicular mirror M.

2.3.1.2. Stereographic projection

In order to derive more complex relationships between point groups, it is useful to employ a planar representation, e.g. a stereographic projection (fig. 2.13a) which maps a sphere of radius R onto its equatorial plane E, the centre of the projection being either the north or the south pole of the sphere. If we make use of both projections, every point of the sphere can be mapped inside the equatorial circle. The angles remain unchanged in a stereographic projection: if two lines on the sphere intersect at C at an angle α, their projected lines will intersect with angle α on the projection of C.

Projections of the north hemisphere (including the equator) will be denoted by crosses (x), of the south hemisphere by circled dots (⊙). A straight line is represented by the projections of its two intersections with the sphere; a plane that contains the centre is similarly projected by two circular arcs, symmetric with each other with respect to its line of intersection with the equatorial plane (fig. 2.13b). When we use a stereogram to describe point symmetry properties, the north-south axis is usually the highest order symmetry axis and the centre of the sphere is chosen as the centre of symmetry, if any. Dotted lines denote constructions; only projections are shown by continuous lines.

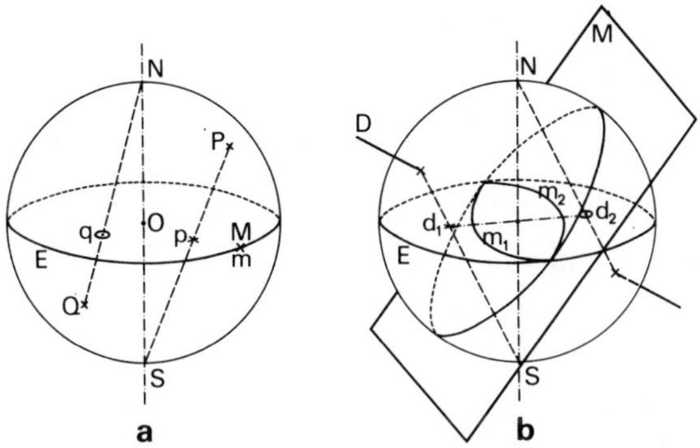

Fig. 2.13. Stereographic projection
a) Triangles SOp and SPN are similar so that $\frac{Sp}{2R} = \frac{R}{SP}$. The stereographic projection is defined by the inversion $\overline{Sp}.\overline{SP} = 2R^2$
b) Diameter D is represented by the projections d_1 and d_2 of both intersections with the sphere. Plane M is described by two circular arcs m_1 and m_2.

2.3.1.3. Equivalence relationships

To begin with, we suppose there is at most one inverse symmetry element.

a) An odd-order n-fold inverse axis is equivalent to a direct n-fold axis and a centre of symmetry. n iterations of the inverse symmetry will give

$$(\overline{A}_n)^n = (A_n \cdot C)\cdot(A_n \cdot C) \cdots (A_n \cdot C) = (A_n)^n \cdot C^n$$

$$(\overline{A}_n)^n = C^n \quad \text{for } (A_n)^n = I \quad \text{(identity)}$$

This is a symmetry operation for the crystal; if n is odd (n = 2p + 1) it reduces to an inversion

$$C^n = C^{2p} \cdot C = C \quad (C^{2p} = I)$$

The equation $(\overline{A}_{2p+1})^{2p+1} = C$ thus establishes the existence of a centre of symmetry. Moreover, the dot product

$$\overline{A}_n \cdot C = A_n \cdot C^2 = A_n$$

is identical to a direct axis. So an odd inverse axis is equivalent to the combination direct axis - centre of symmetry

$$\overline{A}_{2p+1} \equiv A_{2p+1}C.$$

This is shown, for n = 3, in the stereogram 2.14.

N.B. In the above notation $A_{2p+1}C$, both elements are independent, whereas in the definition $\overline{A}_n = A_n \cdot C$, the dot represents the succession of the two operations.

b) An even order n-fold inverse axis implies the existence of a colinear direct n/2-fold axis

$$(\overline{A}_n)^2 = (A_n \cdot C) \cdot (A_n \cdot C) = (A_n)^2.$$

If n = 2p, $(A_n)^2$ is the rotation of angle

$$2 \cdot \frac{2\pi}{2p} = \frac{2\pi}{p} \Rightarrow (\overline{A}_{2p})^2 = A_p.$$

The case n = 4 is depicted in fig. 2.14b.

c) If, furthermore, n/2 = p is odd, there is a mirror perpendicular to the axis. Since p is odd, $C^p = C$ and the operation

$$(\overline{A}_{2p})^p = (A_{2p})^p \cdot C^p = A_2 \cdot C = \overline{A}_2$$

is identical to a perpendicular mirror M.

The above property (b) is also true, so that

$$\overline{A}_{2p} \equiv \frac{A_p}{M} \quad \text{if p is odd.}$$

The bar indicates that the mirror is perpendicular to the axis. This property is shown in fig.2.14c for n=6.

In the figures, symmetry axes are denoted by symbols:

- direct axes diad (2-fold) (⬭), triad (3-fold) (△), tetrad (4-fold) (□), hexad (6-fold) (⬡);

- inverse axes: $\overline{1}$(○), $\overline{3}$ (◬), $\overline{4}$ (▨), $\overline{6}$ (⬡), which takes care of the above results. When we add the inverse diad axis, represented by the equivalent perpendicular mirror, we have the symbols for crystal symmetry elements §2.3.

The occurrence of several symmetry elements at the same time usually generates other elements. We give here some of the most common examples.

d) An even n-fold direct axis and a centre of symmetry, imply a mirror, perpendicular to the axis.

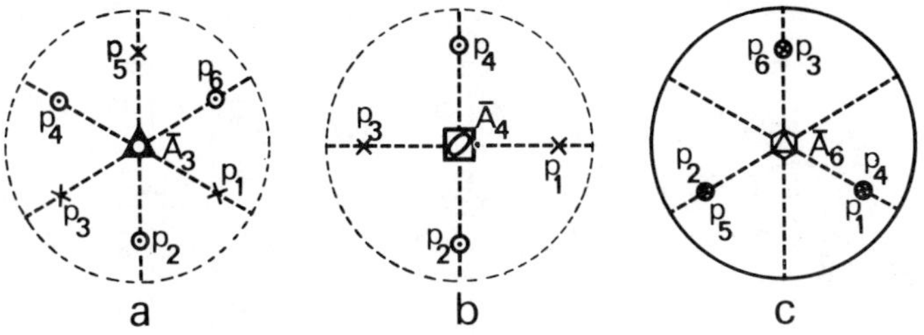

Fig. 2.14. Properties of inverse rotation axes
The sequence p_2, p_3 ... obtained from the starting point p_1 shows that:
a) an inverse triad axis is equivalent to a direct triad axis plus a centre of symmetry.
b) an inverse tetrad axis includes a direct diad axis.
c) an inverse 6-fold axis is equivalent to a direct triad axis plus a perpendicular mirror.

This theorem follows immediately from

$$(A_{2p})^p \cdot C = A_2 \cdot C = \bar{A}_2 \equiv M$$

which is also written

$$A_{2p}C \Rightarrow \frac{A_{2p}}{M} C.$$

e) An n-fold direct axis and a diad axis perpendicular to it generate n perpendicular diad axes intersecting at angle π/n:

Let p_2 and p_3 be the transforms of the original point p_1, from the rotation about the diad axis A_2 and the rotation of $2\pi/n$ (fig. 2.15a). p_3 is then obtained from p_2 by a rotation about the diad A'_2 intersecting with A_2 at an angle π/n to A_2. By repeating the procedure, we obtain the n diad axes.

If n is odd ($n = 2p + 1$), two successive diad axes correspond to each other through the rotation $(A_n)^p$, of angle $p\frac{2\pi}{n} = \pi - \frac{\pi}{n}$. All these diad axes are equivalent and denoted by the same letter A'_2:

$$A_{2p+1}A'_2 \Rightarrow A_{2p+1}(2p + 1)A'_2.$$

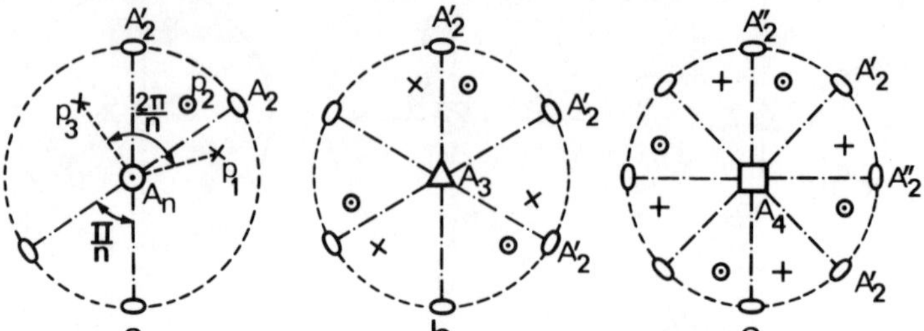

Fig. 2.15. Association of a diad axis and a direct n-fold axis.

a) It implies n diad axes.

b) If n is odd, all diad axes are equivalent.

c) If n is even, alternate axes are equivalent.

If on the other hand n is even (n = 2p), the angle $\pi - 2\pi/n = (p - 1)2\pi/n$ is only a multiple of $2\pi/n$ so that only alternate axes are equivalent, and there are two sets of p equivalent axes.

$$A_{2p}A_2 \Rightarrow A_{2p}pA'_2pA''_2.$$

The above distinction is evident in fig. 2.15b and c for n = 3 and n = 4.

f) An n-fold direct axis and a mirror containing the axis generate n mirrors intersecting at π/n.

The proof is the same (see fig. 2.16), and in the same way, all mirrors are equivalent for odd n

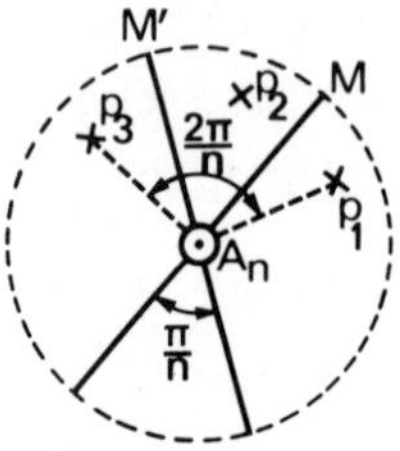

Fig. 2.16. A direct n-fold axis contained in a mirror implies n mirrors.

$$A_{2p+1}M' \Rightarrow A_{2p+1}(2p + 1)M'.$$

and they are of two types for even n

$$A_{2p}M' \Rightarrow A_{2p}pM' \; pM''.$$

g) An n-fold inverse axis and a diad axis generate n equivalent diad axes and n mirrors, perpendicular to the diad axes.

If follows from a, e and d that

$$\overline{A}_{2p+1}A_2 \equiv A_{2p+1}C\;A'_2 \Rightarrow A_{2p+1}\frac{(2p+1)A'_2}{(2p+1)M'}\;C.$$

2.3.2. Lattice Point Groups

Because of the lattice periodicity, only a few angles of rotation are allowed. This selection of allowed rotation axes yields rapid determination of the possible symmetry classes.

2.3.2.1. Symmetry elements of lattices

Lattice periodicity implies the following properties:

a) Every straight line that contains a lattice point and is parallel to an n-fold axis is itself an n-fold axis.

The $2\pi/n$ rotation about A_n maps site P on to site N (fig. 2.17a). The axis A'_n obtained from A_n through the translation $\vec{NP}$, is also equivalent to A_n through the $2\pi/n$ rotation about P. So every site P is located on an n-fold axis. This is also true for inverse axes (fig. 2.17b). In particular, every plane that contains a site and is parallel to a mirror is itself a mirror.

b) Every axis that contains a site (P) is a row. If N_2, N_3 ... N_n correspond to site N_1 through the $2\pi/n$ rotation around A_n (fig. 2.18), then the lattice vector

$$\vec{R} = \sum_{i=1}^{n} \vec{PN}_i = n\vec{PO} + \sum_{i=1}^{n} \vec{ON}_i$$

is parallel to A_n since $\sum_{i=1}^{n} \vec{ON}_i = \vec{0}$. Thus A_n is a lattice row.

c) Every site is a centre of symmetry.

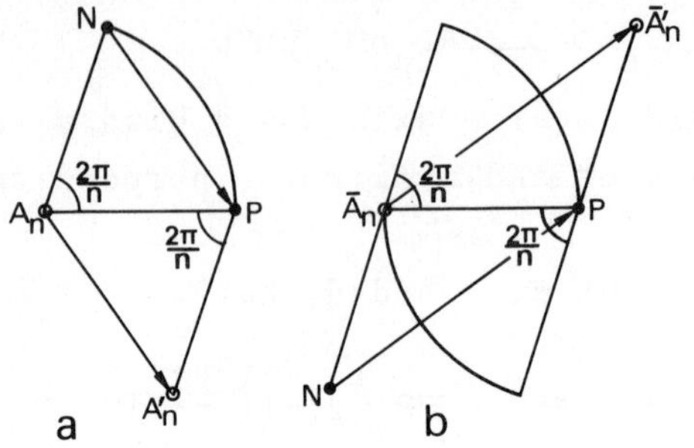

Fig. 2.17. Any straight line which contains a lattice site and is parallel to a direct (a) or inverse (b) axis is itself an axis of symmetry.

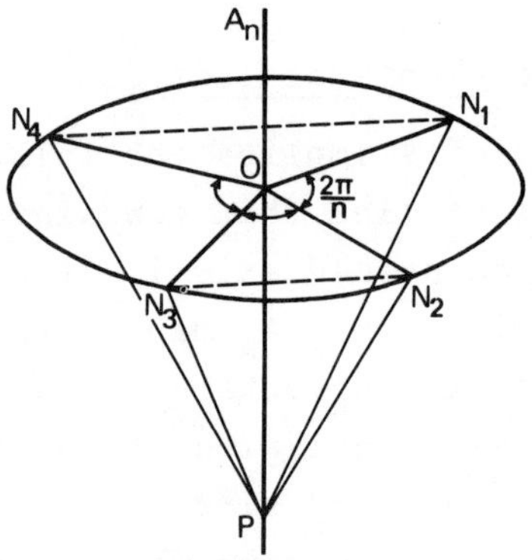

Fig. 2.18. Any symmetry axis that contains a lattice site is a lattice row.

d) The occurrence of an n-fold axis with $n > 2$, implies n perpendicular diad axes.

Let OP_1 be an elementary lattice translation (fig. 2. 19). P_1 goes to P_2 through a $2\pi/n$ rotation about the A_n axis that contains the site O. From the symmetry about O, P_1 goes to P_3 and P_2 goes to P_4. The sites P_1, P_2 and P_4, P_3 correspond to each other by the π-rotation about the row OC which is parallel to P_2P_1. Let both A and B be the nearest neighbours of site O on the axis: the unit cell constructed on O, B, P_3, P_4 is obtained from the OAP_1P_2 unit cell through a π rotation about OC, which then appears to be a diad axis of the lattice.

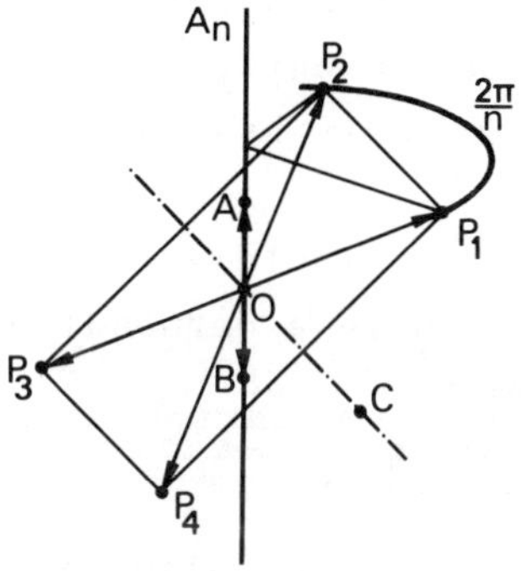

Fig. 2.19. An n-fold axis with n > 2 implies n perpendicular diad axes.

e) <u>A lattice may only have 1-fold, 2-fold, 3-fold, 4-fold, and 6-fold axes (direct or inverse)</u>.

Let us refer again to Fig. 2.18 and assume N_1 is the closest site to the axis; the distance a = N_2N_3 is then the period of the row N_2N_3. Since $\overrightarrow{N_1N_4}$ and $\overrightarrow{N_2N_3}$ are parallel, N_1N_4 is a multiple of N_2N_3:

$$\overrightarrow{N_1N_4} = p\overrightarrow{N_2N_3} \quad \text{(p integer)}$$

This equation can also be written:

$$\overrightarrow{N_1N_2} + \overrightarrow{N_2N_3} + \overrightarrow{N_3N_4} = p\overrightarrow{N_2N_3}$$

Taking the scalar product with $\overrightarrow{N_2N_3}$

$$a^2\left(2 \cos \frac{2\pi}{n} + 1\right) = pa^2$$

since $$(\overrightarrow{N_2N_3}, \overrightarrow{N_1N_2}) = (\overrightarrow{N_3N_4}, \overrightarrow{N_2N_3}) = \frac{2\pi}{n} .$$

Allowed values for n must satisfy the equation

$$2 \cos \frac{2\pi}{n} + 1 = p \qquad \text{p integer.}$$

The absolute value of a cosine is always less than one so p can only take the values -1, 0, 1, 2, 3; the corresponding values for n are

$$n = 2, 3, 4, 6, 1.$$

The above proof is also valid for an inverse axis, since every site is a centre of symmetry, the regular circular pattern $N_1N_2\ldots N_n$ is again obtained.

2.3.2.2. <u>The 7 crystal systems</u>

It is now possible to count the point groups that meet the foregoing requirements.

The simplest class only has one centre of symmetry C.

If we include now a 2-fold axis, because of the centre C, we must also have a perpendicular mirror M, which leads to a symmetry

$$\frac{A_2}{M}\ C.$$

If there is more than one diad axis, there are three orthogonal A_1, A'_2, A''_2 and the centre of symmetry implies three mirrors M, M', M'' perpendicular to A_2, A'_2, A''_2:

$$\frac{A_2}{M}\ \frac{A'_2}{M'}\ \frac{A''_2}{M''}\ C.$$

If the lattice has a triad (3-fold) axis A_3 it should have three axes A'_2 and three mirrors M' (because of C)

$$A_3\ \frac{3A'_2}{3M'}\ C.$$

If the lattice has a tetrad (4-fold) axis A_4, there are only two pairs of equivalent diad axes

$$\frac{A_4}{M}\ \frac{2A'_2}{2M'}\ \frac{2A''_2}{2M''}\ C.$$

With a six-fold axis, we obtain two sets of three diad axes

$$\frac{A_6}{M}\ \frac{3A'_2}{3M'}\ \frac{3A''_2}{3M''}\ C.$$

The six classes listed so far are the only ones that can be constructed with only one axis of order greater than 2. It is possible to prove that the only class with more than one axis of order greater than 2 reflects in fact the complete cubic symmetry (fig. 2.20):

- four axes A_3 along the diagonals;
- three axes A_4 perpendicular to the sides;
- six axes A'_2 containing the midpoints of opposite edges;

- a centre, which will then generate mirrors perpendicular to the A_4 and A'_2 axes. Hence the symmetry:

$$\frac{3A_4}{3M}\ 4A_3\ \frac{6A'_2}{6M'}\ C.$$

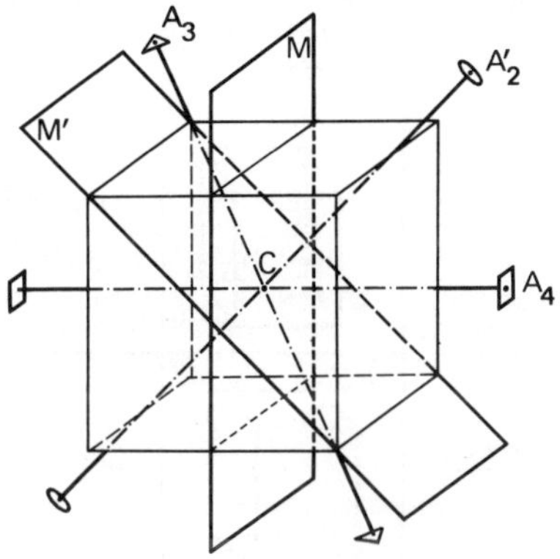

Fig. 2.20. The symmetry elements of the cube

$$\frac{3A_4}{3M}\ 4A_3\ \frac{6A'_2}{6M'}\ C.$$

If we classify crystals according to their lattice point group we define seven crystal systems in the following way: all crystals of a given crystal system have the same lattice point group, which itself is listed in the seven above classes. A crystal system is also characterized by the shape of its cell generating all lattices of the same symmetry. This shape ranges from the most general parallelepiped (triclinic system) to the cube (cubic system), including the right prism of arbitrary base (monoclinic system), the orthogonal parallelepiped (orthorhombic system), the rhombohedron with six identical diamond sides (trigonal system), the right prism with square base (tetragonal system) and the right prism with diamond base of angle $\gamma = 120^o$ (hexagonal system). Table 2.21 shows the basic properties of the seven crystal systems: lattice point group, shape of the cell, angles α β γ between cell edges, lengths a, b, c, of these edges.

SYSTEM	SYMMETRY	UNIT CELL	ANGLES	EDGES
TRICLINIC	C	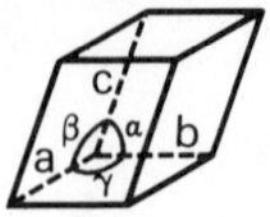	$\alpha\neq\beta\neq\gamma\neq 90^\circ$	$a\neq b\neq c$
MONOCLINIC	$\frac{A_2}{M}C$	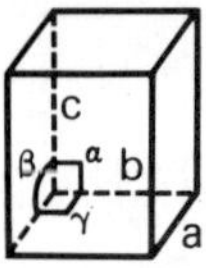	$\alpha=\gamma=90^\circ$ $\beta>90^\circ$	$a\neq b\neq c$
ORTHORHOMBIC	$\frac{A_2}{M}\frac{A_2'}{M'}\frac{A_2''}{M''}C$	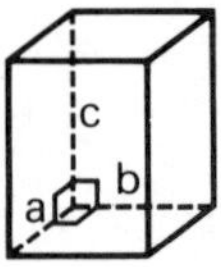	$\alpha=\beta=\gamma=90^\circ$	$a\neq b\neq c$
TRIGONAL	$A_3\frac{3A_2'}{3M'}C$	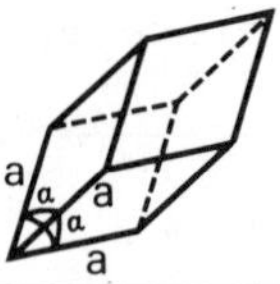	$\alpha=\beta=\gamma\neq 90^\circ$	$a=b=c$
TETRAGONAL	$\frac{A_4}{M}\frac{2A_2'}{2M'}\frac{2A_2''}{2M''}C$	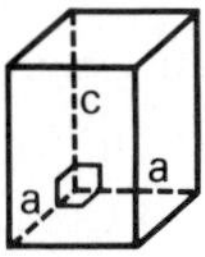	$\alpha=\beta=\gamma=90^\circ$	$a=b\neq c$
HEXAGONAL	$\frac{A_6}{M}\frac{3A_2'}{3M'}\frac{3A_2''}{3M''}C$	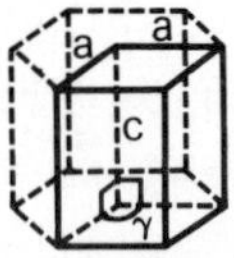	$\alpha=\beta=90^\circ$ $\gamma=120^\circ$	$a=b\neq c$
CUBIC	$\frac{3A_4}{3M}4A_3\frac{6A_2'}{6M'}C$	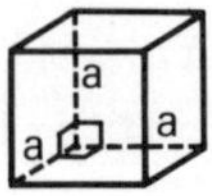	$\alpha=\beta=\gamma=90^\circ$	$a=b=c$

Fig. 2.21. The seven crystal systems.

2.3.2.3. The fourteen Bravais lattices

Generally, different lattices may have the same point group so that crystals of the same class may have different lattices. Besides the seven lattices, known as the primitive (P) lattices which have the seven parallelepipeds of fig. 2.21 as unit cells, there are other lattices obtained by adding other sites onto the cell. Since all sites must have the same environment, the only places where we can add sites are the cell centre and the centres of its faces; in fact, if we want to have a lattice, the point distributions at the corners (A) and the interior (B) of the cell ought to be invariant through any lattice translation, including AA type and BB type translations, whatever the location of the B sites in the cell. If B_o goes to B_1 by translation $\overrightarrow{A_oB_o}$, B_1 has to be a cell corner (fig. 2.22).

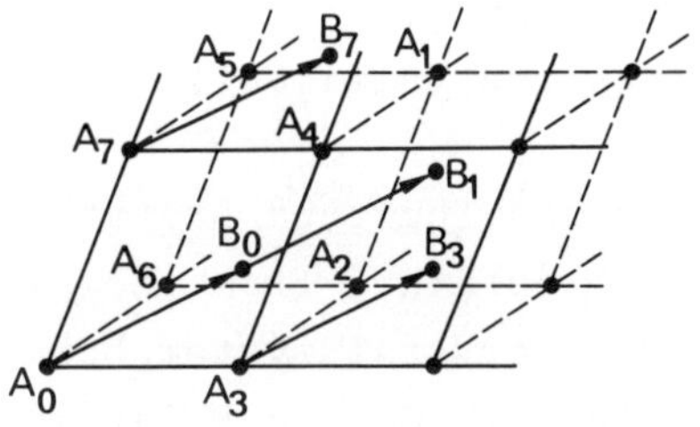

Fig. 2.22. Generation of a new lattice from a primitive lattice. The extra site B_o must be a cell centre or a face centre, in order that the structure be invariant under translation $\overrightarrow{A_oB_o}$.

Now if B_1 is identical to A_1, B_o is the cell centre which generates a body centred lattice (I). If B_1 is one of the corners A_2 A_4 A_5, B_o is a face centre, which generates a face centred lattice (C) where two opposite cell faces are centred. If B_o is the midpoint of an edge, this leads to no change except for a primitive lattice with one cell edge reduced by a factor 2.

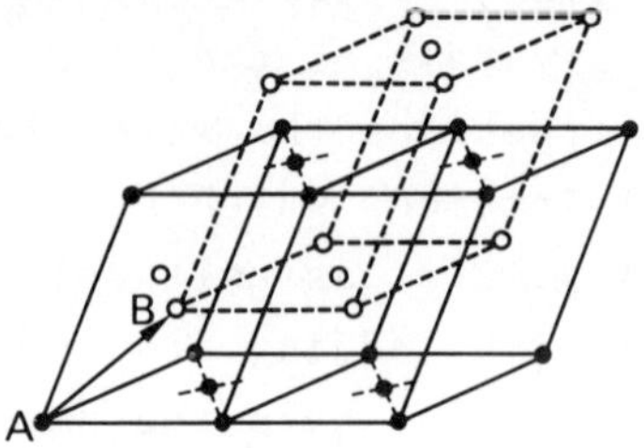

Fig. 2.23. Face centred lattice (F). The point distribution which is obtained on inserting a site at every face centre is decomposed into two C-type lattices, the sites of which (dots · and circles o) are exchanged by translation $\vec{AB}$.

We may also consider the case where all faces of the primitive cell are centred: this yields two C-type lattices, the sites of which are differentiated (fig. 2.22) by dots (·) and circles (o). Both lattices are equivalent through a half face diagonal translation and together they are the face centred lattice (F) derived from the primitive lattice.

Fig. 2.24 shows the fourteen Bravais lattices. The three types (I, C, F) obtained from the primitive lattice do not appear in every crystal system because of redundancy. Thus triclinic I, C, and F lattices are merely primitive triclinic lattices with smaller cells.

Only P and rectangular face centred C-type monoclinic lattices should be considered since the other C-type lattice is simply a P-type monoclinic lattice with a smaller β angle (fig. 2.25a) and the I or F type lattices are also C-type lattices (fig. 2.25b and c).

On the other hand, all orthorhombic lattices are distinct. The only trigonal lattice is the primitive one, which is often denoted by R (rhombohedric): The I lattice is a R-lattice with a smaller cell (fig. 2.26a); no C type exists since all faces of a rhombohedron are equivalent and they should all be centred, and the F type lattice is an R-lattice with a smaller cell (fig. 2.26b).

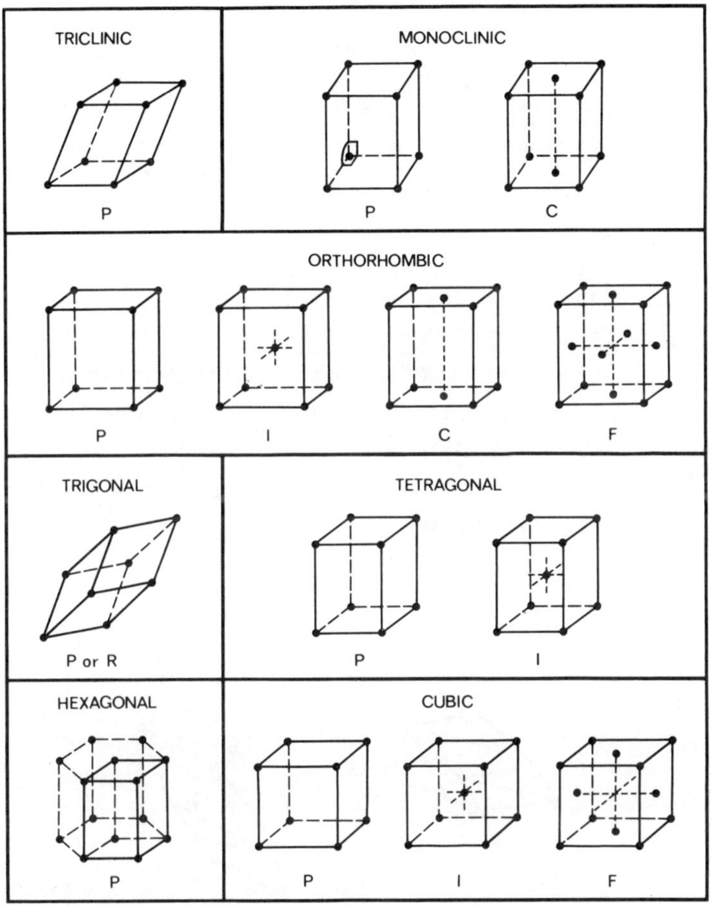

Fig. 2.24. The fourteen Bravais lattices.

Tetragonal lattices can only be of the P or I type: All rectangle faces are equivalent, so only the C-lattice has its square faces centred; fig. 2.27a shows that it is another P-lattice with a smaller cell. Similarly, the F lattice is equivalent to an I quadratic lattice (fig. 2.27b).

For hexagonal lattices, the I, C, and F type would lose their 6-fold axis. Therefore only a P type lattice can exist.

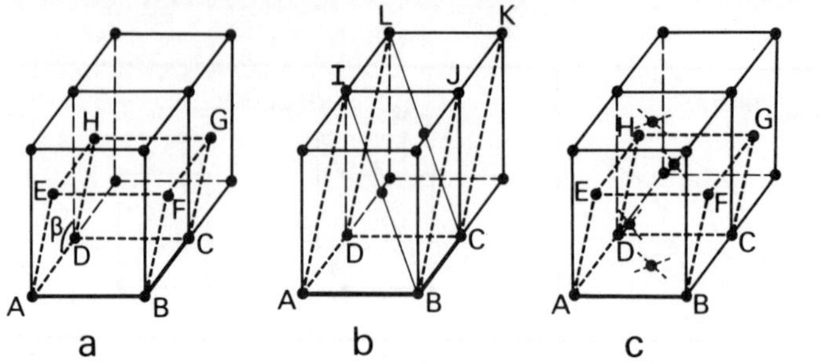

Fig. 2.25. Monoclinic system.
a) The oblique face centred lattice is a P monoclinic lattice with cell ABCDEFGH;
b) the I lattice is a C monoclinic lattice with cell ABCDIJKL;
c) The F lattice is a C monoclinic lattice with cell ABCDEFGH.

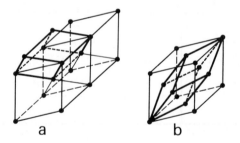

Fig. 2.26. Trigonal system. There is only one lattice.

The cubic system comprises the P, I, and F type lattices. The C lattice cannot exist because the six faces are equivalent and should be centred simultaneously.

These fourteen Bravais lattices are the only possible ways of distributing an infinity of points in space, all of them with the same environment.

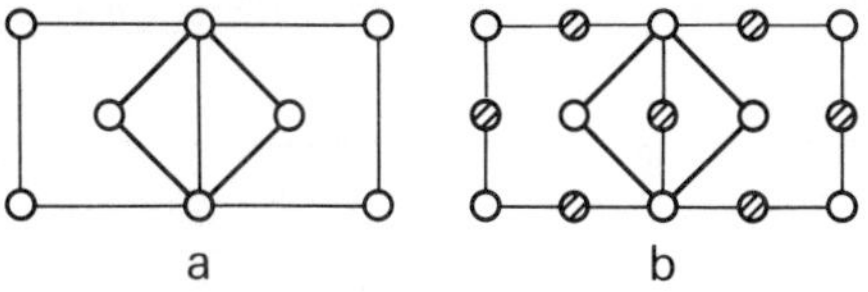

Fig. 2.27. Tetragonal system.
a) The C lattice is a P lattice with a smaller cell;
b) The F lattice is an I lattice.

2.3.3. The Point Groups of Crystals

Adding a group of atoms at each lattice site can only lower the crystal symmetry so that the latter is at most equal to the lattice symmetry. In fact a crystal may have neither a centre of symmetry nor the n diad axes enforced by the presence of an n-fold axis ($n > 2$). The crystal point groups are thus derived from the corresponding lattice groups by removing these optional elements. These classes are listed in Fig. 2.28, according to their crystal system, with the indication of their symmetry elements (above), their stereographic projection and the Hermann-Mauguin notation (below). The first column indicates the seven lattice classes of § 2.3.2.2. Crystals which have the same point group as their lattice are called holohedric. Others are merihedric crystals.

The class that has no symmetry element, obtained by removing the centre of symmetry from the triclinic holohedry ($\bar{1}$) is denoted by 1.

By removing the centre of symmetry from the monoclinic lattice point group $\frac{A_2}{M}$ C, denoted by 2/m, we can only keep the diad axes (class 2) or the mirror (class m).

Similarly, from the orthorhombic holohedry $\frac{A_2}{M}\,\frac{A'_2}{M'}\,\frac{A''_2}{M''}$ C, denoted by mmm, we obtain the point group 222 (on keeping the three diad axes $A_2A'_2A''_2$) or 2mm (on keeping the axis A_2 and the mirrors M', M'' which contain the axis).

The trigonal lattice point group $A_3\,\frac{3A'_2}{3M'}$ C first yields group $A_3 3A'_2$ (denoted by 32) and $A_3 3M'$ (denoted by 3m)

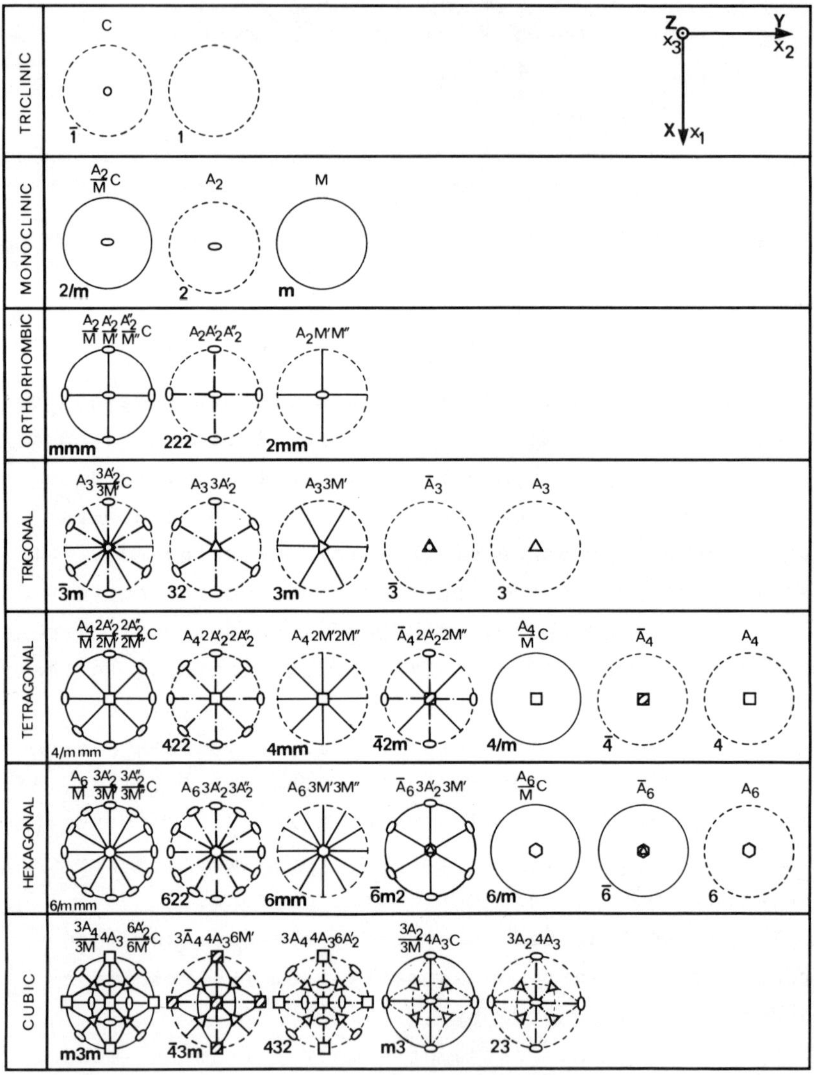

Fig. 2.28. The 32 point groups of crystals.

when one removes the centre, then $A_3C = \bar{A}_3$, denoted by $\bar{3}$, when one removes the three diad axes A'_2 and, accordingly, the three M' mirrors, and lastly the class A_3 denoted by 3, when C is suppressed.

Withdrawing the centre of symmetry from the tetragonal lattice $\frac{A_4}{M}\ \frac{2A'_2}{2M'}\ \frac{2A''_2}{2M''}\ C$ or $\frac{\bar{A}_4}{M}\ \frac{2A'_2}{2M'}\ \frac{2A''_2}{2M''}\ C$ gives

- three point groups when one keeps
    - all diad axes $A_4 2A'_2 2A''_2$ (class 422),
    - all mirrors $A_4 2M' 2M''$ (class 4mm),
    - alternatively a diad axis and a mirror $\bar{A}_4 2A'_2 2M''$ (class $\bar{4}2m$)
- two classes A_4(4) and $\bar{A}_4$($\bar{4}$) when all these elements are removed. The only tetragonal class that has a centre of symmetry is $\frac{A_4}{M}$ C, denoted by 4/m.

The seven classes of the hexagonal system are obtained in a similar fashion.

The cubic symmetry $\frac{3A_4}{3M}\ 4A_3\ \frac{6A'_2}{6M'}\ C$ or $\frac{3\bar{A}_4}{3M}\ 4A_3\ \frac{6A'_2}{6M'}\ C$ when the centre is removed, gives two classes $3A_4 4A_3 6A'_2$ (denoted by 432), and $3\bar{A}_4 4A_3 6M'$ (denoted by $\bar{4}3m$). The minimum symmetry that has the four triad axes of the cubic system is $3A_2 4A_3$ (class 23). Adding the centre of symmetry implies the mirrors M perpendicular to the binary axes. This symmetry class $\frac{3A_2}{3M}\ 4A_3C$ is denoted by m3.

Every crystal system can be characterized by the occurrence of selected symmetry elements:

triclinic	one 1-fold axis
monoclinic	one 2-fold axis
orthorhombic	three 2-fold axes
trigonal	one 3-fold axis
tetragonal	one 4-fold axis
hexagonal	one 6-fold axis
cubic	three direct 2-fold axes and four direct 3-fold axes.

When the nature of the axis is not specified, it may be direct or inverse.

All crystalline species belong to one of the 32 classes listed in table 2.28. Moreover, for each class, there are natural and artificial crystal specimens.

Crystals are anisotropic as far as most physical properties are concerned; so physical constants data make sense only if the orientation of the co-ordinate axes with respect to the symmetry elements is specified. We shall always refer to the orthonormal system shown upright in fig. 2.28, and replace the notation X, Y, Z (usual in crystallography) by x_1, x_2, x_3, which is easier for tensor analysis.

2.4. SPATIAL SYMMETRY

The above considerations which led to the 32 point groups, affect only symmetry of macroscopic properties of crystals. But a comprehensive description of the crystalline medium should also account for some further symmetry operations through which the crystal is invariant. These include:

- screw axis $A_{n,\vec{t}}$: $2\pi/n$ rotation and translation $\vec{t}$ parallel to the axis,
- glide symmetry planes $M_{\vec{t}}$: symmetry with respect to a plane M and translation $\vec{t}$ parallel to the plane.

Inverse axes of order 1, 3, 4, 6 which keep one point invariant, cannot be associated with a translation.

2.4.1. Screw Axes

A screw axis cannot be distinguished from an ordinary axis as far as the point group is concerned. The only allowed values for n are 2, 3, 4 and 6. The result of n operations of the $A_{n,\vec{t}}$ type is a translation $n\vec{t}$ parallel to the axis. Since this must leave the crystal invariant, $n\vec{t}$ should be a multiple of the elementary vector $\vec{R}_{uvw}$ for the [u,v,w] row, parallel to the axis:

$$\vec{t} = \frac{m}{n}\vec{R}_{uvw} \qquad \text{m integer}$$

One can choose $t < R_{uvw}$, i.e. $m < n$, so that the translation values are

$$\vec{t} = \vec{0},\ \vec{t}_1 = \frac{1}{n}\vec{R}_{uvw},\ \vec{t}_2 = \frac{2}{n}\vec{R}_{uvw},\ \ldots,\ \vec{t}_{n-1} = \frac{n-1}{n}\vec{R}_{uvw}.$$

The symbols for the screw axis of the spatial symmetry are listed below, together with the abbreviated notation n_m.

For example, 4_1 represents a $2\pi/4$ rotation associated with an $\vec{R}/4$ translation along the axis (Fig. 2.29).

$2\ \mathbf{0}$. $2_1\ \mathbf{0}$

$3\ \triangle$. $3_1\ \triangle$. $3_2\ \triangle$

$4\ \square$. $4_1\ \square$. $4_2\ \square$. $4_3\ \square$

$6\ \bigcirc$. $6_1\ \Diamond$. $6_2\ \Diamond$. $6_3\ \Diamond$. $6_4\ \Diamond$. $6_5\ \Diamond$

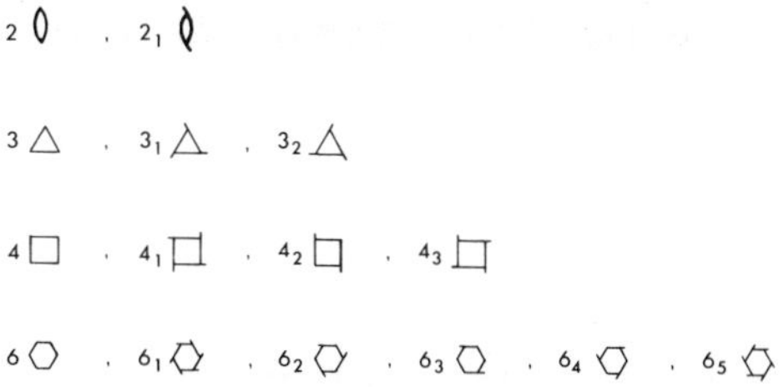

One case of such a tetrad axis is provided by the diamond type structure (germanium, silicon) shown in fig. 2.30.

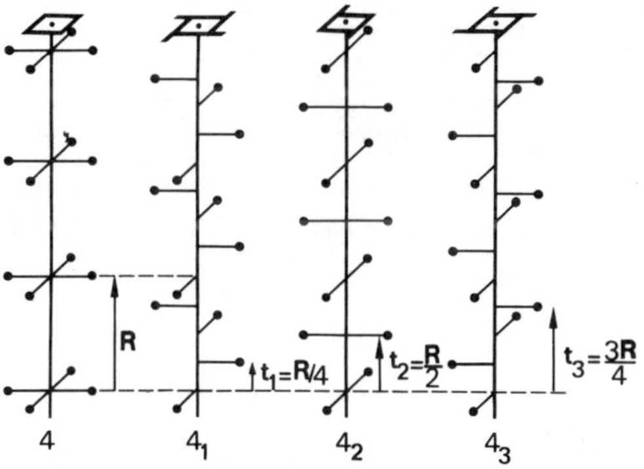

Fig. 2.29. The four tetrad screw axes.

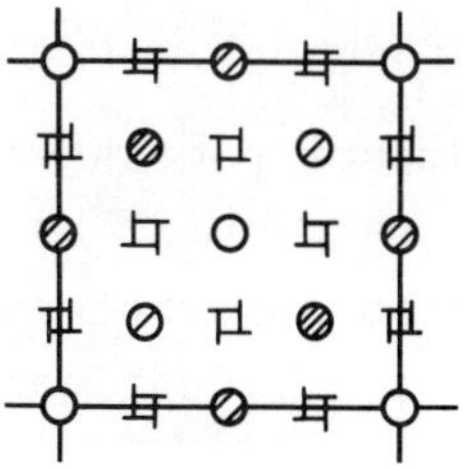

Fig. 2.30. Screw axes of the diamond-type structure.

2.4.2. Glide Planes

The $2\vec{t}$ translation, which arises from two $M_{\vec{t}}$, must be a translation operation of the lattice plane parallel to M:

$$2\vec{t} = m_1\vec{a}_1 + m_2\vec{a}_2$$

where $\vec{a}_1$ and $\vec{a}_2$ are two base vectors in this lattice plane, and m_1, m_2 are integers. Since nothing is changed by removing an elementary translation $\vec{a}_1$ or $\vec{a}_2$, m_1 and m_2 only take the values 0 and 1 so that the only possible translations are

$$\vec{t} = 0,\ \vec{t} = \frac{1}{2}\vec{a}_1,\ \vec{t} = \frac{1}{2}\vec{a}_2,\ \vec{t} = \frac{1}{2}(\vec{a}_1 + \vec{a}_2).$$

One example is depicted in fig. 2.31.

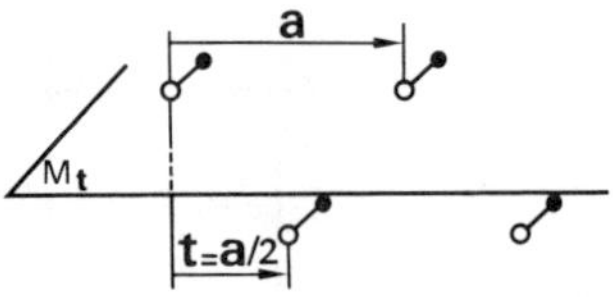

Fig. 2.31. Glide symmetry plane. The translation $\vec{t}$ is half the period $\vec{a}$.

2.4.3. Space Groups

It has been shown by Schönflies and Fedorov that every crystal can be assigned to one of 230 groups, called the space groups, when taking all possible lattices, screw axes and glide planes into account.

2.5. EXAMPLES OF STRUCTURES

The concept of compact packing is often involved in material structure theory as well as in applications. It seems useful therefore to describe the two possible ways of stacking dense layers of identical spheres.

2.5.1. Close-packed Structures

Once the first layer of spheres (with centres denoted by A) has been built as shown in Fig. 2.32, the spheres of the second layer are located in such a way that their centres are projected on the B interstices of the first layer. For the third layer, two possibilities arise

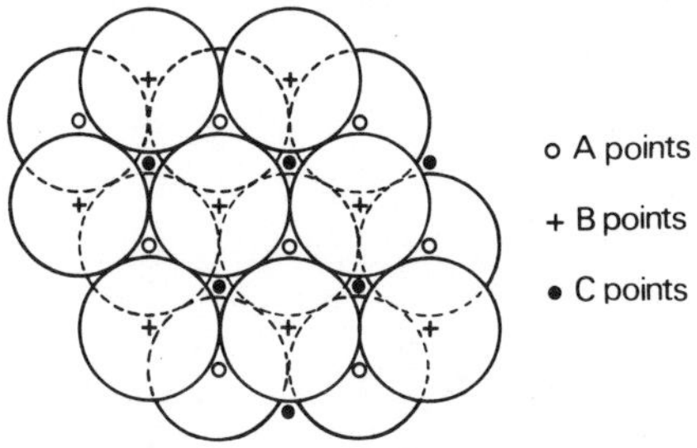

Fig. 2.32. Close packing of identical spheres.

depending on whether the C or A interstices are used. In the former case, the layers are ordered with period ABC. This structure, a close-packed cubic structure, is generated by a face centred cubic cell (fig. 2.33b), the triad axis of which is perpendicular to the layers. The latter case, the ABAB ordered structure, is a close-packed hexagonal structure; its cell is twice as large as the hexagonal cell (fig. 2.34). The ratio of the cell edges is $c/a = 2\sqrt{2/3} \simeq 1.633$ (see exercise 2.11).

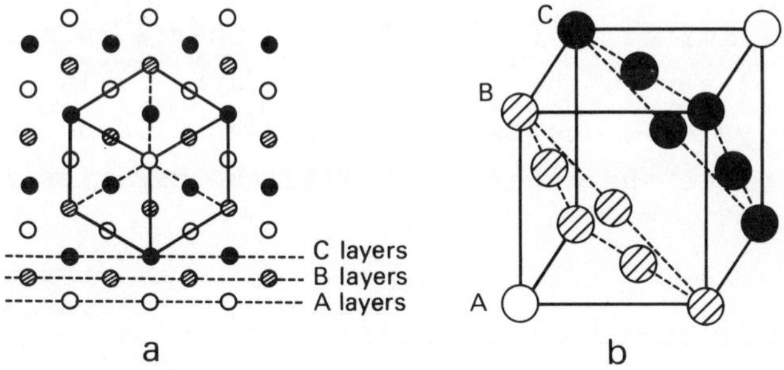

Fig. 2.33. Close-packed cubic structure.
a) Packing (A, B, C) of spheres and cubic cell seen from a diagonal;
b) The fcc lattice cell of this packing. There are octahedral cavities at the cube centre and at the midpoint of every edge, tetrahedral cavities at the mid-points between the centre and the corners.

It would be instructive (and perhaps amusing) for the reader who has not already done so to attempt to pack (at least once) billiard balls (or tennis balls) in these two ways. The major problem arises in preventing the balls from rolling away.

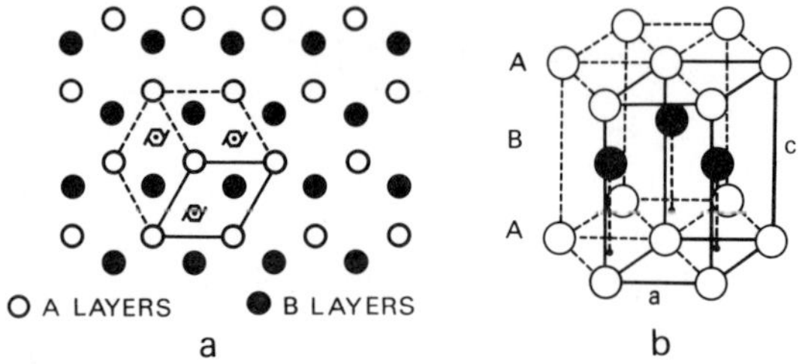

Fig. 2.34. Hexagonal close-packed structure.
a) Piling (ABAB) of spheres and hexagonal cell;
b) 2-fold hexagonal cell. The ratio c/a is $2\sqrt{2/3} = 1.633$.

Close-packed structures have two kinds of cavities between layers: tetrahedral cavities generated by one sphere put on three contiguous spheres, and octahedral cavities delimited by three adjacent spheres in a layer and three others in the neighbouring layer. In some diatomic crystals, one type of atom may build a compact structure while the other type fills the cavities in between.

2.5.2. The Structure of Some Useful Materials

Before going into any detail concerning materials used for elastic wave propagation, we must first emphasize the fact that the crystal structure of a solid material is generally not unique. It depends on its present (or previous) temperature. One says that there are several allotropes of the material. Physical properties may then be greatly altered while the chemical formula remains constant. For example a material may or may not be piezoelectric, depending on whether its temperature is above or below a threshold value. Even before discussing piezoelectric phenomena in Chapter 6 (i.e. the appearance of an electric polarization due to a strain, or vice versa), we have to say that only 20 crystal classes, out of 32, allow piezoelectricity. Among these 20 classes, which share a lack of a centre of symmetry, only 10 allow pyroelectricity (see exercise 3.4): crystals of these 10 classes exhibit spontaneous electric polarization. This steady polarization is normally compensated for by a free charge distribution inside or at the surface of the crystal, but it can be observed, when the temperature is varied, through a charge flow in an external circuit, since the spontaneous polarization magnitude is temperature dependent. If the direction of the polarization is to be altered, or reversed, by the occurrence of an external electric field, the pyroelectric crystal is then said to be <u>ferroelectric</u>. A hysteresis loop (polarization vs electric field) can be plotted. When the temperature is greater than a threshold value, the Curie temperature, any steady polarization vanishes. In such a state (called a paraelectric state), the graph of polarization vs electric field is a straight line. The dielectric constant of these materials, which is often large, is very much dependent, especially near the Curie point, on external influences (thermal or mechanical). This behaviour change can be accounted for by an alteration in crystal structure at the Curie point. The paraelectric phase structure is more symmetric than the ferroelectric structure. As an example, barium titanate ($BaTiO_3$), one of the most frequently analyzed ferroelectric materials, has a cubic (m3m) structure when the temperature θ is above 120^oC. For lower temperatures, the cell becomes quadratic ($10^oC < \theta < 120^oC$), then orthorhombic ($-70^oC < \theta < 10^oC$) and then rhombohedral ($\theta < -70^oC$).

It is possible to separate ferroelectric materials into two groups by referring to their chemical structure [1]. In the former group, ferroelectricity is linked to the hydrogen bonds (Seignette salt: double sodium and potassium tartrate, alcaline phosphates such as PO_4KH_2 ...). In the latter, it arises from the deformation of an oxygen octahedron based structure. $BaTiO_3$, $LiNbO_3$ (lithium niobate), $LiTaO_3$ (lithium tantalate) belong to this group. Before describing the structure of lithium niobate (utilized in the field of elastic waves because of its outstanding piezoelectric properties), we shall refer to cadmium sulphide (CdS) which has been used as a piezoelectric semiconductor for electron-elastic wave interaction studies; zinc oxide (ZnO), it is the main component of thin film transducers; sapphire (Al_2O_3) chosen for high frequency ($f > 1,000$ MHz) elastic wave propagation because of its low attenuation coefficient and price; quartz, used because its properties are nearly independent of temperature.

a) Cadmium sulphide (CdS). Zinc oxide (ZnO). Cadmium sulphide and zinc oxide crystals belong to the hexagonal system, with symmetry 6mm. The size of the hexagon is given by:

CdS: $a = 4.13$ Å $\quad c = 6.69$ Å $\Rightarrow c/a = 1.62$

ZnO: $a = 3.24$ Å $\quad c = 5.19$ Å $\Rightarrow c/a = 1.60$.

In both cases, the ratio c/a is very close to the compact hexagonal structure value (1.63). The anions (sulphur, oxygen) make a close-packed structure; half of its tetrahedral cavities are occupied by cations (cadmium, zinc); the cation diameter being about half the anion diameter. This wurtzite type structure is depicted in fig. 2.35. The 6-fold axis is a polar axis.

b) Sapphire (Al_2O_3). Sapphire crystals belong to the trigonal system with point group $\bar{3}m$. The oxygen atoms form a close-packed structure ABAB ... The aluminium atoms occupy two out of three octahedral cavities between the A and B layers so that the ratio Al_2/O_3 is satisfied (Fig. 2.36). The axis perpendicular to the oxygen layers is now only triad and the rhombohedral cell parameters are:

$$a = 3.51 \text{ Å}, \quad \alpha = 85^{\circ}46'.$$

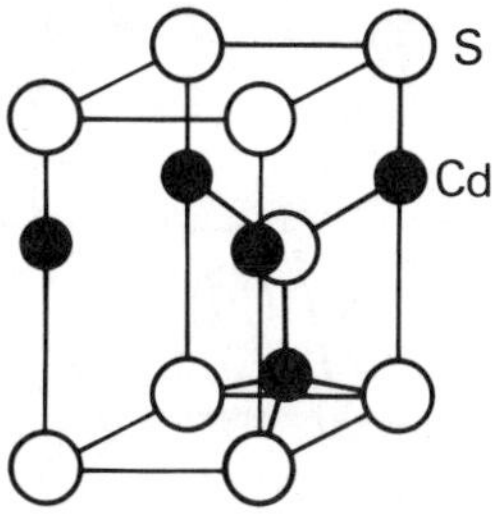

Fig. 2.35. Wurtzite type structure (CdS, ZnO).

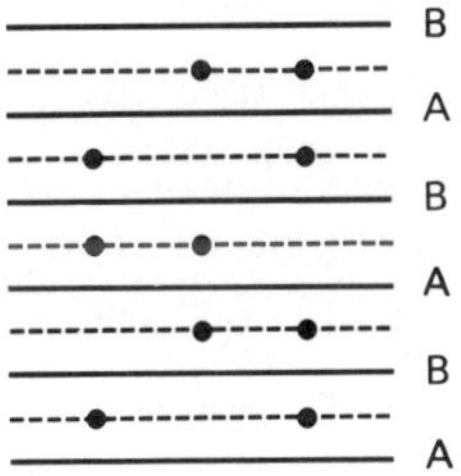

Fig. 2.36. Sapphire structure. Insertion of the aluminium atoms (•) between the oxygen layers (ABAB ...).

c) Quartz (SiO_2). Silica may have many forms: crystalline (quartz, tridymite, cristobalite) or amorphous (silica glass). Furthermore, every crystalline form has two allotropes α and β. The α (trigonal) quartz is stable below 573°C while β (hexagonal) quartz is stable between 573° and 870°C.

We are concerned here with piezoelectric α quartz, which belongs to class 32. Since there is no mirror, it has two enantiomorphic forms: one is dextrorotatory (right α quartz) the other is levorotatory (left α quartz). The triad axis Z is the optical axis; the three diad axes (X), perpendicular to the edges of the hexagonal prism, called the electric axes, are oriented. Fig. 2.37, a projection of the Si and O atoms on a plane perpendicular to the screw axis Z ($\vec{t} = \vec{c}/3$), gives an idea of the structure.

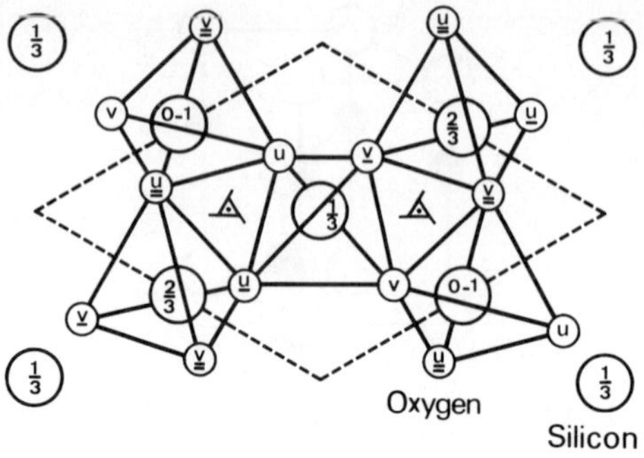

Fig. 2.37. α Quartz structure. The structure made of SiO_4 tetrahedrons linked by their corners (SiO_2 formula) has a screw axis (3_1) perpendicular to the paper plane. Marks in the circles stand for the vertical altitude of atoms, the length of the c edge being taken as unity. Silicon atoms are at height 0, 1/3, 2/3, oxygen atoms at u, $\underline{u}$ = u + 1/3, $\underline{\underline{u}}$ = u + 2/3, v, $\underline{v}$, $\underline{\underline{v}}$. X-ray diffraction provides u = .11 and v = .22 [2].

d) Lithium niobate ($LiNbO_3$). Lithium tantalate ($LiTaO_3$). Barium titanate ($BaTiO_3$). Lithium niobate is of class 3m. The rhombohedral cell (a = 5.492 Å, $\alpha = 55^\circ 33'$) contains two $LiNbO_3$ groups. We have already said this crystal is ferroelectric and its structure involves oxygen octahedrons. The start point is a close-packed hexagonal stack of such ions [3]. As in the case of sapphire two out of three cavities are occupied by metal ions (Li and Nb). Denoting an empty site by Ø, the occupation sequence along the triad axis is: Nb, Li, Ø, Nb, Li, Ø, Nb ... (fig. 2.38). Niobium and lithium ions have slightly different diameters so that the octahedron frame is distorted and no cation is accurately centred.

Another way of representing this structure is to start from the ideal perovskite-type structure, corresponding to the paraelectric state ($\theta > 1{,}200^\circ C$). The oxygen ions make an octahedral frame which can be obtained from Fig. 2.33b

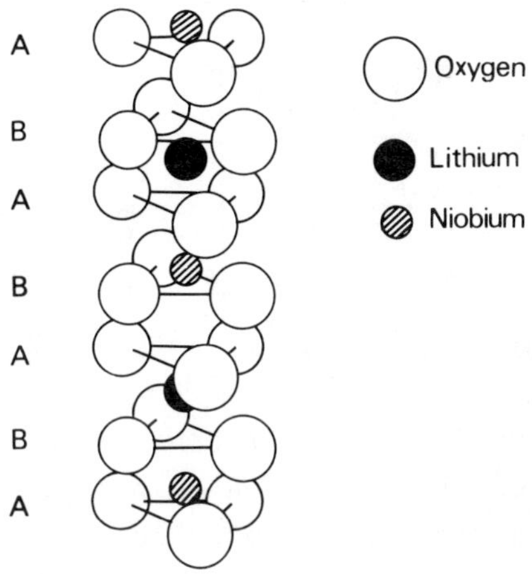

Fig. 2.38. Lithium niobate structure (Fig. 2 of Reference 4).

by removing the ions located at the corners of the cube (Fig. 2.39). The larger ions (A) of compound ABO_3 occupy the cube corners between two octahedrons, the smaller ions (B) being at the octahedron centres. The point group is cubic: m3m. In order to obtain the room temperature ferroelectric structure, one has to distort the paraelectric structure slightly. In the case of lithium niobate (A = Li, B = Nb), metal ions move and create a polarization along a preferred direction lowering the symmetry down to 3m. The above argument also applies to $BaTiO_3$ and $KNbO_3$, but the perovskite structure deformation occurs along the A_4 axis in $BaTiO_3$ which has a 4mm symmetry[1] at room temperature, and along axis A_2 in $KNbO_3$ which thus has 2mm symmetry.

As far as lithium tantalate ($LiTaO_3$) is concerned, all we have said about $LiNbO_3$ is still valid, the Nb ion being replaced by the Ta ion. The rhombohedral cell parameters are a = 5.47 Å and $\alpha = 56^{o}10'$.

---

1. Barium titanate is most often encountered as a ceramic. Single crystal growth is difficult.

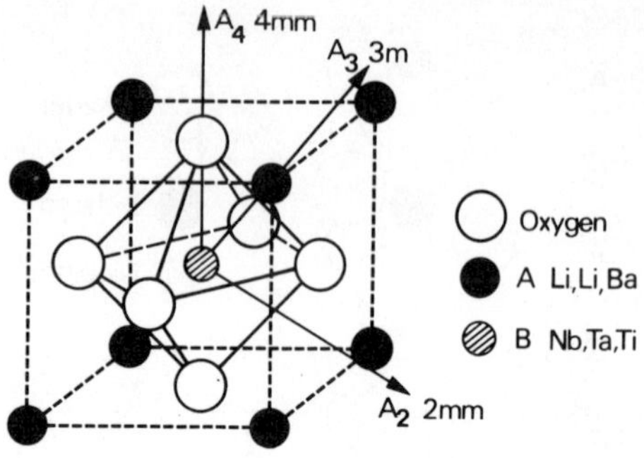

Fig. 2.39. Perovskite type structure ABO_3. The octahedron formed by six oxygen anions contains a B ion. Octahedrons are linked together by their corners. The A ions are at the corners of the cube containing the octahedrons.

REFERENCES

[1] L. Eyraud. 'Diélectriques solides anisotropes et ferroélectricité', chap. 6. Paris: Gauthier-Villars (1967).

[2] K. Schubert. 'Kristallstrukturen zweikomponentiger phasen', p. 200, fig. 6. Berlin: Springer Verlag (1964).

[3] H.D. Megaw. 'Ferroelectricity in crystals', chap. 5, paragr. 5. London: Methuen and Co. (1957).

[4] M. Di Domenico, Jr. and S.H. Wemple. "Oxygen-octahedra ferroelectrics", J. Appl. Phys., 40, 720 (1969).

BIBLIOGRAPHY

Th. Kahan. 'Théorie des groupes en physique classique et quantique', tome 2, p. 49-144: "Les groupes en cristallographie" by H. Curien. Paris: Dunod (1971).

D. Weigel. 'Algèbre et géométrie cristalline et moléculaire'. Paris: Masson et Cie (1972).

M.J. Buerger. 'Elementary Crystallography'. New York: John Wiley & Sons (1956).

F.C. Phillips. 'An introduction to Crystallography'.
London: Longmans (1971).

EXERCISES

2.1. Find the Miller indices of the planes ACGE, BDHF, EBD, and of the rows AC, AG in the cubic lattice shown in fig. 2.40.

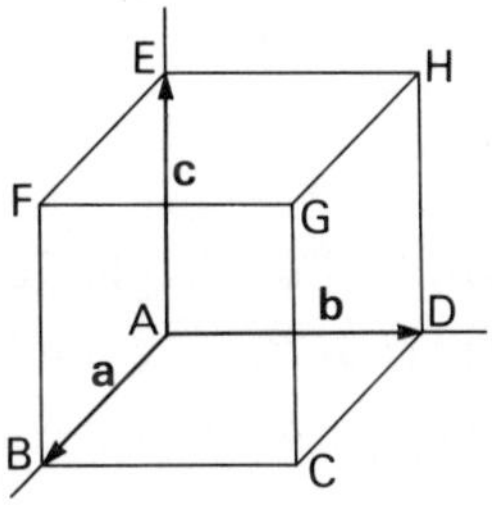

Fig. 2.40.

Solution. $(1,\bar{1}, 0)$, $(1,1,0)$, $(1,1,1)$, $[110]$, $[111]$.

2.2. Show that the standard cubic cell of a cubic face centred lattice is 4-fold by comparing its volume to the volume of the rhomboedron unit cell.

Solution. The cube edge is taken as the unit length. Using the notations of fig. 2.9b, the components of the rhomboedron basis vectors are

$$\vec{a}(\tfrac{1}{2}, \tfrac{1}{2}, 0), \; \vec{b}(\tfrac{1}{2}, 0, \tfrac{1}{2}), \; \vec{c}(0, \tfrac{1}{2}, \tfrac{1}{2}) \rightarrow V = |(\vec{a},\vec{b},\vec{c})| = \frac{1}{4}$$

2.3. Calculate the mass density of aluminium, whose crystal lattice is c. f. c., with cell edge a = 4.04Å Al = 27. The Avogadro number is $N = 6.023.10^{23}$.

Solution. There are four atoms for cell of volume $a^3 \rightarrow \rho = 4M/Na^3 = 2.72\ 10^3$ kg/m^3.

2.4. In fluorine (CaF_2), the ions Ca^{++} constitute a c.f.c. lattice of cell edge a = 5.45 Å. Calculate the mass density of fluorine: Ca = 40, F = 19.

Solution. The cubic cell contains four molecules $CaF_2 \rightarrow \rho = 3.20\ 10^3 kg/m^3$.

2.5. Calculate the equidistance of the planes in the families (100), (110), (111) for the three cubic lattices. For each lattice, which is the plane containing the highest site density?

Solution. The planes of highest density (i.e. of largest equidistance) are underlined.

Simple cubic: $\underline{d_{100}} = a$, $d_{110} = \frac{a\sqrt{2}}{2}$, $d_{111} = \frac{a\sqrt{3}}{3}$

Body centred cubic:

$$d_{100} = \frac{a}{2}, \underline{d_{110}} = \frac{a\sqrt{2}}{2}, d_{111} = \frac{a\sqrt{3}}{6}$$

Face centred cubic:

$$d_{110} = \frac{a}{2}, d_{110} = \frac{a\sqrt{2}}{2}, \underline{d_{111}} = \frac{a\sqrt{3}}{3}$$

2.6. Calculate the angle α for the rhomboedron cell of the c.f.c. lattice.

Solution. Calculate the scalar product $\vec{a}.\vec{b} \rightarrow \alpha = 60^o$.

2.7. Find the equation of the lattice plane family (hkℓ) in the reference frame Oxyz defined by the basis vectors $\vec{a}$, $\vec{b}$, $\vec{c}$.

Solution. Counting on from the origin, the nth plane in the family intersects the axes at points

$$A(x_n = \frac{na}{h}, 0, 0),\ B(0, y_n = \frac{nb}{k}, 0),\ (0, 0, z_n = \frac{nc}{\ell}).$$

Consequently the equation is

$$\frac{x}{x_n} + \frac{y}{y_n} + \frac{z}{z_n} = 1 \quad \text{or} \quad h\frac{x}{a} + k\frac{y}{b} + \ell\frac{z}{c} = n.$$

2.8. In a cubic lattice, what is the distance between two neighbouring planes of the family (h,k,ℓ)?

Solution. According to the solution of the above exercise, the equation of the first lattice plane is

$$hx + ky + \ell z = a$$

its distance from the origin is

$$d_{hk\ell} = \frac{a}{\sqrt{h^2+k^2+\ell^2}} .$$

2.9. Calculate the volume V of a cell in terms of the edges a, b, c, and of the angles $\alpha = (\vec{b}, \vec{c})$, $\beta = (\vec{a}, \vec{c})$ $\gamma = (\vec{a}, \vec{b})$. In order to put V into the form $V = abcF(\alpha, \beta, \gamma)$ use the expression:

$$V^2 = \begin{vmatrix} \vec{a}.\vec{a} & \vec{a}.\vec{b} & \vec{a}.\vec{c} \\ \vec{b}.\vec{a} & \vec{b}.\vec{b} & \vec{b}.\vec{c} \\ \vec{c}.\vec{a} & \vec{c}.\vec{b} & \vec{c}.\vec{c} \end{vmatrix}$$

Give $F(\alpha, \beta, \gamma)$ for each crystal system.

Solution.

$$V^2 = \begin{vmatrix} a^2 & ab\cos\gamma & ac\cos\beta \\ ab\cos\gamma & b^2 & bc\cos\alpha \\ ac\cos\beta & bc\cos\alpha & c^2 \end{vmatrix}$$

$$= a^2b^2c^2 \begin{vmatrix} 1 & \cos\gamma & \cos\beta \\ \cos\gamma & 1 & \cos\alpha \\ \cos\beta & \cos\alpha & 1 \end{vmatrix}$$

Hence

$$V^2 = abc\sqrt{1+2\cos\alpha\cos\beta\cos\gamma-\cos^2\alpha-\cos^2\beta-\cos^2\gamma}$$
$$= abcF(\alpha, \beta, \gamma).$$

Triclinic system:

$$F = \sqrt{1+2\cos\alpha\cos\beta\cos\gamma - \cos^2\alpha - \cos^2\beta - \cos^2\gamma}$$

Monoclinic system: $F = \sin\beta$.

Orthorhombic, tetragonal, cubic systems: $F = 1$.

Trigonal system: $F = \sqrt{1+2\cos^3\alpha - 3\cos^2\alpha}$

Hexagonal system: $F = \sqrt{3}/2$.

2.10. What is the point symmetry class of a regular tetrahedron?

Solution. Fig. 2.41 shows four 3-fold axes orthogonal to the faces, three inverse 4-fold axes orthogonal to each pair of opposite edges and six mirrors. The point symmetry of a regular tetrahedron is the same as that of class $\bar{4}3m$ in the cubic system.

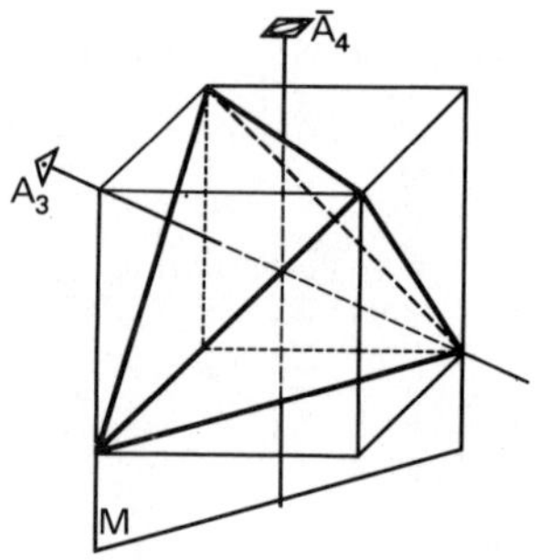

Fig. 2.41. Symmetry elements of a regular tetrahedron.

2.11. Calculate the ratio c/a of the hexagonal cell in stacking ABAB (fig. 2.34b).

Solution. Three spheres A and a sphere B form a tetrahedron of edge $a = 2r$, of height $2r\sqrt{2/3} = c/2$. Hence $c/a = 2\sqrt{2/3} = 1.633$.

2.12. What is the compactness factor f_r (the ratio of the volume occupied by the spheres to the total volume) for the two close-packed structures?

Solution. For the ABC stacking (fig. 2.33) the cubic cell edge is such that $a\sqrt{2} = 4r$, r being the sphere radius. Since the cell contains four spheres:

$$f_r = \frac{4 \times \frac{4}{3}\pi r^3}{(2\sqrt{2}r)^3} = \frac{\pi}{3\sqrt{2}} = 0.74$$

For the ABAB stacking, the hexagonal cell of volume $V = ca^2 \sqrt{3}/2 = 8\sqrt{2}r^3$ contains two spheres:

$$f_r = \frac{2 \times \frac{4}{3}\pi r^3}{8\sqrt{2}r^3} = \frac{\pi}{3\sqrt{2}} = 0.74.$$

It is not surprising to find the same compactness factor since both structures are made of adjacent spheres.

2.13. How many tetrahedron and octahedron cavities are there per cubic cell of the close-packed structure ABC?

Solution. Fig. 2.33b reveals 8 tetrahedron cavities (one per summit) and 4 octahedron cavities (one at the centre of the cube and 12 at the midpoint of each edge, shared by four cubes).

Chapter 3

Representation of the Physical Properties of Crystals by Tensors

In the previous chapter, we regarded the crystal structure as being obtained from the atomic group, repeated in space through the set of the translations generated by three basis vectors. This triple periodic array has symmetry elements from which it is possible to define the 32 crystal classes, from a macroscopic point of view. There is an intuitive feeling that physical and chemical properties of crystals depend not only upon the nature of the atoms, but also on their geometric relationships, i.e. on their symmetry. Therefore, whatever the atoms, all crystals of the same point group exhibit the same kind of behaviour with respect to identically oriented physical actions. This is revealed by tensor analysis, the mathematical tool for studying physical properties of crystals. It should be recalled that this concept was introduced, in the late nineteenth century, by the physicist W. Voigt, in order to describe the mechanical stress field of a solid. Tensor analysis classifies physical quantities according to the transformation properties of their components when the reference frame is changed. When such a change of axes is a symmetry element of the crystal, the identity of physical properties in both sets of co-ordinates provides relations between the components of the tensors that describe those physical properties, and finally reduces the number of independent components. We shall just take, as an example of this reduction, the second rank dielectric tensor. More involved cases, with third or fourth rank tensors, are dealt with in Chapter VI and IV, devoted to piezoelectricity and static elasticity. Since we shall have to derive eigenvalues and eigenvectors of second rank tensors we shall first recall some fundamental results.

3.1. RELATIONSHIP BETWEEN CAUSE AND EFFECT IN A CRYSTAL

The concept of the tensor arises as soon as one tries to set up linear equations between cause and effect in anisotropic media. In a crystal, a cause, oriented in a given direction, will usually give rise to an effect in another direction (fig. 3.1). In a dielectric material, for example, an electric field $\vec{E}$ creates a polarization $\vec{P}$ which is not parallel to $\vec{E}$.

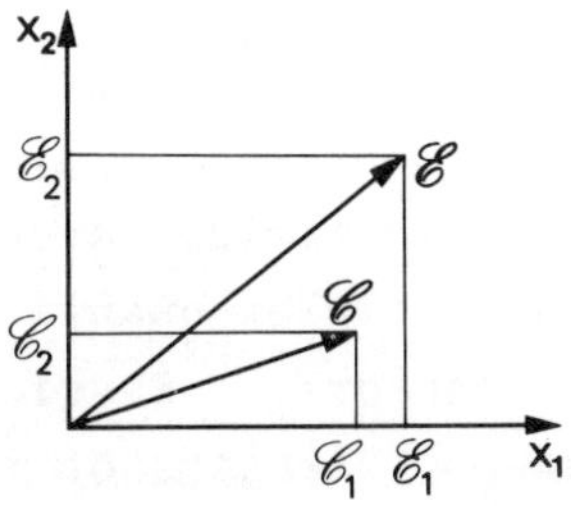

Fig. 3.1. Cause and effect relationship in a crystal. An anisotropic medium exhibits directions of 'lower resistance', a cause $\mathscr{C}$ generates an effect $\mathscr{E}$ ($\mathscr{C}$) which is usually not parallel to $\mathscr{C}$ If we restrict the expansion of $\mathscr{E}$ ($\mathscr{C}$) to the first order terms (linear domain) we obtain (for a vector cause and a vector effect)

$$\mathscr{E}_1 = A_{11}\mathscr{C}_1 + A_{12}\mathscr{C}_2$$

$$\mathscr{E}_2 = A_{21}\mathscr{C}_1 + A_{22}\mathscr{C}_2$$

If the cause $\mathscr{C}$ and the effect $\mathscr{E}$ are represented by vectors, the most general linear equations between the components $\mathscr{E}_1$, $\mathscr{E}_2$, $\mathscr{E}_3$ and $\mathscr{C}_1$, $\mathscr{C}_2$, $\mathscr{C}_3$ (in the same coordinate axes), will involve nine coefficients A_{ij}:

$$\mathscr{E}_1 = A_{11}\mathscr{C}_1 + A_{12}\mathscr{C}_2 + A_{13}\mathscr{C}_3$$

$$\mathscr{E}_2 = A_{21}\mathscr{C}_1 + A_{22}\mathscr{C}_2 + A_{23}\mathscr{C}_3$$

$$\mathscr{E}_3 = A_{31}\mathscr{C}_1 + A_{32}\mathscr{C}_2 + A_{33}\mathscr{C}_3$$

or, in short,

$$\mathscr{E}_i = A_{ij}\mathscr{C}_j \qquad i,j = 1,\ 2,\ 3$$

where we sum over the 3 values of the repeated index j, called a dummy index (Einstein convention). The nine components A_{ij} make a second rank tensor; similarly, an ordinary vector such as $\mathscr{E}_i$ or $\mathscr{C}_i$ is considered to be a first rank tensor and a scalar number such as temperature is a tensor of rank 0.

The $\mathscr{C}_i$ or $\mathscr{E}_i$ tensors and the A_{ij} tensor have different natures. The first two describe physical quantities. The last, A_{ij}, characterizes the medium: it describes the vector response $\mathscr{E}_i$ of the crystal, when undergoing the vector action $\mathscr{C}_i$. A scalar quantity such as temperature or energy is represented by a tensor of rank 0; a vector quantity (electric field, force, ...) by a first rank tensor; other more complex quantities (strain, stress) by second rank tensors. Physical properties of crystals can be measured by tensors of rank 0 (specific heat), 1 (pyroelectricity), 2 (permittivity, conductivity, thermal expansion), 3 (piezoelectricity), 4 (elasticity) ... As an example, the relationship between electric field and electric displacement:

$$D_i = \varepsilon_{ij}E_j \tag{3.1}$$

defines the permittivity tensor ε_{ij}.

The nine numbers A_{ij} are not, in fact, sufficient to describe the corresponding physical property. We must also specify the reference frame because in another reference frame the components of the same tensor have different values A'_{ij}. Both sets of values represent the same physical property, which does not depend upon the system of coordinates. Therefore, a relation between these coefficients should exist, and involve the change of reference frame. Precisely these transformation properties of tensor components are used to define tensors in § 3.3. Let us first look at reference frame changes with matrices.

3.2. CHANGE OF ORTHONORMAL REFERENCE FRAME

A change of reference frame is entirely defined by the nine coefficients α_i^k of the relations giving the new basis vectors $\vec{e}_1'$, $\vec{e}_2'$, $\vec{e}_3'$ in terms of the old ones $\vec{e}_1$, $\vec{e}_2$, $\vec{e}_3$:

$$\vec{e}_i' = \alpha_i^k \, \vec{e}_k \tag{3.2}$$

We have deliberately omitted the symbol Σ, the summation over the dummy index k being automatically understood.

The coefficients α_i^k can be displayed in a table, called the α matrix, the i-th line of which consists of the components α_i^k of $\vec{e}_i'$ on the $\vec{e}_k$ basis:

$$\alpha = \begin{pmatrix} \alpha_1^1 & \alpha_1^2 & \alpha_1^3 \\ \alpha_2^1 & \alpha_2^2 & \alpha_2^3 \\ \alpha_3^1 & \alpha_3^2 & \alpha_3^3 \end{pmatrix}$$

Conversely the old basis vectors $\vec{e}_k$ can be written in terms of the new ones $\vec{e}_j'$, through the elements β_k^j of a matrix β:

$$\vec{e}_k = \beta_k^j \, \vec{e}_j' \tag{3.3}$$

The conversion between α_i^k and β_k^j is made by inserting $\vec{e}_k$ into (3.2)

$$\vec{e}_i' = \alpha_i^k \, \beta_k^j \, \vec{e}_j'$$

or, using the Kronecker symbol $\delta_{ij} = 0$ if $i \neq j$ and $\delta_{ij} = 1$ if $i = j$, so that $\vec{e}_i' = \delta_{ij}\vec{e}_j'$:

$$(\alpha_i^k \, \beta_k^j - \delta_{ij})\vec{e}_j' = 0. \tag{3.4}$$

The $\vec{e}_j'$ terms constitute a basis so there cannot exist a non trivial linear equation between them and all coefficients in (3.4) have to be zero:

$$\alpha_i^k \, \beta_k^j = \delta_{ij} \tag{3.5}$$

According to the rules of matrix multiplications, β is the inverse of the transformation matrix α. The solution of the system (3.5) yields the components β^j_k:

$$\beta^j_k = \frac{M^k_j}{|\alpha|}$$

where M^k_j is the cofactor of the determinant $|\alpha|$, obtained on removing line j and column k.

The coordinate systems we shall use are all orthonormal.

$$\vec{e}'_i . \vec{e}'_j = \delta_{ij} \quad \text{and} \quad \vec{e}_k . \vec{e}_\ell = \delta_{k\ell}$$

In such a case, β is easily derived from α: the first scalar product is given by:

$$\vec{e}'_i = \alpha^k_i \, \vec{e}_k \quad \text{and} \quad \vec{e}'_j = \alpha^\ell_j \, \vec{e}_\ell$$

$$\alpha^k_i \, \alpha^\ell_j (\vec{e}_k . \vec{e}_\ell) = \delta_{ij}$$

Hence, using $\vec{e}_k . \vec{e}_\ell = \delta_{k\ell}$

$$\alpha^k_i \, \alpha^k_j = \delta_{ij} .$$

Comparing the above equation with (3.5), yields

$$\beta^j_k = \alpha^k_j .$$

The β matrix, the inverse of the α matrix for the change of orthonormal reference frame, is obtained by permuting lines and columns of α. This is the transposed matrix α^t. The α matrix, the inverse of which is identical to its transposed matrix, is said to be orthogonal.

Let x_k be the coordinates of a point M, i.e. the components of the vector $\overrightarrow{OM}$ on basis $\vec{e}_k$, and let x'_i be the coordinates of the same point with a basis $\vec{e}'_i$:

$$\vec{x} = x_k \vec{e}_k = x'_i \vec{e}'_i .$$

Substituting for $\vec{e}_k$ from (3.3):

$$\vec{x} = x_k \beta^j_k \vec{e}'_j = x'_i \vec{e}'_i .$$

This expansion is unique, so there is a relation between the new coordinates x'_i and the old coordinates x_k:

$$x'_i = \beta^i_k x_k$$

which is similar to 3.2 if the reference frames are orthonormal ($\beta^i_k = \alpha^k_i$):

$$x'_i = \alpha^k_i x_k. \tag{3.6}$$

3.3. DEFINITION OF A TENSOR

Physical quantities may be distinguished according to their transformation properties when the reference frame is changed. We shall always restrict ourselves to the case of orthonormal reference frames, for which the transformation laws of the coordinates (3.6) and of the basis vectors (3.2) are identical. By definition:

a) A scalar is a quantity that does not depend on the reference frame. It is also called an invariant or a zero rank tensor. The invariance of a scalar function of the coordinates $f(x_1, x_2, x_3)$ is expressed by

$$f(x_1, x_2, x_3) = f(x'_1, x'_2, x'_3).$$

Temperature, energy, and electrostatic potential are examples of a scalar quantity.

b) Any set of three numbers A_i having the same transformation law (under a change of reference frame) as the basis vectors $\vec{e}_i$, is a first rank tensor or a vector:

$$A'_i = \alpha^k_i A_k.$$

For example, the components of any vector of the space spanned by the three basis vectors $\vec{e}_i$ constitute a first rank tensor (cf. 3.6). The three derivatives $\partial f/\partial x_i$ also constitute a vector, the gradient of the scalar function $f(x_i)$: in the new frame, the gradient has components

$$\frac{\partial f}{\partial x'_i} = \frac{\partial f}{\partial x_k} \cdot \frac{\partial x_k}{\partial x'_i}$$

which can be derived from the old components $\partial f/\partial x_k$, with coefficients $\partial x_k/\partial x'_i$. Inverting (3.6), yields

$$x_k = \beta^i_k x'_i = \alpha^k_i x'_i$$

so that the components of the gradient transform as

$$\frac{\partial f}{\partial x'_i} = \alpha_i^k \frac{\partial f}{\partial x_k}.$$

c) Any set of nine numbers A_{ij} with the same transformation rule as the product of two vectors is a second rank tensor:

$$A'_{ij} = \alpha_i^k \alpha_j^\ell A_{k\ell}.$$

These coefficients are arranged in a square table, written between brackets in order to distinguish it from the matrix for the change of reference frame

$$A_{ij} = \begin{bmatrix} A_{11} & A_{12} & A_{13} \\ A_{21} & A_{22} & A_{23} \\ A_{31} & A_{32} & A_{33} \end{bmatrix}$$

Despite this similarity, matrices and tensors are quite different in their fundamental meaning:

- the components α_i^j make a connexion between two reference frames;

- the tensor A_{ij} is a physical (or mathematical) quantity represented in one frame at a time by nine numbers.

An example of a second rank tensor is given by the derivatives $\partial A_i/\partial x_k$ of the components of a vector (the proof is identical to that for the gradient).

Starting from a tensor A_{ij} one can construct an invariant by summing all diagonal terms. Taking the summation over repeated indices into account, this quantity $A_{11} + A_{22} + A_{33}$, called the trace of the tensor, is written A_{ii}. In a change of frame,

$$A'_{ii} = \alpha_i^k \alpha_i^\ell A_{k\ell} = A_{\ell\ell}$$

since $\alpha_i^k \alpha_i^\ell = \delta_{k\ell}$. In particular, the trace of the tensor $A_{ij} = A_i B_j$, i.e. the invariant $A_i B_i$, is the scalar product of vectors A_i and B_j. The product

$$A_i A_i = (A_1)^2 + (A_2)^2 + (A_3)^2,$$

denoted by A_i^2, represents the square of the length of the vector A_i.

d) The definition can be generalized without any difficulty: a tensor of rank r is a set of 3^r components, denoted by r indices, which transform as follows:

$$A'_{...ijk...} = ...\alpha_i^{\ell}\alpha_j^{m}\alpha_k^{n}...A_{...\ell mn...} \qquad (3.7)$$

The construction of the trace of a tensor is a particular case of the law of contraction of indices. The quantity $A_{...ijj\ell...}$ (summed over j), derived from the tensor $A_{...ijk\ell...}$ of rank r, is a tensor of rank (r-2). In a frame change, the new components

$$A'_{...ijj\ell...} = ...\alpha_i^{m}\alpha_j^{n}\alpha_j^{p}\alpha_\ell^{q} ...A_{...mnpq...}$$

can be written (since $\alpha_j^n\alpha_j^p = \delta_{np}$)

$$A'_{...ijj\ell...} = ...\alpha_i^{m}\alpha_\ell^{q}...A_{...mnnq...}$$

In order to check whether a set of numbers does constitute a tensor the following rule may be useful: the linear relationship between a tensor $A_{...ij...}$ of rank m and a tensor $B_{...k\ell...}$ of rank n defines a tensor $C_{...ijk\ell...}$ of rank (m+n):

$$A_{...ij...} = C_{...ijk\ell..}B_{...k\ell...} \qquad (3.8)$$

<u>Proof</u>. In a change of reference frame:

$$A'_{...pq...} = ...\alpha_p^{i}\alpha_q^{j}...A_{...ij...} \qquad (3.9)$$

and

$$B_{...k\ell...} = ...\beta_k^{r}\beta_\ell^{s}...B'_{...rs...} \qquad (3.10)$$

(the matrix β should be used to express the old components in terms of the new ones). Making use of $\beta_k^r = \alpha_r^k$, and inserting 3.10 into (3.8), and (3.8) into (3.9), yields

$$A'_{...pq...} = (...\alpha_p^{i}\alpha_q^{j}\alpha_r^{k}\alpha_s^{\ell}...C_{..ijk\ell..})B'_{..rs..}$$

The factor in brackets is in fact $C'_{..pqrs..}$, so that the terms $C_{..ijk\ell..}$ transform like the components of a tensor of rank (m+n):

$$C'_{...pqrs...} = ...\alpha_p^{i}\alpha_q^{j}\alpha_r^{k}\alpha_s^{\ell}...C_{..ijk\ell..}$$

Consequently, the Kronecker δ_{ij} is a second rank tensor because

$$A_i = \delta_{ij}A_j.$$

In elasticity theory, the rigidity coefficients $c_{ijk\ell}$, which connect the stress tensor T_{ij} and the strain tensor $S_{k\ell}$, constitute a fourth rank tensor:

$$T_{ij} = c_{ijk\ell}S_{k\ell}.$$

3.4. REDUCTION OF THE NUMBER OF INDEPENDENT TENSOR COMPONENTS ENFORCED BY CRYSTAL SYMMETRY

Let $A_{...ijk...}$ be the components (in the reference frame $Ox_1x_2x_3$) of a tensor representing a physical property of a crystal, and $A'_{...ijk...}$ the components of the same tensor in the same reference frame, but for a new crystal orientation, obtained by the transformation $\mathcal{S}$. If we wish to express the terms $A'_{...ijk...}$ in terms of $A_{...pqr...}$, we must first note that it is equivalent to maintaining the crystal in its original position, and applying the inverse operation $\mathcal{S}^{-1}$ to the reference frame (fig. 3.2).

Therefore, if α is the matrix for $\mathcal{S}^{-1}$, then from 3.7

$$A'_{...ijk...} = ...\alpha_i^p\alpha_j^q\alpha_k^r...A_{...pqr...}$$

If now $\mathcal{S}$ belongs to the point group of the crystal, the new orientation of the crystal with respect to the $Ox_1x_2x_3$ reference frame cannot on the basis of macroscopic measurements be distinguished from the first orientation, so that

$$A'_{...ijk...} = A_{...ijk...}$$

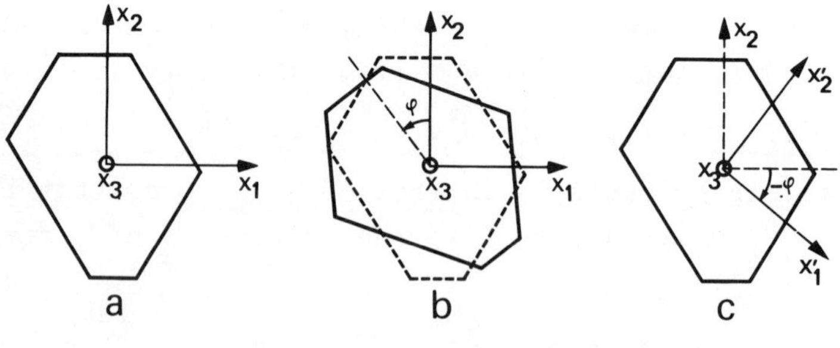

Fig. 3.2. Equivalence between a crystal rotation and a frame transformation

a) crystal and reference frame in the original position;

b) ϕ rotation of the crystal about the x_3 axis;

c) $(-\phi)$ rotation of the reference frame.

The invariance of physical properties with respect to symmetry operations thus implies the equations

$$A_{...ijk...} = ...\ \alpha_i^p \alpha_j^q \alpha_k^r ... A_{...pqr...} \qquad (3.11)$$

which reduce the number of independent tensor components. Since the inverse of any symmetry operation of the crystal ($\mathcal{S} = A_n$ or $\overline{A}_n$) is also a symmetry operation ($\mathcal{S}^{-1} = \mathcal{S}^{n-1}$), α is the reference frame transformation matrix for one of the point groups of the crystal.

3.4.1. Matrices for Symmetry Group Elements in Crystals

The symmetry elements of crystals are the direct or inverse axis, the mirror plane and the centre of symmetry (2.3.1.1.). In a rotation of ϕ about Ox_3 (fig. 3.3.a), the new basis vectors are given in terms of the old ones by

$$\vec{e}_1' = (\cos\phi)\vec{e}_1 + (\sin\phi)\vec{e}_2$$

$$\vec{e}_2' = -(\sin\phi)\vec{e}_1 + (\cos\phi)\vec{e}_2$$

$$\vec{e}_3' = \vec{e}_3 .$$

The reference frame transformation matrix for the rotation of ϕ about Ox_3 is then

$$\alpha_{x_3,\phi} = \begin{pmatrix} \cos\phi & \sin\phi & 0 \\ -\sin\phi & \cos\phi & 0 \\ 0 & 0 & 1 \end{pmatrix}. \qquad (3.12)$$

The matrix denoting symmetry with respect to the origin is

$$\alpha_c = \begin{pmatrix} -1 & 0 & 0 \\ 0 & -1 & 0 \\ 0 & 0 & -1 \end{pmatrix}$$

From the equation $\overline{A}_n = A_n.C$, we derive the reference frame transformation matrix for an n-fold inversion-rotation:

$$\overline{\alpha}_{x_3,2\pi/n} = \alpha_{x_3,2\pi/n} \cdot \alpha_c$$

For a symmetry plane perpendicular to the x_3-axis

$$\alpha_{M\perp x_3} = \begin{pmatrix} 1 & 0 & 0 \\ 0 & 1 & 0 \\ 0 & 0 & -1 \end{pmatrix}.$$

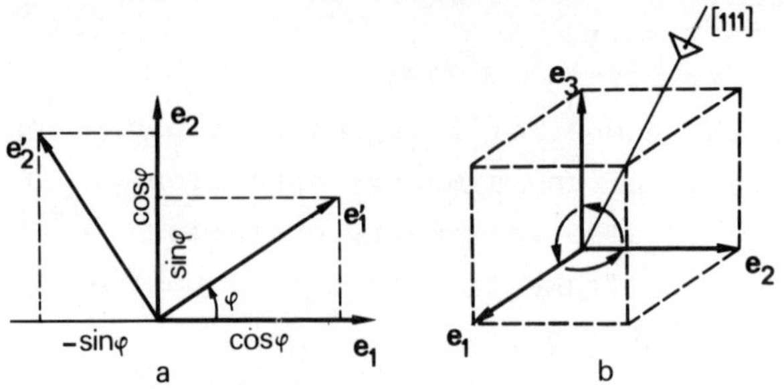

Fig. 3.3. Axis rotation (a) about the x_3-axis, (b) about the [111] direction.

A $2\pi/3$ (triad) rotation about the main diagonal (along [111]) of the cube generated by the three basis vectors (fig. 3.3.b) gives rise to a circular permutation of the axes:

$$\vec{e}_{1'} = \vec{e}_2, \quad \vec{e}_{2'} = \vec{e}_3, \quad \vec{e}_{3'} = \vec{e}_1$$

and is represented by the matrix:

$$\alpha_{[111],2\pi/3} = \begin{pmatrix} 0 & 1 & 0 \\ 0 & 0 & 1 \\ 1 & 0 & 0 \end{pmatrix}.$$

3.4.2. Effect of a Centre of Symmetry

In the case of an inversion the general invariance condition (3.11) reduces, since matrix α is diagonal ($\alpha_i^j = -\delta_{ij}$) to

$$A_{...ijk...} = ...\alpha_i^i \alpha_j^j \alpha_k^k ... A_{...ijk...}$$

For a tensor of rank n:

$$A_{...ijk...} = (-1)^n A_{...ijk...}$$

For odd n ($(-1)^n = -1$), the above equation shows that all components vanish. Physical properties represented by odd rank tensors do not appear in crystals belonging to any of the eleven centro-symmetric classes. This is especially true for piezoelectricity (n = 3).

For even n, equation (3.13) is of no assistance. A centre of symmetry has no influence on physical properties represented by even rank tensors. Moreover, there is no distinction between the axes - direct or inverse - so that the classes of a same system which share one or several - direct or inverse - axes (§ 2.3.3) exhibit a minimal reduction of the number of independent components. These remarks are illustrated by the example of the dielectric tensor.

3.4.3. Reduction of the Number of Independent Dielectric Constants

From thermodynamic arguments, quite similar to those used for the elasticity tensor (§ 4.4), we know that the dielectric tensor, defined by (3.1), is symmetric: $\varepsilon_{ij} = \varepsilon_{ji}$. The invariance condition (3.11) is thus written:

$$\varepsilon_{ij} = \alpha_i^k \alpha_j^\ell \varepsilon_{k\ell} \tag{3.14}$$

This is a tensor of even rank, so there is no reduction in the number of dielectric constants (6) for the triclinic system crystals, which have at most a centre of symmetry

$$[\varepsilon_{ij}] = \begin{bmatrix} \varepsilon_{11} & \varepsilon_{12} & \varepsilon_{13} \\ \varepsilon_{12} & \varepsilon_{22} & \varepsilon_{23} \\ \varepsilon_{13} & \varepsilon_{23} & \varepsilon_{33} \end{bmatrix} \quad \text{triclinic}$$

If the Ox_3 axis of the orthonormal reference frame is taken parallel to the diad axis of monoclinic system crystals, the frame transformation matrix is diagonal:

$$\alpha_{A_2 || x_3} = \begin{pmatrix} -1 & 0 & 0 \\ 0 & -1 & 0 \\ 0 & 0 & 1 \end{pmatrix} \quad \text{or} \quad \alpha_{M \perp x_3} = \begin{pmatrix} 1 & 0 & 0 \\ 0 & 1 & 0 \\ 0 & 0 & -1 \end{pmatrix}$$

Equation 3.14

$$\varepsilon_{ij} = \alpha_i^i \alpha_j^j \varepsilon_{ij}$$

cancels out coefficients if i or j equals 3 ($\alpha_i^i \alpha_3^3 = -1$ if $i \neq 3$), which reduces the number of coefficients to 4:

$$[\varepsilon_{ij}] = \begin{bmatrix} \varepsilon_{11} & \varepsilon_{12} & 0 \\ \varepsilon_{12} & \varepsilon_{22} & 0 \\ 0 & 0 & \varepsilon_{33} \end{bmatrix} \quad \text{monoclinic}$$

For the orthorhombic system, three diad (direct or inverse) axes constitute the reference frame. The foregoing argument, used for each axis, shows that all off-diagonal coefficients vanish:

$$[\varepsilon_{ij}] = \begin{bmatrix} \varepsilon_{11} & 0 & 0 \\ 0 & \varepsilon_{22} & 0 \\ 0 & 0 & \varepsilon_{33} \end{bmatrix} \qquad \text{orthorhombic}$$

When the A_n or $\overline{A}_n$ axis ($n > 2$) of trigonal, tetragonal and hexagonal crystals is along Ox_3, the frame transformation matrix is given by 3.12, with $\phi = 2\pi/n < \pi$. Applying 3.14 to ε_{13} and ε_{23} yields

$$\begin{cases} \varepsilon_{13} = \alpha_1^k \alpha_3^3 \varepsilon_{k3} = \alpha_1^1 \varepsilon_{13} + \alpha_1^2 \varepsilon_{23} & \text{for} \quad \alpha_1^3 = 0 \\ \varepsilon_{23} = \alpha_2^k \alpha_3^3 \varepsilon_{k3} = \alpha_2^1 \varepsilon_{13} + \alpha_2^2 \varepsilon_{23} & \text{for} \quad \alpha_2^3 = 0 \end{cases}$$

or

$$\begin{cases} \varepsilon_{13}(1 - \cos\phi) - \varepsilon_{23}\sin\phi = 0 \\ \varepsilon_{13}\sin\phi + \varepsilon_{23}(1 - \cos\phi) = 0 \end{cases}$$

ϕ being non-zero, $\varepsilon_{13} = \varepsilon_{23} = 0$. For ε_{11} we have

$$\varepsilon_{11} = \alpha_1^k \alpha_1^l \varepsilon_{k\ell} = \alpha_1^1 \alpha_1^2 (\varepsilon_{12} + \varepsilon_{21}) + \alpha_1^1 \alpha_1^1 \varepsilon_{11} + \alpha_1^2 \alpha_1^2 \varepsilon_{22}$$

or

$$\varepsilon_{11}\sin^2\phi = (\varepsilon_{12} + \varepsilon_{21})\cos\phi\sin\phi + \varepsilon_{22}\sin^2\phi.$$

Dividing by $\sin^2\phi$:

$$\varepsilon_{11} - \varepsilon_{22} = (\varepsilon_{12} + \varepsilon_{21})\operatorname{cotan}\phi \qquad (3.15)$$

The same computation, with ε_{12}, gives

$$\varepsilon_{11} - \varepsilon_{22} = -(\varepsilon_{12} + \varepsilon_{21})\tan\phi \qquad (3.16)$$

Since $\phi \neq \pi$, equations (3.15) and (3.16) together imply

$$\varepsilon_{12} + \varepsilon_{21} = 0; \qquad \varepsilon_{22} = \varepsilon_{11}$$

and $\varepsilon_{12} = -\varepsilon_{21} = 0$ because the tensor is symmetric.

Crystals of the trigonal, tetragonal and hexagonal systems have only two independent dielectric constants: ε_{11} and ε_{33}

$$[\varepsilon_{ij}] = \begin{bmatrix} \varepsilon_{11} & 0 & 0 \\ 0 & \varepsilon_{11} & 0 \\ 0 & 0 & \varepsilon_{33} \end{bmatrix} \qquad \begin{matrix} \text{trigonal} \\ \text{tetragonal} \\ \text{hexagonal.} \end{matrix}$$

For cubic system crystals, the minimum symmetry is $4A_3$ and $3A_2$ (these diad axes are taken as the reference frame), only diagonal terms exist as for orthorhombic crystals, but now the rotation about the triad axis [111] creates a circular permutation of the indices:

$$(1,2,3) \rightarrow (2,3,1) \implies \varepsilon_{11} = \varepsilon_{22} = \varepsilon_{33} = \varepsilon$$

$$[\varepsilon_{ij}] = \begin{bmatrix} \varepsilon & 0 & 0 \\ 0 & \varepsilon & 0 \\ 0 & 0 & \varepsilon \end{bmatrix} \qquad \text{cubic}$$

A cubic crystal is an isotropic dielectric medium; the electric field and displacement are parallel: $D_i = \varepsilon E_i$.

3.5. EIGENVECTORS AND EIGENVALUES OF A SECOND RANK TENSOR

When solving many problems in physics, the linear equation approximation involves systems of three (or n) homogeneous linear equations (i = 1, 2, 3) with three (or n) unknown quantities u_1, u_2, u_3:

$$C_{ij}u_j = 0 \qquad (3.17)$$

where the terms C_{ij} are fixed numbers. For a non zero solution to exist, the determinant of the C_{ij} coefficients has to vanish

$$|C_{ij}| = \begin{vmatrix} C_{11} & C_{12} & C_{13} \\ C_{21} & C_{22} & C_{23} \\ C_{31} & C_{32} & C_{33} \end{vmatrix} = 0$$

This is the compatibility condition for the equations (3.17). An important example of such linear homogeneous systems is provided by the equation for the eigenvectors u_j and eigenvalues λ of a tensor A_{ij}:

$$A_{ij}u_j = \lambda u_i \qquad (3.18)$$

Since $u_i = \delta_{ij}u_j$, the system can be put into the standard form, analogous to (3.17)

$$(A_{ij} - \lambda\delta_{ij})u_j = 0 \qquad (3.19)$$

The eigenvalues λ are determined by the compatibility condition:

$$|A_{ij} - \lambda\delta_{ij}| = \begin{vmatrix} A_{11}-\lambda & A_{12} & A_{13} \\ A_{21} & A_{22}-\lambda & A_{23} \\ A_{31} & A_{32} & A_{33}-\lambda \end{vmatrix} = 0. \tag{3.20}$$

This equation, known as the secular equation, is of degree 3 in λ:

$$(A_{11}-\lambda)(A_{22}-\lambda)(A_{33}-\lambda) + A_{21}A_{32}A_{13} + A_{12}A_{23}A_{31}$$
$$- (A_{22}-\lambda)A_{31}A_{13} - (A_{11}-\lambda)A_{23}A_{32} - (A_{33}-\lambda)A_{12}A_{21} = 0.$$

There are generally three distinct roots $\lambda^{(1)}$, $\lambda^{(2)}$, $\lambda^{(3)}$, which are the eigenvalues (or principal values) of the tensor A_{ij}. Every eigenvalue $\lambda^{(k)}$ is assigned to an eigenvector $u^{(k)}$, the components of which satisfy the three (non independent) equations.[1]

$$A_{ij}u_j^{(k)} = \lambda^{(k)}u_i^{(k)} \tag{3.21}$$

In fact, for any scalar μ, $\mu u_j^{(k)}$ is also an eigenvector corresponding to the eigenvalue $\lambda^{(k)}$, so it is better to refer to eigen directions - or principal directions - of a tensor.

Using as basis vectors the eigenvectors (supposed to be of unit length, $u_i^{(k)} = \delta_{ik}$), equation (3.21) is written in terms of the new components A'_{ij}

$$A'_{ij}\delta_{jk} = \lambda^{(k)}\delta_{ik}$$

or

$$A'_{ik} = \lambda^{(k)}\delta_{ik}$$

and the tensor A'_{ij} is diagonal

$$A'_{ij} = \begin{bmatrix} \lambda^{(1)} & 0 & 0 \\ 0 & \lambda^{(2)} & 0 \\ 0 & 0 & \lambda^{(3)} \end{bmatrix}$$

1. The index k in $\lambda^{(k)}$ is written in brackets in order to distinguish it from the index denoting a component. The Einstein summation convention does not apply to indices in brackets.

This form shows that the trace of the tensor, i.e. the invariant A_{ii}, is equal to the sum of the eigenvalues

$$A_{ii} = \lambda^{(1)} + \lambda^{(2)} + \lambda^{(3)}. \qquad (3.22)$$

If two eigenvalues, say $\lambda^{(2)}$ and $\lambda^{(3)}$, are equal, any linear combination of the eigenvectors $u_i^{(2)}$ and $u_i^{(3)}$:

$$u_i = \mu_2 u_i^{(2)} + \mu_3 u_i^{(3)}$$

is also an eigenvector:

$$A_{ij}u_j = \mu_2 A_{ij}u_j^{(2)} + \mu_3 A_{ij}u_j^{(3)}$$

or

$$A_{ij}u_j = \lambda^{(2)}(\mu_2 u_i^{(2)} + \mu_3 u_i^{(3)}) = \lambda^{(2)}u_i.$$

For a twofold (degenerate) eigenvalue, we have not only a principal direction but also a principal plane defined by the vectors $u_i^{(2)}$ and $u_i^{(3)}$.

Symmetric tensors. We shall now prove two basic results for real symmetric tensors ($A_{ij} = A_{ji}$)

1) The eigenvalues are real.

2) Principal directions linked to distinct eigenvalues are mutually orthogonal.

The first point emerges from equation (3.18) by constructing the contracted product

$$u_i^* A_{ij} u_j = \lambda u_i^* u_i \qquad (3.23)$$

which becomes, after a circular permutation of the dummy indices i, j on the left hand side:

$$u_j^* A_{ji} u_i = \lambda u_i^* u_i$$

and for a symmetric tensor ($A_{ji} = A_{ij}$)

$$u_j^* A_{ij} u_i = \lambda u_i^* u_i.$$

The tensor being real, we can compare the complex conjugate equation

$$u_j A_{ij} u_i^* = \lambda^* u_i u_i^*$$

to equation (3.23) and deduce that the eigenvalues are real ($\lambda^* = \lambda$).

In order to prove the second theorem, we write the equation for the eigenvectors $u_i^{(1)}$ and $u_i^{(2)}$:

$$A_{ij}u_j^{(1)} = \lambda^{(1)}u_i^{(1)}$$

$$A_{ij}u_j^{(2)} = \lambda^{(2)}u_i^{(2)}$$

and we construct the scalar product $u_i^{(1)}u_i^{(2)}$ in both cases:

$$A_{ij}u_j^{(1)}u_i^{(2)} = \lambda^{(1)}u_i^{(1)}u_i^{(2)}$$

$$A_{ij}u_j^{(2)}u_i^{(1)} = \lambda^{(2)}u_i^{(2)}u_i^{(1)}$$

We subtract the second equation from the first:

$$A_{ij}u_j^{(1)}u_i^{(2)} - A_{ij}u_j^{(2)}u_i^{(1)} = [\lambda^{(1)} - \lambda^{(2)}]u_i^{(1)}u_i^{(2)}$$

and make a circular permutation of i and j in the left hand side:

$$(A_{ji} - A_{ij})u_i^{(1)}u_j^{(2)} = [\lambda^{(1)} - \lambda^{(2)}]u_i^{(1)}u_i^{(2)}$$

We now deduce that the scalar product $u_i^{(1)}u_i^{(2)}$ vanishes if the tensor is symmetric ($A_{ji} = A_{ij}$) and if the eigenvalues are distinct ($\lambda^{(1)} - \lambda^{(2)} \neq 0$).

Two equal eigenvalues are linked to a principal plane, orthogonal to the third eigenvector - where one can always find a set of two orthogonal vectors. In all cases, the principal directions of a symmetric tensor can be arranged in an orthonormal reference frame.

Determining the eigenvalues and the eigenvectors is particularly simple when at least two non diagonal coefficients are zero, say A_{13} and A_{23}. The expansion of the determinant 3.20 yields

$$(A_{33} - \lambda)[(A_{11} - \lambda)(A_{22} - \lambda)-(A_{12})^2] = 0$$

The root $\lambda^{(3)} = A_{33}$ comes out at once. The other two eigenvalues $\lambda^{(1)}$ and $\lambda^{(2)}$ are the roots of an equation of degree 2:

$$\lambda^{(k)} = \frac{A_{11}+A_{22}}{2} - \frac{(-1)^k}{2} \sqrt{(A_{22}-A_{11})^2+4(A_{12})^2} \text{ with } k = 1,2. \tag{3.24}$$

The system (3.21)

$$\begin{cases} [A_{11}-\lambda^{(k)}]u_1^{(k)} + A_{12}u_2^{(k)} = 0 \\ A_{12}u_1^{(k)} + [A_{22}-\lambda^{(k)}]u_2^{(k)} = 0 \\ [A_{33}-\lambda^{(k)}]u_3^{(k)} = 0 \end{cases}$$

shows that the Ox_3 axis is the principal direction for the eigenvalue $\lambda^{(3)} = A_{33}$

$$u_1^{(3)} = u_2^{(3)} = 0; \quad u_3^{(3)} \text{ arbitrary}$$

For the other two eigenvalues (k = 1,2), $\lambda^{(k)} \neq A_{33} \Longrightarrow u_3^{(k)} = 0$ and

$$\frac{u_2^{(k)}}{u_1^{(k)}} = \frac{\lambda^{(k)}-A_{11}}{A_{12}}$$

or

$$\frac{u_2^{(k)}}{u_1^{(k)}} = \frac{A_{22}-A_{11}}{2A_{12}} -(-1)^k \sqrt{\left(\frac{A_{22}-A_{11}}{2A_{12}}\right)^2+1}.$$

The principal directions, in the Ox_1x_2 plane are defined by the angles β_1 and $\beta_2 = \beta_1 + \pi/2$ with Ox_1:

$$\tan \beta_k = \frac{u_2^{(k)}}{u_1^{(k)}} = a -(-1)^k \sqrt{a^2+1}$$

where

$$a = \frac{A_{22}-A_{11}}{2A_{12}}$$

or by the double angle

$$\tan 2\beta_k = \frac{2[\ a-(-1)^k\sqrt{a^2+1}]}{1-[a^2+a^2+1-2a(-1)^k\sqrt{a^2+1}]} = -\frac{1}{a}.$$

Inserting the expression for a and allowing for an extra π in $2\beta_k$, both orthogonal directions are given by

$$\tan 2\beta = \frac{2A_{12}}{A_{11}-A_{22}} \qquad (3.25)$$

3.6. TENSOR REPRESENTATION OF SURFACE ELEMENTS

We shall now have to calculate the integral of tensors along a curve, on a surface, in a volume.

In the first case, the infinitesimal element is the vector dx_i and the integral of a tensor $A_{\ldots i \ldots}$ is written

$$\int_c A_{\ldots i \ldots} dx_i .$$

For a surface integral, the infinitesimal element is the parallelogram based on the vectors $d\vec{x}^{(1)}$ and $d\vec{x}^{(2)}$. Its projection on the co-ordinate plane $x_j x_k$ has an area:

$$ds_{jk} = dx_j^{(1)} dx_k^{(2)} - dx_k^{(1)} dx_j^{(2)} .$$

These quantities constitute an antisymmetric second rank tensor:

$$[ds_{jk}] = \begin{bmatrix} 0 & ds_{12} & ds_{13} \\ -ds_{12} & 0 & ds_{23} \\ -ds_{13} & -ds_{23} & 0 \end{bmatrix}$$

the three independent coefficients of which can be denoted by an index i : $ds_i = ds_{jk}$ so that the (ijk) permutation is even:

$$ds_1 = ds_{23} = dx_2^{(1)} dx_3^{(2)} - dx_3^{(1)} dx_2^{(2)}$$

$$ds_2 = ds_{31} = dx_3^{(1)} dx_1^{(2)} - dx_1^{(1)} dx_3^{(2)}$$

$$ds_3 = ds_{12} = dx_1^{(1)} dx_2^{(2)} - dx_2^{(1)} dx_1^{(2)} .$$

The transformation law 3.6 can be applied to the components ds_i only for changes of axis system that remains right-handed (i.e. for rotations). If there is an inversion of the reference frame (symmetry with respect to a point or to a plane), the sign should be changed (see exercise 3.5). The pseudo-vector, or axial vector is simply the vector product of the two vectors $d\vec{x}^{(1)}$ and $d\vec{x}^{(2)}$. From the properties of the vector product we know that $d\vec{s} = d\vec{x}^{(1)} \wedge d\vec{x}^{(2)}$ is perpendicular to the parallelogram and that its length is equal to its area ds. Therefore, instead of the antisymmetric tensor ds_{jk}, we take the pseudo-vector ds_i as the integration element. If $ds_i = 1_i ds$ and 1_i is the unit vector perpendicular to the surface, the integral of a tensor $A_{\ldots i \ldots}$ on the surface s is

$$\int_s A_{\ldots i \ldots} 1_i ds .$$

If s is a closed surface, this integral is converted into a volume integral with the aid of the Green's theorem. The vector expression of this theorem

$$\int_s \vec{A}.d\vec{s} = \int_v (\mathrm{div}\ \vec{A})dv$$

is, in tensor notation

$$\int_s A_i ds_i = \int_s A_i l_i ds = \int_v \frac{\partial A_i}{\partial x_i} dv$$

v is the volume within the surface s. In the general case:

$$\int_s A_{..i..} l_i ds = \int_v \frac{\partial A_{..i..}}{\partial x_i} dv. \tag{3.26}$$

It must be remembered that

$$\frac{\partial A_{..i..}}{\partial x_i} = \frac{\partial A_{..1..}}{\partial x_1} + \frac{\partial A_{..2..}}{\partial x_2} + \frac{\partial A_{..3..}}{\partial x_3} .$$

BIBLIOGRAPHY

L. Brillouin. 'Les tenseurs en mécanique et en élasticité'. Paris: Masson et Cie (1949).

E. Bauer. 'Champs de vecteurs et de tenseurs'. Paris: Masson et Cie (1955).

J.F. Nye. 'Physical properties of crystals', chap. I and II. Oxford (1964).

EXERCISES

3.1. Show that the determinant of an orthogonal matrix α is equal to $\pm$ 1.

Solution. The determinant of the inverse matrix $\beta = \alpha^t$ is $1/|\alpha| \Longrightarrow |\alpha|^2 = 1$.

3.2. Write down the matrices for a rotation of angle ϕ about the x_1 and x_2 axes.

Solution.

$$\alpha_{x_1,\phi} = \begin{pmatrix} 1 & 0 & 0 \\ 0 & \cos\phi & \sin\phi \\ 0 & -\sin\phi & \cos\phi \end{pmatrix} \quad \alpha_{x_2,\phi} = \begin{pmatrix} \cos\phi & 0 & \sin\phi \\ 0 & 1 & 0 \\ -\sin\phi & 0 & \cos\phi \end{pmatrix}$$

3.3. Let $\vec{v}$ and $\vec{u}$ be two vectors, $\vec{u}$ having unit length, such that

$$v_i = A_{ij} u_j \quad \text{with} \quad A_{ij} = A_{ji}$$

What is the surface defined by the locus of the end of $\vec{v}$ when $\vec{u}$ is rotated?

Solution. A_{ij}, being a symmetrical tensor, has real eigenvalues and orthogonal eigendirections. Taking these as the reference frame, and making use of $u_i^2 = 1$,

$$\frac{v_1^2}{[\lambda^{(1)}]^2} + \frac{v_2^2}{[\lambda^{(2)}]^2} + \frac{v_3^2}{[\lambda^{(3)}]^2} = 1.$$

The locus of the end of $\vec{v}$ describes an ellipsoid.

3.4. A medium is said to be pyroelectric when a temperature variation $\Delta\theta$ generates an electric polarization $P_i = p_i \Delta\theta$. Which crystal classes can exhibit pyroelectricity?

Solution. The pyroelectric tensor p_i, which is zero for the eleven centro-symmetrical classes $\bar{1}, 2/m, mmm, \bar{3}, \bar{3}m, 4/m, 4/mmm, 6/m, 6/mmm, m3, m3m$ is

$$p_i = [p_1 \ p_2 \ p_3] \quad \text{for class 1.}$$

Since any symmetry axis along the x_3-direction will enforce $p_1 = p_2 = 0$, and any mirror plane orthogonal to the x_3-axis leads to $p_3 = 0$:

$$p_i = [p_1 \ p_2 \ 0] \quad \text{for class m}$$

$$p_i = [0 \ \ 0 \ \ p_3] \quad \text{for classes 2,3,4,6,2mm, 3m,4mm,6mm}$$

and $p_i = 0$ for all other classes. In summary pyroelectricity may arise only in those classes having at most one direct axis, without a perpendicular mirror plane.

3.5. Show that the three independent components ($A_1 = A_{23}$, $A_2 = A_{31}$, $A_3 = A_{12}$) of an antisymmetric tensor A_{ij} transform as the components of a vector if the sign of the reference frame is conserved.

Solution. Equation (3.7):

$$A'_{jk} = \alpha^q_j \alpha^r_k A_{qr}$$

together with $A_{rq} = -A_{qr}$, leads to

$$A'_{jk} = \sum_{q<r} (\alpha^q_j \alpha^r_k - \alpha^r_j \alpha^q_k) A_{qr}$$

Let us assume $j < k$ and $q < r$. Then $M^p_i = (-1)^{i+p}(\alpha^q_j \alpha^r_k - \alpha^r_j \alpha^q_k)$ is the cofactor of the determinant $|\alpha|$ (if $i \neq j$ and k, $p \neq q$ and r). By definition, $A'_i = A'_{jk}$ if the permutation is even

$$A'_i = (-1)^{i+1} A'_{jk} \qquad \text{when } j < k$$

and

$$A_p = (-1)^{p+1} A_{qr} \qquad \text{when } q < r$$

Then, we get

$$A'_i = \sum_p M^p_i A_p \ .$$

For an orthogonal transformation, the inverse matrix is

$$\beta^i_p = \frac{M^p_i}{|\alpha|} = \alpha^p_i \ .$$

Hence

$$A'_i = |\alpha| (\alpha^p_i A_p)$$

with $|\alpha| = \pm 1$ depending on whether or not the transformation changes the sign of the reference frame.

Chapter 4
Static Elasticity

This chapter is devoted to the definition of stress and strain and to the examination of the constants that characterize elastic properties of crystals. These constants are the components of a tensor of rank 4. But in no case may there be more than 21, owing to the symmetry of the strain and stress tensors and because of thermodynamic requirements. All 21 constants are necessary for the triclinic system; fortunately, the symmetry of other crystals drastically reduces this number. Three constants are sufficient for any cubic crystal.

4.1. STRAINS

Before dealing with deformations of solids in general, we begin with a simple case, the stretching of a string.

Let L be a straight string, with one end O fixed to a rigid steady support (fig. 4.1). Under the action of a force F, applied to the free end, the string lengthens. The deformation lasts as long as the traction holds. The relative deformation is (L'-L)/L where L' is the new length of the string. It is not necessary that all parts of the string be stretched in the same way. We must define the deformation of a small element, and then the deformation of the string[1] in the vicinity of a point.

---

1. In fact the deformation does not merely consists of stretching, since the string diameter decreases when the string is stretched. The relative change in diameter variation is proportional to the relative change in length (see exercise 4.3).

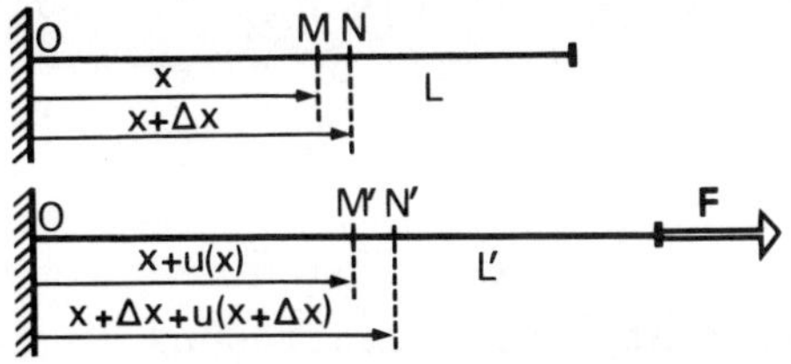

Fig. 4.1. Deformation of a string. Under the action of force $\vec{F}$, the string stretches. Points M and N, which define a portion Δx of the string, are displaced by different amounts, u(x) and u(x+Δx), and go to M' and N'. The strain S at point M is

$$S = \lim_{\Delta x \to 0} \frac{u(x+\Delta x)-u(x)}{\Delta x} = \frac{du}{dx}.$$

Let us now consider the section that was originally between M and N at x and x + Δx. When the force is exerted on the string, these points come to M' and N' at x + u(x) and x + u(x+Δx) + Δx respectively, and the relative deformation of this section MN is

$$\frac{\overline{M'N'}-\overline{MN}}{\overline{MN}} = \frac{u(x+\Delta x)-u(x)}{\Delta x}$$

By definition, the strain S of the string in the vicinity of a point M, located at x when at rest, is the limit of the above ratio when Δx goes to 0, i.e. the derivative of u with respect to x:

$$S = \lim_{\Delta x \to 0} \frac{u(x+\Delta x) - u(x)}{\Delta x} = \frac{du}{dx}. \qquad (4.1)$$

It must be noted that strain is a dimensionless quantity and that the origin of the x-axis is necessary only to define the point M under consideration.

In the example of the string under tension, its different points may move by different amounts, but all in the same direction so that the strain reduces to a lengthening (ignoring a reduction in radius). For a solid of the most general shape, under external forces, there is no reason for two points, even neighbouring points, to move in the same direction. Angular distortions appear, as well as variations in length. Given an origin O, any point of the medium is located by the components x_1, x_2, x_3 (i.e. the co-ordinates) of the vector $\vec{x}$. All points of the solid move under the influence of external forces. So our point is now determined by the vector $\vec{x}'$, of components x'_i, so that

$$\vec{x}' = \vec{x} + \vec{u}.$$

The displacement $\vec{u}$ is a continuous function of the co-ordinates x_k. Two neighbouring points M and N, originally separated by the vector $d\vec{x} = \vec{x}_N - \vec{x}_M$, of components dx_i, move away from or towards each other at the same time as the orientation of MN changes (fig. 4.2), since their displacements $\vec{u}_M$ and $\vec{u}_N$ are different:

$$\vec{u}_N = \vec{u}_M + d\vec{u} \quad \text{with} \quad d\vec{u} = \frac{\partial \vec{u}}{\partial x_i} dx_i \qquad (4.2)$$

(the summation from 1 to 3 over the dummy index i is understood). The square of the distance between the two points, now separated by the vector

$$d\vec{x}' = \vec{x}'_N - \vec{x}'_M = d\vec{x} + d\vec{u}$$

is

$$(d\vec{x}')^2 = (d\vec{x})^2 + 2(d\vec{x}).(d\vec{u}) + (d\vec{u})^2.$$

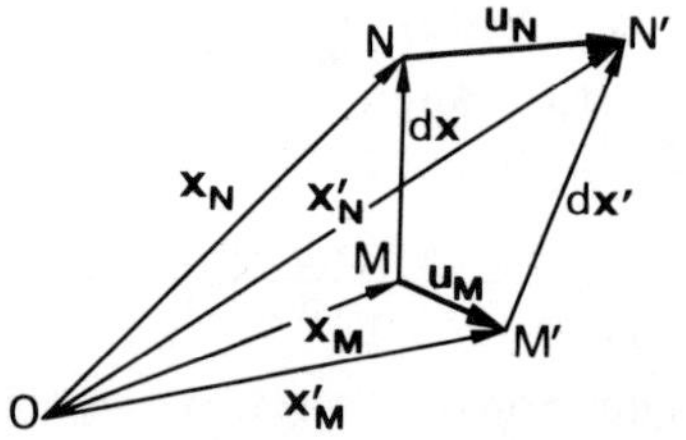

Fig. 4.2. Strains in a solid. Neighbouring points M and N are displaced by different amounts in different directions.

Expanding the scalar products

$$(d\vec{x}')^2 - (d\vec{x})^2 = 2dx_i du_i + du_k\, du_k$$

and substituting for du_i and du_k from (4.2), we obtain:

$$(d\vec{x}')^2 - (d\vec{x})^2 = 2\frac{\partial u_i}{\partial x_j} dx_i dx_j + \frac{\partial u_k}{\partial x_i}\frac{\partial u_k}{\partial x_j} dx_i dx_j. \qquad (4.3)$$

Permuting the dummy indices i and j leaves the value of the sum unchanged

$$\frac{\partial u_i}{\partial x_j}\, dx_i dx_j = \frac{\partial u_j}{\partial x_i}\, dx_j dx_i \qquad (4.4)$$

The equation (4.3) can be written in a more symmetrical form[1]

$$(d\vec{x}')^2 - (d\vec{x})^2 = \left(\frac{\partial u_i}{\partial x_j} + \frac{\partial u_j}{\partial x_i} + \frac{\partial u_k}{\partial x_i}\frac{\partial u_k}{\partial x_j}\right) dx_i dx_j$$

or

$$(d\vec{x}')^2 - (d\vec{x})^2 = 2S_{ij}dx_i dx_j$$

with

$$S_{ij} = \frac{1}{2}\left(\frac{\partial u_i}{\partial x_j} + \frac{\partial u_j}{\partial x_i} + \frac{\partial u_k}{\partial x_i}\frac{\partial u_k}{\partial x_j}\right). \qquad (4.5)$$

The nine quantities S_{ij} constitute a second rank tensor because the sum $2S_{ij}dx_i dx_j$, equal to the difference of the squared lengths of MN before and after the deformation, is an invariant; this is the strain tensor.

In nearly all interesting cases, deformations are small, i.e. the relative variation of distances inside the solid is small:

$$\frac{\partial u_i}{\partial x_j} \ll 1.$$

In this case $\partial u_k/\partial x_i \cdot \partial u_k/\partial x_j$ is a second order infinitesimal and is negligible compared to the first order term $\partial u_i/\partial x_j$ in equation 4.5. The strain tensor components reduce to:

$$\boxed{S_{ij} = \frac{1}{2}\left(\frac{\partial u_i}{\partial x_j} + \frac{\partial u_j}{\partial x_i}\right)} \qquad (4.6)$$

Only six components are distinct, since S_{ij} is a symmetric tensor:

$$S_{ij} = S_{ji}.$$

The value of the components depends on the point M where the deformation is observed. The state of strain in a medium is thus a tensor field, consisting of symmetric second rank tensors.

---

1. Equation 4.4 shows that the antisymmetric tensor $R_{ij} = 1/2\,(\partial u_i/\partial x_j - \partial u_j/\partial x_i)$ is not involved in the change of length It describes a rotation of the solid, which induces no strain. It is therefore logical to express the strain by means of a symmetric tensor. However see § 8.3.2.

At any given point M, the tensor S_{ij} can be reduced to its principal axes or eigenvectors. However, these depend on M, but are orthogonal for every M, S_{ij} being symmetric (§ 3.5). In the frame of the principal axes $M\xi_1\xi_2\xi_3$, all components vanish but the diagonal components $S^{(1)}$, $S^{(2)}$, $S^{(3)}$ and the strain reduces to

$$\frac{\partial u_1}{\partial \xi_1} = S^{(1)} \qquad \frac{\partial u_2}{\partial \xi_2} = S^{(2)} \qquad \frac{\partial u_3}{\partial \xi_3} = S^{(3)}$$

i.e. to three independent extensions along the three principal axes:

$$d\xi_i \rightarrow d\xi_i' = (1 + S^{(i)})d\xi_i \qquad i = 1,2,3.$$

The axes being orthogonal, the volume element

$$dV = d\xi_1 d\xi_2 d\xi_3$$

becomes, after the deformation:

$$dV' = d\xi_1' d\xi_2' d\xi_3' = (1+S^{(1)})(1+S^{(2)})(1+S^{(3)})dV$$

or, to a first order approximation:

$$dV' = (1 + S^{(1)} + S^{(2)} + S^{(3)})dV$$

The expansion $\Delta = (dV' - dV)/dV$ is thus equal to the sum of the eigenvalues, which is itself equal (3.22) to the trace $S_{ii} = S_{11} + S_{22} + S_{33}$ of the strain tensor at this point:

$$\Delta = S^{(1)} + S^{(2)} + S^{(3)} = S_{ii}. \qquad (4.7)$$

The S_{ij} terms with $i \neq j$ involve displacements du_j perpendicular to the element dx_i, and thus correspond to shear deformations. Let us assume, for example, that the only non-diagonal component is S_{12}. Two infinitesimal elements MN_1 and MN_2, originally oriented along the axes 1 and 2, do not remain orthogonal (fig. 4.3). During the deformation, M is displaced by the amount $\vec{u}$, N_1 by $\vec{u} + d\vec{u}^{(1)}$, N_2 by $\vec{u} + d\vec{u}^{(2)}$, with

$$d\vec{u}^{(1)} \begin{cases} \dfrac{\partial u_1}{\partial x_1} dx_1 \\[2ex] \dfrac{\partial u_2}{\partial x_1} dx_1 \\[2ex] 0 \end{cases} \qquad\qquad d\vec{u}^{(2)} \begin{cases} \dfrac{\partial u_1}{\partial x_2} dx_2 \\[2ex] \dfrac{\partial u_2}{\partial x_2} dx_2 \\[2ex] 0 \end{cases}$$

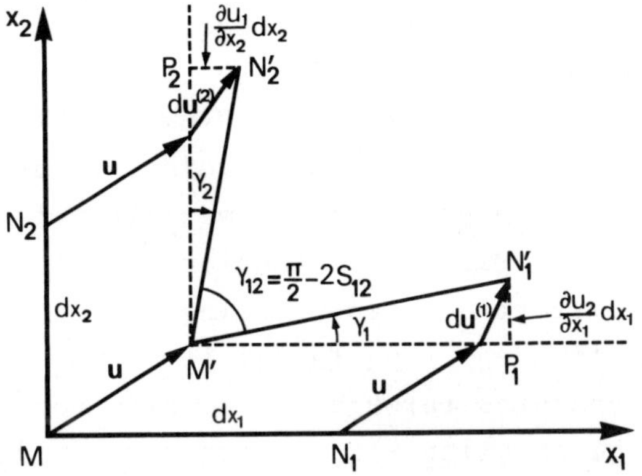

Fig. 4.3. Physical meaning of the shear stress S_{12}. Two elements MN_1, MN_2, originally orthogonal, become "closer" to each other during the deformation, by an angle equal to $2S_{12}$.

for $\frac{\partial u_3}{\partial x_1} = \frac{\partial u_3}{\partial x_2} = 0$ ($S_{13} = S_{23} = 0$, by hypothesis).

After the deformation, the angle γ_{12} between the two segments is

$$\gamma_{12} = \frac{\pi}{2} - \gamma_1 - \gamma_2$$

with

$$\tan \gamma_1 = \frac{\overline{P_1N_1'}}{\overline{M'P_1}} = \frac{\frac{\partial u_2}{\partial x_1}dx_1}{dx_1 + \frac{\partial u_1}{\partial x_1}dx_1} = \frac{\frac{\partial u_2}{\partial x_1}}{1+\frac{\partial u_1}{\partial x_1}}$$

Since the deformations are small: $\partial u_i/\partial x_j \ll 1$

$$\tan \gamma_1 \simeq \gamma_1 \simeq \frac{\partial u_2}{\partial x_1} \quad \text{and} \quad \gamma_2 \simeq \frac{\partial u_1}{\partial x_2} .$$

so that

$$\gamma_{12} = \frac{\pi}{2} - \left(\frac{\partial u_2}{\partial x_1} + \frac{\partial u_1}{\partial x_2}\right) = \frac{\pi}{2} - 2S_{12}. \tag{4.8}$$

The variation of the angle between dx_i and dx_j, initially orthogonal, due to the shear strain S_{ij}, is equal to $-2S_{ij}$.

4.2. STRESS

External forces are necessary to produce deformations of a solid. These forces may be exerted on the surface, by mechanical contact, or inside the solid by a force field. The effects of this field in the medium can be measured either by a force density per unit volume (gravitational field) or by a torque per unit volume (electric field in a polar crystal). Mechanical tensions, or stresses, appear in the distorted solid, tending to restore it to its rest state and thus ensuring the equilibrium of the medium. This stress propagates through the bonding forces between atoms, the range of which - a few interatomic distances - is very small at a macroscopic scale. Therefore, the medium surrounding any given volume acts on it through the boundary surface.[1] We intend to define the stress at a given point of a surface perpendicular to a coordinate axis, then of any surface, and finally we establish the equilibrium conditions of a solid under stress.

4.2.1. Definition of the Stress Tensor

In an orthonormal reference frame, let ΔF_i be the i-th component of the force $\Delta\vec{F}$ exerted on the surface element Δs_k (perpendicular to the k-axis) by the medium in the positive direction (fig. 4.4). The stress T_{ik} is defined as the limit, as Δs_k tends to zero, of the ratio $\Delta F_i/\Delta s_k$

$$T_{ik} = \lim_{\Delta s_k \to 0} \left(\frac{\Delta F_i}{\Delta s_k}\right). \qquad (4.9)$$

T_{ik} is the i-th component of the force exerted on a unit surface perpendicular to the k-axis.

---

1. The strain, in some (piezoelectric) media, may induce fields with a range greater than a few interatomic distances. The analysis is then no longer rigorous. However, the conclusions we are going to draw can (to the best of our knowledge) be applied to such materials.

From the law that action and reaction are equal and opposite, the same surface element is subjected to a force $\Delta\vec{F}' = -\Delta\vec{F}$ from the medium in the negative x_k direction.

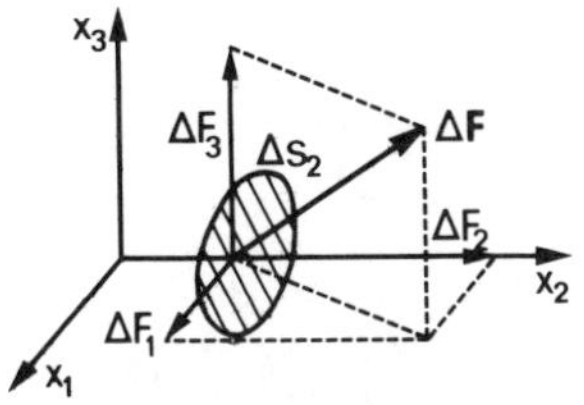

Fig. 4.4. Surface element orthogonal to a co-ordinate axis. On a surface element Δs_2, orthogonal to the x_2-axis, three stresses are acting, namely T_{12}, T_{22}, T_{32}. For example

$$T_{12} = \lim_{\Delta s_2 \to 0} \left(\frac{\Delta F_1}{\Delta s_2}\right).$$

In order to specify the force per unit area $\vec{T}(\vec{\ell})$ exerted on a given element ds, by the medium located on the same side as the unit normal vector $\vec{\ell}(\ell_1,\ell_2,\ell_3)$, we have to look at the projections of ds on the three co-ordinate planes. The tetrahedron of fig. 4.5 is subjected to a net force, from the surface medium and the external force field with density $\vec{\mathfrak{f}}^{ext}$, given by:

$$d\vec{F}' = \vec{T}(\vec{\ell})ds + d\vec{F}'^{(1)} + d\vec{F}'^{(2)} + d\vec{F}'^{(3)} + \vec{\mathfrak{f}}^{ext}dv \quad (4.10)$$

where $d\vec{F}^{(k)}$ is the force exerted on the triangle, perpendicular to the k axis, of area ds_k.

Projecting on to the i-th axis, and taking the definition of T_{ik} into account:

$$dF_i'^{(k)} = - dF_i^{(k)} = - T_{ik}ds_{(k)}.$$

The parentheses indicate that there is no summation over k. The equation (4.10) now becomes:

$$dF_i' = T_i(\vec{\ell})ds - (T_{i1}ds_1 + T_{i2}ds_2 + T_{i3}ds_3) + \mathfrak{f}_i^{ext}dv.$$

Since $ds_k = \ell_k ds$, the net force on the tetrahedron can be put into the form:

$$dF_i' = [T_i(\vec{\ell}) - T_{ik}\ell_k]ds + f_i^{ext}dv$$

which leads to an infinite density of force per unit volume (dv is negligible compared to ds), unless

$$\boxed{T_i(\vec{\ell}) = T_{ik}\ell_k.} \qquad (4.11)$$

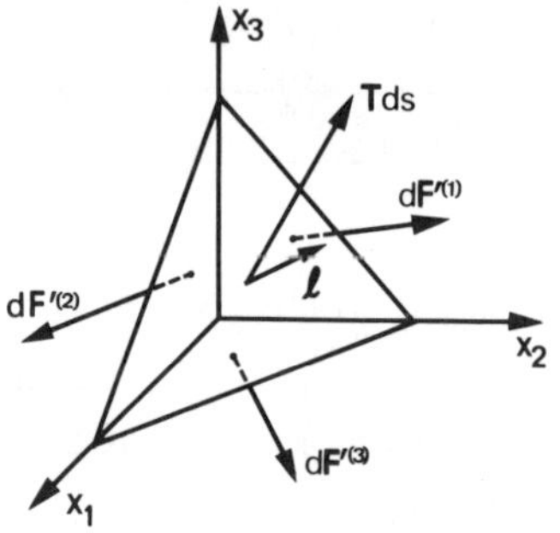

Fig. 4.5. Surface element with any orientation. The mechanical stress $\vec{T}$ depends on the orientation $\vec{\ell}$ of the surface element

$$T_i(\vec{\ell}) = T_{ik}\ell_k.$$

This equation, with the summation over the dummy index k, provides the mechanical stress $\vec{T}(\vec{\ell})$ on a surface of orientation $\vec{\ell}$. The nine quantities T_{ik}, which define the stress state of the distorted medium, constitute a second rank tensor, the stress tensor, because $T_i(\vec{\ell})$ and ℓ_k are the components of the vectors $\vec{T}(\vec{\ell})$ and $\vec{\ell}$.

4.2.2. Equilibrium Conditions

For a body in static mechanical equilibrium, the net force and torque on any given volume must be zero. Let us assume that the stress comes from forces applied to the surface.

From the principle of action and reaction the net force arising from the interactions between different parts of a volume v inside the solid must be zero; so the force $\vec{F}$ on this volume is the sum of the forces exerted by the surrounding medium. Since they are all applied on the boundary surface s, the force $\vec{F}$ is the integral over s of the mechanical stress $\vec{T}(\vec{\ell})$:

$$\vec{F} = \int_s \vec{T}(\vec{\ell})ds$$

or, for each component F_i, after substituting for $T_i(\vec{\ell})$ from (4.11) in terms of the stress tensor:

$$F_i = \int_s T_{ik}\ell_k ds.$$

This surface integral can be converted into a volume integral, through the generalized Green's theorem (equation 3.26):

$$F_i = \int_v \frac{\partial T_{ik}}{\partial x_k} dv.$$

The integrand can thus be regarded as the density of force f_i per unit volume of the strained medium:

$$f_i = \frac{\partial T_{ik}}{\partial x_k} . \qquad (4.12)$$

When the solid is subjected to forces exerted on its surface only, the static equilibrium condition $F_i = 0$ implies $f_i = 0$ or:

$$\frac{\partial T_{ik}}{\partial x_k} = 0. \qquad (4.13)$$

If there is a density of external force, e.g. due to a gravitational field $\vec{g}$, the equilibrium condition becomes:

$$\frac{\partial T_{ik}}{\partial x_k} + \rho g_i = 0 \qquad (4.14)$$

where ρ is the density.

The torque of the stress $\vec{T}$ about the origin is represented by the vector product $\vec{x} \wedge \vec{T}$. In tensor analysis, it is an antisymmetric second rank tensor $T_j x_i - T_i x_j$. From the law of action and reaction we know that the volume v undergoes a net torque that can be calculated by an integration over the surface s:

$$M_{ij} = \int_s (T_j x_i - T_i x_j)ds.$$

The surface integral

$$M_{ij} = \int_s (T_{jk}x_i - T_{ik}x_j)\ell_k ds$$

obtained by using the expression (4.11) for T_i and T_j, is converted into a volume integral

$$M_{ij} = \int_v \frac{\partial}{\partial x_k} (T_{jk}x_i - T_{ik}x_j)dv.$$

The expansion of this equation

$$M_{ij} = \int_v \left[\frac{\partial T_{jk}}{\partial x_k} x_i - \frac{\partial T_{ik}}{\partial x_k} x_j + T_{jk} \frac{\partial x_i}{\partial x_k} - T_{ik} \frac{\partial x_j}{\partial x_k}\right]dv$$

can be simplified, by using the equilibrium condition (4.13) and the independence of the co-ordinates x_i:

$$M_{ij} = \int_v (T_{jk}\delta_{ik} - T_{ik}\delta_{jk})dv$$

or

$$M_{ij} = \int_v (T_{ji} - T_{ij})dv. \qquad (4.15)$$

The volume v being arbitrary, the cancellation of all torque components requires

$$T_{ij} = T_{ji} \quad , \quad \forall \text{ i and j} \qquad (4.16)$$

The stress tensor is symmetric; this reduces the number of independent components to six:

T_{11}, T_{22}, T_{33}: stress normal to the sides

T_{12}, T_{23}, T_{13}: shear stress.

If there is a torque $\vec{G}$ per unit volume, the torque on the volume v is

$$M_{ij} + \int_v G_k dv$$

where k is such that the (ijk) permutation is even. The equilibrium condition thus reads:

$$T_{ji} - T_{ij} + G_k = 0.$$

The stress tensor is no longer symmetric. Such a case is in practice encountered only for polar crystals in an electric field (e.g. between two electrodes, in order to make a transducer). We assume, however, that for a small torque the symmetry of the tensor T_{ij} remains unchanged.

External forces acting on the solid surface determine the mechanical stress at every point of this surface. If the normal vector $\vec{\ell}$ points outwards, the density $\vec{p}$ of the forces applied on the surface is equal to the mechanical stress $\vec{T}(\vec{\ell})$, (where $T(\ell)$ has been defined as the force exerted on a unit surface area by the outside medium). The limiting condition to be satisfied at any point of the strained medium is

$$p_i = T_i(\vec{\ell}) = T_{ik}\ell_k. \tag{4.17}$$

4.3. RELATIONSHIP BETWEEN STRESS AND STRAIN. ELASTIC CONSTANTS

By definition, a medium is said to be elastic if it returns to its initial state after the external forces are removed. This return to the initial state is due to the internal stress. In a non piezoelectric elastic material, there is a one to one correspondence between stress and strain. It is known from experiments that the elastic behaviour of most substances is adequately described (for small deformations) by the first order term in the Taylor expansion of the equation:

$$T_{ij}(S_{k\ell}) = T_{ij}(0) + \left(\frac{\partial T_{ij}}{\partial S_{k\ell}}\right)_{S_{k\ell}=0} S_{k\ell} + \frac{1}{2}\left(\frac{\partial^2 T_{ij}}{\partial S_{k\ell}\partial S_{mn}}\right)_{\substack{S_{k\ell}=0\\ S_{mn}=0}} S_{k\ell}S_{mn} + \ldots$$

or, since $T_{ij}(0) = 0$

$$T_{ij} = c_{ijk\ell}S_{k\ell} \tag{4.18}$$

where

$$c_{ijk\ell} = \left(\frac{\partial T_{ij}}{\partial S_{k\ell}}\right)_{S_{k\ell}=0} \tag{4.19}$$

The coefficients $c_{ijk\ell}$, that describe the most general linear relationship between the two second rank tensors T_{ij} and $S_{k\ell}$, are the components of a fourth rank tensor called the elastic stiffness tensor. This proportionality between stress and strain was first enunciated in the seventeenth century by Hooke, for the simple case of a stretched string.

A fourth rank tensor has $3^4 = 81$ components. But since T_{ij} and $S_{k\ell}$ are symmetric tensors, the elastic constants defined by 4.19 remain unchanged under a permutation of i and j or k and ℓ:

$$c_{ijk\ell} = c_{jik\ell} \quad ; \quad c_{ijk\ell} = c_{ij\ell k}. \tag{4.20}$$

Hooke's law (4.18) can be written in terms of the displacements:

$$T_{ij} = \frac{1}{2} c_{ijk\ell} \frac{\partial u_k}{\partial x_\ell} + \frac{1}{2} c_{ijk\ell} \frac{\partial u_\ell}{\partial x_k} .$$

$c_{ijk\ell} = c_{ij\ell k}$, so both sums are equal:

$$\boxed{T_{ij} = c_{ijk\ell} \frac{\partial u_\ell}{\partial x_k}} \qquad (4.21)$$

Taking the above symmetries into account, we are left with 36 independent elastic constants instead of 81: an unordered pair of indices (i,j) takes only six distinct values, which we number from 1 to 6 in the following way:

$$(11)\leftrightarrow 1 \qquad (22)\leftrightarrow 2 \qquad (33)\leftrightarrow 3$$

$$(23)=(32)\leftrightarrow 4 \quad (31)=(13)\leftrightarrow 5 \quad (12)=(21)\leftrightarrow 6. \quad (4.22)$$

The independent elastic moduli are thus labelled by only two indices α and β, ranging from 1 to 6, and can be arranged in a 6 x 6 table:

$$c_{\alpha\beta} = c_{ijk\ell}$$

with $\alpha \leftrightarrow (ij)$, $\beta \leftrightarrow (k\ell)$ according to (4.22). For example

$$c_{14} = c_{1123} = c_{1132}$$

$$c_{56} = c_{1312} = c_{1321} = c_{3121} = c_{3112}.$$

This so-called matrix notation can be extended to the strain and stress tensors, so that we put Hooke's law into the following form:

$$T_\alpha = c_{\alpha\beta} S_\beta \qquad \alpha,\beta = 1,2,\ldots 6. \qquad (4.23)$$

But if we are to have the same convention for stress:

$$T_\alpha = T_{ij} \text{ with } \alpha \leftrightarrow (ij) \text{ according to 4.22} \quad (4.24)$$

we must set:

$$S_1 = S_{11} \quad S_2 = S_{22} \quad S_3 = S_{33} \quad S_4 = 2\,S_{23} \quad S_5 = 2S_{13} \quad S_6 = 2S_{12} \qquad (4.25)$$

in order that equation (4.23) properly describes Hooke's law. For example, the expansion of T_{11} given by 4.18:

$$T_{11} = c_{11k\ell} S_{k\ell}$$

or

$$T_{11} = c_{1111}S_{11} + c_{1122}S_{22} + c_{1133}S_{33} + 2c_{1123}S_{23} + 2c_{1113}S_{13} + 2c_{1112}S_{12}$$

is identical to that of T_1 given by 4.23:

$$T_1 = c_{11}S_1 + c_{12}S_2 + c_{13}S_3 + c_{14}S_4 + c_{15}S_5 + c_{16}S_6$$

only if the equations 4.25 are satisfied.

Hooke's law can be inverted, and the strain expressed in terms of the stress:

$$S_{ij} = s_{ijk\ell}T_{k\ell}. \qquad (4.26)$$

The compliance constants $s_{ijk\ell}$ constitute a fourth rank tensor which exhibits the same symmetry properties as the $c_{ijk\ell}$ tensor:

$$s_{ijk\ell} = s_{jik\ell} \quad \text{and} \quad s_{ijk\ell} = s_{ij\ell k}.$$

Similarly, the solution of the system 4.23 for the S_α provides

$$S_\alpha = s_{\alpha\beta}T_\beta \qquad (4.27)$$

where the $s_{\alpha\beta}$ matrix is the inverse of the $c_{\alpha\beta}$ matrix.

$$s_{\alpha\beta} = (c_{\alpha\beta})^{-1}$$

or

$$s_{\alpha\beta}c_{\beta\gamma} = \delta_{\alpha\gamma}$$

$\delta_{\alpha\gamma}$ being the Kronecker symbol of dimension 6.

If we wish to relate the $s_{\alpha\beta}$ terms to the $s_{ijk\ell}$ terms, we expand S_{13} according to 4.26:

$$S_{13}=s_{1311}T_{11}+s_{1322}T_{22}+s_{1333}T_{33}+2s_{1323}T_{23}+2s_{1313}T_{13}+2s_{1312}T_{12}.$$

From the correspondence rule 4.24, on comparing S_5 to the expansion of $2S_{13}$ given by 4.27, we obtain

$$s_{51} = 2s_{1311} \quad s_{52} = 2s_{1322} \quad s_{53} = 2s_{1333}$$

$$s_{54} = 4s_{1323} \quad s_{55} = 4s_{1313} \quad s_{56} = 4s_{1312}.$$

More generally, if p is the number of indices greater than three in the pair (α,β):

$$s_{\alpha\beta} = 2^p s_{ijk\ell}. \qquad (4.28)$$

The following thermodynamic considerations will also reduce the number of independent elastic constants.

4.4. ELASTIC ENERGY IN A DISTORTED MEDIUM. MAXWELL'S EQUATIONS

The energy provided by the external forces during the deformation is stored in the medium as elastic potential energy since, as soon as the forces are removed, it is released by the internal stress which makes the solid return to its initial state. During the deformation, the work done by the external forces for a variation $d\vec{u}$ of the displacement $\vec{u}$ consists of two parts, arising from the forces per unit mass $\vec{g}$ and the external forces $\vec{p}$ acting on the surface:

$$\delta W = \int_v \rho\vec{g}.d\vec{u}\,dv + \int_s \vec{p}.d\vec{u}\;ds$$

or

$$\delta W = \int_v \rho g_i du_i dv + \int_s p_i du_i ds.$$

If we assume that there is always an equilibrium, this transformation is thermodynamically reversible and the equilibrium equations 4.14 and 4.17 are valid. If we replace p_i by $T_{ik}\ell_k$:

$$\delta W = \int_v \rho g_i du_i dv + \int_s T_{ik} du_i \ell_k ds$$

This last term is transformed into a volume integral

$$\delta W = \int_v \rho g_i du_i dv + \int_v \frac{\partial}{\partial x_k} (T_{ik} du_i) dv.$$

which can be put into the form

$$\delta W = \int_v (\rho g_i + \frac{\partial T_{ik}}{\partial x_k}) du_i dv + \int_v T_{ik}\; d(\frac{\partial u_i}{\partial x_k}) dv \qquad (4.29)$$

The quantity $(\rho g_i + \partial T_{ik}/\partial x_k)$ vanishes at any instant, because of the equilibrium condition 4.14. The work done by the external forces thus reduces to

$$\delta W = \int_v T_{ik}\; d(\frac{\partial u_i}{\partial x_k}) dv = \int_v T_{ki}\; d(\frac{\partial u_k}{\partial x_i}) dv$$

or, since $T_{ki} = T_{ik}$:

$$\delta W = \frac{1}{2} \int_v T_{ik}\; d(\frac{\partial u_i}{\partial x_k} + \frac{\partial u_k}{\partial x_i}) dv = \int_v T_{ik} dS_{ik} dv.$$

During a change in strain dS_{ik}, the energy per unit volume $\delta\mathcal{W}$ provided by the external forces is equal to

$$\boxed{\delta\mathcal{W} = T_{ik} dS_{ik}.}$$

We shall now use script letters for thermodynamic quantities related to a unit volume. Thus the internal energy variation per unit volume is

$$d\mathcal{U} = \delta\mathcal{W} + \delta\mathcal{Q} \tag{4.30}$$

$\delta\mathcal{Q}$ being the heat received per unit volume. According to the first law of thermodynamics, $\mathcal{U}$ is a function of state and $d\mathcal{U}$ is an exact differential while $\delta\mathcal{W}$ and $\delta\mathcal{Q}$ separately are not. For a reversible transformation, the second law of thermodynamics provides

$$\delta\mathcal{Q} = \theta d\sigma$$

where θ is the absolute temperature and σ the entropy per unit volume of the medium.

The equation 4.30, which reads

$$d\mathcal{U} = \theta d\sigma + T_{ik} dS_{ik} \tag{4.31}$$

shows that the internal energy of a distorted medium is a function of the entropy and of the strains:

$$\mathcal{U} = \mathcal{U}(\sigma, S_{ik}).$$

Therefore

$$T_{ik} = \left(\frac{\partial\mathcal{U}}{\partial S_{ik}}\right)_\sigma$$

The index σ specifies the partial derivation at constant entropy.

Inserting similar expressions for T_{ij} and $T_{k\ell}$ into 4.19:

$$c_{ijk\ell} = \left(\frac{\partial T_{ij}}{\partial S_{k\ell}}\right) \qquad \text{and} \qquad c_{k\ell ij} = \left(\frac{\partial T_{k\ell}}{\partial S_{ij}}\right)$$

we obtain

$$c^{(\sigma)}_{ijk\ell} = \left(\frac{\partial^2\mathcal{U}}{\partial S_{ij}\partial S_{k\ell}}\right)_\sigma = c^{(\sigma)}_{k\ell ij}. \tag{4.32}$$

Exchanging the first two indices with the last two does not change the value of isentropic ($\sigma = C^{te}$) elastic moduli. Precisely these moduli are involved in elastic wave propagation, where the vibration is so fast that there is no time for a thermal exchange with an external medium. Thus $\delta\mathcal{Q} = 0 \Longrightarrow d\sigma = 0$ if the transformation is reversible. The symmetry equation, 4.32, one of Maxwell's equations, is also valid under other thermodynamic conditions. It is sufficient to consider the relevant function of state, e.g. the free energy $\mathcal{F} = \mathcal{U} - \theta\sigma$ for isothermal transformations; thus taking 4.31 into account:

$$d\mathcal{F} = -\sigma d\theta + T_{ik} dS_{ik}. \tag{4.33}$$

The free energy depends on the temperature θ and strains S_{ik}:

$$\mathcal{F} = \mathcal{F}(\theta, S_{ik})$$

so that

$$T_{ik} = \left(\frac{\partial \mathcal{F}}{\partial S_{ik}}\right)_\theta$$

and for isothermal elastic constants:

$$c_{ijk\ell}^{(\theta)} = \left(\frac{\partial^2 \mathcal{F}}{\partial S_{ij}\partial S_{k\ell}}\right)_\theta = c_{k\ell ij}^{(\theta)}.$$

Within the domain of validity of Hooke's law, the change in internal energy given by 4.31:

$$d\mathcal{U} = \theta d\sigma + c_{ijk\ell}^{(\sigma)} S_{k\ell} dS_{ij}$$

can be written, after permuting the dummy indices (ij) and (kℓ)

$$d\mathcal{U} = \theta d\sigma + \frac{1}{2}(c_{ijk\ell}^{(\sigma)} S_{k\ell} dS_{ij} + c_{k\ell ij}^{(\sigma)} S_{ij} dS_{k\ell})$$

and, making use of the Maxwell equation 4.32:

$$d\mathcal{U} = \theta d\sigma + \frac{1}{2} c_{ijk\ell}^{(\sigma)} d(S_{ij}S_{k\ell}).$$

After integration, the internal energy of the strained medium becomes:

$$\mathcal{U}(\sigma, S_{ik}) = \mathcal{U}_o(\sigma) + \frac{1}{2} c_{ijk\ell}^{(\sigma)} S_{ij} S_{k\ell}$$

$\mathcal{U}_o(\sigma)$ is the internal energy of the undeformed medium $\mathcal{U}_o(\sigma) = \mathcal{U}(\sigma, S_{ik} = 0)$. The free energy per unit volume is obtained similarly from 4.33:

$$\mathcal{F}(\theta, S_{ik}) = \mathcal{F}_o(\theta) + \frac{1}{2} c_{ijk\ell}^{(\theta)} S_{ij} S_{k\ell}.$$

Thus the internal energy or the mechanical free energy, which we called the elastic potential energy Φ, without specifying the thermodynamic conditions, is a quadratic function of the strains:

$$\boxed{\Phi = \frac{1}{2} c_{ijk\ell} S_{ij} S_{k\ell}.} \qquad (4.34)$$

This quantity stands for the internal or free energy variation per unit volume, depending on whether the transformation is adiabatic or isothermal.

4.5. RESTRICTIONS IMPOSED BY CRYSTAL SYMMETRY ON THE NUMBER OF INDEPENDENT ELASTIC MODULI

The Maxwell equation 4.32 has been derived from thermodynamics and is thus valid in any medium. In matrix notation, it reads

$$c_{\alpha\beta} = c_{\beta\alpha}.$$

So the 6 x 6 table of the coefficients $c_{\alpha\beta}$ is symmetric with respect to the main diagonal.

$$(c_{\alpha\beta}) = \begin{vmatrix} c_{11} & c_{12} & c_{13} & c_{14} & c_{15} & c_{16} \\ c_{12} & c_{22} & c_{23} & c_{24} & c_{25} & c_{26} \\ c_{13} & c_{23} & c_{33} & c_{34} & c_{35} & c_{36} \\ c_{14} & c_{24} & c_{34} & c_{44} & c_{45} & c_{46} \\ c_{15} & c_{25} & c_{35} & c_{45} & c_{55} & c_{56} \\ c_{16} & c_{26} & c_{36} & c_{46} & c_{56} & c_{66} \end{vmatrix} \quad \text{triclinic} \quad (4.35)$$

The above property reduces the number of independent components to 21 (6 + 5 + 4 + 3 + 2 + 1). Crystals of the triclinic system, the richest from this point of view since a centre of symmetry does not impose any constraints, have 21 independent elastic constants. Fortunately, the point groups of other crystal systems lead to a much lower number. We first show, in the other extreme, that for an isotropic solid (i.e. the highest degree of symmetry), mechanical properties are fully described by two coefficients.

4.5.1. Isotropic Solid

The physical constants of an isotropic medium, by definition, do not depend on the choice of the orthonormal reference frame. In particular, the elastic tensor $c_{ijk\ell}$ should be invariant under all transformations of the reference frame (rotation, symmetry with respect to a point or a plane). And only a scalar or the unit tensor δ_{ij} is invariant in all these orthogonal transformations. Therefore, every component $c_{ijk\ell}$ has to be expressed in terms of components of the unit tensor. Moreover, because of the symmetry $\delta_{ij} = \delta_{ji}$, there are only three distinct combinations containing the four indices $ijk\ell$:

$$\delta_{ij}\delta_{k\ell}; \ \delta_{ik}\delta_{j\ell} \ ; \ \delta_{i\ell}\delta_{jk}$$

So the tensor $c_{ijk\ell}$ has the form

$$c_{ijk\ell} = \lambda\delta_{ij}\delta_{k\ell} + \mu_1\delta_{ik}\delta_{j\ell} + \mu_2\delta_{i\ell}\delta_{jk}$$

where λ, μ_1, μ_2, are constants. Furthermore, the condition $c_{ijk\ell} = c_{jik\ell}$ implies $\mu_1 = \mu_2 = \mu$. Other symmetry conditions are satisfied so that

$$c_{ijk\ell} = \lambda\delta_{ij}\delta_{k\ell} + \mu(\delta_{ik}\delta_{j\ell} + \delta_{i\ell}\delta_{jk}). \quad (4.36)$$

Thus the mechanical properties of an isotropic medium are described by two independent constants, e.g. the Lamé coefficients λ and μ.

Giving to (ij) and (kℓ) all values from 1 to 6 we obtain:

$$\begin{aligned} c_{11} &= c_{22} = c_{33} = \lambda + 2\mu \\ c_{12} &= c_{23} = c_{13} = \lambda \\ c_{44} &= c_{55} = c_{66} = \mu = \frac{c_{11}-c_{12}}{2} \end{aligned} \quad (4.37)$$

The other twelve moduli vanish, since they have an odd number of distinct indices (e.g.: $c_{25} = c_{2213}$). If everything is expressed in terms of c_{11} and c_{12}, the matrix $c_{\alpha\beta}$ looks as follows:

$$(c_{\alpha\beta}) = \begin{vmatrix} c_{11} & c_{12} & c_{12} & 0 & 0 & 0 \\ c_{12} & c_{11} & c_{12} & 0 & 0 & 0 \\ c_{12} & c_{12} & c_{11} & 0 & 0 & 0 \\ 0 & 0 & 0 & \frac{c_{11}-c_{12}}{2} & 0 & 0 \\ 0 & 0 & 0 & 0 & \frac{c_{11}-c_{12}}{2} & 0 \\ 0 & 0 & 0 & 0 & 0 & \frac{c_{11}-c_{12}}{2} \end{vmatrix} \quad \text{isotropic} \quad (4.38)$$

and in an isotropic medium, Hooke's law reduces to the pair of Lamé equations:

- for normal stresses (T_{11}, T_{22}, T_{33}):

$$T_{ii} = c_{iik\ell}S_{k\ell} = (\lambda\delta_{k\ell} + 2\mu\delta_{ik}\delta_{i\ell})S_{k\ell}$$

or

$$T_{ii} = \lambda(S_{11} + S_{22} + S_{33}) + 2\mu S_{ii}; \quad (4.39)$$

- for shear stress (T_{ij} with $i \neq j$):

$$T_{ij} = c_{ijk\ell}S_{k\ell} = \mu(\delta_{ik}\delta_{j\ell} + \delta_{i\ell}\delta_{jk})S_{k\ell}$$

or

$$T_{ij} = 2\mu S_{ij}. \quad (4.40)$$

The equations (4.39) and (4.40) can simultaneously be written as

$$T_{ij} = \lambda\Delta\delta_{ij} + 2\mu S_{ij} \qquad (4.41)$$

where $\Delta = S_{11} + S_{22} + S_{33}$ stands for the expansion $\Delta v/v$. The equation 4.40 shows that $\mu = T_{ij}/2S_{ij}$ is the ratio between the shear stress T_{ij} and the change in the angle between the two directions i and j which were initially orthogonal (fig. 4.3).

4.5.2. Crystals

The reference frame being orthonormal, the general invariance condition 3.11 reads, for the stiffness tensor:

$$c_{ijk\ell} = \alpha_i^p\alpha_j^q\alpha_k^r\alpha_\ell^s c_{pqrs}. \qquad (4.42)$$

As we have already shown, crystals of the triclinic system have 21 elastic constants which are shown in the table (4.35).

If the Ox_3-axis is taken along the diad (direct or inverse) axis of monoclinic crystals, the matrix α for the frame change is diagonal:

$$\alpha = \pm\begin{pmatrix} -1 & 0 & 0 \\ 0 & -1 & 0 \\ 0 & 0 & 1 \end{pmatrix}$$

and (4.42) becomes

$$c_{ijk\ell} = \alpha_i^i\alpha_j^j\alpha_k^k\alpha_\ell^\ell c_{ijk\ell}.$$

This implies that all coefficients with an odd number of the index 3 (for which $\alpha_i^i\alpha_j^j\alpha_k^k\alpha_\ell^\ell = -1$) vanish. The table 4.43 shows 13 elastic constants, for which the set of indices includes 0, 2, or 4 times the index 3:

$$(c_{\alpha\beta}) = \begin{vmatrix} c_{11} & c_{12} & c_{13} & 0 & 0 & c_{16} \\ c_{12} & c_{22} & c_{23} & 0 & 0 & c_{26} \\ c_{13} & c_{23} & c_{33} & 0 & 0 & c_{36} \\ 0 & 0 & 0 & c_{44} & c_{45} & 0 \\ 0 & 0 & 0 & c_{45} & c_{55} & 0 \\ c_{16} & c_{26} & c_{36} & 0 & 0 & c_{66} \end{vmatrix} \quad \text{monoclinic} \qquad (4.43)$$

Crystals of the orthorhombic system are characterized by three diad axes, chosen as the reference frame. The above argument can be applied to every axis and leads to the table (4.44) where non vanishing coefficients must have indices which are repeated an even number of times. We have thus 9 independent constants.

$$(c_{\alpha\beta}) = \begin{vmatrix} c_{11} & c_{12} & c_{13} & 0 & 0 & 0 \\ c_{12} & c_{22} & c_{23} & 0 & 0 & 0 \\ c_{13} & c_{23} & c_{33} & 0 & 0 & 0 \\ 0 & 0 & 0 & c_{44} & 0 & 0 \\ 0 & 0 & 0 & 0 & c_{55} & 0 \\ 0 & 0 & 0 & 0 & 0 & c_{66} \end{vmatrix} \quad \text{orthorhombic} \tag{4.44}$$

Crystals of the <u>cubic</u> system have at least four axes A_3 and three direct diad axes chosen as the reference frame. Non vanishing coefficients are the same as above (orthorhombic system). The $2\pi/3$ rotation about the triad axis [111] will replace x_2 by x_1, x_3 by x_2, x_1 by x_3. The terms $c_{ijk\ell}$ should be invariant under a circular permutation of the indices

$$(123) \rightarrow (231) \rightarrow (312).$$

This implies the following equations:

$$c_{1111} = c_{2222} = c_{3333}$$

and $\quad c_{1122} = c_{2233} = c_{3311}\ ; \quad c_{1212} = c_{2323} = c_{3131}.$

We are left with three independent elastic moduli (c_{11}, c_{12}, c_{44}) for the cubic system. The Table 4.44 becomes:

$$(c_{\alpha\beta}) = \begin{vmatrix} c_{11} & c_{12} & c_{12} & 0 & 0 & 0 \\ c_{12} & c_{11} & c_{12} & 0 & 0 & 0 \\ c_{12} & c_{12} & c_{11} & 0 & 0 & 0 \\ 0 & 0 & 0 & c_{44} & 0 & 0 \\ 0 & 0 & 0 & 0 & c_{44} & 0 \\ 0 & 0 & 0 & 0 & 0 & c_{44} \end{vmatrix} \quad \text{cubic} \tag{4.45}$$

<u>Crystals with an n-fold axis A_n ($n > 2$).</u> Crystals of the trigonal, tetragonal, hexagonal systems have only one direct or inverse axis of order greater than 2. The rotation matrix α is no longer diagonal (the A_n axis points in the x_3 direction)

$$\alpha = \begin{pmatrix} \cos\phi & \sin\phi & 0 \\ -\sin\phi & \cos\phi & 0 \\ 0 & 0 & 1 \end{pmatrix} \quad \text{with} \quad \phi = \frac{2\pi}{n} \neq \pi. \tag{4.46}$$

The invariance relation 4.42 is now more difficult to analyze for it involves many components at the same time. In order to derive a diagonal matrix for the change of frame, it is sufficient to use as the basis vectors the rotation eigenvectors $\vec{\xi}^{(1)}$, $\vec{\xi}^{(2)}$, $\vec{\xi}^{(3)}$. For this purpose, we have to diagonalize the α matrix, or equivalently to solve the equations (similar to 3.19)

$$(\alpha_i^k - \lambda\delta_{ik})\xi_k = 0. \qquad (4.47)$$

The eigenvalues λ are determined by the condition (analogous to 3.20)

$$\begin{vmatrix} \cos\phi - \lambda & \sin\phi & 0 \\ -\sin\phi & \cos\phi - \lambda & 0 \\ 0 & & 1-\lambda \end{vmatrix} = 0.$$

The expansion of the above determinant

$$[(\lambda - \cos\phi)^2 + \sin^2\phi](1 - \lambda) = 0$$

provides

$$\lambda^{(1)} = e^{i\phi} \qquad \lambda^{(2)} = e^{-i\phi} \qquad \lambda^{(3)} = 1. \qquad (4.48)$$

Each of these values is related to an eigenvector $\vec{\xi}^{(i)}$, the components of which are obtained by solving (4.47):

- for $\lambda^{(1)}$

$$\begin{cases} \xi_1(-\, i \sin\phi) + \xi_2 \sin\phi = 0 \\ (1 - e^{i\phi})\xi_3 = 0 \end{cases}$$

or

$$\xi_2 = i\xi_1, \qquad \xi_3 = 0;$$

- for $\lambda^{(2)}$: $\quad \xi_1 = i\xi_2, \quad \xi_3 = 0,$
- for $\lambda^{(3)}$: $\quad \xi_1 = \xi_2 = 0, \quad$ any ξ_3.

These components are complex, so we must use the hermitian product $\xi_i\xi_i^*$ to calculate unit vectors. This leads to

$$\vec{\xi}^{(1)}\left(\frac{1}{\sqrt{2}}, \frac{i}{\sqrt{2}}, 0\right) \quad \vec{\xi}^{(2)}\left(\frac{i}{\sqrt{2}}, \frac{1}{\sqrt{2}}, 0\right) \quad \vec{\xi}^{(3)}(0,0,1). \qquad (4.49)$$

Let us now denote by $\gamma_{ijk\ell}$ the elastic moduli in the orthonormal frame $\vec{\xi}^{(1)}$, $\vec{\xi}^{(2)}$, $\vec{\xi}^{(3)}$. The invariance relation (4.42) reads:

$$\gamma_{ijk\ell} = \lambda^{(i)}\lambda^{(j)}\lambda^{(k)}\lambda^{(\ell)}\gamma_{ijkl}$$

since the reference frame transformation matrix is diagonal in that frame. If ν_1 and ν_2 are the numbers of the indices 1 and 2 in the permutation $(ijk\ell)$, we have (making use of 4.48)

$$\lambda^{(i)}\lambda^{(j)}\lambda^{(k)}\lambda^{(\ell)} = e^{i(\nu_1-\nu_2)2\pi/n}.$$

So $\gamma_{ijk\ell}$ is different from zero when $(\nu_1-\nu_2)$ is a multiple of the order n of the axis, since the product $\lambda^{(i)}\lambda^{(j)}\lambda^{(k)}\lambda^{(\ell)}$ is then equal to one. This is always true for the five moduli

$$\gamma_{1122}, \quad \gamma_{1212}, \quad \gamma_{3312}, \quad \gamma_{2313}, \quad \gamma_{3333}$$

for which $\nu_1 = \nu_2$. There are no other non vanishing coefficients for hexagonal $(n = 6)$ crystals, which have five independent elastic moduli. On the other hand, crystals of the trigonal system have seven independent elastic constants for which γ_{1113} and γ_{2223} are not zero $(\nu_1 - \nu_2 = \pm 3)$. This is also true for tetragonal $(n = 4)$ crystals, for which γ_{1111} and γ_{2222} are not equal to zero $(\nu_1 - \nu_2 = \pm 4)$.

We go back to the $c_{ijk\ell}$ coefficients through the frame transformation equations

$$c_{ijk\ell} = a_i^p a_j^q a_k^r a_\ell^s \gamma_{pqrs} \tag{4.50}$$

where the matrix (a), which relates $Ox_1x_2x_3$ (constants $c_{ijk\ell}$) to $O\vec{\xi}^{(1)}\vec{\xi}^{(2)}\vec{\xi}^{(3)}$ (constants γ_{pqrs}), has the components (4.49) of the eigenvector $\vec{\xi}^{(i)}$ in the $x_1x_2x_3$ frame on its i-th line:

$$a = \begin{pmatrix} 1/\sqrt{2} & i/\sqrt{2} & 0 \\ i/\sqrt{2} & 1/\sqrt{2} & 0 \\ 0 & 0 & 1 \end{pmatrix}. \tag{4.51}$$

There are few non vanishing γ constants, so the expansion of 4.50 is fairly easy. Furthermore, $a_3^3 = 1$ is the only non vanishing a with an index of 3.

To begin with, in the <u>trigonal</u> system, only γ_{3333}, γ_{3312}, γ_{2313}, γ_{1113}, γ_{2223}, γ_{1122}, γ_{1212} are non-zero. Ordering the c's by decreasing number of indices 3:

- $c_{3333} = \gamma_{3333}$ and consequently $c_{33} \neq 0$;
- c_{3313} and c_{3323} are zero since $\gamma_{3313} = \gamma_{3323} = 0$, i.e. $c_{35} = c_{34} = 0$;
- the coefficients with two indices of 3 (c_{ij33} and c_{i3k3}) can be expressed as a function of $\gamma_{1233} = \gamma_{2133}$ and $\gamma_{1323} = \gamma_{2313}$

$$c_{ij33} = (a_i^1 a_j^2 + a_i^2 a_j^1)\gamma_{1233};$$

since $a_1^1 a_1^2 + a_1^2 a_1^1 = a_2^1 a_2^2 + a_2^2 a_2^1$:

$$c_{1133} = c_{2233} \quad \text{or} \quad c_{13} = c_{23}$$

and since $a_1^1 a_2^2 + a_1^2 a_2^1 = 0$:

$$c_{1233} = 0 \quad \text{or} \quad c_{36} = 0.$$

Similarly the expansion

$$c_{i3k3} = (a_i^1 a_k^2 + a_i^2 a_k^1)\gamma_{1323}$$

leads to $c_{1313} = c_{2323}$ or $c_{55} = c_{44}$

and $c_{2313} = 0$ or $c_{45} = 0$;

- for the coefficients with a single index of 3 equation 4.50 becomes:

$$c_{ijk3} = a_i^1 a_j^1 a_k^1 \gamma_{1113} + a_i^2 a_j^2 a_k^2 \gamma_{2223}$$

from which:

$$c_{22k3} = -\frac{1}{2} a_k^1 \gamma_{1113} + \frac{1}{2} a_k^2 \gamma_{2223} = -c_{11k3}$$

and

$$c_{i223} = -\frac{1}{2} a_i^1 \gamma_{1113} + \frac{1}{2} a_i^2 \gamma_{2223} = -c_{i113}$$

or in matrix notation

$$k = 2 \Longrightarrow c_{24} = -c_{14}, \quad k = 1 \Longrightarrow c_{25} = -c_{15}$$

$$i = 1 \Longrightarrow c_{46} = -c_{15}, \quad i = 2 \Longrightarrow c_{24} = -c_{56}$$

or again $c_{14} = -c_{24} = c_{56}$ and $c_{25} = -c_{15} = c_{46}$;

- the coefficients without an index of 3, c_{1111}, c_{2222}, c_{1112}, c_{2221}, c_{1122}, c_{1212} can be expressed as functions of γ_{1122} and γ_{1212}

$$c_{ijk\ell} = (a_i^1 a_j^1 a_k^2 a_\ell^2 + a_i^2 a_j^2 a_k^1 a_\ell^1)\gamma_{1122} + (a_i^1 a_j^2 + a_i^2 a_j^1)(a_k^1 a_\ell^2 + a_k^2 a_\ell^1)\gamma_{1212} \tag{4.52}$$

from which

$$c_{iiii} = 2(a_i^1 a_i^2)^2(\gamma_{1122} + 2\gamma_{1212}) = -\frac{1}{2}(\gamma_{1122} + 2\gamma_{1212})$$

or

$$c_{11} = c_{22} = -\frac{1}{2}(\gamma_{1122} + 2\gamma_{1212}). \tag{4.53}$$

Let us examine the coefficient c_{1112}:

$$c_{1112} = a_1^1 a_1^2 (a_1^1 a_2^2 + a_1^2 a_2^1)(\gamma_{1122} + 2\gamma_{1212}) = 0$$

since $a_1^1 a_2^2 + a_1^2 a_2^1 = 0$. Similarly it can be shown that $c_{2221} = 0$ and therefore

$$c_{16} = c_{26} = 0.$$

A relation exists between c_{1111}, c_{1122} and c_{1212} since there are only two independent constants without an index of 3. Thus from the expansion of 4.52 there remains only

$$c_{1122} = \frac{1}{2}(\gamma_{1122} - 2\gamma_{1212}) = c_{12}$$

and

$$c_{1212} = -\frac{1}{2}\gamma_{1122} = c_{66}.$$

Taking account of equation 4.53 leads to

$$c_{66} = \frac{c_{11} - c_{12}}{2}$$

The above results are all gathered in the following table:

$$(c_{\alpha\beta}) = \begin{vmatrix} c_{11} & c_{12} & c_{13} & c_{14} & -c_{25} & 0 \\ c_{12} & c_{11} & c_{13} & -c_{14} & c_{25} & 0 \\ c_{13} & c_{13} & c_{33} & 0 & 0 & 0 \\ c_{14} & -c_{14} & 0 & c_{44} & 0 & c_{25} \\ -c_{25} & c_{25} & 0 & 0 & c_{44} & c_{14} \\ 0 & 0 & 0 & c_{25} & c_{14} & \frac{c_{11}-c_{12}}{2} \end{vmatrix} \quad \text{trigonal} \qquad (4.54)$$

In the crystal classes 32, 3m, $\bar{3}$m, diad axes perpendicular to the main axis will enforce additional restrictions: if Ox_1 is one of these diad axes, all $c_{ijk\ell}$ coefficients with an odd number of indices of 1 must vanish (this argument has already been used for monoclinic crystals). There are only six independent elastic constants for $c_{15} = c_{1113} = 0$ in those cases.

For crystals of the tetragonal system, we find the same relations for moduli with four, three and two indices of 3, because the non vanishing γ's involved are the same in both cases. On the other hand, elastic constants with a single index of 3 vanish since $\gamma_{1113} = \gamma_{2223} = 0$

$$c_{14} = c_{24} = c_{15} = c_{25} = c_{46} = c_{56} = 0.$$

We are now left with those constants without any index of 3; they can be expressed in terms of γ_{1111}, γ_{2222}, γ_{1122} and γ_{1212}.

- $c_{ijk\ell} = a_i^1 a_j^1 a_k^1 a_\ell^1 \gamma_{1111} + a_i^2 a_j^2 a_k^2 a_\ell^2 \gamma_{2222} + c_{ijk\ell}^{(3)}$

where $c^{(3)}_{ijk\ell}$ is the expansion of 4.52 in the case n = 3. Taking 4.53 into account, we obtain:

$$c_{iiii} = \frac{1}{4}(\gamma_{1111} + \gamma_{2222}) - \frac{1}{2}(\gamma_{1122} + 2\gamma_{1212})$$

or $c_{11} = c_{22}$. Since $c^{(3)}_{1112}$ and $c^{(3)}_{2221}$ are zero:

$$c_{1112} = \frac{i}{4}(\gamma_{1111} - \gamma_{2222}) = -c_{2221} \quad \text{or} \quad c_{26} = -c_{16}.$$

The four moduli without any index of 3 are c_{11}, c_{12}, c_{16}, c_{66} which together with c_{13}, c_{33}, c_{44}, lead to the table below:

$$(c_{\alpha\beta}) = \begin{vmatrix} c_{11} & c_{12} & c_{13} & 0 & 0 & c_{16} \\ c_{12} & c_{11} & c_{13} & 0 & 0 & -c_{16} \\ c_{13} & c_{13} & c_{33} & 0 & 0 & 0 \\ 0 & 0 & 0 & c_{44} & 0 & 0 \\ 0 & 0 & 0 & 0 & c_{44} & 0 \\ c_{16} & -c_{16} & 0 & 0 & 0 & c_{66} \end{vmatrix} \quad \text{tetragonal (4.55)}$$

For classes 422, 4mm, $\bar{4}$2m, 4/mmm, the Ox_1 axis is parallel to a diad axis orthogonal to Ox_3 so that the coefficient $c_{1112} = c_{16}$ vanishes because of the odd number of indices of 1.

In elasticity theory, the principal axis of <u>hexagonal</u> crystals behaves like a direct 6-fold axis, i.e. like both a diad and a triad axis. The $c_{\alpha\beta}$ matrix thus results from the combination of 4.43 and 4.54, the characteristic matrices for the monoclinic and the trigonal systems:

$$(c_{\alpha\beta}) = \begin{vmatrix} c_{11} & c_{12} & c_{13} & 0 & 0 & 0 \\ c_{12} & c_{11} & c_{13} & 0 & 0 & 0 \\ c_{13} & c_{13} & c_{33} & 0 & 0 & 0 \\ 0 & 0 & 0 & c_{44} & 0 & 0 \\ 0 & 0 & 0 & 0 & c_{44} & 0 \\ 0 & 0 & 0 & 0 & 0 & \frac{c_{11}-c_{12}}{2} \end{vmatrix} \quad \text{hexagonal (4.56)}$$

This tensor is invariant, in any rotation about the 6-fold axis. This property (the proof of which is the object of exercise 4.4) has several consequences:

- planes perpendicular to the main axis are isotropic as far as elastic properties are concerned;
- planes containing the main axis are all equivalent;

- all directions at a given angle to the main axis are equivalent.

The results of the above reduction procedure for the independent elastic moduli are shown in fig. 4.6. The equations, with notation due to K.S. Van Dyke, can be readily applied to the stiffness constants $c_{ijk\ell}$ or $c_{\alpha\beta}$. In an orthonormal reference frame, the procedure for the compliance tensor $s_{ijk\ell}$ is the same as for $c_{ijk\ell}$. But in order to obtain the equations for the $s_{\alpha\beta}$ components, one must take 4.28 into account:

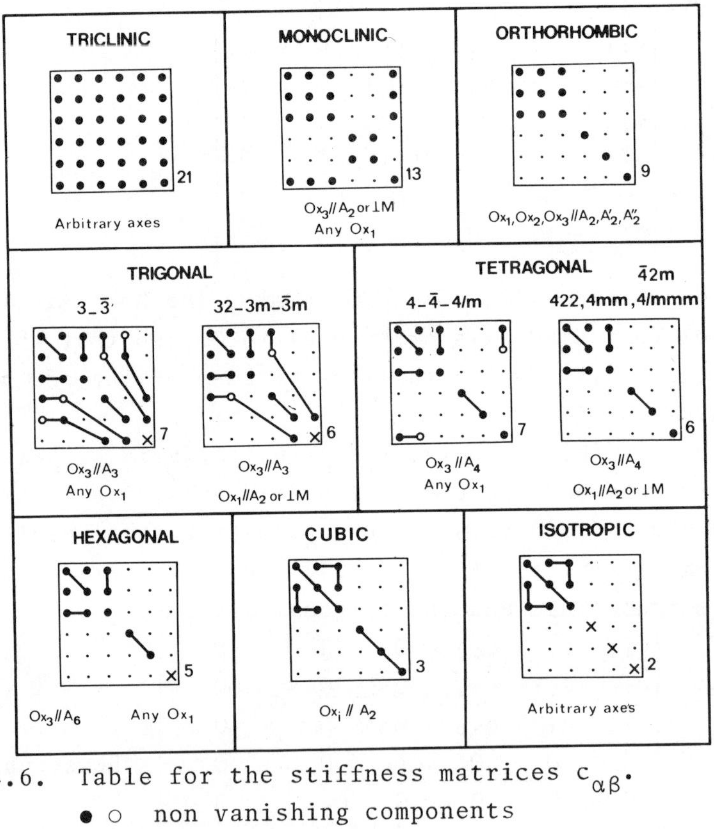

Fig. 4.6. Table for the stiffness matrices $c_{\alpha\beta}$.

● ○ non vanishing components

●—● equal components

• vanishing components

●—○ opposite components

× component equal to $(c_{11} - c_{12})/2$.

The symmetry with respect to the main diagonal is not mentioned. The number of independent constants is shown at the bottom right of each matrix.

Material	Class	Stiffness ($10^{10} N/m^2$)									Density (10^3 kg/m^3)	Ref.
Isotropic cubic system		c_{11}	c_{12}	c_{44}							ρ	
Aluminium (Al)	m3m	10.73	6.08	2.83							2.702	[1]
*Gallium arsenide (GaAs)	$\bar{4}$3m	11.88	5.38	5.94							5.307	[2]
Yttrium aluminium garnet ($Y_3Al_5O_{12}$) YAG	m3m	33.2	11.07	11.50							4.55	[3]
Yttrium iron garnet ($Y_3Fe_5O_{12}$) YIG	m3m	26.9	10.77	7.64							5.17	[3]
*Bismuth germanium oxide ($Bi_{12}GeO_{20}$)	23	12.8	3.05	2.55							9.23	[4]
Gold (Au)	m3m	19.25	16.30	4.24							19.30	[5]
Platinum (Pt)	m3m	34.7	25.10	7.65							21.40	[6]
Silica (SiO_2)	isotropic	7.85	1.61	3.12							2.203	
Silicon (Si)	m3m	16.56	6.39	7.95							2.329	[7]
Tungsten (W)	m3m	52.24	20.44	16.06							19.26	[8]
Hexagonal system		c_{11}	c_{12}	c_{13}	c_{33}	c_{44}						
Beryllium (be)	6/mmm	29.23	2.67	1.4	33.64	16.25					1.848	[9]
*Ceramic PZT-4	trans. isotropic	13.9	7.8	7.4	11.5	2.56					7.5	[25]
*Zinc Oxide (ZnO)	6mm	20.97	12.11	10.51	21.09	4.25					5.676	[10]
*Cadmium sulphide (CdS)	6mm	8.565	5.32	4.62	9.36	1.49					4.824	[11]
Titanium (Ti)	6/mmm	16.24	9.20	6.90	18.07	4.67					4.506	[12]
Tetragonal system		c_{11}	c_{12}	c_{13}	c_{33}	c_{44}	c_{66}	c_{16}				
Indium	4/mmm	4.53	4.00	4.15	4.51	0.651	1.21	0			7.28	[13]
Lead molybdate ($PbMoO_4$)	4/m	10.92	6.83	5.28	9.17	2.67	3.37	1.36			6.95	[14]
Calcium molybdate ($CaMoO_4$)	4/m	14.5	6.6	4.46	12.65	3.69	4.5	1.3			4.255	[15]
*Paratellurite (TeO_2)	422	5.6	5.1	2.2	10.6	2.65	6.6	0			6.00	[16]
Rutile (TiO_2)	4/mmm	27.3	17.6	14.9	48.4	12.5	19.4	0			4.25	[17]
*Barium titanate ($BaTiO_3$)	4mm	27.5	17.9	15.2	16.5	5.43	11.3	0			6.02	[18]
Trigonal system		c_{11}	c_{12}	c_{13}	c_{33}	c_{44}	c_{14}					
Sapphire (Al_2O_3)	$\bar{3}$m	49.7	16.3	11.1	49.8	14.7	-2.35				3.986	[19]
*Lithium niobate ($LiNbO_3$)	3m	20.3	5.3	7.5	24.5	6.0	0.9				4.7	[20]
*Lithium tantalate ($LiTaO_3$)	3m	23.3	4.7	8.0	27.5	9.4	-1.1				7.45	[20]
*α Quartz (SiO_2)	32	8.67	0.70	1.19	10.72	5.79	-1.79				2.648	[21]
*Tellurium (Te)	32	3.27	0.86	2.49	7.22	3.12	1.24				6.25	[22]
Orthorhombic system		c_{11}	c_{12}	c_{13}	c_{22}	c_{23}	c_{33}	c_{44}	c_{55}	c_{66}		
*α Iodic acid (HIO_3)	222	3.01	1.61	1.11	5.8	0.80	4.29	1.69	2.06	1.58	4.64	[23]
*Barium sodium niobate ($Ba_2NaNb_5O_{15}$)	2mm	23.9	10.4	5.0	24.7	5.2	13.5	6.5	6.6	7.6	5.30	[24]

Table 4.7. Elastic stiffness constants.

$$s_{\alpha\beta} = 2^p s_{ijk\ell}$$

where p is the number of indices of value greater than 3 in the pair (α,β). For example, in the trigonal or hexagonal systems and in isotropic media:

$$c_{66} = \frac{c_{11} - c_{12}}{2}, \text{ so } c_{1212} = \frac{c_{1111} - c_{1122}}{2}$$

The same relationship exists between the corresponding $s_{ijk\ell}$:

$$s_{1212} = \frac{s_{1111} - s_{1122}}{2}$$

but, since $s_{66} = 4s_{1212}$, $s_{11} = s_{1111}$, and $s_{12} = s_{1122}$

$$s_{66} = 2(s_{11} - s_{12}).$$

Table 4.7 gives the stiffness constants and the density of a few materials, ordered by their crystal system. For piezoelectric crystals, indicated by a star, we indicate the $c^E_{\alpha\beta}$ constants, as defined in § 6.1.3. Stiffness values are of the order of 10^{11} N/m².

REFERENCES FOR THE ELASTIC CONSTANTS TABLE

[1] J. Vallin and Al. J. Appl. Phys. 35, 1825 (1964).
[2] T.B. Bateman, H.J. McSkimin and J.M. Whelan. J. Appl. Phys. 30, 544 (1959).
[3] E.G. Spencer and Al. J. Appl. Phys. 34, 3059 (1963).
[4] A.J. Slobodnik, Jr., and J.C. Sethares. J. Appl. Phys. 43, 247 (1972).
[5] Y.A. Chang and L. Himmel. J. Appl. Phys. 37, 3567 (1966).
[6] R.E. MacFarlane, J.A. Rayne and C.K. Jones. Phys. Letters 18, 91 (1965).
[7] J.J. Hall. Phys. Rev. 161, 756 (1967).
[8] R. Lowrie and A.M. Gonas. J. Appl. Phys. 38, 4505 (1967).
[9] J.F. Smith and C.L. Arbogast. J. Appl. Phys. 31, 99 (1960).
[10] T.B. Bateman. J. Appl. Phys. 33, 3309 (1962).
[11] J.A. Corll. Phys. Rev. 157, 623 (1967).
[12] E.S. Fischer and C.J. Renken. Phys. Rev. 135, A. 482 (1964).
[13] B.S. Chandrasekhar and J.A. Rayne. Phys. Rev. 124, 1011 (1961).

[14] G.A. Coquin, D.A. Pinnow and A.W. Warner. J. Appl. Phys. 42, 2162 (1971).

[15] W.J. Alton and A.J. Barlow. J. Appl. Phys. 38, 3817 (1967).

[16] Y. Ohmachi and N. Uchida. J. Appl. Phys. 41, 2307 (1970).

[17] R.K. Verma. J. Geophys. Res. 65, 757 (1960).

[18] D. Berlincourt and H. Jaffe. Phys. Rev. 111, 143 (1958).

[19] J. B. Wachtman, W.E. Tefft, D.G. Lam and R.P. Stinchfield. J. Res. Natl. Bur. Std. 64 A, 213 (1960).

[20] A.W. Warner, M. Onoe and G.A. Coquin. J. Acoust. Soc. Am. 42, 1223 (1967).

[21] R. Bechman. Phys. Rev. 110, 1060 (1958).

[22] J.L. Malgrange, G. Quentin and J.M. Thuillier. Phys. Status Solidi 4, 139 (1964).

[23] S. Haussühl. Acta Cryst. A. 24, 697 (1968).

[24] A.W. Warner, G.A. Coquin and J.L. Fink. J. Appl. Phys. 40, 4353 (1969).

[25] H. Jaffe and D.A. Berlincourt. Proc. IEEE, 53, 1372 (1965).

BIBLIOGRAPHY

J.F. Nye. 'Physical properties of crystals', chap. V, VI, VIII. Oxford (1964).

F.I. Fedorov. 'Theory of elastic waves in crystals', chap. I. New York: Plenum Press (1968).

S. Bhagavantam. 'Crystal symmetry and Physical properties', chap. 11. London and New York: Academic-Press (1966).

L. Landau and E. Lifchitz. 'Elastic Theory'. Pergamon.

G. Bruhat and A. Kastler. 'Cours de physique générale. Thermodynamique', chap. IX. Paris: Masson et Cie (1962).

EXERCISES

4.1. A rod, with cylindrical symmetry, is suspended as shown in fig. 4.8. How must the cross section s(x) vary in order that the mean stress due to gravity be constant ($\forall x$)?

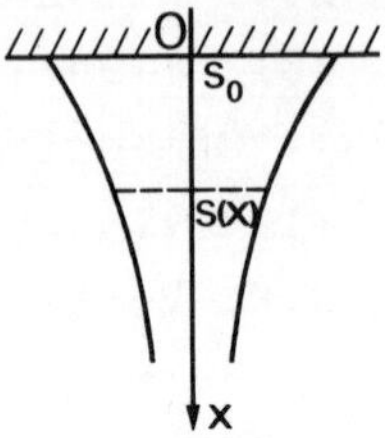

Fig. 4.8.

Solution.

$$T = \frac{\rho g \int_x^\infty s(x)dx}{s(x)} = C^{te} = T_o$$

The derivation yields

$$-\rho g s(x) = T_o \frac{ds}{dx} \Rightarrow s = s_o e^{-\frac{\rho g}{T_o}x}.$$

4.2. Calculate the coefficient of compressibility $\chi = -\Delta/p$ (where Δ is the expansion) of a crystal subject to hydrostatic pressure p.

Solution. With $T_{k\ell} = -p\delta_{k\ell}$, Hooke's law

$$S_{ij} = s_{ijk\ell}T_{k\ell} = -ps_{ijkk}$$

provides the expansion $\Delta = \Delta v/v$ (formula 4.7):

$$\Delta = S_{ii} = -ps_{iikk}$$

and $\chi = s_{iikk} = s_{11} + s_{22} + s_{33} + 2(s_{12}+s_{23}+s_{31})$.

4.3. In terms of the stiffness coefficients, calculate Young's modulus E and Poisson's ratio ν defined, for an isotropic solid, by the strain $\Delta\ell/\ell$ and the relative variation in diameter $\Delta d/d$ of a cylinder, with cross section s, undergoing a uniform stress F/s along the x_1-axis:

$$\frac{\Delta\ell}{\ell} = \frac{1}{E}\frac{F}{s}, \quad \frac{\Delta d}{d} = -\nu\frac{\Delta\ell}{\ell} = -\frac{\nu}{E}\frac{F}{s}.$$

Solution. The strains are $S_{11} = \frac{1}{E}T_{11}$ and $S_{22} = S_{33} = -\frac{\nu}{E}T_{11}$. For an isotropic medium $T_{22} = c_{12}(S_{11}+S_{33}) + c_{11}S_{22} = 0 \Rightarrow \nu = \frac{c_{12}}{c_{11}+c_{12}}$

and $T_{11} = c_{11}S_{11}+c_{12}(S_{22}+S_{33}) \Rightarrow E = c_{11}-2c_{12}\nu = c_{11} - 2\frac{c_{12}^2}{c_{11}+c_{12}}$

4.4. Show that the stiffnesses of hexagonal crystals are invariant with respect to any rotation about the 6-fold axis.

Solution. Let us examine the behaviour of the five independent moduli c_{33}, c_{13}, c_{44}, c_{12}, c_{11} during the reference frame transformation (3.12):

$$c'_{ijk\ell} = \alpha_i^p \alpha_j^q \alpha_k^r \alpha_\ell^s c_{pqrs}$$

- $c'_{3333} = \alpha_3^3 \alpha_3^3 \alpha_3^3 \alpha_3^3 c_{3333} \Longrightarrow c'_{33} = c_{33}$

- $c'_{1133} = \alpha_1^p \alpha_1^q \alpha_3^3 \alpha_3^3 c_{pq33}$

The only non vanishing moduli correspond to $p = q = 1$ or 2:

$$c'_{13} = c_{13} \cos^2\phi + c_{23}\sin^2\phi = c_{13}$$

since $c_{23} = c_{13}$

- $c'_{2323} = \alpha_2^p \alpha_2^r c_{p3r3} \Longrightarrow c'_{44} = c_{44}\cos^2\phi + c_{55}\sin^2\phi = c_{44}$

since $c_{55} = c_{44}$

- $c'_{1122} = \alpha_1^p \alpha_1^q \alpha_2^r \alpha_2^s c_{pqrs}$ with $p,q.r,s \neq 3$,

$$c'_{12} = (c_{11} + c_{22} - 4c_{66})\cos^2\phi\sin^2\phi + c_{12}\cos^4\phi + c_{21}\sin^4\phi$$

and replacing c_{66} by $1/2(c_{11} - c_{12})$

$$c'_{12} = (2\cos^2\phi\sin^2\phi + \sin^4\phi + \cos^4\phi)c_{12} = (\cos^2\phi + \sin^2\phi)^2 c_{12} = c_{12}$$

Similarly $c'_{11} = c_{11}$.

4.5. Calculate the compliance coefficients in terms of the stiffness coefficients for crystals of the quadratic (classes 4/mmm, 422, 4mm, $\bar{4}$2m), hexagonal, and cubic systems.

Solution. Inverting the $c_{\alpha\beta}$ matrix readily provides $s_{44} = 1/c_{44}$ and $s_{66} = 1/c_{66}$.
The other equations can easily be derived from the following transformation of the first three equations 4.23:

$$\begin{cases} (1)\ T_1 = c_{11}S_1 + c_{12}S_2 + c_{13}S_3 \\ (2)\ T_2 = c_{12}S_1 + c_{11}S_2 + c_{13}S_3 \\ (3)\ T_3 = c_{13}S_1 + c_{13}S_2 + c_{33}S_3 \end{cases} \overset{(1)-(2)}{\Longrightarrow} \begin{cases} T_1 - T_2 = (c_{11} - c_{12})(S_1 - S_2) \\ T_2 = c_{12}(S_1 - S_2) + (c_{11} + c_{12})S_2 + c_{13}S_3 \\ T_3 = c_{13}(S_1 - S_2) + 2c_{13}S_2 + c_{33}S_3. \end{cases}$$

The same transformation, performed on equations 4.27, shows that, with variables $\tau_1 = T_1 - T_2$, $\tau_2 = T_2$, $\tau_3 = T_3$, $\sigma_1 = S_1 - S_2$, $\sigma_2 = S_2$, $\sigma_3 = S_3$, the matrix

$$\begin{pmatrix} s_{11}-s_{12} & 0 & 0 \\ s_{12} & s_{11}+s_{12} & s_{13} \\ s_{13} & 2s_{13} & s_{33} \end{pmatrix} \text{ is the inverse of}$$

$$\begin{pmatrix} c_{11}-c_{12} & 0 & 0 \\ c_{12} & c_{11}+c_{12} & c_{13} \\ c_{13} & 2c_{13} & c_{33} \end{pmatrix}$$

Hence

$$s_{11}-s_{12} = \frac{1}{c_{11}-c_{12}} \qquad s_{11}+s_{12} = \frac{c_{33}}{c^2} \qquad s_{13} = -\frac{c_{13}}{c^2}$$

$$s_{33} = \frac{c_{11}+c_{12}}{c^2}$$

where

$$c^2 = (c_{11}+c_{12})c_{33} - 2c_{13}^2.$$

Chapter 5
Dynamic Elasticity

We shall now discuss in detail elastic wave propagation in solids. The reader may already be aware that there is a great variety of elastic waves. Indeed, depending upon the propagation conditions, one currently refers to, amongst others, Rayleigh waves, Bleustein-Gulyaev waves, Lamb waves, Love waves and Stoneley waves. It seems at first useful to emphasise the fact that, basically, only two types of waves exist:

- longitudinal waves, or compression waves (fig. 5.1.a). They are characterized by a particle displacement parallel to the direction of propagation, i.e. a polarization parallel to the wave vector. A longitudinal wave creates a variation in the distance between parallel planes containing given particles, so that the volume occupied by a given number of particles is not a constant.

- transverse waves or shear waves (fig. 5.1.b). The particle displacement is perpendicular to the wave vector and the gliding of parallel planes causes no variation in volume.

These waves propagate in unbounded isotropic solid media (§ 5.1.4). When saying "unbounded", we mean in practice that the size of the medium is much greater than the beam size and that surface effects are negligible.

If now the medium, still assumed to be unbounded, is anisotropic (i.e. crystalline), three waves can propagate in a given direction (as shown in §§ 5.1.1 and 5.1.2) and

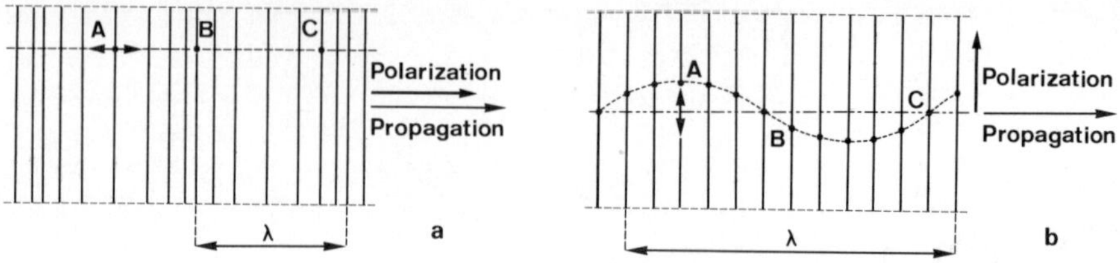

Fig. 5.1. Elastic waves in an unbounded isotropic medium.
a) Longitudinal wave: polarization and wave vector are parallel. Particles A, B, C, move along the same axis. At a given time, the wave looks like a succession of compressions and expansions.
b) Shear wave: polarization and wave vector are orthogonal, particles A, B, C move on both sides of their rest positions. Planes orthogonal to the wave vector glide with respect to each other; their mutual separations remain constant.

in the most general case, none of them is purely longitudinal or transverse. The situation is in fact as follows: for any given direction, there can be three waves:

- a quasi longitudinal wave, so called because the particle vibration and the wave vector make a non-vanishing angle,
- a fast quasi-transverse wave,
- a slow quasi-transverse wave.

The polarizations of these three waves (with different propagation velocities) are always mutually orthogonal. The energy vectors, which show the direction of energy flow for each of these waves, make different angles with the wave vector. It becomes obvious that, currently such propagation conditions are difficult to handle, and it is much better to use specific directions, such as symmetry axes, for which pure modes can propagate, where the energy vector and the propagation vector are parallel as for isotropic media (§§ 5.1.3 and 5.1.5).

When the medium has a finite size, mechanical and electrical (for piezoelectric media) boundary conditions should be satisfied - e.g. vanishing mechanical stress at a free surface, continuous normal component of the electric displacement in the absence of charges - and those waves which can propagate near the boundary are, with a few exceptions, no longer simple waves: they have a longitudinal component and one or two shear component(s).

Let us briefly describe the waves we mentioned above.

At the surface of any semi-infinite medium, a complex wave, called a Rayleigh wave after the scientist who discovered it in 1885, can propagate. In the simplest case of an isotropic medium, this wave consists of a longitudinal wave and a shear wave, with a $\pi/2$ phase shift between them, both contained in the sagittal plane, i.e. the plane which contains the wave vector and the normal to the boundary (fig. 5.2.a and b). The variation with depth is not the same for both components. Since the longitudinal component vanishes and changes sign at a distance of about 0.2λ from the surface, the polarization becomes purely transverse, then elliptic again but with the opposite sense. The particle displacement vanishes completely at a depth 2λ. This is depicted by the curves of fig. 5.42 (§ 5.3.1).

At the surface of a semi-infinite piezoelectric medium, a transverse surface wave, with a polarization parallel to the surface, can propagate (fig. 5.2.c). The penetration depth is smaller, the greater the piezoelectricity (§ 6.2.2.3). The depth to which there is movement for these waves, known as Bleustein-Gulyaev waves, is greater (of the order of 100λ) than for Rayleigh waves.

When the medium is bounded by two parallel planes, Rayleigh waves propagate along each plane as long as their mutual distance is much greater than the wavelength. If the thickness of the medium is of the order of λ, one obtains Lamb waves, which can be either symmetric or antisymmetric (fig. 5.3).

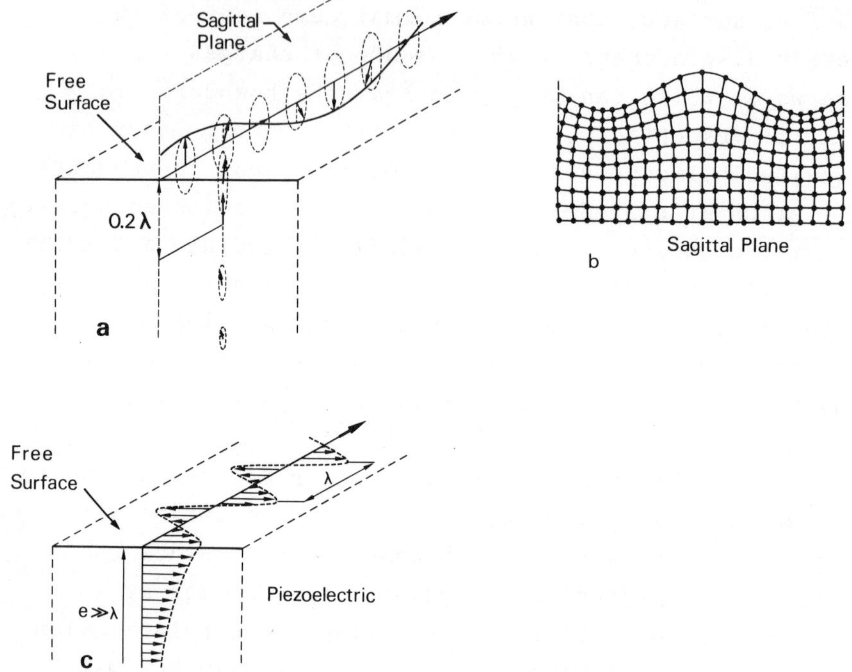

Fig. 5.2. Surface waves

a) and b) Rayleigh wave. In an isotropic medium, the particle displacement, within the sagittal plane, comes from a longitudinal component and a $\pi/2$-phase shifted shear component, the amplitudes of which do not decrease in the same way. The extremity of the polarization vector describes an ellipse, retrograde at the surface (a). The displacement vanishes at a depth 2λ. The wave generates an oscillation of the surface (b, according to fig. 1 of reference 8).

c) Bleustein-Gulyaev wave. The strictly shear polarization is parallel to the surface of the piezoelectric material. The displacement amplitude of the particles decreases exponentially. The surface remains a plane.

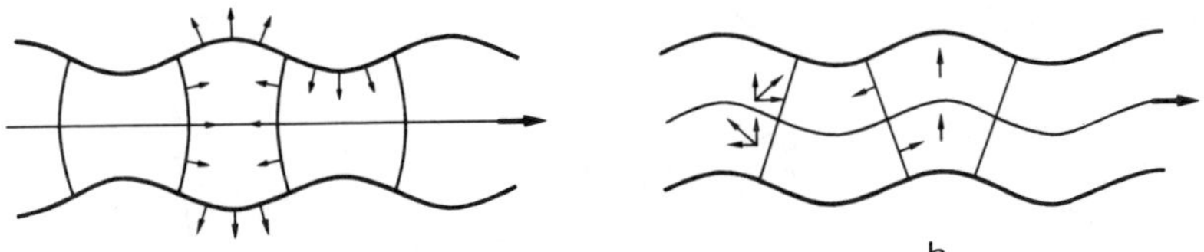

Fig. 5.3. Lamb waves.
a) symmetric: on both sides of the middle plane, the longitudinal components are equal and the shear components have opposite signs.
b) antisymmetric: on both sides of the meridian plane, the longitudinal component changes sign whereas the shear component does not.

The above discussion relates to homogeneous media. In the case of an inhomogeneous medium, e.g. of two adjacent media with different elastic properties, two main situations may be distinguished, depending on the relative thicknesses of these media. The former (shown in fig. 5.4.a) corresponds to the propagation of a so-called Stoneley wave along the boundary between two semi-infinite isotropic media. The displacement is contained in the sagittal plane, and decreases on either side of the boundary. The latter case is shown in fig. 5.4.b: the medium consists of a thin film deposited on a substrate. Such surface waves, with a shear polarization, are called Love waves. Stoneley and Love waves exist only under specific conditions; for example the shear wave velocity in the thin film should be less than that in the substrate (§ 5.2.4).

It may be of interest that Rayleigh, Love and Stoneley discovered the waves named after them whilst studying the propagation of seismic waves.

5.1. ELASTIC WAVES IN AN INFINITE CRYSTAL

Throughout the preceding chapter, it has been assumed that in the strained solid all points remained at rest: the displacement u_i of every point depended only on the

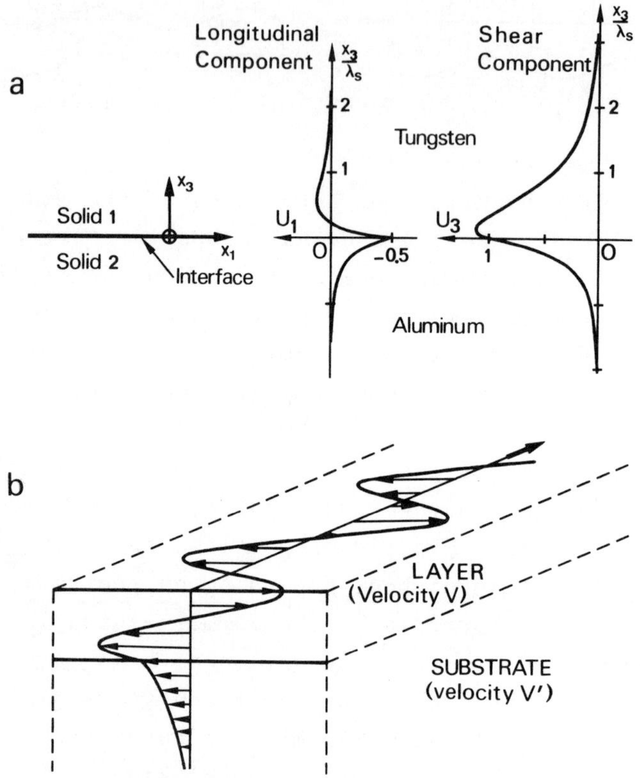

Fig. 5.4. Elastic waves at a boundary between two solids. a) For some pairs of solids, for which the continuity conditions are satisfied, a Rayleigh wave propagates on both sides of the boundary. This boundary wave is called a Stoneley wave. On the right: an example of the decreasing longitudinal and shear components for the case of aluminium-tungsten [1].
b) Love wave. This is a dispersive shear wave propagating in a medium consisting of a layer and a substrate, provided some conditions of symmetry and velocities (V' > V) are fulfilled.

initial co-ordinates x_k. We now suppose that a perturbation travels through the medium, which is locally in motion. The displacement of any point is time dependent:

$$u_i = u_i(x_k, t).$$

5.1.1. Propagation Equation

The equation of motion comes from the fundamental law of dynamics $\vec{F} = m\vec{\gamma}$. It has previously been shown (§ 4.2.2) that the force density per unit volume of stressed material is given by

$$f_i = \frac{\partial T_{ij}}{\partial x_j}$$

We neglect the effect of gravity, and say that this force gives rise to the acceleration $\partial^2 u_i/\partial t^2$ along the i-th axis for the unit volume mass ρ:

$$\rho \frac{\partial^2 u_i}{\partial t^2} = \frac{\partial T_{ij}}{\partial x_j}, \qquad (5.1)$$

Making use of Hooke's Law, i.e. of 4.21:

$$T_{ij} = c_{ijk\ell} \frac{\partial u_\ell}{\partial x_k}.$$

the equation of motion becomes

$$\boxed{\rho \frac{\partial^2 u_i}{\partial t^2} = c_{ijk\ell} \frac{\partial^2 u_\ell}{\partial x_j \partial x_k}.} \qquad (5.2)$$

This is a set of three second order differential equations which constitute the generalization of 1.13, the propagation equation in a fluid:

$$\rho \frac{\partial^2 u}{\partial t^2} = \frac{1}{\chi} \frac{\partial^2 u}{\partial x^2}.$$

for the case of three dimensions and of an anisotropic medium. The general solution for the fluid was

$$u = F\left(t - \frac{x}{V}\right) \quad \text{with } V^2 = \frac{1}{\rho\chi}$$

so, by analogy, we look for a solution in the form of a progressive wave, travelling in the direction of the unit vector $\vec{n}(n_1, n_2, n_3)$ perpendicular to the wave planes $\vec{n}.\vec{x} = C^{te}$:

$$u_i = {}^\circ u_i F(t - \frac{\vec{n}.\vec{x}}{V}) = {}^\circ u_i F(t - \frac{n_j x_j}{V}). \qquad (5.3)$$

In order to calculate the phase velocity V and the wave polarization ${}^\circ u_i$ (i.e. the particle displacement direction), we insert (5.3) into the propagation equation (5.2), denoting by F" the second derivative of F:

$$\frac{\partial^2 u_i}{\partial t^2} = {}^\circ u_i F''$$

and

$$\frac{\partial u_\ell}{\partial x_j} = -{}^\circ u_\ell \frac{n_j}{V} F' \quad \text{or} \quad \frac{\partial^2 u_\ell}{\partial x_j \partial x_k} = {}^\circ u_\ell \frac{n_j n_k}{V^2} F''.$$

so

$$\rho\, {}^\circ u_i F'' = c_{ijk\ell} n_j n_k {}^\circ u_\ell \frac{F''}{V^2}$$

or

$$\rho V^2 \, {}^\circ u_i = c_{ijk\ell} n_j n_k {}^\circ u_\ell . \qquad (5.4)$$

We introduce the second rank tensor

$$\boxed{\Gamma_{i\ell} = c_{ijk\ell} n_j n_k} \qquad (5.5)$$

and the above equation, known as the Christoffel equation, becomes

$$\Gamma_{i\ell} {}^\circ u_\ell = \rho V^2 \, {}^\circ u_i \qquad (5.6)$$

This shows that the polarization ${}^\circ u_i$ is an eigenvector of the $\Gamma_{i\ell}$ tensor with the eigenvalue $\rho V^2 = \gamma$ (we shall assume the results of § 3.5 are known).

Finally, the velocities and polarizations of those plane waves which propagate along a direction $\vec{n}$ in a crystal with a stiffness tensor $c_{ijk\ell}$ are obtained on searching for the eigenvalues and eigenvectors of the tensor $\Gamma_{i\ell} = c_{ijk\ell} n_j n_k$. Generally, for a given propagation direction, there are three velocities, the three roots of the secular equation:

$$|\Gamma_{i\ell} - \rho V^2 \delta_{i\ell}| = 0 \qquad (5.7)$$

which expresses the compatibility of the three homogeneous equations 5.6. Each velocity is related to an eigenvector which defines the displacement direction (i.e. the wave polarization).

5.1.2. General Properties of Elastic Plane Waves

Due to the symmetry properties of the stiffness tensor (equations 4.20 and 4.32), the $\Gamma_{i\ell}$ tensor is symmetric:

$$\Gamma_{\ell i} = c_{\ell jki}n_j n_k = c_{ki\ell j}n_j n_k = c_{ikj\ell}n_j n_k = c_{ijk\ell}n_j n_k = \Gamma_{i\ell}.$$

Therefore, its eigenvalues are real and its eigenvectors are orthogonal (§ 3.5). Furthermore, the eigenvalues $\gamma = \rho V^2$ are positive (this is a necessary condition for the propagation velocity V to be real). In order to prove this, we go back to equation 5.6 which provides (after a contraction with $\circ u_i$):

$$\gamma = \frac{\circ u_i \Gamma_{i\ell} \circ u_\ell}{\circ u_i^2} .$$

γ has the sign of the numerator:

$$\circ u_i \Gamma_{i\ell} \circ u_\ell = \circ u_i n_j c_{ijk\ell} n_k \circ u_\ell .$$

But the elastic energy density (e.g. 4.34) $\Phi = 1/2 c_{ijk\ell} S_{ij} S_{k\ell}$ is positive, whatever the symmetric second rank tensor S_{ij}. In the case of a tensor A_{ij} which can be decomposed into its symmetric part $\overset{+}{A}_{ij} = (A_{ij}+A_{ji})/2$ and its antisymmetric part $\overset{-}{A}_{ij} = (A_{ij}-A_{ji})/2$:

$$A_{ij} = \overset{+}{A}_{ij} + \overset{-}{A}_{ij}$$

the contracted product

$$P = A_{ij} c_{ijk\ell} A_{k\ell} = (\overset{+}{A}_{ij} + \overset{-}{A}_{ij}) c_{ijk\ell} (\overset{+}{A}_{k\ell} + \overset{-}{A}_{k\ell})$$

reduces to

$$P = \overset{+}{A}_{ij} c_{ijk\ell} \overset{+}{A}_{k\ell}$$

for, if the dummy indices i and j are permuted,

$$\overset{-}{A}_{ij} c_{ijk\ell} = \overset{-}{A}_{ji} c_{jik\ell} = -\overset{-}{A}_{ij} c_{ijk\ell} = 0.$$

As a result, $P = A_{ij} c_{ijk\ell} A_{k\ell}$ is positive for any tensor A_{ij}. In particular for $A_{ij} = \circ u_i n_j$:

$$P = \circ u_i n_j c_{ijk\ell} \circ u_\ell n_k = \circ u_i \Gamma_{i\ell} \circ u_\ell > 0 \Longrightarrow \gamma > 0.$$

Since the eigenvalues are real and positive, there are generally <u>three waves</u> propagating in the same direction with different velocities, and mutually <u>orthogonal polarizations.</u> This is depicted in fig. 5.5.

The displacement vector $\vec{u}$ is generally not parallel or perpendicular to the propagation direction $\vec{n}$. The

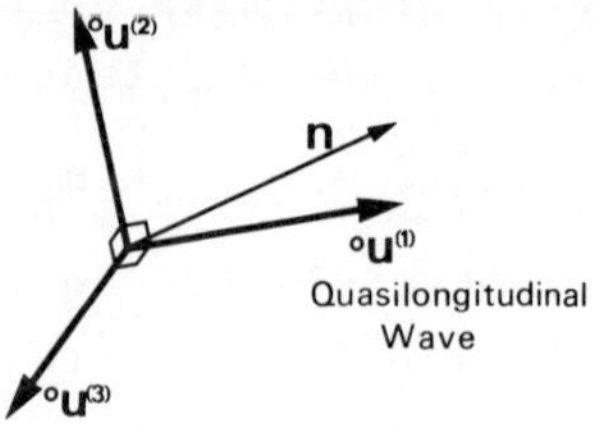

Fig. 5.5. Anisotropic medium. General case. In a crystal three plane waves can propagate in a given direction $\vec{n}$, each with its own velocity. The wave whose displacement $\circ\vec{u}^{(1)}$ is closest to $\vec{n}$ is called quasi-longitudinal. Its velocity is usually greater than that of the other two waves (quasi-shear waves) polarized along $\circ\vec{u}^{(2)}$ and $\circ\vec{u}^{(3)}$. The three vectors $\circ\vec{u}^{(1)}$, $\circ\vec{u}^{(2)}$, $\circ\vec{u}^{(3)}$ are mutually orthogonal.

wave for which $\vec{u}$ is closest to $\vec{n}$ is said to be quasi-longitudinal; the other two are quasi-transverse. Usually, the latter propagate more slowly than the quasi-longitudinal wave. Only for special propagation directions are these waves purely longitudinal or transverse [2,3].

Before we deal with propagation along directions linked to the symmetry elements, we first establish the expressions for the $\Gamma_{i\ell}$ tensor components. Expanding the sum over j and k, one obtains:

$$\Gamma_{i\ell} = c_{i11\ell}n_1^2+c_{i22\ell}n_2^2+c_{i33\ell}n_3^2+(c_{i12\ell}+c_{i21\ell})n_1n_2$$
$$+(c_{i13\ell}+c_{i31\ell})n_1n_3+(c_{i23\ell}+c_{i32\ell})n_2n_3$$

Hence the six components:

$$\Gamma_{11}=c_{11}n_1^2+c_{66}n_2^2+c_{55}n_3^2+2c_{16}n_1n_2+2c_{15}n_1n_3+2c_{56}n_2n_3$$
$$\Gamma_{12}=c_{16}n_1^2+c_{26}n_2^2+c_{45}n_3^2+(c_{12}+c_{66})n_1n_2+(c_{14}+c_{56})n_1n_3+(c_{46}+c_{25})n_2n_3$$
$$\Gamma_{13}=c_{15}n_1^2+c_{46}n_2^2+c_{35}n_3^2+(c_{14}+c_{56})n_1n_2+(c_{13}+c_{55})n_1n_3+(c_{36}+c_{45})n_2n_3$$
$$\Gamma_{22}=c_{66}n_1^2+c_{22}n_2^2+c_{44}n_3^2+2c_{26}n_1n_2+2c_{46}n_1n_3+2c_{24}n_2n_3$$
$$\Gamma_{23}=c_{56}n_1^2+c_{24}n_2^2+c_{34}n_3^2+(c_{46}+c_{25})n_1n_2+(c_{36}+c_{45})n_1n_3+(c_{23}+c_{44})n_2n_3$$
$$\Gamma_{33}=c_{55}n_1^2+c_{44}n_2^2+c_{33}n_3^2+2c_{45}n_1n_2+2c_{35}n_1n_3+2c_{34}n_2n_3.$$
$$\Gamma_{21} = \Gamma_{12} \qquad \Gamma_{31} = \Gamma_{13} \qquad \Gamma_{32} = \Gamma_{23} \qquad (5.8)$$

5.1.3. Propagation Along Directions Linked to the Crystal Symmetry

a) The propagation tensor $\Gamma_{i\ell}$ along the x_3 axis ($n_1 = n_2 = 0$, $n_3 = 1$) is written $\Gamma_{i\ell} = c_{i33\ell}$, or explicitly

$$\Gamma_{x_3} = \begin{bmatrix} c_{55} & c_{45} & c_{35} \\ c_{45} & c_{44} & c_{34} \\ c_{35} & c_{34} & c_{33} \end{bmatrix}$$

When the x_3 axis is the diad axis of monoclinic system crystals (see table 4.43) $c_{35} = c_{34} = 0$:

$$\Gamma_{x_3//A_2} = \begin{bmatrix} c_{55} & c_{45} & 0 \\ c_{45} & c_{44} & 0 \\ 0 & 0 & c_{33} \end{bmatrix}$$

A longitudinal wave propagates with velocity $V_1 = \sqrt{c_{33}/\rho}$ as well as two shear waves with different velocities (§ 3.5).

If Ox_3 is one of the diad axes of the orthorhombic system, c_{45} also vanishes. The shear wave of velocity $V_2 = \sqrt{c_{44}/\rho}$ is polarized along Ox_2; the other wave of velocity $V_3 = \sqrt{c_{55}/\rho}$ is polarized along Ox_1.

For a p-fold axis with $p > 2$, $c_{44} = c_{55}$ and both shear waves have the same velocity; the polarization may be anywhere in a plane perpendicular to x_3 (two eigenvalues being identical). The shear waves are said to be degenerate. Such an axis, along which a shear wave may propagate with any polarization, is called an acoustic axis. Directions off the symmetry axis may also fulfil this condition (equality of the shear wave velocities).

The above results can also be obtained by symmetry arguments. When $\vec{n}$ is along the A_p axis, the whole system - crystal and propagation direction - has symmetry A_p. The polarizations of the waves should be invariant under rotations of $2\pi/p$ about A_p, which implies they are purely longitudinal or transverse. Furthermore, if $p > 2$, the displacement vectors for the initial wave ($\vec{u}$) and the transformed wave ($\vec{u}'$) are not parallel (fig. 5.6). Since they correspond to the same velocity, i.e. to the same eigenvalue of the $\Gamma_{i\ell}$ tensor, any vector orthogonal to A_p

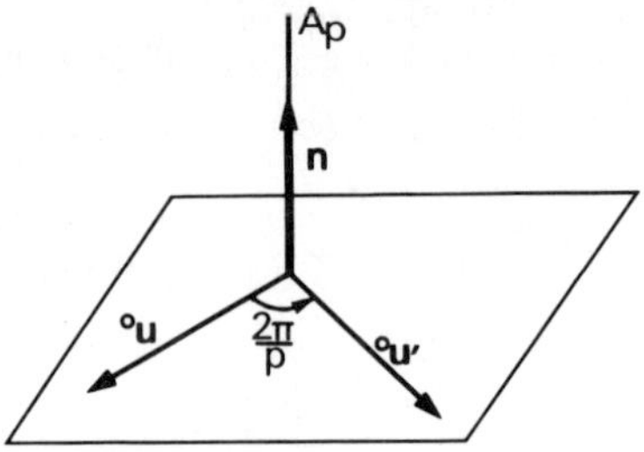

Fig. 5.6. Propagation along a symmetry axis. A_p (p > 2). Any vector orthogonal to A_p is decomposed into the vectors $°\vec{u}$ and $°\vec{u}'$ derived from $\vec{u}$ by a $2\pi/p$ rotation, and is an eigenvector of Γ with the same eigenvalue as $°\vec{u}$ and $°\vec{u}'$.

decomposes along $\vec{u}$ and $\vec{u}'$, and is an eigenvector of $\Gamma_{i\ell}$ for the same eigenvalue: shear waves all travel with the same velocity and an arbitrary polarization.

b) Let us now concentrate on propagation within a symmetry plane, or in a plane perpendicular to an A_p axis (with even p). In both cases, a diad axis exists. Elastic properties and the propagation direction are invariant under inversion (§ 3.4.2). If we take the equivalence $A_2C = \overline{A}_2 = M$ into account, we see that the system - crystal plus wave - is symmetric about the plane perpendicular to A_p. Polarizations can only be parallel to A_p (shear wave S) or in the plane perpendicular to A_p (quasi longitudinal waves QL and quasi shear waves QS) (fig. 5.7). The above argument also proves that the Γ_{13} and Γ_{23} components of the propagation tensor in the x_1x_2 plane (orthogonal to A_p, itself along x_3) vanish:

$$\Gamma_{i\ell} = \begin{bmatrix} \Gamma_{11} & \Gamma_{12} & 0 \\ \Gamma_{12} & \Gamma_{22} & 0 \\ 0 & 0 & \Gamma_{33} \end{bmatrix} .$$

In the case of a 4-fold or 6-fold axis, the expression for Γ_{33}, taken from 5.8 with $n_3 = 0$

$$\Gamma_{33} = c_{55}n_1^2 + c_{44}n_2^2 + 2c_{45}n_1n_2$$

can be simplified for $c_{45} = 0$ and $c_{44} = c_{55}$ (see 4.55 and 4.56):

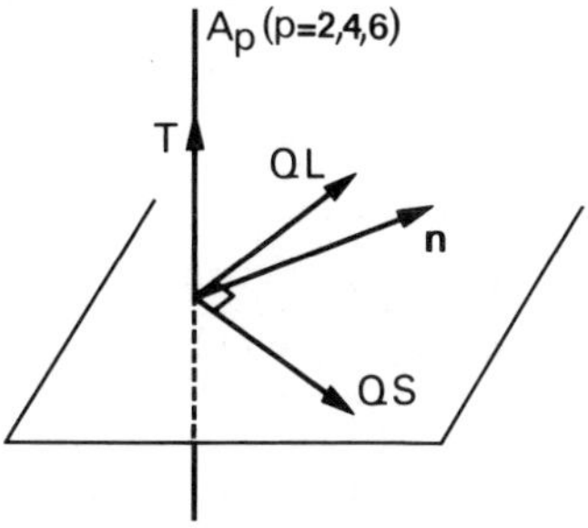

Fig. 5.7. For any direction $\vec{n}$ orthogonal to an A_p axis of even order (direct or inverse), there is a shear wave with A_p polarization.

$$\Gamma_{33} = c_{44}(n_1^2 + n_2^2) = c_{44}.$$

The velocity of the shear wave ($V_3 = \sqrt{\Gamma_{33}/\rho} = \sqrt{c_{44}/\rho}$) does not depend on the propagation direction in a plane perpendicular to a 4-fold or 6-fold axis.

5.1.4. Elastic Waves in an Isotropic Medium

In an isotropic medium, the elastic moduli can be expressed in terms of the Lamé-coefficient λ and μ through 4.36:

$$c_{ijk\ell} = \lambda\delta_{ij}\delta_{k\ell} + \mu(\delta_{ik}\delta_{j\ell} + \delta_{i\ell}\delta_{jk})$$

and the propagation tensor components are:

$$\Gamma_{i\ell} = \lambda\delta_{ij}\delta_{k\ell}n_j n_k + \mu(\delta_{ik}\delta_{j\ell}n_j n_k + \delta_{i\ell}\delta_{jk}n_j n_k)$$

or $$\Gamma_{i\ell} = (\lambda + \mu)n_i n_\ell + \mu\delta_{i\ell}n_k n_k.$$

Given equations 4.37:

$$\lambda + \mu = \frac{c_{11} + c_{12}}{2}, \qquad \mu = \frac{c_{11} - c_{12}}{2}$$

we obtain:

$$\Gamma_{i\ell} = \frac{c_{11} + c_{12}}{2} n_i n_\ell + \frac{c_{11} - c_{12}}{2} n_k n_k \delta_{i\ell}. \qquad (5.9)$$

Since $n_k n_k = \vec{n}^2 = 1$

$$\Gamma_{i\ell} = \frac{c_{11} + c_{12}}{2} n_i n_\ell + \frac{c_{11} - c_{12}}{2} \delta_{i\ell}$$

The equation for the eigenvectors (5.6) is:

$$\frac{c_{11} + c_{12}}{2} n_i n_\ell {}^{o}u_\ell + \frac{c_{11} - c_{12}}{2} {}^{o}u_i = \rho V^2 {}^{o}u_i.$$

or, in vector notation:

$$\frac{c_{11} + c_{12}}{2}(\vec{n}.{}^{o}\vec{u})\vec{n} = (\rho V^2 - \frac{c_{11} - c_{12}}{2})\,{}^{o}\vec{u}$$

There are two solutions:

- $\vec{n}.{}^{o}\vec{u} = 0$, which corresponds to a degenerate shear wave, propagating with velocity $V_T = \sqrt{(c_{11} - c_{12})/2\rho}$;
- ${}^{o}\vec{u}//\vec{n}$ (when $\vec{n}.{}^{o}\vec{u} \neq 0$) which corresponds to a longitudinal wave propagating with velocity $V_L = \sqrt{c_{11}/\rho}$ (since $(\vec{n}.{}^{o}\vec{u})\vec{n} = \vec{n}^2\,{}^{o}\vec{u} = {}^{o}\vec{u}$.)

So, in an isotropic solid, plane waves are longitudinal or transverse; the velocities do not depend on the propagation direction. Furthermore, since c_{12} is positive,

$$\boxed{V_T < \frac{V_L}{\sqrt{2}}} \tag{5.10}$$

for any isotropic solid.

This decomposition into two waves, travelling with velocities V_L and V_T, also applies to any wave (non plane). Let us return to equation 5.2:

$$\rho\frac{\partial^2 u_i}{\partial t^2} = c_{ijk\ell}\frac{\partial^2 u_\ell}{\partial x_j \partial x_k}$$

which can be written:

$$\rho\frac{\partial^2 u_i}{\partial t^2} = \Delta_{i\ell}u_\ell. \tag{5.11}$$

The differential operator

$$\Delta_{i\ell} = c_{ijk\ell}\frac{\partial^2}{\partial x_j \partial x_k}$$

can be derived from $\Gamma_{i\ell}$ by replacing $n_j n_k$ by $\partial^2/\partial x_j \partial x_k$. The same substitution in 5.9 leads to:

$$\Delta_{i\ell} = \frac{c_{11} + c_{12}}{2}\frac{\partial^2}{\partial x_i \partial x_\ell} + \frac{c_{11} - c_{12}}{2}\delta_{i\ell}\Delta$$

where $\Delta = \partial^2/\partial x_k^2$ is the laplacian operator. Equation (5.11) now becomes:

$$\rho\frac{\partial^2 u_i}{\partial t^2} = \frac{c_{11} + c_{12}}{2}\frac{\partial^2 u_\ell}{\partial x_i \partial x_\ell} + \frac{c_{11} - c_{12}}{2}\Delta u_i$$

or, in vector form, with $\partial u_\ell/\partial x_\ell = \text{div}\,\vec{u}$ and $\Delta u_i = (\Delta\vec{u})_i$

$$\rho \frac{\partial^2 \vec{u}}{\partial t^2} = \frac{c_{11} + c_{12}}{2} \overrightarrow{\text{grad}} \text{ div } \vec{u} + \frac{c_{11} - c_{12}}{2} \Delta\vec{u}.$$

Replacing c_{11} and c_{12} by the plane wave velocities $V_L = \sqrt{c_{11}/\rho}$ and $V_T = \sqrt{(c_{11} - c_{12})/2\rho}$:

$$\frac{\partial^2 \vec{u}}{\partial t^2} = V_T^2 \Delta\vec{u} + (V_L^2 - V_T^2) \overrightarrow{\text{grad}} \text{ div } \vec{u}. \qquad (5.12)$$

We can always decompose $\vec{u}$ into a divergenceless vector $\vec{u}_T$ and an irrotational vector $\vec{u}_L$:

$$\vec{u} = \vec{u}_T + \vec{u}_L$$

where div $\vec{u}_T = 0$ and $\overrightarrow{\text{curl}}\ \vec{u}_L = 0$. (5.12) now reads:

$$\frac{\partial^2 \vec{u}_T}{\partial t^2} + \frac{\partial^2 \vec{u}_L}{\partial t^2} = V_T^2 \Delta\vec{u}_T + V_T^2 \Delta\vec{u}_L + (V_L^2 - V_T^2) \overrightarrow{\text{grad}} \text{ div } \vec{u}_L$$

Making use of

$$\overrightarrow{\text{grad}} \text{ div } \vec{u}_L = \Delta\vec{u}_L + \overrightarrow{\text{curl}}\ \overrightarrow{\text{curl}}\ \vec{u}_L = \Delta\vec{u}_L:$$

$$\left(\frac{\partial^2 \vec{u}_T}{\partial t^2} - V_T^2 \Delta\vec{u}_T\right) + \left(\frac{\partial^2 \vec{u}_L}{\partial t^2} - V_L^2 \Delta\vec{u}_L\right) = 0.$$

Both of the expressions in brackets vanish, because they both have a vanishing divergence and curl:

$$\frac{\partial^2 \vec{u}_T}{\partial t^2} - V_T^2 \Delta\vec{u}_T = 0 \quad \text{and} \quad \frac{\partial^2 \vec{u}_L}{\partial t^2} - V_L^2 \Delta\vec{u}_L = 0.$$

The above two equations describe the propagation of the two independent components $\vec{u}_L$ and $\vec{u}_T$, of velocities V_T and V_L. More generally, two waves propagate simultaneously: a dilatation wave (with a relative volume expansion equal to div $\vec{u}_L$) of velocity V_L and a shear wave (div $\vec{u}_T = 0$) of velocity V_T.

5.1.5. Elastic Energy Flux

Whenever an elastic wave propagates, energy is transported. In this section, we show that this transport of energy can be regarded as the flux of a Poynting vector equal to the density of elastic power per unit area, and we calculate, in the case of a plane wave, the components of the energy velocity vector, equal to the quotient of the Poynting vector divided by the energy density per unit volume.

5.1.5.1. Poynting vector

The kinetic energy due to the motion of solid matter appears in the expression 4.29 for the work done by external forces on a solid of volume v_o:

$$\delta W = \int_{v_o} (\rho g_i + \frac{\partial T_{ik}}{\partial x_k}) du_i dv + \int_{v_o} T_{ik} d(\frac{\partial u_i}{\partial x_k}) dv$$

In dynamic elastic theory, the density of force per unit volume $(\rho g_i + \partial T_{ik}/\partial x_k)$ no longer vanishes:

$$\rho g_i + \frac{\partial T_{ik}}{\partial x_k} = \rho \frac{\partial^2 u_i}{\partial t^2} .$$

Transforming the second term as in § 4.4, we obtain

$$\delta W = \int_{v_o} \rho \frac{\partial^2 u_i}{\partial t^2} du_i dv + \int_{v_o} T_{ik} dS_{ik} dv.$$

But du_i stands for the displacement during the time dt:

$du_i = \frac{\partial u_i}{\partial t} dt$ and the first term writes:

$$\int_{v_o} \rho \frac{\partial^2 u_i}{\partial t^2} \frac{\partial u_i}{\partial t} dt\ dv = \int_{v_o} \frac{\partial}{\partial t}\left[\frac{1}{2} \rho (\frac{\partial u_i}{\partial t})^2\right] dt\ dv = \int_{v_o} \frac{\partial \mathscr{E}_c}{\partial t} dt\ dv$$

where
$$\mathscr{E}_c = \frac{1}{2} \rho (\frac{\partial u_i}{\partial t})^2 \qquad (5.13)$$

is the density of kinetic energy per unit volume. The work done by the forces acting on the volume v_o is now

$$\delta W = \int_{v_o} d\mathscr{E}_c dv + \int_{v_o} d\Phi dv = \int_{v_o} d(\mathscr{E}_c + \Phi) dv$$

where $d\Phi = T_{ik} dS_{ik}$ is the potential energy variation per unit volume, i.e. the internal energy ($d\mathscr{U}$) variation for an isentropic transformation, the free energy ($d\mathscr{F}$) for an isothermal transformation. The work δW done by external forces increases the total energy $E_o = \int_{v_o} (\mathscr{E}_c + \Phi) dv$ by the amount $\delta W = dE_o$.

The energy E contained in a volume v inside the medium varies with time; the increase dE is equal to the work done by the mechanical stress $T_i(\vec{\ell})$ on every point of the boundary surface s:

$$dE = \int_s T_i(\vec{\ell})du_i ds$$

or, replacing $T_i(\vec{\ell})$ by $T_{ik}\ell_k$:

$$dE = \int_s T_{ik}\ell_k du_i ds.$$

The variation of the energy contained in v per unit time:

$$\frac{dE}{dt} = \int_s T_{ik}\frac{\partial u_i}{\partial t}\ell_k ds$$

can be put into the form:

$$\frac{dE}{dt} + \int_s P_k \ell_k ds = 0 \qquad (5.14)$$

where we have introduced a vector $\vec{P}$:

$$\boxed{P_k = - T_{ik}\frac{\partial u_i}{\partial t}\;.} \qquad (5.15)$$

Equation 5.14 shows that the energy variation within a volume v arises from the flux of the vector $\vec{P}$ across the boundary surface s. $\vec{P}$ is the Poynting vector; its direction shows the direction of energy transport; its magnitude is equal to the amount of energy traversing unit area per unit time. Denoting by V^e the energy transport velocity and by $\mathscr{E}$ the total energy density per unit volume: $P = \mathscr{E}V^e$. The energy velocity vector, parallel to $\vec{P}$, is given by

$$\vec{V^e} = \frac{\vec{P}}{\mathscr{E}}\;. \qquad (5.16)$$

5.1.5.2. <u>Energy velocity of an elastic plane wave</u>

In the case of an elastic plane wave, the displacement u_i obeys equation 5.3. Inserting

$$\frac{\partial u_i}{\partial t} = {}^o u_i F' \qquad (5.17)$$

into 5.13, the kinetic energy density becomes

$$\mathscr{E}_c = \frac{1}{2}\rho\, {}^o u_i^2 F'^2 \qquad (5.18)$$

and the potential energy density (equation 4.34):

$$\Phi = \frac{1}{2}\, c_{ijk\ell}S_{ij}S_{k\ell} = \frac{1}{2}\, c_{ijk\ell}\frac{\partial u_i}{\partial x_j}\frac{\partial u_\ell}{\partial x_k}$$

together with

$$\frac{\partial u_i}{\partial x_j} = - n_j\, {}^o u_i \frac{F'}{V} \quad \text{and} \quad \frac{\partial u_\ell}{\partial x_k} = - n_k\, {}^o u_\ell \frac{F'}{V} \qquad (5.19)$$

leads to

$$\Phi = \frac{1}{2} c_{ijk\ell} n_j n_k {}^o u_i {}^o u_\ell \frac{F'^2}{V^2}$$

Moreover, the Christoffel equation 5.4. expressed in terms of ${}^o u_i$, provides:

$$c_{ijk\ell} n_j n_k {}^o u_\ell {}^o u_i = \rho V^2 {}^o u_i^2. \tag{5.20}$$

So, the potential energy is given by

$$\Phi = \frac{1}{2} \rho {}^o u_i F'^2. \tag{5.21}$$

On comparing 5.18 and 5.21, we see that the potential energy and the kinetic energy of a plane wave are equal. The total energy density is then

$$\mathscr{E} = \rho {}^o u_i F'^2. \tag{5.22}$$

The Poynting vector

$$P_i = - T_{ij} \frac{\partial u_j}{\partial t} = - c_{ijk\ell} \frac{\partial u_\ell}{\partial x_k} \frac{\partial u_j}{\partial t}$$

can be written (from 5.17 and 5.19)

$$P_i = c_{ijk\ell} {}^o u_j {}^o u_\ell \frac{n_k}{V} F'^2.$$

The energy transport velocity is obtained on dividing P_i by the energy density $\mathscr{E}$:

$$V_i^e = \frac{c_{ijk\ell} {}^o u_j {}^o u_\ell n_k}{\rho {}^o u_i^2 V} \tag{5.23}$$

With a unit displacement vector (${}^o u_i^2 = 1$) this becomes

$$\boxed{V_i^e = c_{ijk\ell} \frac{{}^o u_j {}^o u_\ell n_k}{\rho V}} \tag{5.24}$$

The energy velocity $\vec{V^e}$ shows the direction of the energy transport, i.e. of the acoustic ray. When this ray is orthogonal to the wave planes - i.e. parallel to $\vec{n}$ - the propagation mode is said to be "pure".* These pure mode directions are very important. Crystals used

* Many authors define a "pure mode" as a wave having exactly a longitudinal or shear polarization (see reference [3]). The two definitions are identical only for a longitudinal wave (see exercise 5.2 and following example).

for devices, e.g. for delay lines, are long cylinders or long right parallelepipeds with transducers at both ends (fig. 9.2). If the beam emitted by one transducer is to be detected at the other, the Poynting vector should be parallel to the bar axis, i.e. to the wave vector, perpendicular to the ends of the crystal. The systematic investigation of these pure mode directions has been performed for cubic, hexagonal, and quadratic system crystals [4].

If we now construct the scalar product

$$\overrightarrow{V^e}.\vec{n} = V_i^e n_i = \frac{c_{ijk\ell}\,{}^o u_j\,{}^o u_\ell n_i n_k}{\rho\,{}^o u_i^2 V}$$

and if we take 5.20 into account:

$$\boxed{\overrightarrow{V^e}.\vec{n} = V.} \qquad (5.25)$$

This equation shows that the component of the energy velocity vector in the propagation direction is equal to the phase velocity of the plane wave (fig. 5.8). V^e is always greater than or equal to V.

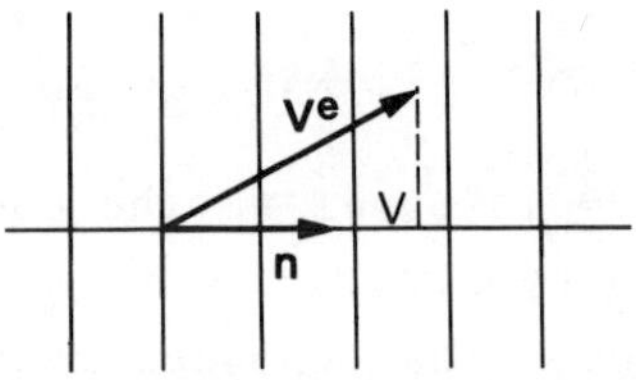

Fig. 5.8. The component of the energy velocity $\overrightarrow{V^e}$ in the propagation direction $\vec{n}$ is equal to the plane wave phase velocity.

It is worth comparing this energy transport velocity to the group velocity V^g which has been defined in section 1.1.3 through equation 1.10:

$$V^g = \frac{\partial\omega}{\partial k}$$

Here, in a three dimensional space, the group velocity components are:

$$V_j^g = \frac{\partial\omega}{\partial k_j} \quad \text{with} \quad k_j = kn_j .$$

Solving the secular equation 5.7

$$|c_{ijk\ell}n_jn_k - \rho V^2\delta_{i\ell}| = 0 \tag{5.26}$$

provides a link between the phase velocity V and the propagation direction n_j:

$$V = f(n_j).$$

Multiplying 5.26 by k^6 (i.e. every line of the determinant by k^2):

$$|c_{ijk\ell}kn_jkn_k - \rho k^2V^2\delta_{i\ell}| = 0$$

we obtain the same link between $\omega = kV$ and $k_j = kn_j$:

$$\omega = f(k_j).$$

Hence

$$V_j^g = \frac{\partial\omega}{\partial k_j} = \frac{\partial V}{\partial n_j} . \tag{5.27}$$

Let us calculate $\partial V/\partial n_j$, through the derivative of V^2 (obtained from equation 5.20):

$$2V\frac{\partial V}{\partial n_j} = 2c_{ijk\ell}\frac{n_k \overset{o}{u}_i \overset{o}{u}_\ell}{\rho \overset{o}{u}{}_i^2}$$

so

$$V_j^g = \frac{\partial V}{\partial n_j} = c_{ijk\ell}\frac{n_k \overset{o}{u}_i \overset{o}{u}_\ell}{\rho \overset{o}{u}{}_i^2 V} \tag{5.28}$$

The group velocity $\vec{V^g}$ is equal to the energy transport velocity $\vec{V^e}$ as given by 5.23.

It should be noted that the equation $\vec{V^e}.\vec{n} = V$ results from Euler's theorem relating to homogeneous functions. If f is a homogeneous function of the variables n_i with rank p:

$$n_i\frac{\partial f}{\partial n_i} = pf .$$

The equation $\omega = Vk = f(kn_i)$ shows that $V = f(n_i)$ is a homogeneous function of degree 1:

$$n_i\frac{\partial V}{\partial n_i} = V \quad\Longrightarrow\quad \vec{n}.\vec{V^e} = V.$$

The same theorem, when applied to $V_i^e = \dfrac{\partial V}{\partial n_i}$ which is a homogeneous function of degree 0, shows that

$$n_k \frac{\partial V_i^e}{\partial n_k} = 0.$$

Example. As an illustration of this concept of energy transport velocity, let us look for the acoustic ray direction for propagation along a triad axis. This is the case of sapphire (class $\bar{3}m$). The $c_{\alpha\beta}$ matrix is given in fig. 4.6. Since $n_k = \delta_{k3}$, the energy transport velocity, derived from equation 5.24, is

$$V_i^e = \frac{c_{ij3\ell}\,{}^o u_j\,{}^o u_\ell}{\rho V} \quad . \tag{5.29}$$

For the longitudinal wave propagating with velocity $V_1 = \sqrt{c_{33}/\rho}$ (§ 5.1.3)

$$ {}^o u_i = \delta_{i3} \quad \Longrightarrow \quad \vec{V^e} \begin{cases} \dfrac{c_{53}}{\rho V_1} = 0 \\ \dfrac{c_{43}}{\rho V_1} = 0 \\ \dfrac{c_{33}}{\rho V_1} = V_1 \end{cases}$$

This is a pure longitudinal mode.

For the degenerate transverse wave with velocity $V_2 = \sqrt{c_{44}/\rho}$, the displacement is arbitrary within the x_1-x_2 plane (fig. 5.9):

$$u_1 = \cos\phi \quad u_2 = \sin\phi \quad u_3 = 0.$$

Expanding equation (5.29), we obtain:

$$\rho V_2 V_i^e = c_{i131}\cos^2\phi + (c_{i132} + c_{i231})\cos\phi\,\sin\phi + c_{i232}\sin^2\phi$$

which provides the energy velocity components:

$$\begin{cases} \rho V_2 V_1^e = c_{15}\cos^2\phi + (c_{14}+c_{56})\cos\phi\,\sin\phi + c_{46}\sin^2\phi \\ \rho V_2 V_2^e = c_{56}\cos^2\phi + (c_{46}+c_{25})\cos\phi\,\sin\phi + c_{24}\sin^2\phi \\ \rho V_2 V_3^e = c_{55}\cos^2\phi + (c_{54}+c_{45})\cos\phi\,\sin\phi + c_{44}\sin^2\phi \end{cases}$$

Since c_{15}, c_{46}, c_{25} and c_{45} vanish, and since $c_{56} = c_{14} = -c_{24}$ as well as $c_{55} = c_{44}$,

$$\begin{cases} \rho V_2 V_1^e = c_{14}\sin 2\phi \\ \rho V_2 V_2^e = c_{14}\cos 2\phi \\ \rho V_2 V_3^e = c_{44} \end{cases} \Longrightarrow \vec{V^e} \begin{cases} V_1^e = \dfrac{c_{14}}{\rho V_2}\sin 2\phi \\ V_2^e = \dfrac{c_{14}}{\rho V_2}\cos 2\phi \\ V_3^e = \dfrac{c_{44}}{\rho V_2} \end{cases}$$

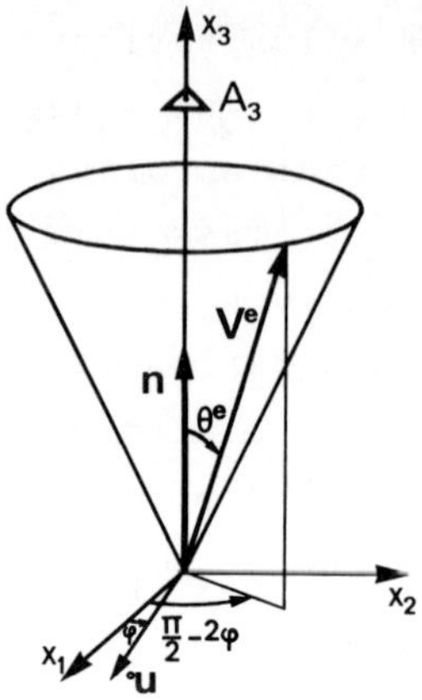

Fig. 5.9. Propagation along a triad axis. The degenerate shear wave has polarization $\overset{o}{\vec{u}}$ and its energy velocity $\vec{V^e}$ is at a fixed angle θ^e with respect to the propagation direction $\vec{n}$ (i.e. the triad axis). For sapphire, $\theta^e = 9^o$.

The angle θ^e between $\vec{V^e}$ and the triad axis does not depend upon ϕ:

$$\tan \theta^e = \frac{|c_{14}|}{c_{44}} \qquad (5.30)$$

As a result, the energy velocity vector lies along a cone, the vertex angle of which is θ^e, when the transverse wave polarization turns in the x_1-x_2 plane (fig. 5.9). For sapphire:

$c_{14} = -2.35\ 10^{10} N/m^2 \quad c_{44} = 14.74\ 10^{10} N/m^2 \Longrightarrow \theta^e = 9^o$.

The triad axis provides the only case of a symmetry axis which is not a pure mode propagation direction. For n = 2, 4 and 6 the c_{14} modulus vanishes so that $\theta^e = 0$.

5.1.6. Characteristic Surfaces

In the same way that the index surface and the wave surface are used in crystal optics there are several characteristic surfaces used to illustrate the phenomena of elastic wave propagation in crystals.

5.1.6.1. Definitions and properties

From the origin O, let us draw a vector which is perpendicular to the wave plane and has length V, where V is the phase velocity. When the propagation direction is varied the end of the vector $\vec{\mathcal{V}} = V\vec{n}$ describes a surface known as the velocity surface. It generally has three sheets, one for the quasi-longitudinal wave (V_1 sheet) and two for the quasi-shear waves (V_2 and V_3 sheets). Normally, the quasi-longitudinal wave travels faster than the quasi-shear waves,* so that the V_1 sheet contains the V_2 and V_3 sheets without there being any intersection. On the other hand, the V_2 and V_3 sheets have some points in common in the directions of the acoustic axes - e.g. along a p-fold axis with $p > 2$. For an isotropic medium, the velocity surface consists of two spheres, of radii V_L and V_T.

The slowness surface (L) is the locus of the ends of the vectors $\vec{L} = \vec{n}/V$. Since $\vec{L}$ and $\vec{\mathcal{V}}$ are parallel, with $LV = 1$, the slowness and velocity surfaces are related by an inversion with respect to O. This slowness surface plays the same role as the index surface in optics, and is a key concept in dealing with reflexion or refraction problems, on which we shall focus our attention in § 5.2. In addition, this surface provides the energy displacement direction. If $\vec{L}$ and $\vec{L} + \vec{dL}$ are the slowness radii for $\vec{n}$ and $\vec{n} + d\vec{n}$ then:

$$\vec{dL} = \frac{\partial \vec{L}}{\partial n_k} dn_k$$

is a tangent to the slowness surface (fig. 5.10). And the contraction of

$$\frac{\partial L_i}{\partial n_k} = \frac{\partial (\frac{n_i}{V})}{\partial n_k} = \frac{\delta_{ik}}{V} - \frac{n_i}{V^2} \frac{\partial V}{\partial n_k}$$

with the energy velocity components V_i^e yields:

* There are a few exceptions, such as TeO_2 (fig. 5.20).

$$V_i^e \frac{\partial L_i}{\partial n_k} = \frac{1}{V} (V_k^e - \frac{V_i^e n_i}{V} \frac{\partial V}{\partial n_k}).$$

Making use of equations 5.25: $V_i^e n_i = V$ and 5.28 $V_k^e = V_k^g = \frac{\partial V}{\partial n_k}$ one is left with

$$V_i^e \frac{\partial L_i}{\partial n_k} = 0$$

and

$$\vec{V^e}.d\vec{L} = V_i^e \frac{\partial L_i}{\partial n_k} dn_k = 0 \text{ for any } d\vec{L}.$$

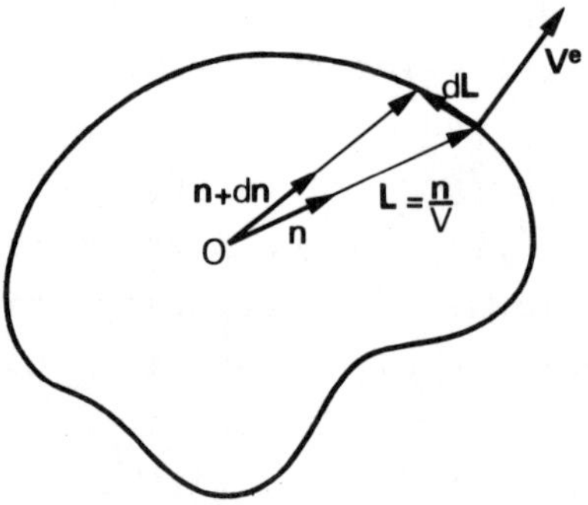

Fig. 5.10. Slowness surface and energy velocity. The energy velocity is always orthogonal to the slowness surface:
$$\vec{V^e}.d\vec{L} = 0.$$

Since this orthogonality relationship is valid for all $d\vec{L}$ vectors in a plane tangent to the slowness surface, the energy velocity is always normal to the slowness surface. As before, this surface consists of 3 sheets but now, the quasi-longitudinal sheet is entirely within both quasi-transverse sheets.

The third surface is the wave surface and contains the end of the energy velocity vector $\vec{V^e}$. The radius vector joining 0 to a point of this wave surface R shows the distance travelled by the elastic energy in unit time. So the wave surface also shows the set of the points reached after unit time interval by a perturbation emitted by a point source. It is also an equi-phase surface, since all of its points start vibrating at the same time. This surface has other remarkable features; the propagation direction of a plane wave with energy velocity $\vec{V^e}$ is normal to the wave surface at the corresponding point. The equations

$$\vec{V^e}.\vec{L} = 1 \quad \text{and} \quad \vec{V^e}.d\vec{L} = 0$$

imply:

$$\vec{L}.d\vec{V^e} = 0$$

or

$$\vec{n}.d\vec{V^e} = 0 \quad \text{since} \quad \vec{L} = \frac{\vec{n}}{V} .$$

So $\vec{n}$ is perpendicular to all $d\vec{V^e}$ vectors within the tangent plane (fig. 5.11).

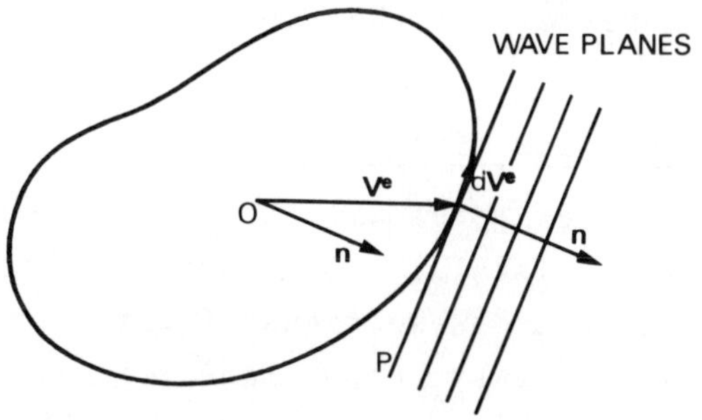

Fig. 5.11. Wave surface and propagation direction. The wave plane P is tangential to the wave surface at the extremity of the energy velocity vector $\vec{V^e}$.

There is a straightforward connexion between the wave surface and the velocity surface (fig. 5.12): from A we draw the plane tangential to the wave surface R. The equation $\vec{V^e}.\vec{n} = V$ shows that the phase velocity V is the projection of $\vec{V^e}$ onto the normal direction $\vec{n}$. Therefore the extremity B of $\vec{\mathcal{V}}$ is the foot of the perpendicular dropped from O onto the plane P, and the velocity surface is the pode of the wave surface with respect to the origin O. Conversely the wave surface is the envelope of the planes perpendicular to the $\vec{\mathcal{V}}$ vectors.

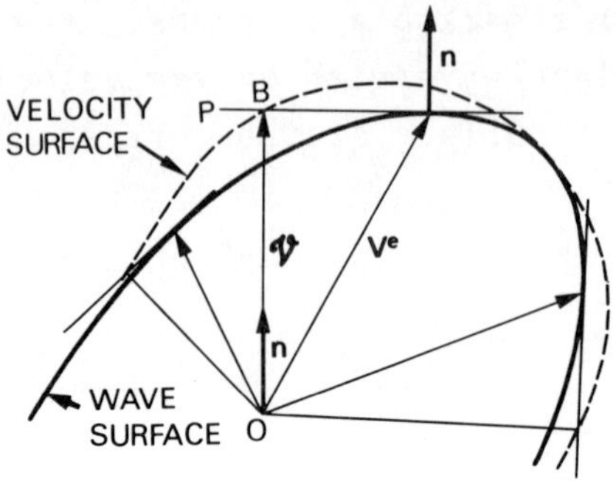

Fig. 5.12. Velocity surface and wave surface. The points of the velocity surface are the orthogonal projections of O onto the tangent planes of the wave surface. Conversely, the wave surface is the envelope of all planes orthogonal to $\vec{\mathcal{V}}$ at the extremity of $\vec{\mathcal{V}}$.

5.1.6.2. Some examples of slowness surfaces

The following examples will describe the slowness surfaces of a few crystals. On a sheet of paper, we can only draw plane cross-sections of this surface. If the section plane is a symmetry plane for elastic properties (a mirror or a plane perpendicular to an even order axis), the surface is symmetric with respect to this plane which thus contains the normal. The propagation direction is defined by two polar angles θ and ϕ, as shown in fig. 5.13.

$$n_1 = \sin\theta\cos\phi \quad n_2 = \sin\theta\sin\phi \quad n_3 = \cos\theta. \quad (5.31$$

It is not generally possible to solve the secular equation analytically. We can do it only for special directions or planes, when in the $\Gamma_{i\ell}$ tensor at least two off diagonal components out of three vanish. The determinant is then factorised as a product of polynomials of degree 1 or 2 in V^2 (§ 3.5).

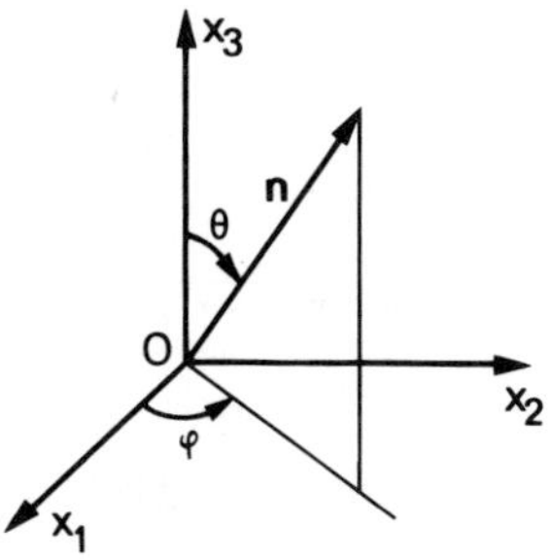

Fig. 5.13. The propagation direction $\vec{n}$ is determined by angles θ and ϕ.

a) Cubic system. The expansion 5.8 of the propagation tensor components $\Gamma_{i\ell}$ is simplified on taking 4.45 into account:

$$\Gamma_{11} = c_{11}n_1^2 + c_{44}(n_2^2 + n_3^2)$$

$$\Gamma_{21} = \Gamma_{12} = (c_{12} + c_{44})n_1n_2$$

$$\Gamma_{31} = \Gamma_{13} = (c_{12} + c_{44})n_1n_3 \qquad (5.32)$$

$$\Gamma_{22} = c_{44}(n_1^2 + n_3^2) + c_{11}n_2^2$$

$$\Gamma_{32} = \Gamma_{23} = (c_{12} + c_{44})n_2n_3$$

$$\Gamma_{33} = c_{44}(n_1^2 + n_2^2) + c_{11}n_3^2.$$

In addition, if the propagation vector is parallel to one face of the cube, say (001) then

$$n_1 = \cos\phi, \quad n_2 = \sin\phi, \quad n_3 = 0 \Longrightarrow \Gamma_{13} = \Gamma_{23} = 0$$

and the $\Gamma_{i\ell}$ tensor is as follows:

$$\Gamma_{i\ell} = \begin{vmatrix} \Gamma_{11} & \Gamma_{12} & 0 \\ \Gamma_{12} & \Gamma_{22} & 0 \\ 0 & 0 & \Gamma_{33} \end{vmatrix}$$

where

$$\Gamma_{11} = c_{11}\cos^2\phi + c_{44}\sin^2\phi \qquad \Gamma_{12} = (c_{12}+c_{44})\cos\phi\,\sin\phi$$

$$\Gamma_{22} = c_{11}\sin^2\phi + c_{44}\cos^2\phi \qquad \Gamma_{33} = c_{44}.$$

For any such propagation direction, there exists a shear wave with x_3 polarization and velocity $V_3 = \sqrt{c_{44}/\rho}$ (for any ϕ).

The other two velocities are obtained on solving the eigenvalue equation for $\gamma = \rho V^2$; then according to 3.24:

$$\gamma_{\frac{1}{2}} = \frac{\Gamma_{11} + \Gamma_{22}}{2} \pm \frac{1}{2} \sqrt{(\Gamma_{11} - \Gamma_{22})^2 + 4(\Gamma_{12})^2}. \quad (5.33)$$

V_1 and V_2 can be expressed in terms of ϕ:

$$2\rho V_1^2 = c_{11}+c_{44}+\sqrt{(c_{11}-c_{44})^2\cos^2 2\phi+(c_{12}+c_{44})^2\sin^2 2\phi} \quad (5.34)$$

$$2\rho V_2^2 = c_{11}+c_{44}-\sqrt{(c_{11}-c_{44})^2\cos^2 2\phi+(c_{12}+c_{44})^2\sin^2\ \phi} \quad (5.35)$$

The polarization is generally neither longitudinal nor transverse. When $\phi = 0$ or $\pi/2$, the velocity $V_1 = \sqrt{c_{11}/\rho}$ corresponds to a longitudinal wave and the other velocity $V_2 = \sqrt{c_{44}/\rho}$, to a shear wave. Otherwise, (5.34) and (5.35) provide the velocities V_1 and V_2 for the quasi-longitudinal and quasi-shear modes.

The slowness curves of silicon are shown in fig. 5.14, using the elastic constants of table 4.7. They exhibit the maximal cubic system symmetry for the stiffness tensor (which is the same for the five classes of this system), is invariant under point group operations. The quasi-shear wave velocity has an extremum for directions [100] and [110]:

$$V_2[100] = \sqrt{c_{44}/\rho} \qquad V_2[110] = \sqrt{c_{11}-c_{12}/2\rho}.$$

The ratio of the above values

$$\frac{V_2[100]}{V_2[110]} = \sqrt{2c_{44}/(c_{11}-c_{12})}$$

characterizes the mechanical crystal anisotropy. The curves for silicon correspond to the most frequent situation where the anisotropy factor:

$$A = \frac{2c_{44}}{c_{11}-c_{12}} \quad (5.36)$$

is greater than 1 (A(si) = 1.565). When A is less than 1, the curve distortion is reversed; this is the case for bismuth-germanium oxide ($Bi_{12}GeO_{20}$), a piezoelectric material (fig. 6.6).

It should be noted that A = 1 for an isotropic medium.

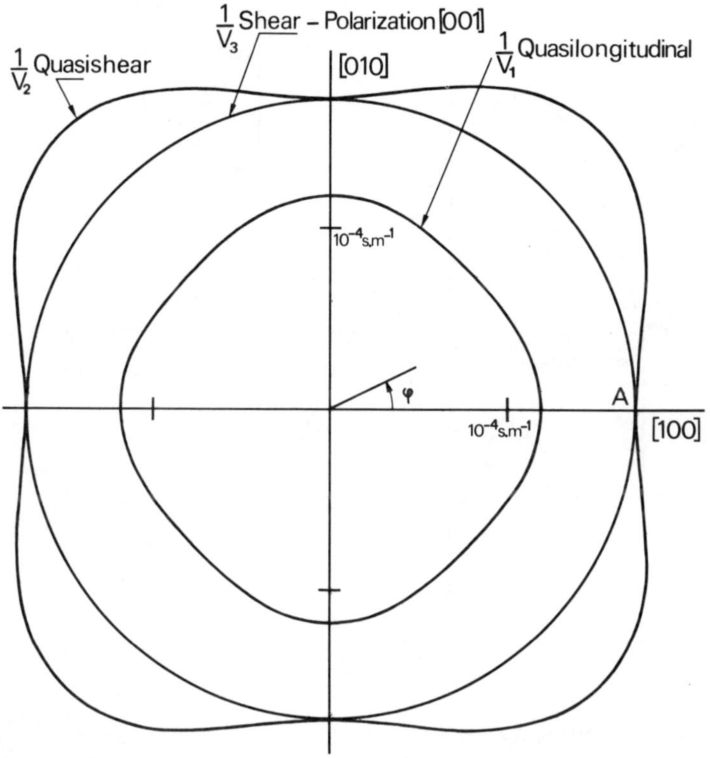

Fig. 5.14. Cubic system. Cross section of the slowness surface of silicon in the (001) plane. Point A corresponds to a velocity of 5,843 m/s.

The polarization pattern also depends on the propagation direction. The quasi-longitudinal and quasi-shear displacement vectors make angles β_1 and β_2 with the x_1 direction, which are given by (3.25):

$$\tan 2\beta = \frac{2\Gamma_{12}}{\Gamma_{11}-\Gamma_{22}}$$

or, in the (001) plane of a cubic crystal

$$\tan 2\beta = \frac{c_{12}+c_{44}}{c_{11}-c_{44}} \tan 2\phi. \tag{5.37}$$

Fig. 5.15 shows the deviation $(\beta_1 - \phi)$ of the quasi-longitudinal displacement vector from the propagation direction in the (001) plane, for silicon. There is a maximum deviation at $\phi_M = 18^\circ 52'$ such that

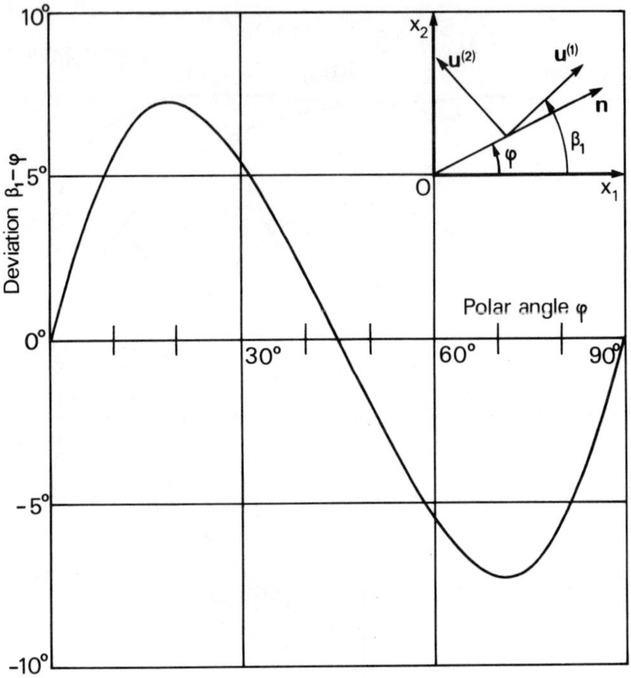

Fig. 5.15. Cubic system. Deviation ($\beta_1 - \phi$) of the quasi-longitudinal displacement vector from the propagation direction in the (001) plane of silicon.

$$\tan 2\phi_M = \sqrt{\frac{c_{11}-c_{44}}{c_{12}+c_{44}}}$$

The maximum value ($7^o16'$) is equal to

$$\beta_1 - \phi_M = \frac{\pi}{4} - 2\phi_M.$$

Let us now focus on propagation in the diagonal plane ($1\bar{1}0$). It is convenient to use a reference frame $Ox_1'x_2'x_3$, derived from $Ox_1x_2x_3$by a $\pi/4$ rotation about Ox_3 (fig. 5.16). The stiffness matrix $c'_{\alpha\beta}$ is analogous to the matrix for a tetragonal system crystal, with $c'_{16} = 0$ since Ox_1' is a diad axis. The expansion (5.8) of the

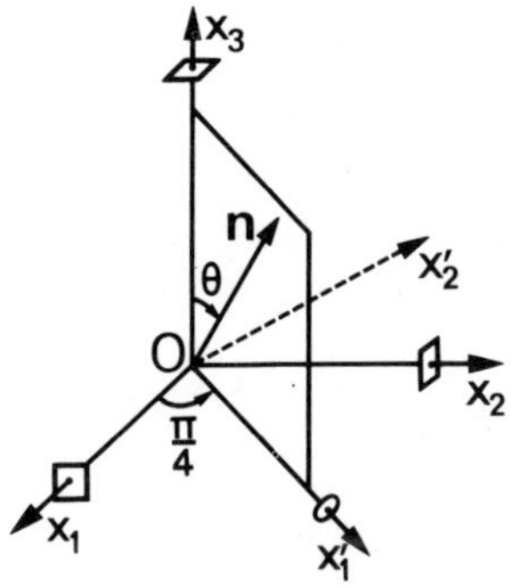

Fig. 5.16. Cubic system. New reference frame $Ox_1'x_2'x_3$ used to calculate the velocities in the diagonal plane $(1\bar{1}0)$.

propagation tensor components $\Gamma_{i\ell}'$, remembering that $n_1' = \sin\theta$, $n_2' = 0$, $n_3' = \cos\theta$ and taking (4.55) into account, reduces to

$$\Gamma_{i\ell}' = \begin{vmatrix} \Gamma_{11}' & 0 & \Gamma_{13}' \\ 0 & \Gamma_{22}' & 0 \\ \Gamma_{13}' & 0 & \Gamma_{33}' \end{vmatrix}$$

where

$$\Gamma_{11}' = c_{11}' \sin^2\theta + c_{44}' \cos^2\theta \qquad \Gamma_{13}' = \frac{c_{13}' + c_{44}'}{2} \sin 2\theta$$

$$\Gamma_{22}' = c_{66}' \sin^2\theta + c_{44}' \cos^2\theta \qquad \Gamma_{33}' = c_{44}'\sin^2\theta + c_{33}'\cos^2\theta.$$

Therefore a shear wave, with Ox_2' polarization i.e. along $[1\bar{1}0]$, propagates with velocity

$$V_3 = \sqrt{\frac{\Gamma_{22}'}{\rho}} = \sqrt{\frac{c_{66}' \sin^2\theta + c_{44}' \cos^2\theta}{\rho}}$$

There are two other waves, one quasi-longitudinal and one quasi-shear, with velocities V_1 and V_2 such that

$$2\rho V_{\frac{1}{2}}^2 = \Gamma_{11}' + \Gamma_{33}' \pm \sqrt{(\Gamma_{11}' - \Gamma_{33}')^2 + 4(\Gamma_{13}')^2}$$

On replacing the Γ's by their values:

$$2\rho V_{\frac{1}{2}}^2 = c_{44}' + c_{11}' \sin^2\theta + c_{33}' \cos^2\theta$$

$$\pm\sqrt{[(c_{11}' - c_{44}')\sin^2\theta + (c_{44}' - c_{33}')\cos^2\theta]^2 + (c_{13}' + c_{44}')^2 \sin^2 2\theta}.$$

But the elastic constants are always given in the standard crystal reference frame and we have to express the coefficients $c'_{\alpha\beta}$ in terms of the coefficients $c_{\alpha\beta}$ through the reference frame transformation matrix

$$\alpha_i^j = \begin{pmatrix} 1/\sqrt{2} & 1/\sqrt{2} & 0 \\ -1/\sqrt{2} & 1/\sqrt{2} & 0 \\ 0 & 0 & 1 \end{pmatrix} .$$

The equation

$$c'_{ijk\ell} = \alpha_i^p \alpha_j^q \alpha_k^r \alpha_\ell^s c_{pqrs}$$

leads to

$$c'_{3333} = c_{3333} \Rightarrow \quad c'_{33} = c_{33} = c_{11}$$

$$c'_{2323} = \alpha_2^1\alpha_2^1 c_{1313} + \alpha_2^2\alpha_2^2 c_{2323} = (\alpha_2^1\alpha_2^1 + \alpha_2^2\alpha_2^2)c_{44} \Rightarrow c'_{44} = c_{44}$$

$$c'_{1133} = \alpha_1^1\alpha_1^1 c_{1133} + \alpha_1^2\alpha_1^2 c_{2233} = (\alpha_1^1\alpha_1^1 + \alpha_1^2\alpha_1^2)c_{13} \Rightarrow c'_{13} = c_{13} = c_{12}$$

$$c'_{1212} = \frac{1}{4} c_{11} + \frac{1}{4} c_{22} - \frac{1}{2} c_{12} \Rightarrow c'_{66} = \frac{c_{11} - c_{12}}{2}$$

$$c'_{1111} = \frac{1}{4} c_{11} + \frac{1}{4} c_{22} + c_{66} + \frac{1}{2} c_{12} \Rightarrow c'_{11} = \frac{c_{11} + c_{12}}{2} + c_{44} .$$

Hence the final expressions for the velocities in terms of the c constants are

- for the shear wave

$$V_3 = \sqrt{\frac{c_{44}}{\rho} \cos^2\theta + \frac{c_{11} - c_{12}}{2\rho} \sin^2\theta} . \qquad (5.38)$$

- for the quasi longitudinal (V_1) and quasi shear (V_2) waves:

$$2\rho V_{\frac{1}{2}}^2 = c_{44} + (c_{44} + \frac{c_{11}+c_{12}}{2})\sin^2\theta + c_{11} \cos^2\theta \qquad (5.39)$$

$$\pm \sqrt{[\frac{c_{11}+c_{12}}{2} \sin^2\theta + (c_{44} - c_{11})\cos^2\theta]^2 + (c_{12}+c_{44})^2 \sin^2 2\theta} .$$

The slowness curves for $1/V_1$, $1/V_2$, $1/V_3$ are plotted in fig. 5.17 for the case of silicon.

The following table gives the expressions for the velocities and the polarizations in some specific propagation directions, as well as numerical data for a few cubic system crystals: silicon, gallium arsenide (GaAs), aluminium, gold and platinum.

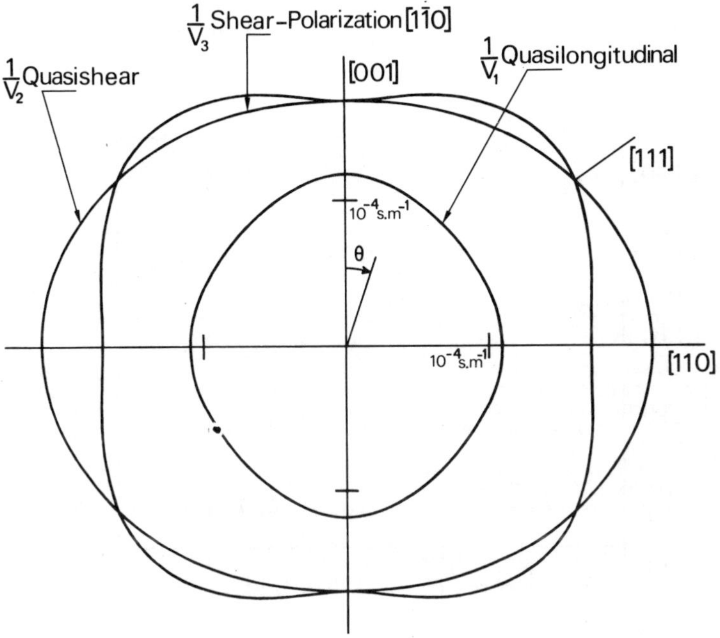

Fig. 5.17. Cubic system. Cross section of the slowness surface of silicon in the (110) plane.

b) Tetragonal system. Making use of 5.8 and 4.55, we first have the Christoffel tensor

$$\begin{aligned}
\Gamma_{11} &= c_{11}n_1^2 + c_{66}n_2^2 + c_{44}n_3^2 + 2c_{16}n_1n_2 \\
\Gamma_{12} &= c_{16}(n_1^2 - n_2^2) + (c_{12} + c_{66})n_1n_2 \\
\Gamma_{13} &= (c_{13} + c_{44})n_1n_3 \\
\Gamma_{22} &= c_{66}n_1^2 + c_{11}n_2^2 + c_{44}n_3^2 - 2c_{16}n_1n_2 \\
\Gamma_{23} &= (c_{13} + c_{44})n_2n_3 \\
\Gamma_{33} &= c_{44}(n_1^2 + n_2^2) + c_{33}n_3^2.
\end{aligned} \qquad (5.40)$$

If the propagation takes place in a plane perpendicular to the tetrad axis ($n_1 = \cos\phi$, $n_2 = \sin\phi$, $n_3 = 0$) the $\Gamma_{i\ell}$ tensor is as follows:

Table 5.18. Cubic system.

Direction of propagation	Polarisation	Velocity	Examples (ms^{-1})				
			Si	GaAs*	Al	Au	Pt
[100]	[100] (L)	$\sqrt{c_{11}/\rho}$	8433	4735	6300	3158	4025
	plan (100) (T)	$\sqrt{c_{44}/\rho}$	5843	3347	3236	1482	1891
[110]	[110] (L)	$\sqrt{\frac{c_{11}+c_{12}+2c_{44}}{2\rho}}$	9134	5242	6450	3376	4189
	[$1\bar{1}0$] (T)	$\sqrt{\frac{c_{11}-c_{12}}{2\rho}}$	4673	2478	2935	874	1498
	[001] (T)	$\sqrt{\frac{c_{44}}{\rho}}$	5843	3347	3236	1482	1891
[111]	[111] (L)	$\sqrt{\frac{c_{11}+2c_{12}+4c_{44}}{3\rho}}$	9360	5401	6496	3447	4242
	plan (111) (T)	$\sqrt{\frac{c_{11}-c_{12}+c_{44}}{3\rho}}$	5085	2798	3039	1114	1639

* Piezoelectricity has been ignored in the calculation of the velocities.

$$\Gamma_{i\ell} = \begin{bmatrix} \Gamma_{11} & \Gamma_{12} & 0 \\ \Gamma_{12} & \Gamma_{22} & 0 \\ 0 & 0 & \Gamma_{33} \end{bmatrix}$$

with

$$\Gamma_{11} = c_{11}\cos^2\phi + c_{66}\sin^2\phi + c_{16}\sin 2\phi$$

$$\Gamma_{12} = c_{16}\cos 2\phi + (c_{12} + c_{66})\frac{\sin 2\phi}{2}$$

$$\Gamma_{22} = c_{66}\cos^2\phi + c_{11}\sin^2\phi - c_{16}\sin 2\phi$$

$$\Gamma_{33} = c_{44}.$$

The secular equation being the same as for the cubic system, there is a shear wave, polarized along x_3, with velocity $V_3 = \sqrt{c_{44}/\rho}$, and two waves polarized in the (001) plane, the velocities of which are given by

$$2\rho V_{\frac{1}{2}}^2 = \Gamma_{11} + \Gamma_{22} \pm \sqrt{(\Gamma_{11} - \Gamma_{22})^2 + 4(\Gamma_{12})^2}$$

Replacing Γ_{11}, Γ_{22}, Γ_{12} by expressions in terms of ϕ, we get

$$2\rho V_{\frac{1}{2}}^2 = c_{11} + c_{66}$$

$$\pm\sqrt{[(c_{11}-c_{66})\cos 2\phi+2c_{16}\sin 2\phi]^2+[(c_{12}+c_{66})\sin 2\phi+2c_{16}\cos 2\phi]^2}$$

or

$$2\rho V_{\frac{1}{2}}^2 = c_{11} + c_{66} \qquad (5.41)$$

$$\pm\sqrt{(c_{11}-c_{66})^2\cos^2 2\phi+(c_{12}+c_{66})^2\sin^2 2\phi+2c_{16}(c_{11}+c_{12})\sin 4\phi+4c_{16}^2}.$$

The velocity V_1 with the + sign, corresponds to the quasi-longitudinal wave, and V_2 (with the - sign) to the quasi-shear wave.

For crystal classes 422, 4mm, $\bar{4}$2m, 4/mmm, the c_{16} modulus vanishes and the above equation becomes

$$2\rho V_{\frac{1}{2}}^2 = c_{11}+c_{66}\pm\sqrt{(c_{11}-c_{66})^2\cos^2 2\phi+(c_{12}+c_{66})^2\sin^2 2\phi}. \qquad (5.42)$$

This is the case of rutile (TiO_2, class 4/mmm), for which we have drawn the slowness curves in fig. 5.19, using the data of (4.7). These curves show a strong anisotropy because of the high value of the ratio

$$A = \frac{2c_{66}}{c_{11} - c_{12}} = 4.$$

The quasi shear wave velocity for the [100] direction is twice that for the [110] direction.

The table below gives the velocities for a few crystals having a vanishing c_{16}: rutile (TiO_2), paratellurite (TeO_2), indium (In).

For classes 4, $\bar{4}$, 4/m, c_{16} does not vanish. Fig. 5.21 shows the slowness curves in the (001) plane for lead molybdate (PbM_0O_4, class 4/m, see table 4.7 for constants). The outline is the same as for rutile, except for a tilt of angle ϕ_0 in the (001) plane. This is by no means a coincidence: this comes from the fact that a change of reference frame may cause c_{16} to vanish. In fact (see exercise 5.5) c'_{16} vanishes in the frame $Ox'_1x'_2x_3$ defined by $(Ox_1, Ox'_1) = \phi_0$ where:

$$\tan 4\phi_0 = \frac{4c_{16}}{c_{11} - c_{12} - 2c_{66}}. \qquad (5.43)$$

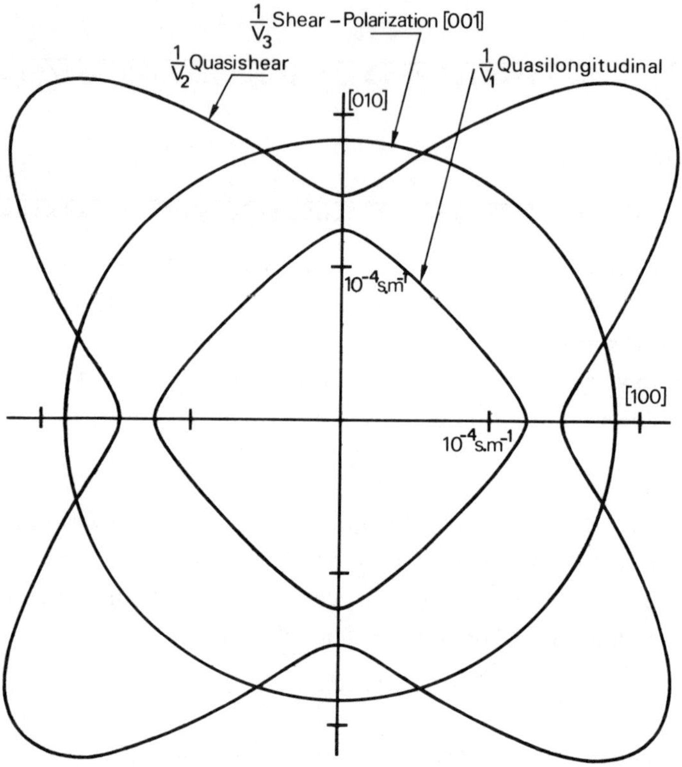

Fig. 5.19. Tetragonal system, classes 422, 4mm, $\bar{4}$2m, 4/mmm. Cross section of the slowness surface of rutile (4/mm) by plane (001).

In the Ox_1' and Ox_2' directions, and along their bisectors, the waves polarized in the (001) plane are longitudinal or transverse. In both cases, these are pure modes. For lead molybdate $\phi_0 = -16^{\circ}$ and the pure mode directions make an angle of -16° or of $29^{\circ} = 45^{\circ} - 16^{\circ}$ with the [100] crystal axis.

In summary, for this last category of crystals, we finally have two methods for calculating the velocities:

- start from the $c_{\alpha\beta}$ in the crystal axis and use the complete equations (5.41).

Table 5.20. Tetragonal system. Classes 422, 4mm, $\bar{4}2m$, 4/mmm.

Direction of Propagation	Polarization	Velocity	Examples (m/s)		
			TiO_2	TeO_2	In
[100]	[100] (L)	$\sqrt{c_{11}/\rho}$	8014	3050	2495
	[010] (T)	$\sqrt{c_{66}/\rho}$	6756	3317	1290
	[001] (T)	$\sqrt{c_{44}/\rho}$	5424	2100	946
[110]	[110] (L)	$\sqrt{\frac{c_{11}+c_{12}+2c_{66}}{2\rho}}$	9923	4663	2740
	[1$\bar{1}$0] (T)	$\sqrt{\frac{c_{11}-c_{12}}{2\rho}}$	3378	616	603
	[001] (T)	$\sqrt{c_{44}/\rho}$	5424	2100	946
[001]	[001] (L)	$\sqrt{c_{33}/\rho}$	10671	4200	2490
	plan(001)(T)	$\sqrt{c_{44}/\rho}$	5424	2100	946

For paratellurite, the shear wave in the [100] direction is faster than the longitudinal wave.

- rotate the axes so as to cause c_{16} to vanish and use the reduced equations (5.42) with the constants $c'_{\alpha\beta}$ together with the angle ϕ'.

The second procedure has the advantage of providing the pure mode directions and simple expressions for the velocities in those major directions. For example, along Ox'_1, $V_1 = \sqrt{c'_{11}/\rho}$ and $V_2 = \sqrt{c'_{66}/\rho}$. For lead molybdate, this gives $V_1 = 3780\ ms^{-1}$ and $V_2 = 2460\ ms^{-1}$.

c) Hexagonal system. In a crystal belonging to the hexagonal system, all planes that contain the 6-fold axis (vertical planes) are equivalent as far as elastic properties are concerned (see exercise 4.4). The 6-fold axis is thus an axis of revolution for all characteristic surfaces. For the same reason, in a horizontal plane, velocities do not depend on the propagation direction.

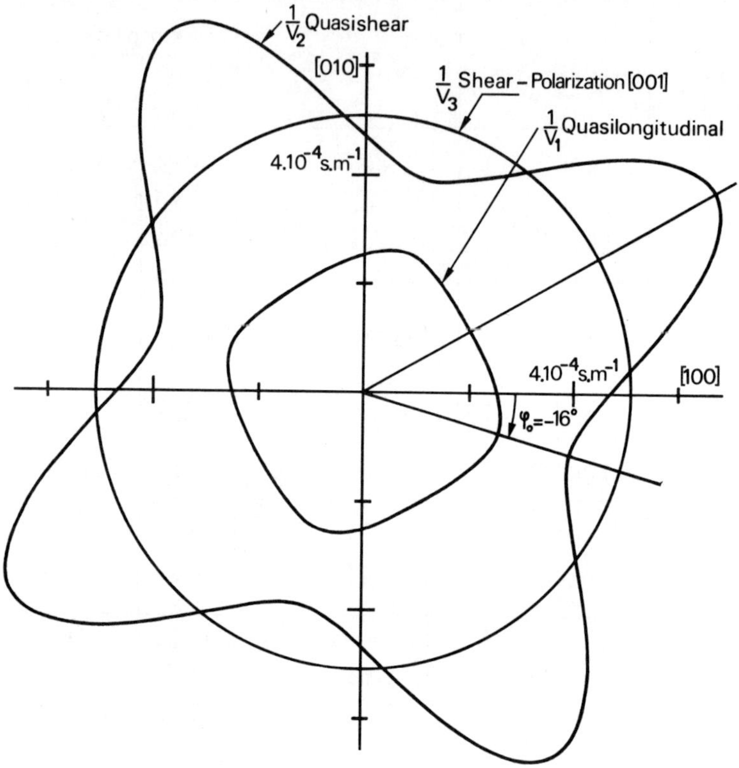

Fig. 5.21. Tetragonal system. Classes 4, $\bar{4}$, 4/m. Cross section of the slowness surface of lead molybdate (4/m) by plane (001).

In order to calculate these, we first write down the $\Gamma_{i\ell}$ tensor for the hexagonal system:

$$\begin{aligned}
\Gamma_{11} &= c_{11}n_1^2 + c_{66}n_2^2 + c_{44}n_3^2 \\
\Gamma_{12} &= (c_{12} + c_{66})n_1n_2 \\
\Gamma_{13} &= (c_{13} + c_{44})n_1n_3 \\
\Gamma_{22} &= c_{66}n_1^2 + c_{11}n_2^2 + c_{44}n_3^2 \qquad \text{with } c_{66} = \frac{c_{11}-c_{12}}{2} \qquad (5.44) \\
\Gamma_{23} &= (c_{13} + c_{44})n_2n_3 \\
\Gamma_{33} &= c_{44}(n_1^2 + n_2^2) + c_{33}n_3^2.
\end{aligned}$$

We choose the [100] direction, for which Γ is diagonal:

$$\Gamma = \begin{vmatrix} c_{11} & 0 & 0 \\ 0 & c_{66} & 0 \\ 0 & 0 & c_{44} \end{vmatrix} .$$

Therefore, three waves propagate in the (001) plane:
- a longitudinal wave of velocity $V_1 = \sqrt{c_{11}/\rho}$;
- a shear wave, polarized in the (001) plane, of velocity $V_2 = \sqrt{c_{66}/\rho}$;
- a shear wave, polarized along the 6-fold axis, of velocity $V_3 = \sqrt{c_{44}/\rho}$.

The slowness curves are three circles with radii $\sqrt{\rho/c_{11}}$, $\sqrt{\rho/c_{66}}$, $\sqrt{\rho/c_{44}}$.

In a vertical plane, the velocities depend on the angle θ between the 6-fold axis (Ox_3) and the propagation direction $\vec{n}$. Let us choose the x_2x_3 plane:

$$n_1 = 0 \quad n_2 = \sin\theta \quad n_3 = \cos\theta \implies \Gamma_{12} = \Gamma_{13} = 0.$$

The propagation tensor is written

$$\Gamma = \begin{vmatrix} \Gamma_{11} & 0 & 0 \\ 0 & \Gamma_{22} & \Gamma_{23} \\ 0 & \Gamma_{23} & \Gamma_{33} \end{vmatrix}$$

with

$$\Gamma_{11} = c_{66}\sin^2\theta + c_{44}\cos^2\theta \qquad \Gamma_{22} = c_{11}\sin^2\theta + c_{44}\cos^2\theta$$

$$\Gamma_{23} = (c_{13} + c_{44})\sin\theta\cos\theta \qquad \Gamma_{33} = c_{44}\sin^2\theta + c_{33}\cos^2\theta \tag{5.45}$$

A shear wave, the polarization of which is orthogonal to the vertical plane, propagates with velocity

$$V_3 = \sqrt{\frac{\Gamma_{11}}{\rho}} \quad \sqrt{\frac{c_{66}\sin^2\theta + c_{44}\cos^2\theta}{\rho}} \quad \text{with} \quad c_{66} = \frac{c_{11}-c_{12}}{2} \tag{5.46}$$

For the other two modes, with polarization in the x_2x_3 plane, the velocities are given by

$$2\rho V_{\frac{1}{2}}^2 = \Gamma_{22} + \Gamma_{33} \pm \sqrt{(\Gamma_{22} - \Gamma_{33})^2 + 4\Gamma_{23}^2}.$$

Inserting the expressions (5.45) for the Γ´s, we obtain the quasi longitudinal wave velocity

$$2\rho V_1^2 = c_{44} + c_{11}\sin^2\theta + c_{33}\cos^2\theta \tag{5.47}$$
$$+ \sqrt{[(c_{11}-c_{44})\sin^2\theta + (c_{44}-c_{33})\cos^2\theta]^2 + (c_{13}+c_{44})^2\sin^2 2\theta}$$

and the quasi shear wave velocity

$$2\rho V_2^2 = c_{44} + c_{11}\sin^2\theta + c_{33}\cos^2\theta \tag{5.48}$$

$$- \sqrt{[(c_{11}-c_{44})\sin^2\theta + (c_{44}-c_{33})\cos^2\theta]^2 + (c_{13}+c_{44})^2\sin^2 2\theta}.$$

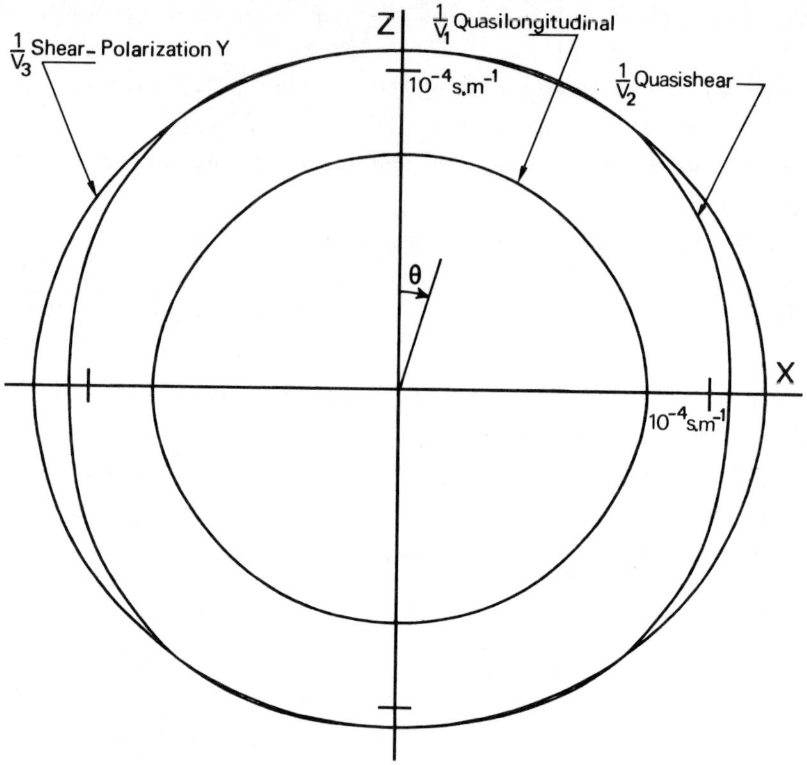

Fig. 5.22. Hexagonal system. Cross section of the slowness surface of beryllium in a meridian plane.

Table 5.23. Hexagonal system.

Direction of Propagation	Polarization	Velocity	Examples (ms^{-1})	
			Be	Ti
Z	Z(L)	$\sqrt{c_{33}/\rho}$	13490	6330
	Plan ⊥ Z(T)	$\sqrt{c_{44}/\rho}$	9380	3220
Perpendicular to the Z axis	Longitudinal	$\sqrt{c_{11}/\rho}$	12580	6000
	Transverse ⊥Z	$\sqrt{(c_{11}-c_{12})/2\rho}$	8480	2795
	Transverse //Z	$\sqrt{c_{44}/\rho}$	9380	3220

The curves in Fig. 5.22, which represent vertical cross sections of the beryllium slowness surface have been calculated from the data of (4.7).

Numerical data of Fig. 5.23 refer to beryllium (Be) and titanium (Ti).

d) Trigonal system. When the c_{25} modulus does not vanish, the calculations are rather involved. But most trigonal crystals currently used in practice (sapphire, quartz, lithium niobate) belong to classes $\bar{3}m$, 32 or 3m, for which $c_{25} = 0$. The stiffness matrix is given in table 4.6 and the propagation tensor components are

$$\Gamma_{11} = c_{11}n_1^2 + c_{66}n_2^2 + c_{44}n_3^2 + 2c_{14}n_2n_3$$

$$\Gamma_{12} = (c_{12} + c_{66})n_1n_2 + 2c_{14}n_1n_3$$

$$\Gamma_{13} = 2c_{14}n_1n_2 + (c_{13} + c_{44})n_1n_3$$

$$\Gamma_{22} = c_{66}n_1^2 + c_{11}n_2^2 + c_{44}n_3^2 - 2c_{14}n_2n_3 \quad \text{with } c_{66} = \frac{c_{11}-c_{12}}{2}$$

$$\Gamma_{23} = c_{14}(n_1^2 - n_2^2) + (c_{13} + c_{44})n_2n_3 \qquad (5.49)$$

$$\Gamma_{33} = c_{44}(n_1^2 + n_2^2) + c_{33}n_3^2.$$

In a direction perpendicular to the triad axis ($n_1 = \cos\phi$, $n_2 = \sin\phi$, $n_3 = 0$), none of the Γ's vanish and the secular equation can be solved only by numerical means, except for the x_1 and x_2 direction.

Along x_1, $n_1 = 1$, $n_2 = n_3 = 0$, we have

$$\Gamma = \begin{vmatrix} c_{11} & 0 & 0 \\ 0 & c_{66} & c_{14} \\ 0 & c_{14} & c_{44} \end{vmatrix}$$

which implies a longitudinal wave of velocity $V = \sqrt{c_{11}/\rho}$ and two shear waves of velocities V_2 and V_3 such that

$$2\rho V_2^2 = c_{66} + c_{44} + \sqrt{(c_{66} - c_{44})^2 + 4c_{14}^2} \qquad (5.50)$$

$$2\rho V_3^2 = c_{66} + c_{44} - \sqrt{(c_{66} - c_{44})^2 + 4c_{14}^2}. \qquad (5.51)$$

Along the x_2 direction, $n_2 = 1$, $n_1 = n_3 = 0$:

$$\Gamma = \begin{vmatrix} c_{66} & 0 & 0 \\ 0 & c_{11} & -c_{14} \\ 0 & -c_{14} & c_{44} \end{vmatrix}$$

and we find

- a shear wave with x_1 polarization and velocity $V_3 = \sqrt{c_{66}/\rho}$;

- a quasi longitudinal wave, with velocity V_1:

$$2\rho V_1^2 = c_{11} + c_{44} + \sqrt{(c_{11} - c_{44})^2 + 4c_{14}^2}; \qquad (5.52)$$

- a quasi shear wave with velocity V_2:

$$2\rho V_2^2 = c_{11} + c_{44} - \sqrt{(c_{11} - c_{44})^2 + 4c_{14}^2}; \qquad (5.53)$$

The slowness curves for sapphire are plotted in fig. 5.24 (numerical data from table 4.7).

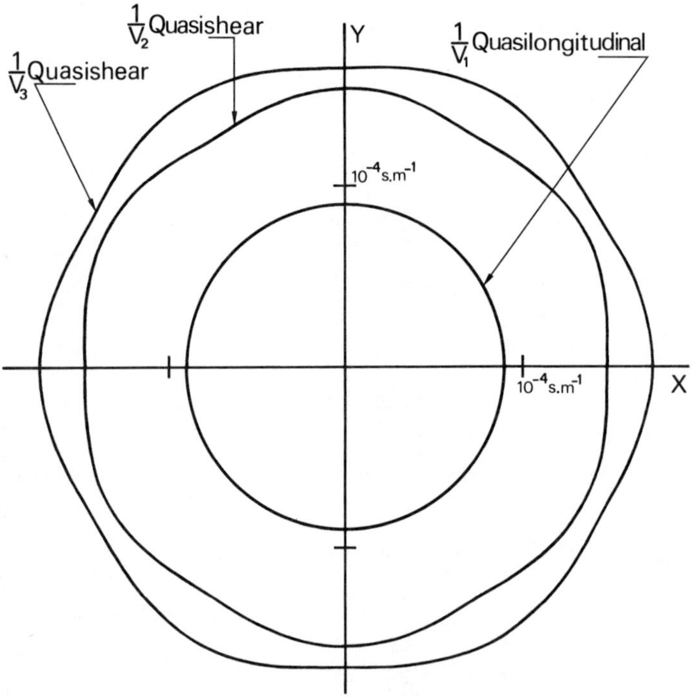

Fig. 5.24. Trigonal system. Classes $\overline{3}$m, 32, 3m. Cross section of the slowness surface of sapphire ($\overline{3}$m) in the XY plane. The normal is not in the XY plane which is not a mirror. Thus the Y axis is not a pure mode direction. Figure 5.25 shows that the energy velocity makes a non-zero angle with the XY cross-section.

If the propagation takes place in the x_2x_3 plane (i.e. a YZ symmetry plane), $n_1 = 0$, $n_2 = \sin\theta$, $n_3 = \cos\theta$. The Γ tensor is as follows:

$$\Gamma = \begin{vmatrix} \Gamma_{11} & 0 & 0 \\ 0 & \Gamma_{22} & \Gamma_{23} \\ 0 & \Gamma_{23} & \Gamma_{33} \end{vmatrix}$$

with

$$\begin{aligned} \Gamma_{11} &= c_{66}\sin^2\theta + c_{44}\cos^2\theta + c_{14}\sin 2\theta \\ \Gamma_{22} &= c_{11}\sin^2\theta + c_{44}\cos^2\theta - c_{14}\sin 2\theta \\ \Gamma_{23} &= -c_{14}\sin^2\theta + \frac{(c_{13} + c_{44})}{2}\sin 2\theta \\ \Gamma_{33} &= c_{44}\sin^2\theta + c_{33}\cos^2\theta \end{aligned} \quad (5.54)$$

therefore, there is a shear wave polarized along x_1 with velocity

$$V_3 = \sqrt{\frac{c_{44}\cos^2\theta + c_{66}\sin^2\theta + c_{14}\sin 2\theta}{\rho}} \quad (5.55)$$

and two waves, quasi-longitudinal and quasi-shear with velocities V_1 and V_2 such that

$$2\rho V_{\frac{1}{2}}^2 = \Gamma_{22} + \Gamma_{33} \pm \sqrt{(\Gamma_{22} - \Gamma_{33})^2 + 4\Gamma_{23}^2}$$

Fig. 5.25, plotted for the case of sapphire, shows that the triad axis is not a pure mode direction for shear waves (see example in § 5.1.5.2); in fact, the normals to both curves intersecting on the x_3 axis do make an angle with Ox_3. However, the X-polarized shear wave is pure in some specific directions of the YZ plane. In order to derive these directions, we first determine the energy velocity components using 5.24,

$$V_i^e = c_{i1k1}\frac{n_k}{\rho V_3} \qquad \text{for} \qquad {}^0u_i = \delta_{i1}$$

or, in terms of θ ($n_1 = 0$, $n_2 = \sin\theta$, $n_3 = \cos\theta$):

$$\vec{V^e} \begin{cases} V_1^e = 0 \\ V_2^e = \dfrac{c_{66}\sin\theta + c_{14}\cos\theta}{\rho V_3} \\ V_3^e = \dfrac{c_{14}\sin\theta + c_{44}\cos\theta}{\rho V_3} \end{cases} \quad (5.56)$$

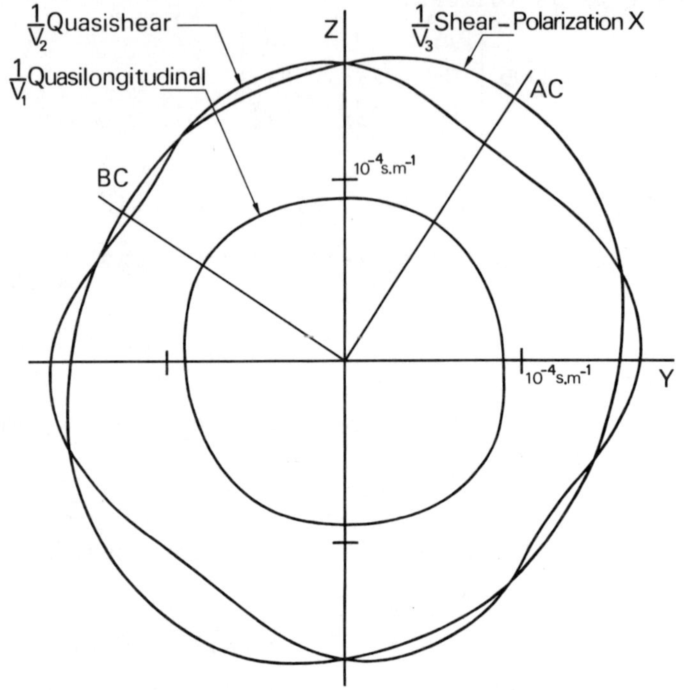

Fig. 5.25. Trigonal system, classes $\bar{3}m$, 32, 3m. Cross section of the slowness surface of sapphire ($\bar{3}m$) in the YZ plane.

The energy velocity and the Z axis are at an angle θ^e such that

$$\tan \theta^e = \frac{V_2^e}{V_3^e} = \frac{c_{66} \sin \theta + c_{14} \cos \theta}{c_{14} \sin \theta + c_{44} \cos \theta} \tag{5.57}$$

and fig. 5.26 shows the deviation of the acoustic ray with respect to the propagation direction in the YZ plane. The shear wave is pure for $\theta - \theta^e = 0$, i.e. for θ_1 and θ_2 given by

$$\tan 2\theta = \frac{2c_{14}}{c_{44} - c_{66}} \ . \tag{5.58}$$

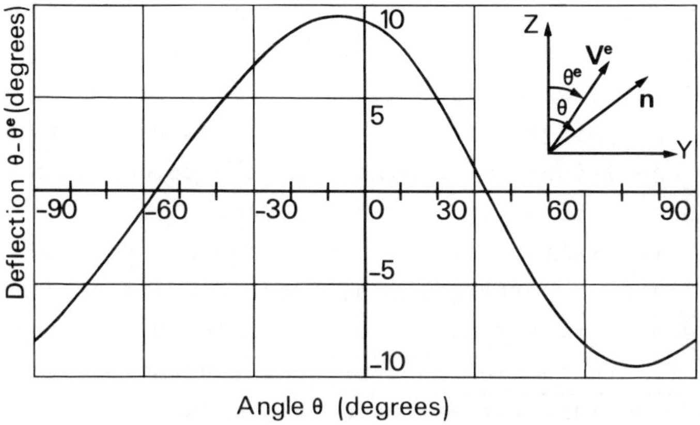

Fig. 5.26. Deviation of the shear wave acoustic ray ($\vec{V^e}$) from the propagation direction $\vec{n}$ in the YZ plane of sapphire.

For sapphire, $\theta_1 = 33^\circ 28'$, $\theta_2 = -56^\circ 32'$. These directions, which are orthogonal, and denoted by AC and BC in fig. 5.25, are of interest since the wave is both pure and transverse.

5.2. REFLEXION AND REFRACTION OF PLANE ELASTIC WAVES

A plane monochromatic elastic wave with a given polarization gives rise, when it encounters a boundary between two crystals, to three waves propagating on each side of the boundary. In this way, an incident quasi-longitudinal wave may generate a quasi longitudinal and two quasi shear (slow and fast) waves (the reflected waves) in the crystal where it propagates, as well as a quasi-longitudinal and two quasi shear waves, the transmitted waves, in the other crystal. The situation is much simpler for isotropic media, because firstly the original wave generates at most two waves per medium, and secondly the waves are either longitudinal or transverse.

5.2.1. Equations of Continuity

The reflexion or refraction problem can be stated as follows: given the incident wave, its propagation direction, polarization and amplitude, given the elastic properties of both crystals, then what are the propagation directions, the polarizations and the amplitudes of the reflected or refracted waves, as well as the directions of energy flow? In principle, the solution is obtained by expanding the propagation equations for both media together with the boundary conditions. In practice, the calculation can be performed only in a few selected cases.

When two solids, with a plane boundary surface, are rigidly linked together, the boundary conditions can be expressed by saying that the displacement u_i and the stresses T_i are continuous everywhere on the boundary. The boundary plane, if it has a unit normal vector $\vec{\ell}$ (ℓ_1, ℓ_2, ℓ_3) has the equation

$$\vec{\ell}.\vec{x} = 0$$

provided the origin of the coordinates is on the boundary. So we should have, at any time, and throughout this plane:

$$u_i^I + \sum_R u_i^R = \sum_T u_i^T \tag{5.59}$$

and

$$T_i^I + \sum_R T_i^R = \sum_T T_i^T \tag{5.60}$$

where the indices I, R, T correspond to the incident, reflected and transmitted waves. For plane sine waves, the continuity of the displacements

$$^{o}u_i^I e^{i(\omega^I t - \vec{k}^I.\vec{x})} + \sum_R {^{o}u_i^R} e^{i(\omega^R t - \vec{k}^R.\vec{x})} = \sum_T {^{o}u_i^T} e^{i(\omega^T t - \vec{k}^T.\vec{x})} \tag{5.61}$$

implies, if it is to be satisfied for all terms:

$$\omega^R = \omega^T = \omega^I \tag{5.62}$$

and throughout the plane $\vec{\ell}.\vec{x} = 0$,

$$\vec{k}^R.\vec{x} = \vec{k}^T.\vec{x} = \vec{k}^I.\vec{x} \tag{5.63}$$

There is no frequency shift in a reflexion or a refraction.

In addition, the equations (5.63), which can be written as

$$(\vec{k}^R - \vec{k}^I).\vec{x} = 0; \qquad (\vec{k}^T - \vec{k}^I).\vec{x} = 0$$

show that the vectors $(\vec{k}^R - \vec{k}^I)$ and $(\vec{k}^T - \vec{k}^I)$ are

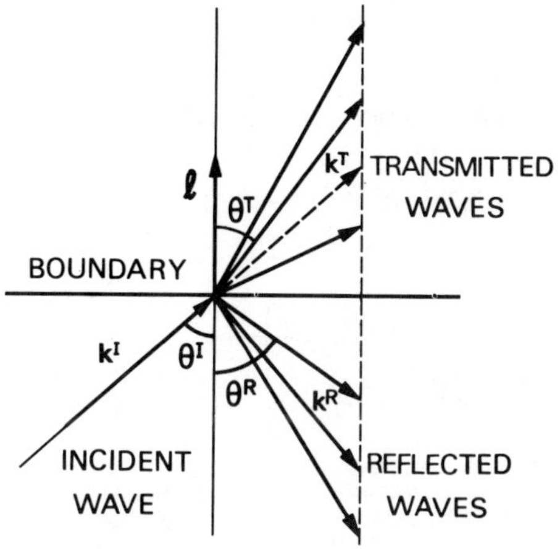

Fig. 5.27. Snell-Descartes Law. The projections of the wave vectors $\vec{k}^R$ and $\vec{k}^T$ (reflected and transmitted waves) onto the boundary are the same as for the incident wave $\vec{k}^I$.

orthogonal to the boundary plane. Therefore, all wave vectors are contained in the plane of incidence, defined by the normal $\vec{\ell}$ and the wave vector $\vec{k}^I$, and all these wave vectors have the same projection on the boundary as $\vec{k}^I$ (fig. 5.27). This condition is no other than the Descartes-Snell's law:

$$k^R \sin \theta^R = k^T \sin \theta^T = k^I \sin \theta^I \quad (5.64)$$

where θ^I, θ^R, θ^T are the incidence, reflexion and refraction angles.

Since all frequencies are equal, the propagation directions are determined by the equations:

$$\frac{\sin \theta^R}{V(\theta^R)} = \frac{\sin \theta^T}{V(\theta^T)} = \frac{\sin \theta^I}{V(\theta^I)} \quad (5.65)$$

or, graphically, by a construction involving the slowness surfaces of both crystals. Fig. 5.28 shows the case of a cubic crystal (silicon) stuck to an isotropic crystal (silica) with an (010) boundary plane. The quasi-longitudinal incident wave propagates in the (001) plane and gives rise to two reflected waves (quasi longitudinal and quasitransverse) and to two transmitted waves in the isotropic medium, one longitudinal and one transverse. No shear wave with a polarization orthogonal to the incident plane can be excited since the incident wave displacement has no component in that direction.

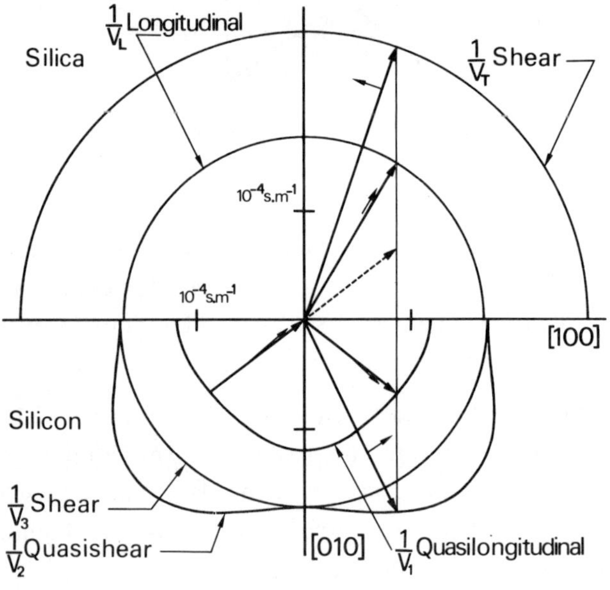

Fig. 5.28. Geometrical construction of the reflected and refracted wave vectors at a silicon-silica boundary. The quasilongitudinal incident wave gives rise to quasilongitudinal and quasi-shear reflected waves, and to longitudinal and shear refracted waves.

This construction also provides the velocity of energy flow for each wave. In view of what was said about slowness surfaces in the above section, we may visualize the various possible cases:

- there is no intersection with any slowness curve, i.e. we have an evanescent wave. This is analytically expressed by the fact that we have an imaginary solution to the transmitted wave dispersion equation:

$$(k^T)^2 = (K^T_{||})^2 + (k^T_{\perp})^2 = (\frac{\omega}{V'})^2$$

The projection of the refracted wave vector on the boundary, $k^T_{||}$, being equal to $k^I_{||}$, the normal component $k^T_{\perp}$ is imaginary if $k^I_{||} > \omega/V'$, i.e. since $k^I_{||} = (\omega/V)\sin\theta^I$:

$$k^T_{\perp} = \frac{i\omega}{V}\sqrt{\sin^2\theta^I - (\frac{V}{V'})^2}. \qquad (5.66)$$

The wave amplitude decreases exponentially in the second medium. This situation appears beyond the critical angle of incidence θ^I_c such that $k^T_{\perp} = 0 \Leftrightarrow \theta^T = \pi/2$:

$$\sin\theta^I_c = \frac{V(\theta^I_c)}{V'(\frac{\pi}{2})}$$

- The acoustic ray, which is orthogonal to the slowness surface, and the wave vector may be on opposite sides of the boundary for the same mode. The energy is then reflected, although the wave vector is refracted.

Before we look at a specific example, let us write down the boundary conditions omitting the propagating terms which are the same throughout the boundary:

$$\boxed{{}^ou^I_i + \sum_R {}^ou^R_i = \sum_T {}^ou^T_i\ .} \qquad (5.67)$$

For mechanical stresses $T_i = T_{ij}\ell_j$ (Hooke's law):

$$T_{ij} = c_{ijk\ell}\frac{\partial u_k}{\partial x_\ell} = -\,ic_{ijk\ell}\,k_\ell\,{}^ou_k e^{i(\omega t-\vec{k}.\vec{x})}$$

so we obtain:

$$\boxed{c_{ijk\ell}\ell_j(k^I_\ell\ {}^ou^I_k + \sum_R k^R_\ell\ {}^ou^R_k) = c'_{ijk\ell}\ell_j\sum_T k^T_\ell\ {}^ou^T_k} \qquad (5.68)$$

where $c'_{ijk\ell}$ are the stiffness coefficients of the second crystal.

5.2.2. An Example: Shear Horizontal Incident Wave

It was shown in section 5.1.3b that a plane orthogonal to a 4-fold or 6-fold axis is isotropic as far as the shear wave polarized along the axis is concerned, i.e. the velocity $V_3 = \sqrt{c_{44}/\rho}$ does not depend upon the propagation direction. So the simplest case of reflexion at a boundary between two crystals - shown in fig. 5.29 - corresponds to

- a medium consisting of two crystals, each having a 4-fold or a 6-fold axis along the x_3 direction, parallel to the boundary plane.

- an incident shear wave with x_3 polarization, (the incidence plane is x_1x_2) called a S.H. wave (shear horizontal).

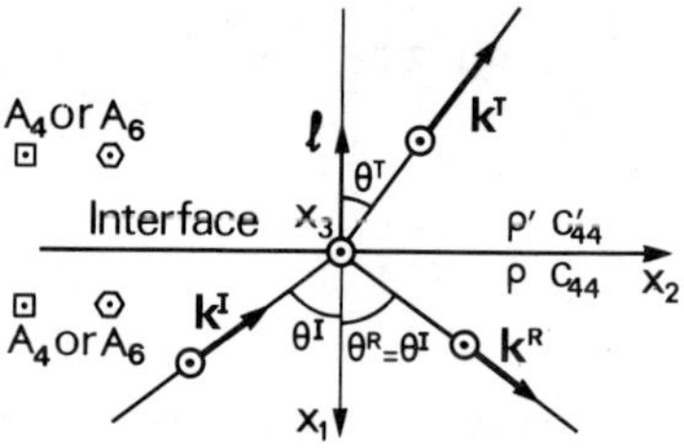

Fig. 5.29. Reflexion and refraction of a shear horizontal (S.H.) wave at a boundary between two crystals having an A_4 or A_6 axis orthogonal to the plane of incidence (or at a boundary between two isotropic solids).

The allowed displacements in both crystals being parallel or perpendicular to the x_3 axis, the incident wave polarization is conserved and the discontinuity can only give rise to two waves, one reflected and one transmitted, having the same shear polarization as the incident wave. These conditions are also fulfilled in the case of two isotropic solids. Since ${}^o u_i = {}^o u \delta_{i3}$ for each wave, the continuity of displacement implies

$$ {}^o u^I + {}^o u^R = {}^o u^T . \qquad (5.69)$$

Equation 5.68 can be written down in another way since $\ell_j = -\delta_{j1}$ and ${}^o u_k = {}^o u \delta_{k3}$:

$$c_{i13\ell}(k_\ell^I\, {}^o u^I + k_\ell^R\, {}^o u^R) = c'_{i13\ell} k_\ell^T\, {}^o u^T$$

or

$$c_{i131}(k_1^I\, {}^o u^I + k_1^R\, {}^o u^R) + c_{i132}(k_2^I\, {}^o u^I + k_2\, {}^o u^R) = c'_{i131} k_1^T\, {}^o u^T + c'_{i132} k_2^T\, {}^o u^T .$$

In crystals having an A_4 or A_6 axis, only i = 3 leads to non-vanishing constants: $c_{3131} = c_{55} = c_{44}$ $c'_{3131} = c'_{44}$ (see 4.55 and 4.56), so

$$c_{44}(k_1^I \, {}^o u^I + k_1^R \, {}^o u^R) = c'_{44} k_1^T \, {}^o u^T. \qquad (5.70)$$

By solving the equations 5.69 and 5.70, we readily obtain the reflexion coefficient:

$$A_R = \frac{{}^o u^R}{{}^o u^I} = \frac{c'_{44} k_1^T - c_{44} k_1^I}{c_{44} k_1^R - c'_{44} k_1^T} \qquad (5.71)$$

and the transmission coefficient

$$A_T = \frac{{}^o u^T}{{}^o u^I} = A_R + 1.$$

Snell's law 5.65 implies $\theta^R = \theta^I$ and

$$\frac{\sin \theta^T}{\sin \theta^I} = \frac{V'_3}{V_3} \qquad (5.72)$$

Substituting $k_1^R = - k_1^I = k^I \cos \theta^I$ and $k_1^T = - k^T \cos \theta^T$, the coefficient A_R is expressed in terms of the angles of incidence and of refraction:

$$A_R = \frac{c_{44} k^I \cos \theta^I - c'_{44} k^T \cos \theta^T}{c_{44} k^I \cos \theta^I + c'_{44} k^T \cos \theta^T} \, .$$

We may introduce the elastic impedances ($Z_3 = \rho V_3$; $Z'_3 = \rho' V'_3$) for shear waves in both crystals:

$$k^I c_{44} = \frac{\omega}{V_3} c_{44} = \frac{\omega \rho V_3^2}{V_3} = \omega Z_3 \quad \text{and} \quad k^T c'_{44} = \omega Z'_3$$

so that

$$A_R = \frac{Z_3 \cos \theta^I - Z'_3 \cos \theta^T}{Z_3 \cos \theta^I + Z'_3 \cos \theta^T} \qquad (5.73)$$

and

$$A_T = A_R + 1 = \frac{2 Z_3 \cos \theta^I}{Z_3 \cos \theta^I + Z'_3 \cos \theta^T} \qquad (5.74)$$

where θ^I and θ^R are related by 5.72.

The variation of A_R and A_T in terms of the angle of incidence θ^I depends on the elastic impedance ratio Z'_3/Z_3 and on the velocity ratio V'_3/V_3. If V'_3/V_3 is smaller than 1, θ^T is less than θ^I and there is no critical angle for the transmitted wave. In addition, if $Z_3'/Z_3 < 1$, the reflexion coefficient A_R vanishes for θ^I such that

$$\frac{\cos\theta^I}{\cos\theta^T} = \frac{Z_3'}{Z_3}$$

and for this direction we have total transmission into the second crystal. The curves of fig. 5.30 correspond to a medium consisting of silica ($V_3' = 3{,}763$ m/s, $Z_3' = 8.29\ 10^6$ kg/m²/s) over silicon ($V_3 = 5{,}843$ m/s, $Z_3 = 13.6\ 10^6$ kg/m²/s). If $V_3' > V_3$, the transmitted wave is evanescent for θ^I greater than the critical angle θ_c where $\sin\theta_c = V_3/V_3'$. The wave vector component k_1^T, normal to the boundary, is then imaginary: $k_1^T = i\chi^T$ and according to 5.71, the reflexion coefficient is a complex number:

$$A_R = \frac{c_{44}k^I \cos\theta^I + ic_{44}'\chi^T}{c_{44}k^I \cos\theta^I - ic_{44}'\chi^T} \ .$$

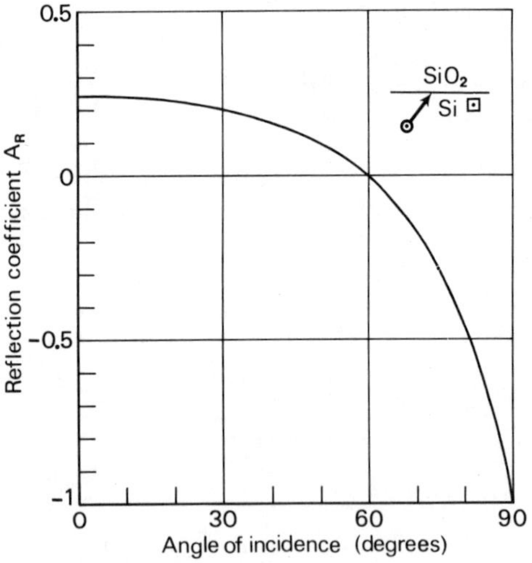

Fig. 5.30. Reflexion coefficient vs angle of incidence for a SH wave at a boundary Si → SiO_2.

Since A_R has unit modulus , we have a total reflexion of the incident wave beyond the critical angle of incidence. The ratio A_T (transmitted wave displacement to incident wave displacement at the boundary) is no longer the transmission coefficient since the transmitted wave amplitude is not a constant, instead it decreases exponentially in the second medium. Fig. 5.31 shows this with the same materials, but in reverse order, i.e. silicon on silica.

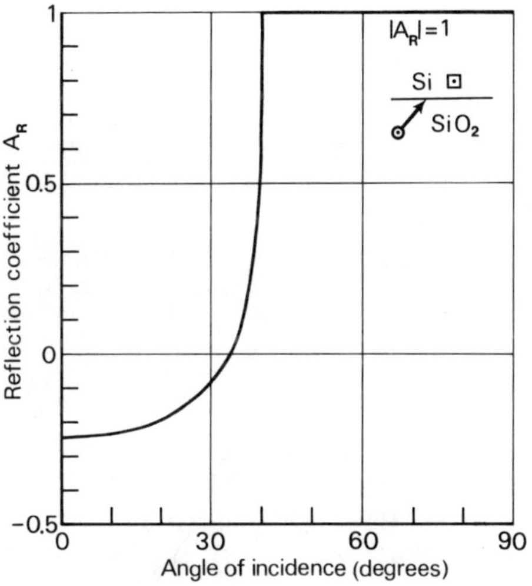

Fig. 5.31. Reflexion coefficient vs. angle of incidence for a SH wave at a boundary $SiO_2 \rightarrow Si$. Beyond the threshold value $\theta_c = 40^o$, there is total reflexion: $|A_R| = 1$.

5.2.3. Reflexion at a Free Surface

For a free surface boundary between a solid and air or any other low pressure fluid, there are no transmitted waves and the only condition that has to be satisfied is that the mechanical stress vanish at the surface:

$$T_{ij}\ell_j = 0$$

or

$$c_{ijk\ell}\ell_j(k_\ell^I \, {}^o u_k^I + \sum_R k_\ell^R \, {}^o u_k^R) = 0. \qquad (5.75)$$

On comparing with 5.68, we see that this is equivalent to equating to zero the stiffness coefficients $c'_{ijk\ell}$, of the second medium. In the above example $A_R = 1$ when $Z'_3 = 0$; i.e. a shear wave polarized along the free surface is totally reflected.

At the free surface of an isotropic solid, or of a solid with an A_6 axis orthogonal to the plane of incidence, a longitudinal wave is reflected into a longitudinal and a shear wave, denoted by the indices (*) L and T (fig. 5.32). Since the material is isotropic in the plane of incidence:

$$k^L = k^I = \frac{\omega}{V_L} \quad \text{and} \quad \frac{k^T}{k^I} = \frac{V_L}{V_T}$$

and Snell's law provides:

$$\theta^L = \theta^I \quad \text{and} \quad \frac{\sin\theta^T}{\sin\theta^I} = \frac{V_T}{V_L}\,. \qquad (5.76)$$

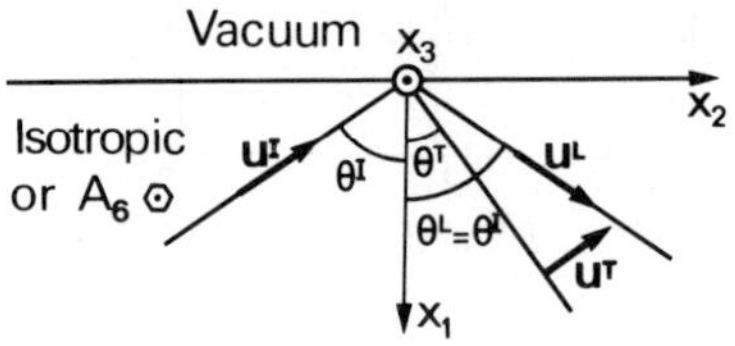

Fig. 5.32. Reflexion of a longitudinal plane wave at the free surface of an isotropic solid or of a hexagonal crystal where A_6 is orthogonal to the plane of incidence.

In the expansion of the vanishing stress condition (5.75), on the x_2x_3 plane,

$$c_{i1k\ell}(k^I_\ell\,{}^ou^I_k + k^L_\ell\,{}^ou^L_k + k^T_\ell\,{}^ou^T_k) = 0 \qquad (5.77)$$

---

* Strictly speaking, one would use RL and RT for "reflected longitudinal" and "reflected transverse". Omitting the R is not misleading since there is no transmitted wave in this example.

the indices ℓ and k can only be equal to 1 or 2, because all vectors $\vec{k}$ and ${}^{o}\vec{u}$ are within the x_1x_2 plane, with components:

$$\vec{k}^I \begin{cases} -k^I \cos\theta^I \\ k^I \sin\theta^I \end{cases} \qquad \vec{k}^L \begin{cases} k^I \cos\theta^I \\ k^I \sin\theta^I \end{cases} \qquad \vec{k}^T \begin{cases} k^T \cos\theta^T \\ k^T \sin\theta^T \end{cases}$$

$$ {}^{o}\vec{u}^I \begin{cases} -{}^{o}u^I \cos\theta^I \\ {}^{o}u^I \sin\theta^I \end{cases} \qquad {}^{o}\vec{u}^L \begin{cases} {}^{o}u^L \cos\theta^I \\ {}^{o}u^L \sin\theta^I \end{cases} \qquad {}^{o}\vec{u}^T \begin{cases} -{}^{o}u^T \sin\theta^T \\ {}^{o}u^T \cos\theta^T. \end{cases}$$

For i = 1, the expression 5.77 involves $c_{1111} = c_{11}$ and and $c_{1122} = c_{12}$ only:

$$c_{11}[k^I({}^{o}u^I + {}^{o}u^L)\cos^2\theta^I - \frac{{}^{o}u^T k^T}{2}\sin 2\theta^T]$$
$$+ c_{12}[k^I({}^{o}u^I + {}^{o}u^L)\sin^2\theta^I + \frac{{}^{o}u^T k^T}{2}\sin 2\theta^T] = 0.$$

For i = 2, only $c_{2112} = c_{2121} = c_{66}$ do not vanish:

$$c_{66}[k^I(-{}^{o}u^I + {}^{o}u^L)\sin 2\theta^I + {}^{o}u^T k^T \cos 2\theta^T] = 0.$$

For i = 3, all terms of 5.77 vanish. The two quantities $A_L = {}^{o}u^L/{}^{o}u^I$ and $A_T = {}^{o}u^T/\,{}^{o}u^I$ satisfy the following equations:

$$\begin{cases} (1+A_L)k^I(c_{11}\cos^2\theta^I + c_{12}\sin^2\theta^I) - A_T k^T(\frac{c_{11}-c_{12}}{2})\sin 2\theta^T = 0 \\ (A_L - 1)k^I \sin 2\theta^I + A_T k^T \cos 2\theta^T = 0. \end{cases} \qquad (5.78)$$

Let us now transform the factor

$$F = c_{11}\cos^2\theta^I + c_{12}\sin^2\theta^I = c_{11} + (c_{12}-c_{11})\sin^2\theta^I$$

by introducing the longitudinal and shear waves velocities: $V_L = \sqrt{c_{11}/\rho}$, $V_T = \sqrt{(c_{11}-c_{12})/2\rho}$:

$$F = \rho V_L^2 - 2\rho V_T^2 \sin^2\theta^I$$

or, making use of 5.76:

$$F = \rho V_L^2(1 - 2\sin^2\theta^T) = \rho V_L^2 \cos 2\theta^T.$$

Inserting this into (5.78) and replacing k^T/k^I by V_L/V_T, we obtain:

$$\begin{cases} A_L\left(\frac{V_L}{V_T}\right)\cos 2\theta^T - A_T \sin 2\theta^T = -\left(\frac{V_L}{V_T}\right)\cos 2\theta^T \\ A_L \sin 2\theta^I + A_T\left(\frac{V_L}{V_T}\right)\cos 2\theta^T = \sin 2\theta^I. \end{cases}$$

the solution of which yields the longitudinal wave reflexion coefficient:

$$A_L = \frac{{}^{o}u^{L}}{{}^{o}u^{I}} = \frac{\sin 2\theta^I \sin 2\theta^T - \left(\frac{V_L}{V_T}\right)^2 \cos^2 2\theta^T}{\sin 2\theta^I \sin 2\theta^T + \left(\frac{V_L}{V_T}\right)^2 \cos^2 2\theta^T} \qquad (5.79)$$

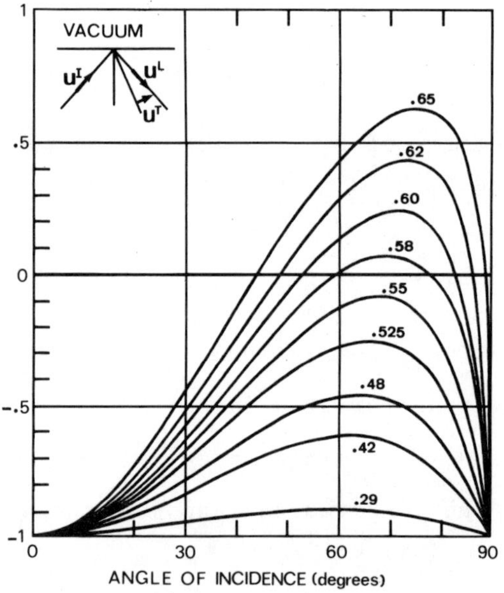

Fig. 5.33. Reflexion coefficient vs angle of incidence for a longitudinal wave at the free surface of an isotropic solid, for various values of the parameter V_T/V_L. For silica $V_T/V_L = .63$ (figure 3 of reference 5).

together with the conversion factor for the shear wave:

$$A_T = \frac{{}^o u^T}{{}^o u^I} = \frac{2(\frac{V_L}{V_T})\cos 2\theta^T \sin 2\theta^I}{\sin 2\theta^I \sin 2\theta^T + (\frac{V_L}{V_T})^2 \cos^2 2\theta^T}$$

$$= \frac{V_L}{V_T}(A_L + 1) \text{ cotan } 2\theta^T \qquad (5.80)$$

θ^T and θ^I being related by (5.76). Since $V_T/V_L < 1/\sqrt{2}$, the angle θ^T is less than $\pi/4$ and A_T is positive, whatever the angle of incidence. The set of curves in fig. 5.33 shows that, on the other hand, A_L vanishes for two values of the angle of incidence if V_T/V_L is greater than 0.565. The incident longitudinal wave is then entirely reflected into a single shear wave. This is used for the conversion of longitudinal waves into shear waves.

5.2.4. Love Waves

These waves are surface waves with shear polarization, which propagate in a medium consisting of a layer and a substrate with different elastic properties (fig. 5.34). The thin layer, rigidly linked to the substrate, acts as a one dimensional guide with non identical sides. The presence of the substrate removes the cutoff frequency the guide would normally have if it were alone (§ 1.1.2.2). It can be predicted that the propagation velocity, between the substrate velocity and the layer velocity, depends on the frequency. This dispersive effect is the basic feature of Love waves.

Let us now look at fig. 5.34. Since the crystals have an A_4 or A_6 symmetry axis in the x_3 direction, the x_1x_2 plane is isotropic for a shear wave with displacement u_3 and wave vector $\vec{k}(k_1, k_2, 0)$. The resultant wave, propagating along the x_2 direction, generates shear stresses only:

$$T_{31} = c_{44}\frac{\partial u_3}{\partial x_1} \quad \text{and} \quad T_{32} = c_{44}\frac{\partial u_3}{\partial x_2}.$$

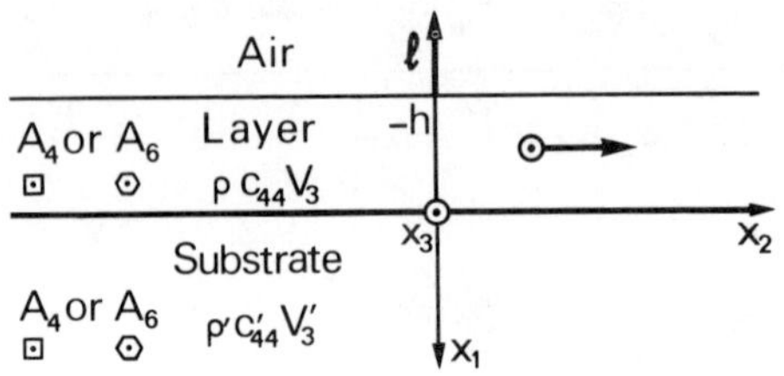

Fig. 5.34. Love waves. The SH wave propagates along the x_2 axis, undergoing alternating reflexions at the free surface of the layer and the boundary with the substrate.

At the free surface $x_1 = -h$, setting the mechanical stress to zero $T_i = T_{ij}\ell_j = -T_{i1}$ has no effect except on T_{31}, the expression of which results, as in § 1.1.2.2, from the superposition of the incident and reflected waves:

$$T_{31} = A\, e^{i[\omega t - k_1(x_1+h) - k_2 x_2]} - A\, e^{i[\omega t + k_1(x_1+h) - k_2 x_2]}$$

or, if we take $T_0 = -2iA$:

$$T_{31} = T_0 \sin k_1(x_1+h) e^{i(\omega t - k_2 x_2)} \quad \text{for } -h < x_1 < 0. \quad (5.81)$$

The shear wave velocity in the layer is $V_3 = \sqrt{c_{44}/\rho}$ so that the wave vector components k_1, k_2, $k_3 = 0$ obey the following equation:

$$k^2 = k_1^2 + k_2^2 = \left(\frac{\omega}{V_3}\right)^2 \quad (5.82)$$

and, since k_1 is real,

$$k_2 < \frac{\omega}{V_3} \quad . \quad (5.83)$$

A Love wave being a surface wave,* the stress T_{31} created in the substrate

$$T_{31} = T'_0\, e^{i(\omega t - k'_1 x_1 - k'_2 x_2)} \quad \text{for} \quad x_1 > 0$$

should decrease for high values of x_1; it is thus necessary that

---

* It should be noted that the boundary condition $T_{31} = 0$ is also fulfilled when u_3 does not depend on x_1. This solution corresponds to a shear horizontal wave of constant amplitude, propagating in the layer or in the substrate (velocity V_3 or V'_3).

$$k_1' = -i\chi_1 \quad \text{with} \quad \chi_1 > 0.$$

Furthermore, the continuity of mechanical stress at the boundary $x_1 = 0$ implies $k_2' = k_2$. T_{31} is then expressed by

$$T_{31} = T_0' \, e^{-\chi_1 x_1} \, e^{i(\omega t - k_2 x_2)} \quad \text{for} \quad x_1 > 0.$$

χ_1 and k_2 satisfy the equation

$$k_1'^2 + k_2'^2 = \left(\frac{\omega}{V_3'}\right)^2 = k_2^2 - \chi_1^2 \qquad (5.84)$$

where $V_3' = \sqrt{c_{44}'/\rho'}$ is the shear wave velocity in the substrate. If χ_1 has to be real,

$$k_2 > \frac{\omega}{V_3'} \qquad (5.85)$$

The inequalities 5.83 and 5.85 are compatible only if

$$\boxed{V_3' > V_3\ .}$$

So, the existence of Love waves requires that the shear wave velocity in the substrate is greater than in the layer. The phase velocity of the Love wave is then situated between the above two values:

$$V_3 \text{ (layer)} < V_\phi < V_3' \text{ (substrate)}.$$

The continuity of mechanical stress at the boundary $x_1 = 0$ implies

$$T_0 \sin k_1 h = T_0'.$$

The shear stress is equal to

$$\begin{cases} T_{31} = T_0' \dfrac{\sin k_1 (x_1 + h)}{\sin k_1 h} \, e^{i(\omega t - k_2 x_2)} & \text{for} \quad -h < x_1 < 0, \\[2ex] T_{31} = T_0' \, e^{-\chi_1 x_1} \, e^{i(\omega t - k_2 x_2)} & \text{for} \quad x_1 > 0. \end{cases}$$

The displacement u_3 is obtained by integrating Hooke's equation in the layer and in the substrate: $T_{31} = c_{44} \dfrac{\partial u_3}{\partial x_1}$ and $T_{31} = c_{44}' \dfrac{\partial u_3}{\partial x_1}$, together with $^{o}u = -T_0'/c_{44}'\chi_1$ yield:

$$\begin{cases} u_3 = {}^{o}u \, \dfrac{c_{44}'}{c_{44}} \cdot \dfrac{\chi_1}{k_1} \cdot \dfrac{\cos k_1 (x_1 + h)}{\sin k_1 h} \, e^{i(\omega t - k_2 x_2)} & \text{for } -h < x_1 < 0, \\[2ex] u_3 = {}^{o}u \, e^{-\chi_1 x_1} \, e^{i(\omega t - k_2 x_2)} & \text{for } x_1 > 0. \end{cases}$$

The continuity of displacement at $x_1 = 0$ provides

$$\chi_1 = \frac{c_{44}}{c'_{44}}\, k_1 \tan k_1 h. \qquad (5.86)$$

and u_3 can be put into the following form:

$$\begin{cases} u_3 = {}^0u\, \dfrac{\cos k_1(x_1+h)}{\cos k_1 h}\, e^{i(\omega t - k_2 x_2)} & \text{for } -h < x_1 < 0, \\ u_3 = {}^0u\, e^{-\chi_1 x_1}\, e^{i(\omega t - k_2 x_2)} & \text{for } x_1 > 0. \end{cases} \qquad (5.87)$$

By making use of 5.86 and of

$$k_1^2 + \chi_1^2 = \left(\frac{\omega}{V_3}\right)^2 - \left(\frac{\omega}{V'_3}\right)^2 \qquad (5.88)$$

(which is a consequence of 5.82 and 5.84), we obtain χ_1, k_1 and k_2 for all frequencies:

$$k_2 = \left[\left(\frac{\omega}{V_3}\right)^2 - k_1^2\right]^{1/2} = \left[\left(\frac{\omega}{V'_3}\right)^2 + \chi_1^2\right]^{1/2} \qquad (5.89)$$

The set of equations 5.86, 5.88 and 5.89 contains an implicit definition of the dispersion equation $\omega(k_2)$ for Love waves. The determination of k_1 and χ_1 is then achieved graphically from fig. 5.35, by using the dimensionless variables $k_1 h$, $\chi_1 h$ and $\omega h/V_3$. The solutions are given by the intersections between the curve

$$\chi_1 h = \frac{c_{44}}{c'_{44}}\,(k_1 h)\ \tan(k_1 h)$$

and the circle of radius

$$R = \frac{\omega h}{V_3}\left[1 - \left(\frac{V_3}{V'_3}\right)^2\right]^{\frac{1}{2}}$$

centered at the origin, and which corresponds to $R^2 = (k_1 h)^2 + (\chi_1 h)^2$. Every intersection defines a mode. There are p modes if $(p-1)\pi < R < p\pi$, i.e. if $(p-1)\omega_0 < \omega < p\omega_0$ with

$$\frac{\omega_0 h}{V_3} = \frac{\pi}{\left[1-\left(\frac{V_3}{V'_3}\right)^2\right]^{\frac{1}{2}}}$$

For the first three Love modes in a silica layer on silicon, the x_1 dependence of the displacement u_3 is depicted in fig. 5.36. The p-mode has (p-1) nodal planes in the layer. At a given frequency, the penetration depth is greater for high orders, because the damping

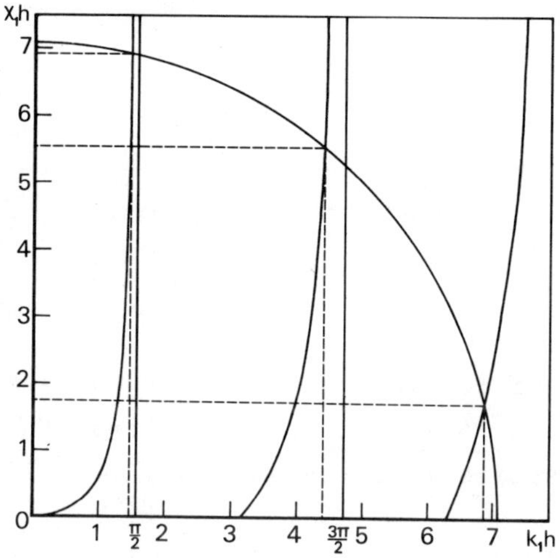

Fig. 5.35. Graphical solution of equations 5.86 and 5.88, showing the existence of several allowed modes (silica layer on a silicon substrate).

factor χ_1 will decrease. The dependence of the particle displacement upon the frequency (fig. 5.37) can easily be interpreted through the graphical construction of fig. 5.35: for a given mode, χ_1 increases with ω, as does the maximum value ${}^0u/\cos k_1 h$ of the displacement amplitude in the layer. Therefore, the fraction of the energy which is carried in the layer, with velocity $V_3 < V_3'$, increases with ω. On the other hand, at low frequencies, $(\omega h < V_3)$ almost all the energy propagates in the substrate with velocity V_3'. Since the width of the layer is negligible compared to $1/\chi_1$, the layer is no longer of any importance. These results appear in the dispersion curves of Fig. 5.38, plotted in reduced co-ordinates $\omega h/V_3$ and $k_2 h$. The branches for the various modes stem from equally spaced points on a line of slope V_3'/V_3 (corresponding to $\omega = V_3' k_2$). For the cut-off frequencies $\omega_p = p\omega_0$, the value $\chi_1 = 0$ will imply $\omega = V_3' k_2$ for the corresponding mode. As the frequency increases, the curves tend closer and closer to the straight line $\omega = V_3 k_2$. Consequently (see exercise 5.7) the phase velocity $V_\phi = \omega/k_2$ and the group velocity $V_g = d\omega/dk_2$ depend on the normalized frequency fh/V_3 as shown in fig. 5.39.

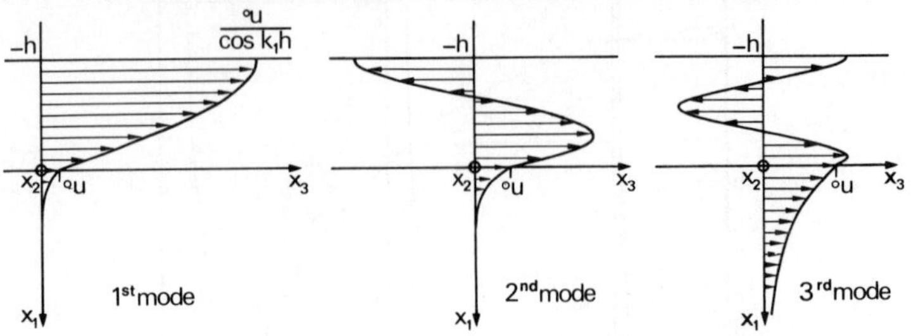

Fig. 5.36. Particle displacements for the same frequency ($\omega = 2.25\ \omega_0$) and for the first three modes in a silica layer on silicon.

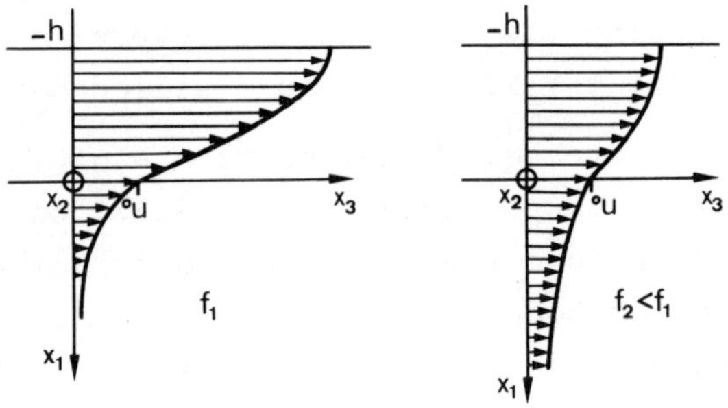

Fig. 5.37. Particle displacements for the first mode at two different frequencies f_1 and $f_2 < f_1$.

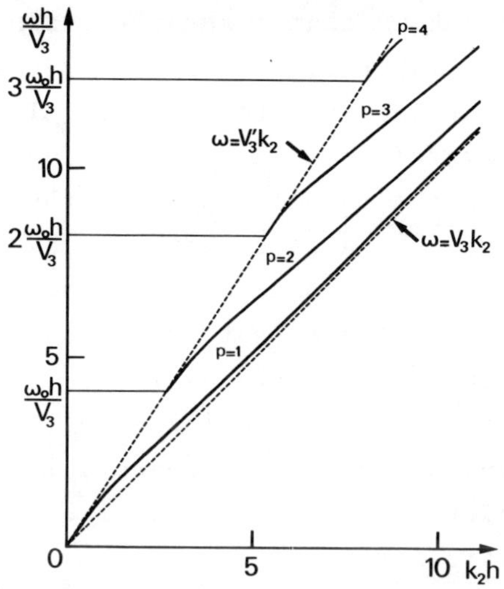

Fig. 5.38. Love waves. In reduced coordinates, dispersion curves $\omega(k_2)$ for the first modes in a silica layer on silicon.

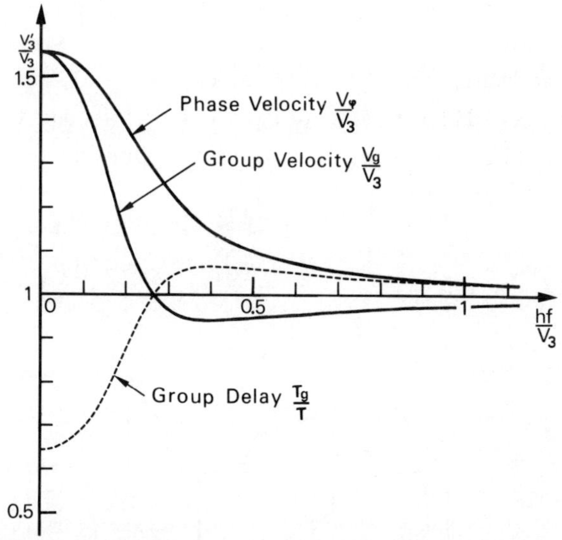

Fig. 5.39. Love waves (silica layer on silicon). Phase and group velocities vs frequency. The dotted curve shows the group delay $\tau_g = \ell/V_g$ in units of the delay $\tau = \ell/V_3$ of the SH wave in the layer.

5.3. SURFACE ELASTIC WAVES. RAYLEIGH WAVES

We have just shown that a shear horizontal wave can propagate within a layer or its vicinity. This wave (a Love wave) is a surface wave for the overall medium (layer plus substrate), but the propagation exhibits dispersion, due to the inhomogeneity of the medium. It is worth asking whether a wave, without any layer and dispersive effect, can propagate along the surface of a homogeneous medium. The answer is yes but the wave, as Rayleigh [6] showed as early as 1885, no longer has linear polarization. This complex wave involves both a longitudinal and a shear displacement, in order to satisfy the boundary condition at the free surface (vanishing mechanical stress).

In a bounded medium, a propagating elastic wave should satisfy the propagation equation 5.2

$$\rho \frac{\partial^2 u_i}{\partial t^2} = c_{ijk\ell} \frac{\partial^2 u_\ell}{\partial x_j \partial x_k}$$

on the one hand, and the boundary condition 4.17

$$p_i = T_{ij} \ell_j$$

on the other hand.

If the medium, in the half space $x_3 > 0$, is unbounded in the x_1 and x_2 directions (fig. 5.40), we must express the fact that the surface $x_3 = 0$ undergoes no external force: $p_i = 0$, or, since $\ell_j = \delta_{j3}$

$$T_{i3} = c_{i3k\ell} \frac{\partial u_k}{\partial x_\ell} = 0 \quad \text{for} \quad x_3 = 0. \qquad (5.90)$$

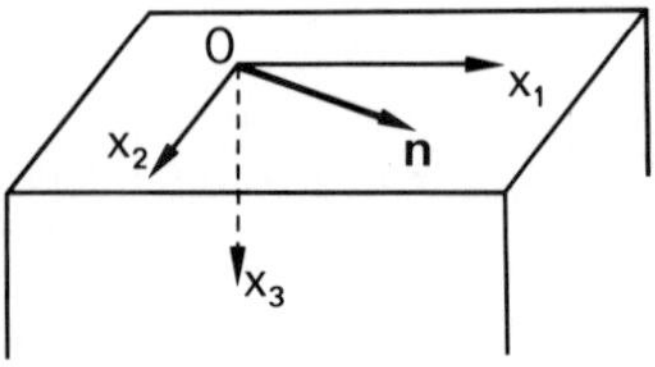

Fig. 5.40. Reference frame for Rayleigh waves.

Let us now look for solutions, if any, that satisfy the above restriction and represent surface waves, which means that the amplitude should decrease with the depth x_3. We consider a displacement of the form

$$u_k = {}^o u_k \; e^{-\chi x_3} \; e^{i\omega[t-(n_1 x_1 + n_2 x_2)/V]} \text{with } \mathrm{Re}[\chi] > 0$$

the propagation direction of which is defined, within the free surface plane, by the cosines n_1 and n_2, and the amplitude of which decreases exponentially from the surface. In order to recover some standard expression, we define

$$\chi = i\frac{\omega}{V} n_3$$

where n_3 is not a third cosine for $\vec{n}$ but an unknown quantity which has to be determined [7]. By substituting

$$u_k = {}^o u_k \; e^{i\omega(t-n_\ell x_\ell/V)}$$

in the propagation equation, we obtain the Christoffel equation as in § 5.1.1:

$$\rho V^2 \; {}^o u_i = c_{ijk\ell} n_j n_k \; {}^o u_\ell .$$

The determinant

$$|c_{ijk\ell} n_j n_k - \rho V^2 \delta_{i\ell}| = 0 \qquad (5.91)$$

has to vanish. This yields a cubic equation in n_3^2, where the velocity V acts as a parameter. Since the coefficients of all powers of n are real, there are three pairs of complex conjugate solutions and we must select those having a negative imaginary part, say $n_3^{(1)}$, $n_3^{(2)}$, $n_3^{(3)}$. The others would give a negative $\mathrm{Re}[\chi]$ value and thus an exponentially increasing amplitude. For each value $n_3^{(r)}$ there is an eigenvector ${}^o u_k^{(r)}$ and a displacement

$$u_k^{(r)} = {}^o u_k^{(r)} \; e^{i\omega(t-n_\ell{}^{(r)} x_\ell/V)} . \qquad (5.92)$$

The general solution is a linear combination of the three displacements corresponding to the same velocity V:

$$u_k = \sum_{r=1}^{3} A_r u_k^{(r)} . \qquad (5.93)$$

The coefficients A_r and the velocity V are determined by inserting this expression into the boundary condition (5.90) at $x_3 = 0$:

$$\sum_{r=1}^{3} c_{i3k\ell} A_r \left(\frac{\partial u_k^{(r)}}{\partial x_\ell}\right)_{x_3=0} = 0$$

and from

$$\left(\frac{\partial u_k^{(r)}}{\partial x_\ell}\right)_{x_3=0} = {}^o u_k^{(r)} \left(-\frac{i\omega}{V}\right) n_\ell^{(r)} \, e^{i\omega[t-(n_1x_1+n_2x_2)/V]}$$

we get

$$\sum_{r=1}^{3} c_{i3k\ell} n_\ell^{(r)} \, {}^o u_k^{(r)} A_r = 0 \qquad i = 1, 2, 3. \qquad (5.94)$$

The Rayleigh wave velocity is a solution of the secular equation which expresses the compatibility of these three homogeneous equations.

If the procedure is straightforward, its application to anisotropic media rapidly leads to intricate calculations which require numerical solutions. So we deal first with the case of an isotropic medium.

5.3.1. Isotropic Medium

We first make use of the plane wave velocities in an unbounded isotropic medium: $V_L = \sqrt{c_{11}/\rho}$ and $V_T = \sqrt{(c_{11}-c_{12})/2\rho}$. Equation 5.9 for the propagation tensor components

$$\Gamma_{i\ell}' = \frac{c_{11} + c_{12}}{2} n_i n_\ell + \frac{c_{11} - c_{12}}{2} n_k^2 \delta_{i\ell}$$

becomes

$$\Gamma_{i\ell} = \rho[(V_L^2 - V_T^2) n_i n_\ell + V_T^2 n_k^2 \delta_{i\ell}].$$

Similarly, the Christoffel equation 5.6

$$\Gamma_{i\ell} \, {}^o u_\ell = \rho V^2 \, {}^o u_i$$

gives

$$(V_L^2 - V_T^2)(n_\ell \, {}^o u_\ell) n_i = (V^2 - V_T^2 n_k^2) \, {}^o u_i \, . \qquad (5.95)$$

Two cases should be considered:

- $n_\ell \, {}^o u_\ell \neq 0$. Equation 5.95 implies that ${}^o u_i$ is proportional to n_i: ${}^o u_i = B n_i$, and

$$(n_\ell {}^o u_\ell) n_i = n_\ell^2 B n_i = n_\ell^2 {}^o u_i \quad .$$

Equating the coefficients of ${}^o u_i$ on the two sides of (5.95) we obtain

$$(V_L^2 - V_T^2) n_\ell^2 = V^2 - V_T^2 n_k^2$$

or

$$V^2 = V_L^2 n_k^2 \qquad (5.96)$$

$$- n_\ell {}^o u_\ell = 0 \Rightarrow \quad V^2 = V_T^2 n_k^2 . \qquad (5.97)$$

Since the medium is isotopic,, we may specify the propagation direction to be along the co-ordinate axis, e.g. along the x_1 axis:

$$n_1 = 1, \qquad n_2 = 0.$$

The solutions 5.96 and 5.97 of the eigenvalue equation will determine n_3 in terms of the propagation velocity V:

- the first solution

$$V^2 = V_L^2 (1 + n_3^2)$$

implies $V < V_L$ for n_3 should be imaginary

$$n_3^L = \pm i(1 - V^2/V_L^2)^{1/2}$$

- the second solution

$$V^2 = V_T^2 (1 + n_3^2)$$

provides

$$n_3^T = \pm i(1 - V^2/V_T^2)^{1/2} \quad \text{with } V < V_T .$$

In both cases, the plus sign must be discarded since it leads to a negative $\chi = i\omega n_3/V$ and to an exponentially increasing amplitude with x_3 which is not physically acceptable. Our eigenvectors will be

- for

$$n_3^{(1)} = - i(1 - V^2/V_L^2)^{1/2} :$$

$$ {}^o\vec{u}^{(1)} \quad \begin{cases} {}^o u_1^{(1)} = 1 \\ {}^o u_2^{(1)} = 0 \\ {}^o u_3^{(1)} = n_3^{(1)} \end{cases}$$

the components of which satisfy ${}^o u_i^{(1)} = B n_i$ with $B = 1$;

- for $n_3^{(2)} = n_3^{(3)} = -i(1 - V^2/V_T^2)^{1/2}$, the two independent vectors

$$
{}^{o}\vec{u}^{(2)} \begin{cases} {}^{o}u_1^{(2)} = 0 \\ {}^{o}u_2^{(2)} = 1 \\ {}^{o}u_3^{(2)} = 0 \end{cases} \quad \text{and} \quad {}^{o}\vec{u}^{(3)} \begin{cases} {}^{o}u_1^{(3)} = -n_3^{(3)} \\ {}^{o}u_2^{(3)} = 0 \\ {}^{o}u_3^{(3)} = 1 \end{cases}
$$

the components of which satisfy the equation

$$n_\ell {}^{o}u_\ell = 0 \quad \text{or} \quad {}^{o}u_1 + n_3 {}^{o}u_3 = 0.$$

For each i, we now expand the boundary condition 5.94:

$$c_{i3k\ell} \sum_{r=1}^{3} n_\ell^{(r)} \, {}^{o}u_k^{(r)} \, A_r = 0$$

and we make use of the stiffness coefficients (4.38) for an isotropic solid. We get

$$i = 1 \qquad c_{1313} \sum_r (n_1^{(r)} \, {}^{o}u_3^{(r)} + n_3^{(r)} \, {}^{o}u_1^{(r)}) A_r = 0$$

$$i = 2 \qquad c_{2323} \sum_r (n_2^{(r)} \, {}^{o}u_3^{(r)} + n_3^{(r)} \, {}^{o}u_2^{(r)}) A_r = 0$$

$$i = 3 \qquad \sum_r (c_{3311} n_1^{(r)} \, {}^{o}u_1^{(r)} + c_{3322} n_2^{(r)} \, {}^{o}u_2^{(r)} + c_{3333} n_3^{(r)} \, {}^{o}u_3^{(r)}) A_r = 0.$$

The above equations can be simplified because $n_1^{(r)} = 1$ and $n_2^{(r)} = 0$. If we define $n_3^{(r)} = q_r$:

$$
\begin{cases} \sum_r ({}^{o}u_3^{(r)} + q_r \, {}^{o}u_1^{(r)}) A_r = 0 \\ \sum_r q_r \, {}^{o}u_2^{(r)} A_r = 0 \\ \sum_r (c_{12} {}^{o}u_1^{(r)} + c_{11} q_r \, {}^{o}u_3^{(r)}) A_r = 0. \end{cases} \tag{5.98}
$$

Knowing the components ${}^{o}u_k^{(r)}$ of the eigenvectors and expanding the sum over r, we finally have

$$
\begin{cases} 2q_1 A_1 + 0 \,.\, A_2 + (1 - q_3^2) A_3 = 0 \\ 0.A_1 + q_2 A_2 + 0.A_3 = 0 \\ (c_{12} + c_{11} q_1^2) A_1 + 0.A_2 + (-c_{12} + c_{11}) q_3 A_3 = 0. \end{cases} \tag{5.99}
$$

In this system, the second equation implies $A_2 = 0$. The other two can be solved if the determinant vanishes:

$$2q_1q_3(c_{11} - c_{12}) - (1 - q_3^2)(c_{12} + c_{11}q_1^2) = 0$$

Replacing c_{11} and c_{12} by their expressions in terms of the velocities V_L and V_T, also $q_1 = n_3^{(1)}$ and $q_3 = n_3^{(3)}$ by their values.

$$c_{11} - c_{12} = 2\rho V_T^2$$

$$1 - q_3^2 = 2 - \frac{V^2}{V_T^2}$$

$$c_{12} + c_{11}q_1^2 = \rho(V_L^2 - 2V_T^2) - \rho V_L^2\left(1 - \frac{V^2}{V_L^2}\right) = -\rho V_T^2\left(2 - \frac{V^2}{V_T^2}\right)$$

we obtain

$$4\left(1 - \frac{V^2}{V_L^2}\right)^{1/2}\left(1 - \frac{V^2}{V_T^2}\right)^{1/2} = \left(2 - \frac{V^2}{V_T^2}\right)^2 \quad . \qquad (5.100)$$

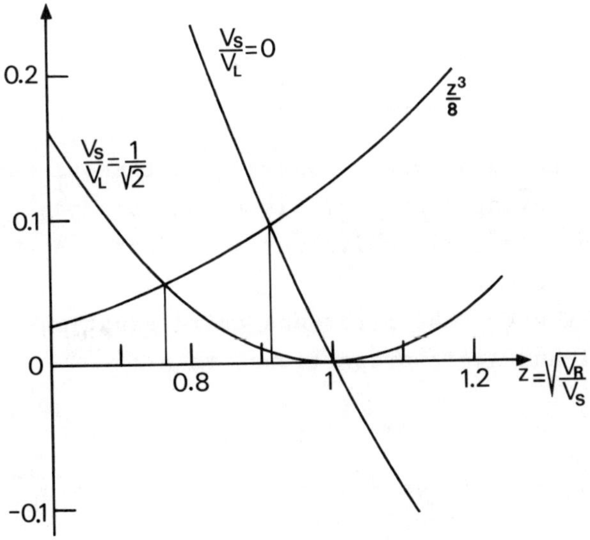

Fig. 5.41. Graphical solution of equation 5.101, giving the Rayleigh wave velocity V_R in the limiting cases $V_T/V_L = 0$ and $V_T/V_L = 1/\sqrt{2}$.

When expanded, this equation turns out to be cubic in V^2. Let us define $z = (V/V_T)^2$:

$$z^3 - 8z^2 + 8z[3 - 2(\frac{V_T}{V_L})^2] - 16[1 - (\frac{V_T}{V_L})^2] = 0$$

or

$$z^3 - 8(z - 1)[z - 2(1 - \frac{V_T^2}{V_L^2})] = 0. \quad (5.101)$$

The quantity z can be graphically determined by the intersection between the curve $y_1 = z^3/8$ and the parabola $y_2 = (z - 1)[z - 2(1 - V_T^2/V_L^2)]$. The curves of fig. 5.41, which have been drawn for the extremum values $V_T/V_L = 0$ and $V_T/V_L = 1/\sqrt{2}$, show that there is only one root in the interval [0, 1], which varies between .912 and .764. This solution defines the Rayleigh wave velocity through the ratio $V_R/V_T = \sqrt{z_R}$, varying from .955 to .874 when V_T/V_L varies from 0 to $1/\sqrt{2}$. The Viktorov formula [8] is a useful approximation to V_R:

$$\frac{V_R}{V_T} = \frac{0.718 - (\frac{V_T}{V_L})^2}{0.75 - (\frac{V_T}{V_L})^2} .$$

The fact that V_R is always less than V_T can be interpreted by saying that there is no matter upon the free surface and this is equivalent to a decrease in the stiffness coefficients.

For $V = V_R$, the first and third equations in 5.99 are identical and A_3 is proportional to A_1:

$$A_3 = - \frac{2q_1}{1 - q_3^2} A_1.$$

The components of the displacement $u_k = \sum_r A_r u_k^{(r)}$

$$\begin{cases} u_1 = A_1(u_1^{(1)} - \frac{2q_1}{1-q_3^2} u_1^{(3)}) \\ u_2 = 0 \\ u_3 = A_1(u_3^{(1)} - \frac{2q_1}{1 - q_3^2} u_3^{(3)}) \end{cases}$$

can be rewritten, using 5.92 and the expressions for the terms ${}^{o}u_k^{(r)}$:

$$\begin{cases} u_1 = A_1\left(e^{-i(q_1\omega/V_R)x_3} + \frac{2q_1q_3}{1-q_3^2}e^{-i(q_{3\omega}/V_R)x_3}\right)e^{i\omega(t-x_1/V_R)} \\ u_2 = 0 \\ u_3 = A_1q_1\left(e^{-i(q_1\omega/V_R)x_3} - \frac{2}{1-q_3^2}e^{-i(q_3\omega/V_R)x_3}\right)e^{i\omega(t-x_1/V_R)}. \end{cases} \tag{5.102}$$

This involves the constants

$$\chi_1 = iq_1\frac{\omega}{V_R} = \frac{2\pi}{\lambda_R}\left(1 - \frac{V_R^2}{V_L^2}\right)^{1/2}$$

$$\chi_3 = iq_3\frac{\omega}{V_R} = \frac{2\pi}{\lambda_R}\left(1 - \frac{V_R^2}{V_T^2}\right)^{1/2} \tag{5.103}$$

where λ_R is the Rayleigh wavelength, and the factor

$$a = -\frac{2q_1q_3}{1 - q_3^2} = \frac{2\left(1 - \frac{V_R^2}{V_L^2}\right)^{1/2}\left(1 - \frac{V_R^2}{V_T^2}\right)^{1/2}}{2 - \frac{V_R^2}{V_T^2}}$$

which, because of 5.100, is equal to

$$a = 1 - \frac{V_R^2}{2V_T^2} = \frac{1 - q_3^2}{2} \tag{5.104}$$

Finally, the displacement components are

$$\begin{cases} u_1 = A_1\left(e^{-\chi_1x_3} - a\,e^{-\chi_3x_3}\right)e^{i\omega(t - x_1/V_R)} \\ u_2 = 0 \\ u_3 = -iA_1\left(1 - \frac{V_R^2}{V_L^2}\right)^{1/2}\left(e^{-\chi_1x_3} - \frac{1}{a}e^{-\chi_3x_3}\right)e^{i\omega(t - x_1/V_R)}. \end{cases} \tag{5.105}$$

Since $u_2 = 0$, the particle displacement takes place in the plane that contains the propagation direction (x_1) and the normal to the free surface (x_3), which is called a sagittal plane. The imaginary quantity (i) stands for a $\pi/2$ phase shift between u_1 and u_3, which we also find on the real parts

$$\begin{cases} Re[u_1] = U_1(x_3) \cos \omega \left(t - \frac{x_1}{V_R}\right) \\ \\ Re[u_3] = U_3(x_3) \sin \omega \left(t - \frac{x_1}{V_R}\right). \end{cases} \quad (5.106)$$

We see that a Rayleigh wave will cause an elliptic particle movement, whatever the depth x_3. The x_3 variation of the amplitude $U_1(x_3)$ and $U_3(x_3)$ are given by

$$\begin{cases} U_1(x_3) = A_1(e^{-\chi_1 x_3} - a\, e^{-\chi_3 x_3}) \\ \\ U_3(x_3) = A_1\left(1 - \frac{V_R^2}{V_L^2}\right)^{1/2} \left(e^{-\chi_1 x_3} - \frac{1}{a}\, e^{-\chi_3 x_3}\right) \end{cases} \quad (5.107)$$

Since the variations are different the shape of the ellipse changes and the polarization, negative at the surface, may become positive. The curves of 5.42, plotted for the case of silica (V_R = 3410 m/s) are an illustration of the U_1 (longitudinal) and U_3 (shear) variations which can be extended to many isotropic media: at the surface, the shear component is roughly 1.5 times the longitudinal component; the latter changes sign at a depth of about .2 λ_R.

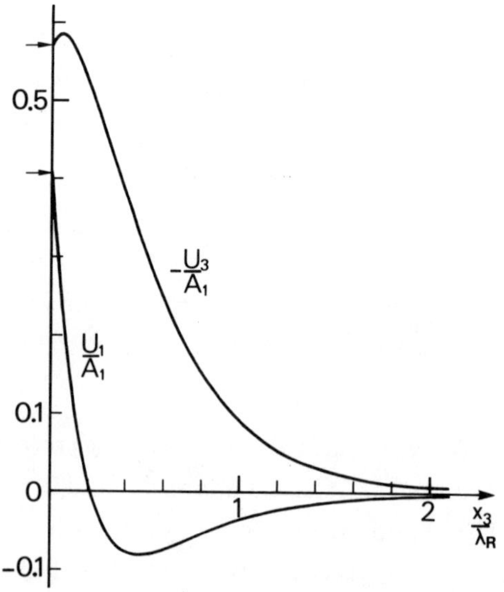

Fig. 5.42. Rayleigh waves on an isotropic solid (silica). Amplitudes of the longitudinal (U_1) and shear (U_3) displacements vs depth.

5.3.2. Anisotropic Medium

In a crystal the equations can only be solved numerically. Several departures from the isotropic case show up.

- The surface wave velocity depends on the direction of propagation in the free surface plane except if this plane is orthogonal to a six-fold axis (CdS, ZnO).

- If the sagittal plane is not a symmetry plane for the elastic properties, the displacement vector has three orthogonal components instead of two and it runs along an ellipse, the plane of which is at an angle to the sagittal plane.

- The roots $n_3^{(r)}$ of the secular equation 5.91 are no longer purely imaginary, but complex. Consequently the amplitudes of the displacement components go to zero and oscillate, as depicted in fig. 5.43.

- Generally, the energy flux is not parallel to the propagation direction as for bulk waves. The Rayleigh mode is pure only at the surface of selected planes and for specific directions which, in practice, are very important.

- In some isolated directions, on the surface of anisotropic media, there exists a Rayleigh type solution, the phase velocity of which is greater than the slow shear wave velocity. However, as soon as one tilts the propagation direction from the above position, the wave is no longer strictly a surface wave. A component appears which gives rise to an inward energy radiation. This solution, which does not satisfy the vanishing displacement condition at infinite depth, is called a pseudo surface wave [9], or a leaky surface wave.

The example we have chosen to illustrate the general method described at the beginning of § 5.3, will indicate how intricate the calculations may be, even for the simplest anisotropic case.

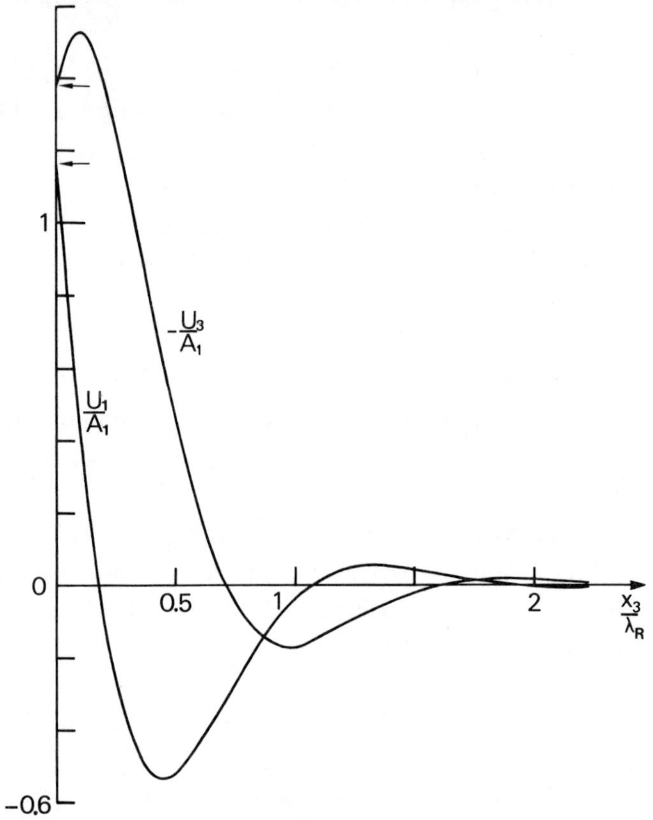

Fig. 5.43. Rayleigh waves on a cubic crystal (GaAs) in the [100] direction of the (001) plane. The amplitudes of the longitudinal (U_1) and shear (U_3) components follow a damped oscillation (see equations 5.125).

Propagation of a Rayleigh wave along the [100] direction of the (001) plane in a cubic crystal

Since $n_1 = 1$ and $n_2 = 0$, the expressions (5.32) for the propagation tensor components in a cubic crystal become, with $n_3 = q$:

$$\Gamma_{11} = c_{11} + c_{44}q^2 \qquad \Gamma_{12} = 0 \qquad \Gamma_{13} = (c_{13} + c_{44})q$$

$$\Gamma_{22} = c_{44}(1 + q^2) \qquad \Gamma_{23} = 0 \qquad \Gamma_{33} = c_{44} + c_{11}q^2.$$

Γ_{12} and Γ_{23} vanish, so the secular equation can be factorised out:

$$(\Gamma_{22} - \rho V^2)[(\Gamma_{11} - \rho V^2)(\Gamma_{33} - \rho V^2) - \Gamma_{13}^2] = 0.$$

The root $q_2 = n_3^{(2)}$ is given by

$$c_{44}(1 + q_2^2) = \rho V^2$$

or, introducing the degenerate shear wave velocity $V_T = \sqrt{c_{44}/\rho}$,

$$q_2^2 = \frac{V^2}{V_T^2} - 1.$$

In order that q_2 be imaginary, the Rayleigh wave velocity should be less than V_T:

$$q_2 = - i(1 - \frac{V^2}{V_T^2})^{1/2}$$

The other two roots $q_1 = n_3^{(1)}$ and $q_3 = n_3^{(3)}$ are the solutions of the quartic equation.

$$(c_{11} - \rho V^2 + c_{44}q^2)(c_{44} - \rho V^2 + c_{11}q^2) - (c_{12} + c_{44})^2 q^2 = 0$$

which can also be written as

$$c_{11}c_{44}(q^2)^2 + q^2[(c_{11} - \rho V^2)c_{11} + (c_{44} - \rho V^2)c_{44} - (c_{12} + c_{44})^2] + (c_{11} - \rho V^2)(c_{44} - \rho V^2) = 0 \tag{5.108}$$

The x_2-components of ${}^o\vec{u}^{(1)}$ and ${}^o\vec{u}^{(3)}$ vanish. The other two components ${}^ou_1$ and ${}^ou_3$ satisfy the equation

$$(\Gamma_{11} - \rho V^2)\,{}^ou_1 + \Gamma_{13}\,{}^ou_3 = 0.$$

Taking ${}^ou_1 = 1$, we have

$${}^ou_3 = -\frac{\Gamma_{11} - \rho V^2}{\Gamma_{13}}$$

or, for each eigenvector (${}^o\vec{u}^{(1)}$ and ${}^o\vec{u}^{(3)}$):

$$p_1 = {}^ou_3^{(1)} = - \frac{c_{11} - \rho V^2 + c_{44}q_1^2}{(c_{12}+c_{44})q_1}$$

(5.109)

and

$$p_3 = {}^ou_3^{(3)} = - \frac{c_{11} - \rho V^2 + c_{44}q_3^2}{(c_{12}+c_{44})q_3}$$

The above results yield

$$\overset{o}{\vec{u}}^{(1)}\ (1,\ 0,\ p_1) \qquad \overset{o}{\vec{u}}^{(2)}\ (0,\ 1,\ 0) \qquad \overset{o}{\vec{u}}^{(3)}\ (1,\ 0,\ p_3)$$

and the boundary condition expansion is identical to that of the isotropic medium (5.98) since the non vanishing components of ${}^{o}u_k^{(r)}$ and $c_{ijk\ell}$ are the same:

$$\begin{cases} \sum_r ({}^{o}u_3^{(r)} + q_r\, {}^{o}u_1^{(r)}) A_r = 0 \\ \sum_r q_r\, {}^{o}u_2^{(r)} A_r = 0 \\ \sum_r (c_{12}\, {}^{o}u_1^{(r)} + c_{11} q_r\, {}^{o}u_3^{(r)}) A_r = 0 \end{cases}$$

Given the values for ${}^{o}u_k^{(r)}$:

$$\begin{cases} (p_1 + q_1)A_1 + 0.A_2 + (p_3 + q_3)A_3 = 0 \\ 0.A_1 + q_2 A_2 + 0.A_3 = 0 \\ (c_{12} + c_{11} q_1 p_1)A_1 + 0.A_2 + (c_{12} + c_{11} q_3 p_3)A_3 = 0. \end{cases} \tag{5.110}$$

The second equation implies $A_2 = 0$. The other two are compatible only if the corresponding determinant is zero:

$$(p_1 + q_1)(c_{12} + c_{11} p_3 q_3) - (p_3 + q_3)(c_{12} + c_{11} p_1 q_1) = 0 \tag{5.111}$$

Starting from (5.109) and, defining $c = c_{11} - \rho V^2$, we obtain

$$p_r q_r = -\frac{c + c_{44} q_r^2}{(c_{12} + c_{44})} \qquad r = 1 \text{ or } 3 \tag{5.112}$$

and

$$p_r + q_r = \frac{(c_{12}+c_{44})q_r^2 - c - c_{44} q_r^2}{(c_{12} + c_{44}) q_r} = -\frac{c - c_{12} q_r^2}{(c_{12}+c_{44}) q_r} \tag{5.113}$$

After a substitution into 5.111 and some rearrangement:

$$(c - c_{12}\, q_3^2)[c_{12}(c_{12} + c_{44}) - c c_{11} - c_{11} c_{44} q_1^2] q_1$$

$$-(c - c_{12} q_1^2)[c_{12}(c_{12} + c_{44}) - c c_{11} - c_{11} c_{44} q_3^2] q_3 = 0$$

or

$$c_{12} c_{11} c_{44} q_1^2 q_3^2 (q_1 - q_3) - c c_{11} c_{44} (q_1^3 - q_3^3)$$

$$+ c_{12}[c_{12}(c_{12}+c_{44}) - c c_{11}] q_1 q_3 (q_1 - q_3)$$

$$+ c[c_{12}(c_{12}+c_{44}) - c c_{11}](q_1 - q_3) = 0$$

We divide by $(q_1 - q_3)$ and put all products q_1q_3 on the right hand side:

$$c_{12}c_{11}c_{44}q_1^2q_3^2 - cc_{11}c_{44}(q_1^2 + q_3^2) + c[c_{12}(c_{12} + c_{44}) - cc_{11}]$$
$$= (c_{12} + c_{44})(cc_{11} - c_{12}^2)q_1q_3. \qquad (5.114)$$

The product $q_1^2q_3^2$ and the sum $q_1^2 + q_3^2$ of the roots of equation (5.108) are given by

$$c_{11}c_{44}q_1^2q_3^2 = c(c_{44} - \rho V^2)$$

and

$$-c_{11}c_{44}(q_1^2 + q_3^2) = cc_{11} + c_{44}(c_{44} - \rho V^2) - (c_{12} + c_{44})^2.$$

Inserting all this into the left hand side M of 5.114, the latter becomes:

$$M = c_{12}c(c_{44} - \rho V^2)$$
$$+c[cc_{11}+c_{44}(c_{44}- \rho V^2)-(c_{12}+c_{44})^2 + c_{12}(c_{12}+c_{44})-cc_{11}]$$

or

$$M = c[(c_{44} - \rho V^2)(c_{12} + c_{44}) - c_{44}(c_{12} + c_{44})]$$

and finally

$$M = - \rho V^2c(c_{12} + c_{44}).$$

Squaring the right hand side N of 5.114 provides

$$N^2 = (c_{12} + c_{44})^2(cc_{11} - c_{12}^2)^2 \frac{c(c_{44}-\rho V^2)}{c_{11}c_{44}}$$

From $M^2 = N^2$, we get

$$(\rho V^2)^2cc_{11} = (cc_{11} - c_{12}^2)^2 (1 - \frac{\rho V^2}{c_{44}}).$$

Defining

$$R = \frac{\rho V^2}{c_{11}} = (\frac{V}{V_L})^2 \qquad (5.115)$$

where V_L is the longitudinal wave velocity along the [100] direction, we have an equation for R (of degree 3) which gives the Rayleigh wave velocity $V_R = V_LR^{\frac{1}{2}}$. Making use of $c = c_{11} - \rho V^2 = c_{11}(1 - R)$, this becomes

$$\boxed{R^2(1 - R) = [1 - (\frac{c_{12}}{c_{11}})^2 - R]^2(1 - \frac{c_{11}}{c_{44}}R).} \qquad (5.116)$$

This equation was first established by Stoneley [10].

The particle displacement is deduced from 5.92 and 5.93:

$$u_k = \sum_{r=1}^{3} A_r \, {}^{o}u_k^{(r)} \; e^{-i\omega \, n_3^{(r)}/V_R \, x_3} \; e^{i\omega(t - x_1/V_R)}$$

and, since $A_2 = 0$ and ${}^{o}u_1^{(1)} = {}^{o}u_1^{(3)} = 1$, this gives, for each component:

$$\begin{cases} u_1 = A_1\left(e^{-i2\pi q_1 \, x_3/\lambda_R} + \dfrac{A_3}{A_1} e^{-i2\pi q_3 \, x_3/\lambda_R}\right) e^{i\omega(t - x_1/V_R)} \\ u_2 = 0 \\ u_3 = A_1\left(p_1 e^{-i2\pi q_1 \, x_3/\lambda_R} + \dfrac{A_3}{A_1} p_3 e^{-i2\pi q_3 \, x_3/\lambda_R}\right) e^{i\omega(t - x_1/V_R)} \end{cases} \qquad (5.117)$$

The particle displacement takes place in the sagittal plane, i.e. the symmetry plane (010), as in the isotropic case. Let us assume that the roots q^2 of 5.108 are complex and consequently conjugate:

$$q^2 = Q \, e^{\pm i2\theta} \quad \text{with } Q > 0 \quad \text{and } 0 < \theta < \frac{\pi}{2} .$$

Only the roots q_1 and q_3 that have a negative imaginary part are acceptable:

$$q_1 = -\sqrt{Q} \, e^{i\theta} = -g - ih \quad \text{and} \quad q_3 = \sqrt{Q} \, e^{-i\theta} = g - ih = -q_1^* \qquad (5.118)$$

with

$$g = \sqrt{Q} \cos\theta \qquad \text{and} \qquad h = \sqrt{Q} \sin\theta .$$

The equations 5.109 show that $p_3 = -p_1^*$. If we define

$$p_1 = r \, e^{-i\eta} \implies p_3 = -r \, e^{i\eta}$$

we have

$$r e^{-i\eta} = \frac{c \, e^{-i\theta} + c_{44} Q \, e^{i\theta}}{(c_{12} + c_{44}) \sqrt{Q}}$$

or

$$r^2 = \frac{(c + c_{44}Q)^2 \cos^2\theta + (c - c_{44}Q)^2 \sin^2\theta}{(c_{12} + c_{44})^2 Q} \qquad (5.119)$$

and

$$\tan\eta = \frac{c - c_{44}Q}{c + c_{44}Q} \tan\theta \qquad (5.120)$$

The ratio A_3/A_1 is given by 5.110, which reduces to the first equation when $V = V_R$:

$$\frac{A_3}{A_1} = -\frac{p_1 + q_1}{p_3 + q_3}$$

or, making use of 5.113:

$$\frac{A_3}{A_1} = -\frac{c - c_{12}q_1^2}{c - c_{12}q_3^2} \cdot \frac{q_3}{q_1}$$

Replacing q_1 and q_3 by their expressions, using the fact that A_3/A_1 has unit modulus, we obtain

$$\frac{A_3}{A_1} = \frac{c\,e^{-i\theta} - c_{12}Q\,e^{i\theta}}{c\,e^{i\theta} - c_{12}Q\,e^{-i\theta}} = e^{-i2\alpha} \qquad (5.121)$$

where α is the phase angle of the denominator

$$\tan\alpha = \frac{c + c_{12}Q}{c - c_{12}Q}\tan\theta. \qquad (5.122)$$

Since 5.120 and 5.122 show that the angles η and α are related, one can prove that $\eta = 2\alpha - \pi/2$. Thus

$$p_1 = ir\,e^{-i2\alpha} \qquad p_3 = ir\,e^{i2\alpha} \qquad (5.123)$$

The longitudinal and shear components of the displacement (u_1 and u_3) are obtained by inserting 5.118, 5.123 and 5.121 into 5.117:

$$\begin{cases} u_1 = A_1\,e^{-2\pi h\,x_3/\lambda_R}\,(e^{i2\pi g\,x_3/\lambda_R} + e^{-i2\alpha}e^{-i2\pi g\,x_3/\lambda_R}) \\ \qquad\qquad e^{i\omega(t - x_1/V_R)} \\ u_3 = iA_1 r\,e^{-2\pi h\,x_3/\lambda_R}\,(e^{-i2\alpha}\,e^{i2\pi g\,x_3/\lambda_R} + e^{-i2\pi g\,x_3/\lambda_R}) \\ \qquad\qquad e^{i\omega(t - x_1/V_R)} \end{cases}$$

or

$$\begin{cases} u_1 = 2A_1\,e^{-2\pi h\,x_3/\lambda_R}\cos\left(2\pi g\frac{x_3}{\lambda_R} + \alpha\right)\exp i\left[\omega\left(t - \frac{x_1}{V_R}\right) - \alpha\right] \\ u_3 = i2A_1 r\,e^{-2\pi h\,x_3/\lambda_R}\cos\left(2\pi g\frac{x_3}{\lambda_R} - \alpha\right)\exp i\left[\omega\left(t - \frac{x_1}{V_R}\right) - \alpha\right] \end{cases}$$

The real parts of the displacement are

$$\begin{cases} \mathrm{Re}[u_1] = U_1(x_3)\cos\left[\omega\left(t - \frac{x_1}{V_R}\right) - \alpha\right] \\ \mathrm{Re}[u_3] = U_3(x_3)\sin\left[\omega\left(t - \frac{x_1}{V_R}\right) - \alpha\right] \end{cases} \qquad (5.124)$$

They exhibit a $\pi/2$ phase shift so that the particles move along an ellipse, the axes of which are parallel and orthogonal to the surface. The longitudinal amplitude

$$U_1(x_3) = 2A_1\ e^{-2\pi h\ x_3/\lambda_R} \cos\left(2\pi g \frac{x_3}{\lambda_R} + \alpha\right) \qquad (5.125a)$$

and the shear amplitude

$$U_3(x_3) = -2A_1 r\ e^{-2\pi h\ x_3/\lambda_R} \cos\left(2\pi g \frac{x_3}{\lambda_R} - \alpha\right) \qquad (5.125b)$$

oscillate with a spatial period λ_R/g equal to the result of dividing the wave length by the real part of the roots q_1 and q_3. They show an exponential damping which depends on the imaginary parts of the roots. The curves of Fig. 5.43 show the variations of the amplitudes U_1/A_1 and U_3/A_1 in terms of the normalized depth x_3/λ_R, for the case of gallium arsenide.* With the elastic moduli of 4.7, the Stoneley equation is satisfied for R = .3305, i.e.

$$V_R = 2{,}720 \text{ m/s}.$$

The roots q_1 and q_3 are complex:

$$g = 0.5613 \qquad h = 0.4017$$

and the equations 5.119 and 5.122 yield

$$r = 1.185 \qquad \alpha = 54^\circ\ 23'.$$

The curves for the depth dependence of the displacements exhibit oscillations since the solutions $n_3^{(r)}$ are complex whereas in the isotropic case they are purely imaginary. For the cubic system crystals, this happens when the anisotropy factor $A = 2c_{44}/(c_{11} - c_{12})$ is greater than 1. For GaAs, A = 1.83. Also for silicon A = 1.565:

$$V_R = 4{,}917 \text{ m/s} \qquad g = .4808 \qquad h = .4556$$
$$r = 1.226 \qquad \alpha = 58^\circ\ 1'$$

whose displacement components are sketched in fig. 5.43.

In practice, Rayleigh waves are generated directly at the surface of the crystal by an electrical field (§ 7.2). Useful materials are thus piezoelectric. Therefore we shall describe the conditions of propagation for the major crystals principally employed (quartz, lithium niobate) in section 6.2.2 of the following chapter entitled "Piezoelectricity".

---

* The piezoelectric properties of this medium have no influence on Rayleigh wave propagation in the [100] direction of the plane (001).

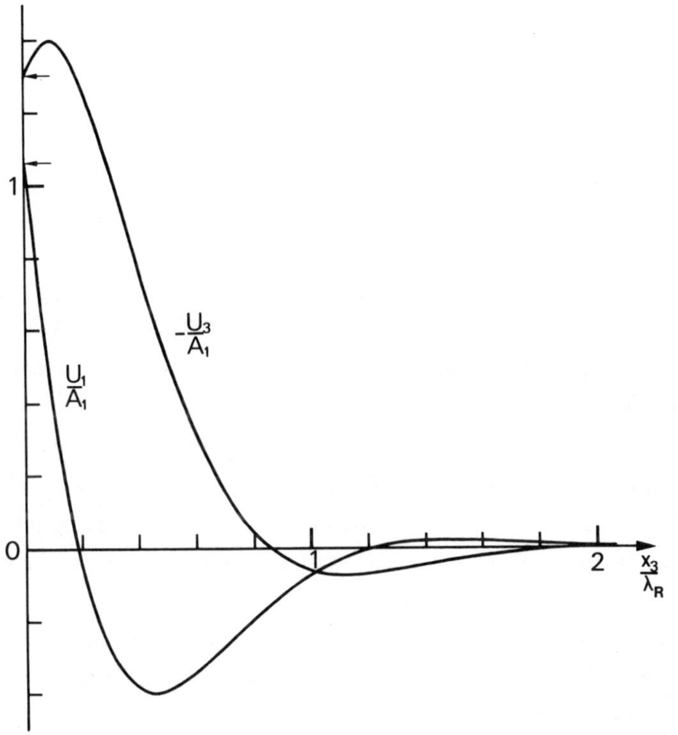

Fig. 5.44. Rayleigh waves on a silicon crystal in the [100] direction of the (001) plane. Amplitudes of the longitudinal (U_1) and shear (U_3) displacements.

REFERENCES

[1] G.W. Farnell and E.L. Adler. Elastic wave propagation in thin layers, fig. 25. In Physical Acoustics (W.P. Mason and R.N. Thurston, eds) vol. 9, p. 35-127. New York: Academic Press (1972).

[2] F.E. Borgnis. Specific directions of longitudinal wave propagation in anisotropic media, Phys. Rev., 98, p. 1000-1005 (1955).

[3] K. Brugger. Pure modes for elastic waves in crystals, J. Appl. Phys., 36, p. 759-768 (1965).

[4] A. Levelut. Propagation de l'énergie acoustique dans les cristaux (systèmes cubique, hexagonal et quadratique) Acta Cryst., A 25, p. 553-563 (1969).

[5] D.L. Arenberg. Ultrasonic solid delay lines, J. Acoust. Soc. Am., 20, p. 1-25 (1948).

[6] Lord Rayleigh. On waves propagated along the plane surface of an elastic solid, Proc. London Math. Soc., 17, p. 4-11 (1885).

[7] G.W. Farnell. Properties of elastic surface waves, p. 114. In Physical Acoustics (W.P. Mason and R.N. Thurston, eds), vol. 6, p. 109-166. New York: Academic Press (1970).

[8] I.A. Viktorov. Rayleigh and Lamb waves, p. 3. New York: Plenum Press (1967).

[9] T.C. Lim and G.W. Farnell. Character of pseudo surface waves on anisotropic crystals, J. Acoust. Soc. Am., 45, p. 845-851 (1969).

[10] R. Stoneley. The propagation of surface elastic waves in a cubic crystal, Proc. Roy. Soc., A 232, p. 447-458 (1955).

BIBLIOGRAPHY

<u>Plane waves in an unbounded crystal. Reflexion. Refraction</u>

F.I. Federov. Theory of elastic waves in crystals, chap. 3, 4, 8. New York: Plenum Press (1968).

B.A. Auld. Acoustic fields and waves in solids, chap. 7 vol. I, chap. 9 vol. II. New York: Wiley-Interscience (1973).

L. Landau and E. Lifchitz. Théorie de l'élasticité, chap. 3. Moscow: Mir (1967).

M.R. Redwood. Mechanical wave guides, chap. 2. London: Pergamon Press (1960).

M.J.P. Musgrave. The propagation of elastic waves in crystals and other anisotropic media, Reports Prog. in Phys., 22, p. 74-96 (1959).

G.W. Farnell. Elastic waves in trigonal crystals, Can. J. Phys., 39, p. 65-80 (1961).

Propagation in a thin layer. Love waves

G.W. Farnell and E.L. Adler. Reference [1].
P. Tournois and C. Lardat. Love wave dispersive delay lines for wide band pulse compression, IEEE Trans. Sonics Ultrason., vol. SU-16, p. 107-117 (1969).

Rayleigh waves

G.W. Farnell. Reference [7].
I.A. Viktorov. Reference [8].
K. Kajimura, R. Inaba and N. Mikoshiba. Experimental evidence for elliptic particule motion accompanied by elastic surface waves, Appl. Phys. Lett., 19, 182 (1971).
M.W. Lawrence and L.W. Davies. Surface motion measurements on surface elastic waves, Appl. Phys. Lett., 20, 328 (1972).

EXERCISES

5.1. In the case of a plane sine wave of amplitude a, calculate the mean power $\mathcal{P}$ carried by a beam of cross sectional area A.

Solution. For a pure mode:

$$\mathcal{P} = \langle P\rangle A = V\langle\mathcal{E}\rangle A$$

and for a sine wave F(t) = cos ωt, equation 5.22 provides

$$\langle\mathcal{E}\rangle = \rho a^2\omega^2\langle\sin^2 \omega t\rangle = \frac{1}{2}\rho a^2\omega^2$$

or, in terms of the elastic impedance per unit area $Z = \rho V$ as defined in exercise 1.3:

$$\mathcal{P} = \frac{1}{2} Z\omega^2 a^2 A.$$

5.2. Show that any longitudinal wave is a pure wave.

Solution. Since $^{o}u_j = n_j$, equation 5.24 gives the energy velocity:

$$V_i^e = c_{ijk\ell} \frac{n_j n_k {}^o u_\ell}{\rho V} = \frac{\Gamma_{i\ell} {}^o u_\ell}{\rho V} .$$

Making use of the Christoffel equation

$$\Gamma_{i\ell} {}^o u_\ell = \rho V^2 {}^o u_i ,$$

we obtain

$$V_i^e = V {}^o u_i = V n_i$$

so that the energy velocity is parallel to the direction of propagation.

5.3. Show that, if there is a pure mode, of polarization $^o\vec{u}$ and velocity V, in the $\vec{n}$ direction, another pure mode can propagate at the same velocity in the $^o\vec{u}$ direction with polarization $\vec{n}$. Show that in the YZ plane of a trigonal crystal, only two pure directions can be found for the shear wave with X-polarization (denoted AC and BC in fig. 5.25).

Solution. From equation 5.24, we see that the pure mode condition $V_i^e = Vn_i$ reads

$$c_{ijk\ell} {}^o u_j {}^o u_\ell n_k = \gamma n_i \qquad \gamma = \rho V^2$$

or, after a permutation of k and ℓ (since $c_{ij\ell k} = c_{ijk\ell}$):

$$c_{ijk\ell} {}^o u_j {}^o u_k n_\ell = \gamma n_i .$$

On comparing with the Christoffel equation 5.4, we notice that $\vec{n}$ is the polarization of a wave which propagates in the $^o\vec{u}$ direction with velocity V. Such a mode is pure because 5.4 ensures

$$c_{ijk\ell} n_j n_\ell {}^o u_k = \gamma {}^o u_i .$$

For a pure mode, polarization and propagation direction can be permuted.

As an example, in a crystal with symmetry $\bar{3}m$, 3m, 32, the X-polarized shear wave is pure in two orthogonal directions of the YZ plane, which are the polarizations of the two non degenerate pure waves propagating along the diad axis (A_2 or $\bar{A}_2$), i.e. the X direction.

5.4. Give the polar angle θ^e of the energy velocity direction for a shear wave, the polarization of which is orthogonal to the meridian plane of a hexagonal system crystal. Which value of the angle θ corresponds to the maximum deviation $(\theta - \theta^e)$?

Solution. Using the axes of § 5.1.6.2c:

$$^{o}u_i = \delta_{i1} \qquad n_1 = 0 \qquad n_2 = \sin\theta \qquad n_3 = \cos\theta$$

Equation 5.24 gives

$$V_i^e = c_{i1k1}\,\frac{n_k}{\rho V}$$

or

$$V_1^e = 0, \quad V_2^e = c_{66}\,\frac{\sin\theta}{\rho V}, \quad V_3^e = c_{44}\,\frac{\cos\theta}{\rho V}$$

and

$$\tan\theta^e = \frac{V_2^e}{V_3^e} = \frac{c_{66}}{c_{44}}\tan\theta = k\tan\theta$$

$$\text{with } k = \frac{c_{66}}{c_{44}}.$$

The deviation $\theta - \theta^e$ is a maximum for $d\theta^e/d\theta = 1$, or, from the above equation for

$$\frac{\cos\theta_M}{\cos\theta_M^e} = \sqrt{k}$$

so that

$$\tan\theta_M^e = \sqrt{\frac{k}{\cos^2\theta_M} - 1} = \sqrt{(k-1)+k\tan^2\theta_M}$$

which together with $\tan\theta_M^e = k\tan\theta_M$, yields

$$\tan\theta_M = \frac{1}{\sqrt{k}} = \sqrt{\frac{c_{44}}{c_{66}}}$$

5.5. What is the angle ϕ_0 by which the frame $Ox_1x_2x_3$ should be rotated about Ox_3, to cancel out the c_{16} modulus of a crystal with symmetry 4, $\overline{4}$ or 4/m?

<u>Solution</u>. We know from equation 3.12 the matrix α_i^j corresponding to a rotation of ϕ about the x_3 axis. Accordingly, the expansion of the new modulus c'_{16}

$$c'_{16} = c'_{1112} = \alpha_1^i\alpha_1^j\alpha_1^k\alpha_2^\ell c_{ijk\ell}$$

only contains coefficients $c_{ijk\ell}$ without an index of 3 in table 4.55

$$c_{1111} = c_{2222} = c_{11}, \quad c_{1122} = c_{12}, \quad c_{1212} = c_{66},$$

$$c_{1112} = c_{16}, \quad c_{2212} = -c_{16}.$$

The coefficient of c_{11} is:

$$a_{11} = \alpha_1^1\alpha_1^1\alpha_1^1\alpha_2^1+\alpha_1^2\alpha_1^2\alpha_1^2\alpha_2^2=-\cos^3\phi\sin\phi+\sin^3\phi\cos\phi$$

$$= \frac{\sin 2\phi}{2}(-\cos 2\phi) = -\frac{\sin 4\phi}{4}$$

For c_{12}, the coefficient is:

$$a_{12} = \alpha_1^1\alpha_1^1\alpha_1^2\alpha_2^2+\alpha_1^2\alpha_1^2\alpha_1^1\alpha_2^1=\cos^3\phi\sin\phi-\sin^3\phi\cos\phi$$

$$= \frac{\sin 4\phi}{4}$$

For c_{66} it is

$$a_{66} = (\alpha_1^1\alpha_1^2+\alpha_1^2\alpha_1^1)(\alpha_1^1\alpha_2^2+\alpha_1^2\alpha_2^1)=\sin 2\phi\cos 2\phi$$

$$= \frac{\sin 4\phi}{2}$$

and finally for c_{16}:

$$a_{16} = \alpha_1^1\alpha_1^1(\alpha_1^1\alpha_2^2+\alpha_1^2\alpha_2^1)+\alpha_1^1\alpha_2^1(\alpha_1^1\alpha_1^2+\alpha_1^2\alpha_1^1)-\alpha_1^2\alpha_1^2(\alpha_1^1\alpha_2^2$$
$$+\alpha_1^2\alpha_2^1)-\alpha_1^2\alpha_2^2(\alpha_1^1\alpha_1^2+\alpha_1^2\alpha_1^1)$$

$$a_{16} = \cos^2\phi\cos 2\phi-\frac{\sin^2 2\phi}{2}-\sin^2\phi\cos 2\phi-\frac{\sin^2 2\phi}{2}$$

$$= \cos^2 2\phi - \sin^2 2\phi = \cos 4\phi.$$

The expression

$$c'_{16} = \frac{1}{2}\left(c_{66}-\frac{c_{11}-c_{12}}{2}\right)\sin 4\phi + c_{16}\cos 4\phi$$

shows that the modulus c'_{16} vanishes for ϕ_0 such that

$$\boxed{\tan 4\phi_0 = \frac{4c_{16}}{c_{11}-c_{12}-2c_{66}}.}$$

5.6. What should be the angle by which the frame $Ox_1x_2x_3$ is tilted about Ox_3 to cancel out the c_{25} modulus of crystals having symmetry 3 and $\overline{3}$?

Solution. Since $\alpha_3^\ell = \delta_{3\ell}$, the expansion of the new modulus

$$c'_{25} = c'_{2213} = \alpha_2^i\alpha_2^j\alpha_1^k c_{ijk3}$$

only involves the c_{ijk3} components with i, j, k $\neq$ 3, or, according to table 4.54:

$c_{1113} = -c_{25}$, $c_{2213} = c_{25}$, $c_{1213} = c_{2113} = c_{14}$,

$c_{1123} = c_{14}$, $c_{2223} = -c_{14}$, $c_{1223} = c_{2123} = c_{25}$.

Gathering terms with c_{25} or c_{14}, we have

$c'_{25} = c_{25}(\cos^3\phi - 3\sin^2\phi\cos\phi) + c_{14}(\sin^3\phi - 3\cos^2\phi\sin\phi)$.

The new modulus

$$c'_{25} = c_{25}\cos 3\phi - c_{14}\sin 3\phi$$

vanishes for ϕ_0 such that

$$\boxed{\tan 3\phi_0 = \frac{c_{25}}{c_{14}}\,.}$$

5.7. Find the expressions for the phase velocity $V_\phi = \omega/k_2$ and the group velocity $V_g = d\omega/dk_2$ of Love waves.

Solution. According to 5.82:

$$\frac{V_\phi}{V_3} = \frac{\omega/V_3}{\sqrt{(\frac{\omega}{V_3})^2 - k_1^2}} = \frac{1}{\sqrt{1 - (\frac{k_1V_3}{\omega})^2}}$$

Differentiating equations 5.86, 5.82 and 5.84, we get:

$$(\frac{V_g}{V_3})\,(\frac{V_\phi}{V_3}) = \frac{1+\alpha}{(\frac{V_3}{V'_3})^2+\alpha} \quad \text{with} \quad \alpha = \frac{\chi_1^2}{k_1^2} + \frac{c_{44}}{c'_{44}}\cdot\frac{\chi_1 h}{\cos^2 k_1 h}\,.$$

Chapter 6
Piezoelectricity

Piezoelectricity, an interdependence of elastic and electric properties in certain materials, is intimately related to the study of elastic waves. Most devices that convert mechanical energy into electric energy (or vice versa), known as transducers, involve direct or inverse piezoelectric effects,* (the two major types of transducer currently used to launch surface or bulk elastic waves are described in Chap. VII). Furthermore, the wide utilization of the remarkable features of some crystals (such as their high Q factor) in electronics (for filtering) relies on piezoelectricity. Electromechanical resonators, made typically of quartz, are directly inserted into the circuits, the vibration being maintained by the electric field. This chapter is divided into two main sections; the first deals with static phenomena, the second with elastic wave propagation in piezoelectric solids.

6.1. STATIC PHENOMENA

Included under the heading of static phenomena are the basic Curie principles which were the starting point for the discovery of piezoelectricity, an elementary explanation of electromechanical coupling, various tensor

* There exist also magnetostrictive transducers, but they can be employed only at low frequencies.

expressions for the relationships between mechanical and electrical quantities, and finally the reduction of the number of piezoelectric tensor components.

6.1.1. Curie Symmetry Principle. Application to Piezoelectricity

A material is piezoelectric if a mechanical stress induces an electric polarization (direct effect) or if it undergoes a strain when an electric field is applied (inverse effect). The discovery of the direct piezoelectric effect, made in 1880 by Pierre and Jacques Curie, directly followed the annunciation of principles relating symmetry or asymmetry to various causes and effects. Symmetry conditions were previously dealt with more or less intuitively but Pierre Curie, in order to cast some light on the subject, made the following two statements:

1. A phenomenon must show (at least) the same symmetry as its causes (it may be more symmetric).

2. The asymmetry of a phenomenon is already present in the causes that give rise to it. Every phenomenon is characterized by a maximum symmetry. The symmetry elements of a scalar quantity (such as temperature) are an infinity of isotropy axes A_∞ and of mirrors M perpendicular to those A_∞, together with a centre $(\infty A_\infty/\infty M)C$. A polar vector such as the electric field has one isotropy axis A_∞ and an infinity of mirrors M containing A_∞: $A_\infty \infty M$. The characteristic symmetry of a compression or a dilation, resulting from the action of two opposing forces (fig. 6.1) is

$$\frac{A_\infty}{M} \frac{\infty A_2}{\infty M}, C$$

Both Curie principles can be summed up in the following statement: a phenomenon can exist only in a system which is a subgroup of the characteristic symmetry of this phenomenon.

In the case of the direct piezoelectric effect, for an axial stress on the crystal to generate an electric polarization, it is necessary that the system (crystal-stress) be a subgroup of the polarization characteristic

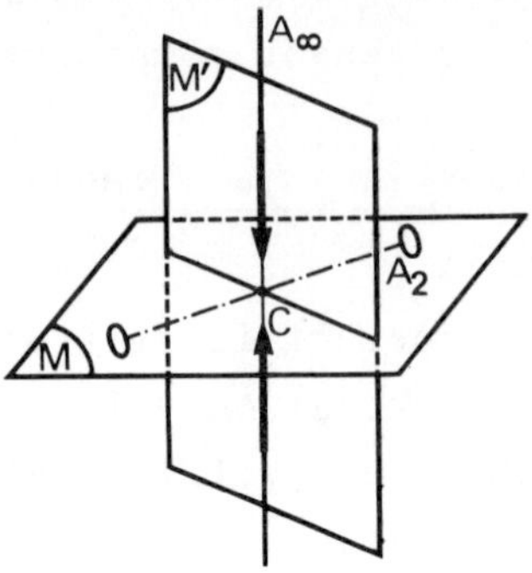

Fig. 6.1. Characteristic symmetry of a compression (or a dilation) resulting from two opposing forces:

$$\frac{A_\infty}{M}\ \frac{\infty A_2}{\infty M'}\ C.$$

symmetry: $A_\infty \infty M$. In particular, the system should have no centre of symmetry; since the stress field does have one, the crystal itself must possess no centre of symmetry, i.e. centrosymmetric crystals cannot be piezoelectric. This deduction was first made by Curie, and later confirmed by experiment: in all the 21 classes that have no centre - except for the cubic class 432 (§ 6.1.4e) - Curie found piezoelectric crystals.

Symmetry principles also predict the direction in which the mechanical stress has to be exerted: in the case of α-quartz, which belongs to class 32, an axial compression (or dilation) along the triad axis (fig. 6.2.a) does not give rise to any polarization since the overall symmetry of the system crystal-stress ($A_3 3A_2$) is not a subgroup of the phenomenon symmetry ($A_\infty \infty M$). On the other hand, the crystal-stress symmetry, for an axial stress along A_2, is compatible with a polarization $\vec{P}$ along the same A_2 (fig. 6.2.b). A compression perpendicular to both A_2 and A_3 (fig. 6.2.c) will also create a polarization along the axis A_2, the only symmetry element shared by the crystal and the stress.

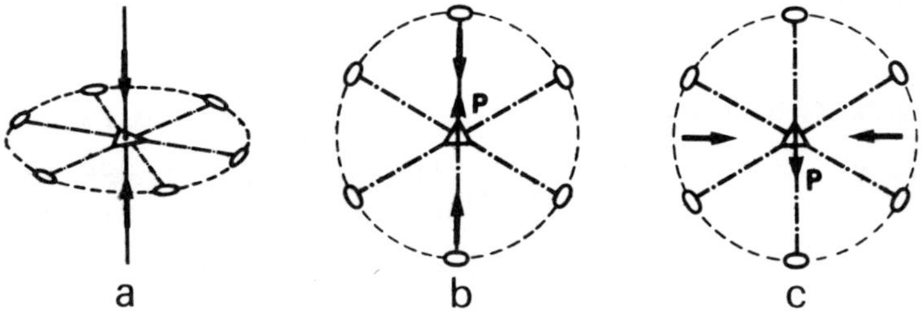

Fig. 6.2. Curie principles applied to quartz. Piezoelectricity can arise if the crystal-compression system is a subgroup of the polarization symmetry $A_\infty \infty M$.

a) A stress along the triad axis generates no effect since the system symmetry ($A_3 3A_2$) is not a subgroup of the polarization.

b) A stress along the diad axis induces a polarization along this axis.

c) A stress orthogonal both to A_3 and one A_2 induces a polarization along A_2.

6.1.2. Physical Mechanism. A One-dimensional Model

In § 2.5.2.a, we have already explained that a cadmium sulphide crystal consists of a succession of layers made respectively of cadmium or sulphur ions. In a direction orthogonal to those layers, i.e. parallel to the 6-fold axis, ions are located in identical rows, corresponding to each other through the screw axis 6_3 (fig. 2.35). Such a row is depicted in fig. 6.3.a.

For external actions (stress, electric field) directed along the 6-fold axis, it is sufficient to consider a single row, because all ions in a layer undergo the same displacement. Let -q and +q be the effective charges of the cadmium and sulphur ions, assumed to be bound by springs. The two nearest neighbours of any single ion are not symmetric with respect to that ion, so the springs on each side have different elastic constants K_1 and K_2. The chain is divided into cells, of length a, containing two dipoles of moment $(q/2)(a-b)$ and $(-q/2)b$ (fig. 6.3.a). The dipole moment of a molecule is

$$p_0 = \frac{q}{2}(a - 2b).$$

At rest, the polarization per unit volume P_0 does not vanish if b is different from a/2 and the medium is polar:

$$P_0 = \frac{nq}{2}(a - 2b)$$

where n is the number of CdS molecules per unit volume.

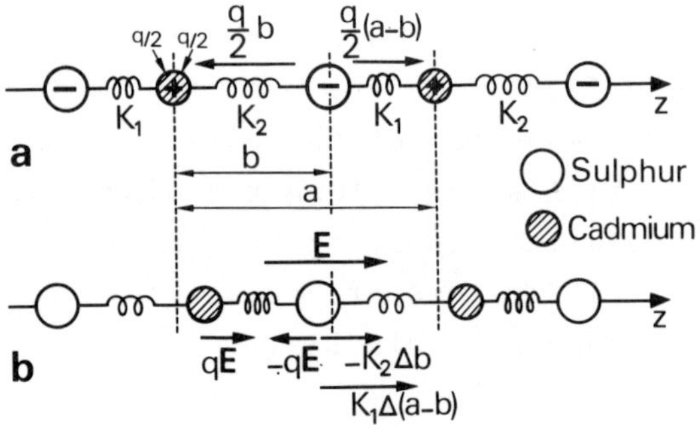

Fig. 6.3. Piezoelectricity in a cadmium sulphide crystal.
a) Pattern of the sulphur and cadmium ions on a row parallel to the 6-fold axis.
b) Strain of the chain under the action of an electric field $\vec{E}$; the Cd and S ions move in opposite directions. The cell length is changed by Δa.

This one dimensional model reveals, quite clearly, the interdependence of mechanical and electrical properties in a piezoelectric material:

- under a stress, the chain will be distorted and the change in a and b will cause a polarization change

$$P = \Delta P_0 = \frac{nq}{2}(\Delta a - 2\Delta b) \qquad (6.1)$$

which is the direct piezoelectric effect.

- an electric field gives rise to a displacement of

the ions, the sign of which depends on the charge; the distance (a - b) will increase, while b decreases. The springs do not have the same stiffness, so a strain results. This is the inverse piezoelectric effect.

Let us look for a relationship between mechanical quantities (stress T and strain S) and electrical quantities (field E and polarization P or displacement $D = \varepsilon_0 E + P$). The static equilibrium of each ion requires (fig. 6.3.b)

$$-qE + K_1\Delta(a - b) - K_2\Delta b = 0. \qquad (6.2)$$

In order to express the mechanical stress T, let us consider a cross-section perpendicular to the axis A_6. Two neighbouring rows, connected through a screw rotation about the axis A_6 are separated by a half period a/2 shift (fig. 2.35). Therefore, a cross section orthogonal to A_6 will involve, alternately, springs of constant K_1 and K_2. In the former case, the right hand side of the chain exerts a force

$$F_1 = K_1\Delta(a - b)$$

on the left hand side and in the latter case the force is

$$F_2 = K_2\Delta b.$$

Let N be the number of rows per unit area of cross section orthogonal to the axis A_6. The mechanical stress, i.e. the force per unit area is

$$T = \frac{N}{2} K_1\Delta(a - b) + \frac{N}{2} K_2\Delta b$$

since there are equal numbers of springs of each type. Since N = na, we get

$$T = \frac{na}{2} [K_1\Delta a + (K_2 - K_1)\Delta b]. \qquad (6.3)$$

The induced polarization is obtained by inserting

$$\Delta b = -\frac{qE}{K_1 + K_2} + \frac{K_1}{K_1 + K_2}\Delta a \qquad (6.4)$$

(which arises from 6.2) into equation 6.1:

$$P = \frac{nq}{2} [\frac{2qE}{K_1 + K_2} + \frac{K_2 - K_1}{K_1 + K_2}\Delta a] . \qquad (6.5)$$

There are two terms; the first is proportional to the electric field and defines the ionic crystal polarizability $\chi_{ion} = nq^2/(K_1 + K_2)$. The second, proportional to the strain $\Delta a/a = S$, reflects the direct piezoelectric effect:

$$P = \chi_{ion}E + eS$$

where e is the piezoelectric constant

$$e = \frac{nq}{2}\left(\frac{K_2 - K_1}{K_2 + K_1}\right)a. \quad (6.6)$$

The overall polarization, taking the electronic polarizability χ_e into account, is

$$P = (\chi_{ion} + \chi_e)E + eS$$

and the electric displacement is

$$D = \varepsilon E + cS \quad (6.7)$$

where $\varepsilon = \varepsilon_0 + \chi_{ion} + \chi_e$ is the dielectric constant.

Inserting 6.4 into 6.3, we obtain the stress

$$T = \frac{na}{2}\left(\frac{2K_2K_1\Delta a}{K_2 + K_1} - \frac{K_2 - K_1}{K_2 + K_1}qE\right).$$

which can be put into the form

$$T = cS - eE \quad (6.8)$$

where $c = na^2K_1K_2/(K_2 + K_1)$ is the stiffness constant and e the piezoelectric constant which we have just calculated. Equation (6.6) shows that, in this one dimensional model, the existence of a piezoelectric effect is linked to the asymmetry in the elastic restoring forces, i.e. $e = 0$ for $K_1 = K_2$. The equation 6.8 can be interpreted in various ways:

- for zero strain (e.g. for a chain with fixed bounds), an electric field generates a stress: $T = -eE$

- for zero stress, an electric field induces a strain $S = eE/c$. Note that from now the constant e is actually a component of a third rank tensor, since the displacement is a first rank tensor (a vector) and the strain a second rank tensor. Taking the x_3 direction along the 6-fold axis A_6, equation 6.7 reads

$$D_3 = \varepsilon_{33}E_3 + e_{333}S_{33}.$$

This model provides a qualitative description of the origin of piezoelectric effects in non symmetric ionic crystals. The mechanism is, in fact, somewhat more involved since piezoelectricity arises in monoatomic crystals such as tellurium Te [1] or selenium Se [2].

In those cases, the electric polarization induced by a strain should be attributed to an electronic distribution change. Polycrystalline materials such as ceramics, become piezoelectric only after undergoing some specific processes. These ceramics, consisting of lead, zirconium and titanium in the case PZT, contain ferroelectric domains (§ 2.5.2) randomly oriented during the growth of the material. Applying a high electric field (about 20,000 V/cm), at a temperature of 100°C, produces the alignment of the domains and gives rise to an elongation; both partly remain at room temperature when the field has been removed. The permanent polarization, and therefore the piezoelectricity, disappear above the Curie temperature, usually located between 200 and 400°C. As far as symmetry is concerned, the polarization direction is an isotropy axis. All planes containing this axis are equivalent and the material is said to be transversely isotropic.

6.1.3. Tensor Formulation of Piezoelectricity

A three dimensional extension of equation 6.7 defines the third rank tensor e_{ijk}

$$D_i = \varepsilon_{ij}E_j + e_{ijk}S_{jk}. \tag{6.9}$$

The piezoelectric constants e_{ijk} are the link between an electrical displacement variation D_i (at constant electric field) and the strain S_{jk}:

$$e_{ijk} = \left(\frac{\partial D_i}{\partial S_{jk}}\right)_E . \tag{6.10}$$

Since $S_{jk} = S_{kj}$, then the tensor e_{ijk} is symmetric with respect to its last two indices:

$$e_{ijk} = e_{ikj}.$$

The number of independent piezoelectric constants is thus reduced from 27 (ordinary third rank tensor) to 18; the pair (j, k) has only 6 distinct values, denoted by the index α according to the convention of 4.22 (see § 4.3)

$$e_{i\alpha} = e_{ijk} \quad i = 1, 2, 3 \quad \alpha = (j,k) = 1, 2, \ldots 6.$$

The piezoelectric moduli are arranged in a 3 x 6 table (three columns, six rows):

$$e_{i\alpha} = \begin{vmatrix} e_{11} & e_{12} & e_{13} & e_{14} & e_{15} & e_{16} \\ e_{21} & e_{22} & e_{23} & e_{24} & e_{25} & e_{26} \\ e_{31} & e_{32} & e_{33} & e_{34} & e_{35} & e_{36} \end{vmatrix} \qquad (6.11)$$

Extending this matrix notation to strains, as given by equation 4.25, equation 6.9 becomes

$$D_i = \varepsilon_{ij}E_j + e_{i\alpha}S_\alpha \ .$$

The inverse piezoelectric coefficient, which relates the stress T_{jk} and the electric field E_i in the generalized equation 6.8, can be deduced from e_{ijk} through thermodynamic arguments. The external forces acting on a piezoelectric medium are both mechanical and electric. The work which has to be provided in order to modify the state of the crystal involves, besides the mechanical term calculated in § 4.4,

$$\delta \mathcal{W}_M = T_{jk}dS_{jk}$$

an electric term equal to

$$\delta \mathcal{W}_E = E_i dD_i .$$

For a reversible transformation, the internal energy variation per unit volume is

$$d\mathcal{U} = \theta d\sigma + T_{jk}\, dS_{jk} + E_i\, dD_i$$

where θ is the absolute temperature and σ the entropy. $\mathcal{U}$ is a function depending on the entropy, on strains, and on the electrical displacement. In order to have as variables σ, S_{jk} and E_i, we introduce the thermodynamic potential

$$\mathcal{G} = \mathcal{U} - E_i D_i .$$

The variation of this function of state

$$d\mathcal{G} = \theta d\sigma + T_{jk}\, dS_{jk} - D_i\, dE_i$$

is an exact differential:

$$T_{jk} = \left(\frac{\partial \mathcal{G}}{\partial S_{jk}}\right)_{\sigma, E_i} \quad \text{and} \quad D_i = -\left(\frac{\partial \mathcal{G}}{\partial E_i}\right)_{\sigma, S_{jk}}$$

Second order derivations do not depend on the order of the variables.

$$\left(\frac{\partial T_{jk}}{\partial E_i}\right)_{\sigma,S} = -\left(\frac{\partial D_i}{\partial S_{jk}}\right)_{\sigma,E} = \left(\frac{\partial^2 \mathcal{G}}{\partial E_i \partial S_{jk}}\right)_{\sigma}$$

This equation shows that, if a strain at constant electric field leads to an electrical displacement:

$$\left(\frac{\partial D_i}{\partial S_{jk}}\right)_{\sigma,E} = e_{ijk},$$

an electric field will also generate a stress in a rigidly bonded material (constant strain)

$$\left(\frac{\partial T_{jk}}{\partial E_i}\right)_{\sigma,S} = -e_{ijk} \tag{6.12}$$

Proportionality coefficients of the two effects are of opposite sign. The inverse piezoelectric effect appears to be a thermodynamic consequence of the direct effect. In the linear domain, the coefficients e_{ijk} are assumed to be constant and the integration of equation 6.12 (for constant σ and S) leads to

$$T_{jk} = -e_{ijk}E_i.$$

If, in addition, the crystal experiences a strain $S_{\ell m}$, then

$$T_{jk} = -e_{ijk}E_i + c^E_{jk\ell m}S_{\ell m}. \tag{6.13}$$

The index E means that the stiffness coefficient of Hooke's law

$$c^E_{jk\ell m} = \left(\frac{\partial T_{jk}}{\partial S_{\ell m}}\right)_{\sigma,E} \tag{6.14}$$

connects strain and stress when the electric field is constant. Since the material is piezoelectric, the mechanical constants depend on the electric conditions, which for this reason must be specified. Similarly, the coefficient ε_{ij} in equation 6.9 is the dielectric constant for constant strain ε^S_{ij}. Equations 6.9 and 6.13 are a first set of equations of state which read, in matrix notation:

$$\boxed{T_\alpha = c^E_{\alpha\beta}S_\beta - e_{i\alpha}E_i} \qquad i, j = 1, 2, 3 \tag{6.15a}$$

$$\boxed{D_i = \varepsilon^S_{ij}E_j + e_{i\alpha}S_\alpha} \qquad \alpha, \beta = 1,\ldots 6 \tag{6.15b}$$

They provide the mechanical stress and the electrical displacement in terms of the independent variables, strain and electric field.

With other pairs of independent variables: D and S, D and T, or E and T, the equations of state of a piezoelectric material have different forms. The thermodynamic potential

$$\mathcal{I} = \mathcal{G} - T_{jk}S_{jk} = \mathcal{U} - E_iD_i - T_{jk}S_{jk}$$

changes by the amount

$$d\mathcal{I} = \theta d\sigma - S_{jk}dT_{jk} - D_i dE_i$$

so that $\mathcal{I}$ is a function of T_{jk} and E_i: $\mathcal{I}(\sigma, T_{jk}, E_i)$ and

$$S_{jk} = -\left(\frac{\partial \mathcal{I}}{\partial T_{jk}}\right)_{\sigma,E} \qquad D_i = -\left(\frac{\partial \mathcal{I}}{\partial E_i}\right)_{\sigma,T}$$

Equating the second order derivations yields

$$\left(\frac{\partial S_{jk}}{\partial E_i}\right)_{\sigma,T} = \left(\frac{\partial D_i}{\partial T_{jk}}\right)_{\sigma,E} = d_{ijk} \tag{6.16}$$

If we recall the definition of the compliance (at constant electric field) $s^E_{jk\ell m}$ and of the permittivity (at constant stress) ε^T_{ij}:

$$s^E_{jk\ell m} = \left(\frac{\partial S_{jk}}{\partial T_{\ell m}}\right)_E \quad \text{and} \quad \varepsilon^T_{ij} = \left(\frac{\partial D_i}{\partial E_j}\right)_T \tag{6.17}$$

and if we integrate, we obtain a second pair of equations of state where the independent variables are the electric field and the stress:

$$\begin{cases} S_{jk} = d_{ijk}E_i + s^E_{jk\ell m}T_{\ell m} \\ D_i = \varepsilon^T_{ij}E_j + d_{ijk}T_{jk} \end{cases}$$

or, in matrix form:

$$\boxed{S_\alpha = s^E_{\alpha\beta}T_\beta + d_{i\alpha}E_i} \tag{6.18a}$$

$$\boxed{D_i = \varepsilon^T_{ij}E_j + d_{i\alpha}T_\alpha} \tag{6.18b}$$

provided that

$$\begin{aligned} d_{i\alpha} &= d_{ijk} \quad \text{for} \quad \alpha \le 3 \\ d_{i\alpha} &= 2d_{ijk} \quad \text{for} \quad \alpha > 3 \end{aligned} \tag{6.19}$$

in order to satisfy the definition of T_α (4.24) and of S_α (4.25).

The piezoelectric ($e_{i\alpha}$, $d_{i\alpha}$) and mechanical ($c^E_{\alpha\beta}$, $s^E_{\alpha\beta}$) coefficients are not independent. Inserting

$$S_\beta = d_{i\beta}E_i + s^E_{\beta\gamma}T_\gamma$$

into 6.15a, yields

$$T_\alpha = (-e_{i\alpha} + c^E_{\alpha\beta}d_{i\beta})E_i + c^E_{\alpha\beta}s^E_{\beta\gamma}T_\gamma = \delta_{\alpha\gamma}T_\gamma$$

whatever E_i and T_γ. Therefore:

$$\boxed{c^E_{\alpha\beta}s^E_{\beta\gamma} = \delta_{\alpha\gamma}} \tag{6.20}$$

and

$$\boxed{e_{i\alpha} = d_{i\beta}c^E_{\beta\alpha}}. \tag{6.21}$$

The former equation shows that, as in the non-piezoelectric case, the $s^E_{\alpha\beta}$ matrix is the inverse of $c^E_{\alpha\beta}$. The latter allows a calculation of $e_{i\alpha}$ from $d_{i\alpha}$, or vice versa through the equation

$$\boxed{d_{i\beta} = e_{i\alpha}s^E_{\alpha\beta}.} \tag{6.22}$$

In order to connect ε^T_{ij} and ε^S_{ij}, we introduce

$$T_\alpha = -e_{j\alpha}E_j + c^E_{\alpha\beta}S_\beta$$

into 6.18b, which gives

$$D_i = (\varepsilon^T_{ij} - d_{i\alpha}e_{j\alpha})E_j + d_{i\alpha}c^E_{\alpha\beta}S_\beta \ .$$

Comparing with 6.15b, results in

$$\boxed{\varepsilon^T_{ij} - \varepsilon^S_{ij} = d_{i\alpha}e_{j\alpha} = d_{i\alpha}c^E_{\alpha\beta}d_{j\beta} \ .} \tag{6.23}$$

The dielectric coefficient ε^S_{ij} of a rigidly bonded solid is very difficult to measure. The above equation provides a method of calculation given the piezoelectric moduli and the usual dielectric coefficients ε^T_{ij} of the stress free solid. The difference between ε^T_{ij} and ε^S_{ij} is important for highly piezoelectric media. For example, in lithium niobate ($LiNbO_3$)

$$\varepsilon^T_{11} = 74.3\ 10^{-11}\ \text{F/m} \qquad \varepsilon^S_{11} = 38.9\ 10^{-11}\ \text{F/m}.$$

Equations 6.15 and 6.18 are the most frequently used in practice. The other two pairs of equations of state where the electric displacement D_i is an independent variable, can be obtained by considering the internal energy $\mathcal{U}(\sigma, S_{jk}, D_i)$ and the enthalpy

$$\mathcal{H}(\sigma, T_{jk}, D_i) = \mathcal{U} - T_{jk}S_{jk}.$$

Thus, in terms of D and S, we have

$$\boxed{\begin{aligned} E_i &= \beta^S_{ij}D_j - h_{ijk}S_{jk} \\ T_{jk} &= c^D_{jk\ell m}S_{\ell m} - h_{ijk}D_i \end{aligned}} \qquad (6.24)$$

$c^D_{jk\ell m}$ is the stiffness tensor at constant electrical displacement and β^S_{ij} the inverse of the ε^S_{ij} tensor. The piezoelectric coefficient h_{ijk} defined by

$$h_{ijk} = -\left(\frac{\partial E_i}{\partial S_{jk}}\right)_D = -\left(\frac{\partial T_{jk}}{\partial D_i}\right)_S = -\left(\frac{\partial^2 \mathcal{U}}{\partial D_i \partial S_{jk}}\right)_\sigma \qquad (6.25)$$

is written, in matrix notation, as $h_{ijk} = h_{i\alpha}$. The piezoelectric coefficient g_{ijk} in the last pair of equations:

$$\boxed{\begin{aligned} E_i &= \beta^T_{ij}D_j - g_{ijk}T_{jk} \\ S_{jk} &= s^D_{jk\ell m}T_{\ell m} + g_{ijk}D_i \end{aligned}} \qquad (6.26)$$

is the opposite of the second derivative of the enthalpy $\mathcal{H}$ with respect to D_i and T_{jk} (the independent variables)

$$g_{ijk} = -\left(\frac{\partial^2 \mathcal{H}}{\partial D_i \partial T_{jk}}\right)_\sigma$$

In matrix notation, $g_{i\alpha} = g_{ijk}$ if $\alpha = 1, 2, 3$ and $g_{i\alpha} = 2g_{ijk}$ for $\alpha > 3$.

6.1.4. Reduction in the Number of the Independent Piezoelectric Coefficients by Crystal Symmetry

The requirement that the piezoelectric tensor components be invariant under a symmetry operation with transformation matrix α leads, according to equation 3.11

$$e_{ijk} = \alpha^\ell_i \alpha^m_j \alpha^n_k e_{\ell mn} \qquad (6.28)$$

Let us discuss the effect of the crystal point group operations.

a) Centre of symmetry. Since $\alpha^k_i = -\delta^k_i$, equation 6.28 gives

$$e_{ijk} = (-1)^3 e_{ijk} \Rightarrow e_{ijk} = 0 \quad \forall\ i, j, k.$$

The piezoelectric tensor vanishes in the eleven centro-symmetric classes:

$\bar{1}$, 2/m, mmm, $\bar{3}$, $\bar{3}$m, 4/m, 4/mmm, 6/m, 6/mmm, m3, m3m.

We recover the result already derived from the Curie symmetry principle (§ 6.1.1).

b) Symmetry plane. The transformation matrix is again diagonal. If the x_3 axis is orthogonal to the mirror,

$$\alpha = \begin{pmatrix} 1 & 0 & 0 \\ 0 & 1 & 0 \\ 0 & 0 & -1 \end{pmatrix}$$

then condition 6.28

$$e_{ijk} = \alpha_i^i \alpha_j^j \alpha_k^k e_{ijk} \tag{6.29}$$

implies that coefficients with one or three indices of 3 vanish. There are only 10 moduli $e_{i\alpha}$ in the class m:

$$e_{i\alpha} = \begin{vmatrix} e_{11} & e_{12} & e_{13} & 0 & 0 & e_{16} \\ e_{21} & e_{22} & e_{23} & 0 & 0 & e_{26} \\ 0 & 0 & 0 & e_{34} & e_{35} & 0 \end{vmatrix} \quad \begin{matrix} \text{class m} \\ (M \perp x_3) \end{matrix} \tag{6.30}$$

c) Direct diad axis. The transformation matrix for rotation about the x_3 axis is

$$\alpha = \begin{pmatrix} -1 & 0 & 0 \\ 0 & -1 & 0 \\ 0 & 0 & 1 \end{pmatrix}$$

Equation 6.29 cancels out those components which have an even number of indices of 3, so that, for class 2, the matrix $e_{i\alpha}$ is the complement of (6.30) for class m:

$$e_{i\alpha} = \begin{vmatrix} 0 & 0 & 0 & e_{14} & e_{15} & 0 \\ 0 & 0 & 0 & e_{24} & e_{25} & 0 \\ e_{31} & e_{32} & e_{33} & 0 & 0 & e_{36} \end{vmatrix} \quad \begin{matrix} \text{class 2} \\ (A_2 // x_3) \end{matrix} \tag{6.31}$$

d) Several diad axes. The non vanishing coefficients of class 222 crystals have an odd number of each index: this is true only for e_{123}, e_{213}, e_{312}:

$$e_{i\alpha} = \begin{vmatrix} 0 & 0 & 0 & e_{14} & 0 & 0 \\ 0 & 0 & 0 & 0 & e_{25} & 0 \\ 0 & 0 & 0 & 0 & 0 & e_{36} \end{vmatrix} \quad \text{class 222} \qquad (6.32)$$

In class 2mm, the mirrors, orthogonal to the x_1 and x_2 axes, cause the coefficients $e_{14} = e_{123}$, $e_{25} = e_{213}$, and $e_{36} = e_{312}$ in table 6.31 to cancel since they have an odd number of indices of 1 and 2:

$$e_{i\alpha} = \begin{vmatrix} 0 & 0 & 0 & 0 & e_{15} & 0 \\ 0 & 0 & 0 & e_{24} & 0 & 0 \\ e_{31} & e_{32} & e_{33} & 0 & 0 & 0 \end{vmatrix} \quad \text{class 2mm} \qquad (6.33)$$

Classes 23 and $\bar{4}3m$ of the cubic system have three diad axes which constitute the reference frame $Ox_1x_2x_3$. Since a $2\pi/3$ rotation about the triad axis [111] generates a permutation of x_1, x_2 and x_3, as a result e_{123}, e_{312} and e_{231} (from the class 222) are equal:

$$e_{i\alpha} = \begin{vmatrix} 0 & 0 & 0 & e_{14} & 0 & 0 \\ 0 & 0 & 0 & 0 & e_{14} & 0 \\ 0 & 0 & 0 & 0 & 0 & e_{14} \end{vmatrix} \quad \text{classes 23 and } \bar{4}3m \qquad (6.34)$$

e) Direct n-fold axis for $n > 2$. In order to determine the form of piezoelectric matrices for crystals having a direct 3 , 4 , or 6-fold axis, we use the same method as in § 4.5.2 for the calculation of the stiffness coefficients. With the same notation, the invariance conditions for the new moduli η_{ijk} read, using as basis the eigenvectors of the rotation matrix 4.46 $\{\vec{\xi}^{(1)}\ \vec{\xi}^{(2)}\ \vec{\xi}^{(3)}\}$

$$\eta_{ijk} = \lambda^{(i)}\lambda^{(j)}\lambda^{(k)}\eta_{ijk}$$

Since the eigenvalues are

$$\lambda^{(1)} = e^{i\ 2\pi/n} \qquad \lambda^{(2)} = e^{-i\ 2\pi/n} \qquad \lambda^{(3)} = 1$$

a component η_{ijk} does not vanish if the difference between ν_1, the number of indices of 1, and ν_2, the number of indices of 2, is a multiple of n for, in this case, the product $\lambda^{(i)}\lambda^{(j)}\lambda^{(k)} = \exp 2i\pi(\nu_1-\nu_2)/n$ is equal to 1. This is always true for η_{123}, η_{213}, η_{312}, η_{333}, for which $\nu_1 = \nu_2$. There are no other non vanishing components for $n = 4$ or 6; but for $n = 3$, the components η_{111} and η_{222}, for which $\nu_1 - \nu_2 = \pm 3$, do not vanish.

We recover the e_{ijk} coefficients through the equation

$$e_{ijk} = a_i^\ell a_j^m a_k^n \eta_{\ell mn} \qquad (6.35)$$

where (a) is the transformation matrix 4.51 which connects the $Ox_1x_2x_3$ frame (constants e_{ijk}) to the eigenbasis $\vec{\xi}^{(1)}\vec{\xi}^{(2)}\vec{\xi}^{(3)}$:

$$a = \begin{pmatrix} 1/\sqrt{2} & i/\sqrt{2} & 0 \\ i/\sqrt{2} & 1/\sqrt{2} & 0 \\ 0 & 0 & 1 \end{pmatrix}$$

We begin with classes 4 and 6, the e_{ijk} tensors of which are the same, since the non vanishing η_{ijk} are already the same. An A_4 or A_6 axis being a diad axis, we need only consider non vanishing components of table 6.31 (for class 2). Constants with a single index of 3, such as e_{3jk} with j, k $\neq$ 3, can be expressed in terms of $\eta_{312} = \eta_{321}$

$$e_{3jk} = (a_j^1 a_k^2 + a_j^2 a_k^1)\eta_{312} \qquad (6.36)$$

In particular

$$e_{3jj} = 2a_j^1 a_j^2 \eta_{312} = i\eta_{312} \qquad \forall j \neq 3 \Rightarrow \; e_{31} = e_{32}$$

and

$$e_{312} = (a_1^1 a_2^2 + a_1^2 a_2^1)\eta_{312} = 0 \Rightarrow \; e_{36} = 0.$$

The expansion of the constants with just one index of 3, such as e_{ij3} with i, j $\neq$ 3

$$e_{ij3} = a_i^1 a_j^2 \eta_{123} + a_i^2 a_j^1 \eta_{213}$$

leads to

$$e_{jj3} = a_j^1 a_j^2(\eta_{123} + \eta_{213}) = \frac{i}{2}(\eta_{123} + \eta_{213}) \quad \forall\, j \neq 3$$

or

$$e_{15} = e_{24}.$$

From 6.35, we also derive e_{123} and e_{213}:

$$e_{123} = \frac{1}{2}(\eta_{123} - \eta_{213})$$

$$e_{213} = \frac{1}{2}(-\eta_{123} + \eta_{213})$$

and see that these moduli are equal and of opposite sign, so that $e_{14} = -e_{25}$. Since $e_{333} = \eta_{333}$ is not zero, the matrix $e_{i\alpha}$ has the following form:

$$e_{i\alpha} = \begin{vmatrix} 0 & 0 & 0 & e_{14} & e_{15} & 0 \\ 0 & 0 & 0 & e_{15} & -e_{14} & 0 \\ e_{31} & e_{31} & e_{33} & 0 & 0 & 0 \end{vmatrix} \quad \text{classes 4 and 6.} \tag{6.37}$$

For classes 422 and 622 which have three orthogonal diad axes, we need only combine 6.32 and 6.37:

$$e_{i\alpha} = \begin{vmatrix} 0 & 0 & 0 & e_{14} & 0 & 0 \\ 0 & 0 & 0 & 0 & -e_{14} & 0 \\ 0 & 0 & 0 & 0 & 0 & 0 \end{vmatrix} \quad \text{classes 422 and 622.} \tag{6.38}$$

Similarly, taking table 6.33 into account (class 2mm), we obtain for classes 4mm and 6mm:

$$e_{i\alpha} = \begin{vmatrix} 0 & 0 & 0 & 0 & e_{15} & 0 \\ 0 & 0 & 0 & e_{15} & 0 & 0 \\ e_{31} & e_{31} & e_{33} & 0 & 0 & 0 \end{vmatrix} \quad \text{classes 4mm and 6mm.} \tag{6.39}$$

The aim of the exercise 6.3 is to prove that in classes 4mm and 6mm the piezoelectric tensor is invariant under any rotation about the x_3 axis. If we bear in mind a similar observation about stiffness constants of hexagonal crystals (see § 4.5.2 and exercise 4.4), it appears that elasto-electric properties of 6mm symmetry crystals are invariant under any rotation about the 6-fold axis. Since this transverse isotropy is also encountered in a plane orthogonal to the direction of polarization of piezoelectric ceramics, we should describe the elastic, electric and piezoelectric properties of these materials by tensors of class 6mm.

The class 432 of the cubic system involves all symmetry elements of the class 422 and has, at most, one piezoelectric coefficient:

$$e_{14} = -e_{25}.$$

But the triad axis along [111] induces a permutation of the indices 1, 2 and 3, so that $e_{123} = e_{231}$ and

$$e_{14} = e_{25} = -e_{25} = 0$$

All components vanish. In spite of the absence of a centre of symmetry, piezoelectricity does not exist in crystals of the class 432.

$$e_{i\alpha} = 0 \quad \text{class 432.}$$

The above equations relating moduli with one or three indices of 3 for classes 4 and 6 are still valid for crystals having a triad axis. On the other hand, coefficients without an index of 3 no longer vanish; they can be expressed in terms of η_{111} and η_{222}:

$$e_{ijk} = a_i^1 a_j^1 a_k^1 \eta_{111} + a_i^2 a_j^2 a_k^2 \eta_{222} \qquad (6.40)$$

Especially

$$e_{i11} = \frac{1}{2}(a_i^1 \eta_{111} - a_i^2 \eta_{222})$$

and

$$e_{i22} = \frac{1}{2}(-a_i^1 \eta_{111} + a_i^2 \eta_{222}) = -e_{i11}$$

When i takes the values 1 and 2 we get

$$e_{12} = -e_{11} \quad \text{and} \quad e_{22} = -e_{21}.$$

Likewise, the expansion 6.40 leads to

$$e_{112} = \frac{1}{2\sqrt{2}}(i\eta_{111} - \eta_{222}) = -e_{222}$$

$$e_{212} = \frac{1}{2\sqrt{2}}(-\eta_{111} + i\eta_{222}) = -e_{111}\ .$$

Hence, the table:

$$e_{i\alpha} = \begin{vmatrix} e_{11} & -e_{11} & 0 & e_{14} & e_{24} & -e_{22} \\ -e_{22} & e_{22} & 0 & e_{15} & -e_{14} & -e_{11} \\ e_{31} & e_{31} & e_{33} & 0 & 0 & 0 \end{vmatrix} \quad \text{class 3} \qquad (6.41)$$

A diad axis along the x_1 axis will cancel out all moduli with an even number of indices of 1 (§ 6.1.4.c) in the above table:

$$e_{15} = e_{24} \quad e_{16} = -e_{22} \quad e_{21} = -e_{22} \quad e_{31} = e_{32} \quad e_{33}$$

so that crystals of symmetry 32 have only two independent piezoelectric coefficients: e_{11} and e_{14}:

$$e_{i\alpha} = \begin{vmatrix} e_{11} & -e_{11} & 0 & e_{14} & 0 & 0 \\ 0 & 0 & 0 & 0 & -e_{14} & -e_{11} \\ 0 & 0 & 0 & 0 & 0 & 0 \end{vmatrix} \quad \text{class 32} \qquad (6.42)$$

A mirror orthogonal to the x_1 axis cancels out all coefficients with one or three indices of 1 in table 6.41, namely e_{14} and e_{11}. The matrix for crystals of class 3m reduces to

$$e_{i\alpha} = \begin{vmatrix} 0 & 0 & 0 & 0 & e_{15} & -e_{22} \\ -e_{22} & e_{22} & 0 & e_{15} & 0 & 0 \\ e_{31} & e_{31} & e_{33} & 0 & 0 & 0 \end{vmatrix} \quad \text{class } 3m \qquad (6.43)$$

f) Inverse principal axis. Since an inverse 6-fold axis is equivalent to 3/m (§ 2.3.1.3.c), the $e_{i\alpha}$ matrix of class $\overline{6}$ crystals is readily derived from 6.30 and 6.41:

$$e_{i\alpha} = \begin{vmatrix} e_{11} & -e_{11} & 0 & 0 & 0 & -e_{22} \\ -e_{22} & e_{22} & 0 & 0 & 0 & -e_{11} \\ 0 & 0 & 0 & 0 & 0 & 0 \end{vmatrix} \quad \text{class } \overline{6} \qquad (6.44)$$

In crystals of class $\overline{6}m2$ the diad axis, which is parallel to the x_1 axis, removes e_{22} from the above table since it has no index of 1 (§ 6.1.4.c):

$$e_{i\alpha} = \begin{vmatrix} e_{11} & -e_{11} & 0 & 0 & 0 & 0 \\ 0 & 0 & 0 & 0 & 0 & -e_{11} \\ 0 & 0 & 0 & 0 & 0 & 0 \end{vmatrix} \quad \text{class } \overline{6}m2 \qquad (6.45)$$

We return to the standard procedure for the case of an inverse tetrad axis. The rotation-inversion matrix $\overline{\alpha}$ is equal to $-\alpha$ and the vectors $\vec{\xi}^{(i)}$ are also the eigenvectors of $\overline{\alpha}$, with eigenvalues $\overline{\lambda}^{(i)} = -\lambda^{(i)}$.

The existence condition of the η_{ijk}

$$\overline{\lambda}^{(i)}\overline{\lambda}^{(j)}\overline{\lambda}^{(k)} = 1 ,$$

equivalent to

$$\lambda^{(i)}\lambda^{(j)}\lambda^{(k)} = e^{i(\nu_1-\nu_2)2\pi/n} = -1$$

is satisfied, for $n = 4$, by η_{113}, η_{223}, η_{311}, η_{322}, for which $\nu_1 - \nu_2 = \pm 2$. Since an $\overline{A}_4$ axis includes a diad axis, we only need to look for the e_{ijk} coefficients of table 6.31. Applying equation 6.35 to e_{ii3}:

$$e_{ii3} = a_i^1 a_i^1 \eta_{113} + a_i^2 a_i^2 \eta_{223}$$

implies $e_{24} = -e_{15}$. Similarly, the expansion

$$e_{3ii} = a_i^1 a_i^1 \eta_{311} + a_i^2 a_i^2 \eta_{322}$$

leads to $e_{32} = -e_{31}$; and the identity of e_{25} and e_{14} comes from

$$e_{ij3} = a_i^1 a_j^1 \eta_{113} + a_i^2 a_j^2 \eta_{223} = e_{ji3} .$$

The coefficient e_{33} being zero ($\eta_{333} = 0$), the $e_{i\alpha}$ matrix is written as

$$e_{i\alpha} = \begin{vmatrix} 0 & 0 & 0 & e_{14} & e_{15} & 0 \\ 0 & 0 & 0 & -e_{15} & e_{14} & 0 \\ e_{31} & -e_{31} & 0 & 0 & 0 & e_{36} \end{vmatrix} \quad \text{class } \bar{4} \qquad (6.46)$$

For the class $\bar{4}2m$, the diad axis along the x_1 direction causes e_{15} and e_{31} to cancel (having an odd number of indices of 1):

$$e_{i\alpha} = \begin{vmatrix} 0 & 0 & 0 & e_{14} & 0 & 0 \\ 0 & 0 & 0 & 0 & e_{14} & 0 \\ 0 & 0 & 0 & 0 & 0 & e_{36} \end{vmatrix} \quad \text{class } \bar{4}2m \qquad (6.47)$$

The above results are valid for any third rank tensor A_{ijk}, symmetric in j and k; therefore, they can also be applied to e_{ijk}, d_{ijk}, h_{ijk}, g_{ijk}. This is still true for the coefficients with two indices $e_{i\alpha} = e_{ijk}$ and $h_{i\alpha} = h_{ijk}$. However if we want to find the relationship between the $d_{i\alpha}$ (or $g_{i\alpha}$) moduli, we must take equation 6.19 into account. For example, in class 32 crystals: $e_{26} = -e_{11}$ but $d_{26} = -2d_{11}$. The table of fig. 6.4 gives the results of the reduction of the number of tensor components for the elastic, electric and piezoelectric properties of crystals and for all the point groups.

Table 6.5 gives the piezoelectric ($e_{i\alpha}$) and dielectric (ε^S_{ij}) coefficients of several frequently used crystals.

Model indicating the layout of elements in table 6.4:

c_{11} c_{12} . . . . c_{12} c_{22} . . . . . . $c^E_{\alpha\beta}$. $(c^D_{\alpha\beta})$.	e_{11} e_{21} e_{31} e_{12} . . $e_{i\alpha}$. $(h_{i\alpha})$.
e_{11} e_{12} . . . . e_{21} $e_{i\alpha}$ e_{31} $(h_{i\alpha})$	ε^S_{11} . . . ε^S_{ij} . (β^S_{ij})

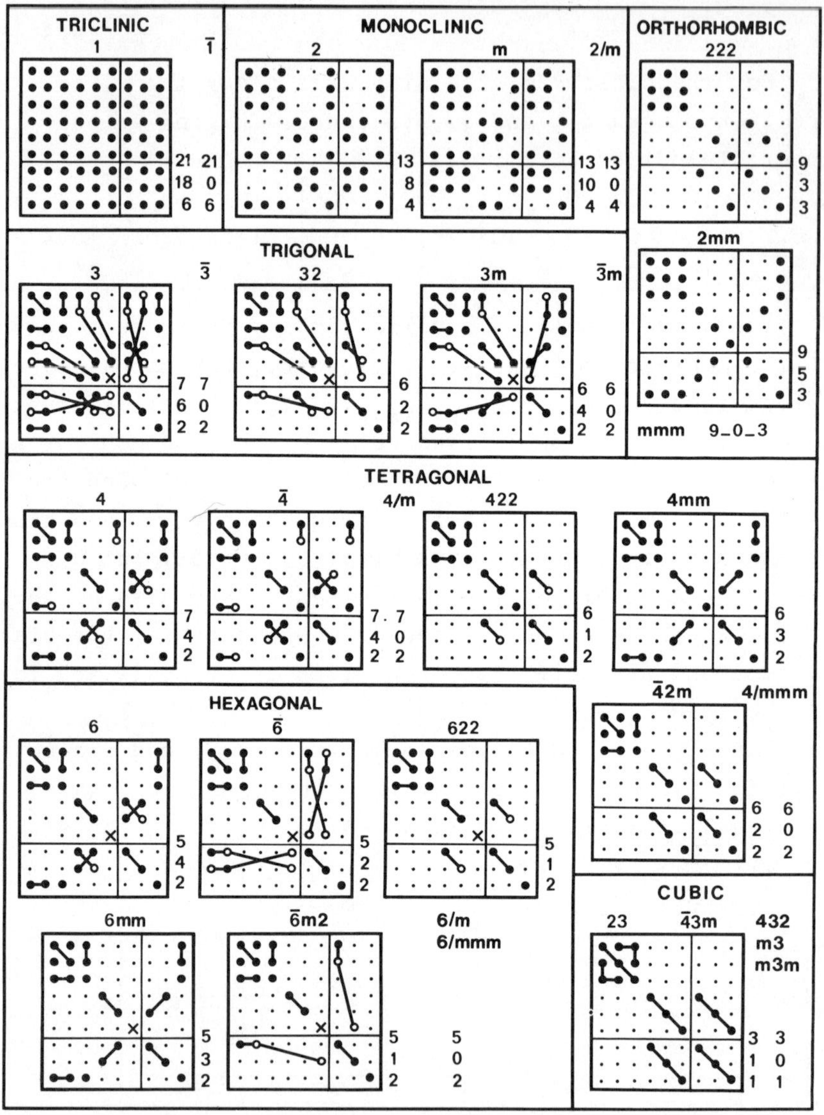

Fig. 6.4. Table of the matrices for elastic, piezoelectric, and dielectric constants, in the crystallographic axes reference frame (fig. 2.28).

○ non vanishing component
● vanishing component
●—● equal components
●—○ opposite components
× component equal to $(c_{11}-c_{12})/2$.

The layout of the matrices is explained on page 263. The symmetry with respect to the main diagonal is not mentioned. The number of independent elastic, piezoelectric and dielectric constants is indicated to the right of each matrix. The $c_{\alpha\beta}$ and ε_{ij} matrices for non-piezoelectric classes are not shown; they are identical to those of the preceding piezoelectric class in the same system.

Material	Class	Piezoelectric constant (C/m^2)						Permittivity (10^{-11}F/m)			Ref.
Cubic system		e_{14}						ε^S			
Gallium arsenide (GaAs)	$\bar{4}3m$	-0.16						9.73			[3]
Bismuth germanium oxide ($Bi_{12}GeO_{20}$)	23	0.99						34.2			[4] chap.4
Hexagonal system		e_{15}	e_{31}	e_{33}				ε^S_{11}	ε^S_{33}		
Ceramic PZT-4	trans. iso-tropic	12.7	-5.2	15.1				650	560		[25] chap.4
Zinc oxide (ZnO)	6mm	-0.59	-0.61	1.14				7.38	7.83		-
Cadmium sulphide (CdS)	6mm	-0.21	-0.24	0.44				7.99	8.44		-
Tetragonal system		e_{14}	e_{15}	e_{31}	e_{33}			ε^S_{11}	ε^S_{33}		
Paratellurite (TeO_2)	422	0.22	0	0	0			20	22		[16] chap.4
Barium titanate ($BaTiO_3$)	4mm	0	21.3	-2.65	3.64			1.744	97		[18] chap.4
Trigonal system		e_{11}	e_{14}	e_{15}	e_{22}	e_{31}	e_{33}	ε^S_{11}	ε^S_{33}		
Lithium niobate ($LiNbO_3$)	3m	0	0	3.7	2.5	0.2	1.3	28.9	25.7		[20] chap.4
Lithium tantalate ($LiTaO_3$)	3m	0	0	2.6	1.6	≃ 0	1.9	36.3	38.1		-
Quartz (SiO_2)	32	0.171	-0.0406	0	0	0	0	3.92	4.10		[21] chap. 4
Orthorhombic system		e_{15}	e_{24}	e_{31}	e_{32}	e_{33}		ε^S_{11}	ε^S_{22}	ε^S_{33}	
Barium sodium niobate ($Ba_2NaNb_5O_{15}$)	2mm	2.8	3.4	-0.4	-0.3	4.3		196	201	28	[24] chap.4

Fig. 6.5. Piezoelectric and dielectric constants.

6.2. ELASTIC WAVES IN A PIEZOELECTRIC SOLID

In a piezoelectric material, the interrelationship of electric and mechanical quantities, which is expressed, under static conditions, by equations such as 6.15, implies a coupling between elastic waves and electromagnetic waves. Specifically equation 6.13 introduces the electric field into the equations of dynamics and equation 6.9 introduces the mechanical strain into Maxwell's equations. In principle, the problem of wave propagation cannot be treated comprehensively if one does not solve simultaneously Newton's equations and Maxwell's equations. The solutions are mixed elasto-electromagnetic waves, i.e. elastic waves of velocity V with an associated electric field and electromagnetic waves (of velocity $v \simeq 10^5$ V) with an associated mechanical strain. For the first type of wave, the magnetic field created by the electric field moving at a velocity V, much smaller than the electromagnetic wave, velocity v, can be neglected. According to Maxwell's equations:

$$\overrightarrow{\text{curl}}\ \vec{E} = -\frac{\partial \vec{B}}{\partial t} \simeq 0$$

and the field $\vec{E}$ is derived from a scalar potential Φ, as in electrostatics:

$$\vec{E} = -\overrightarrow{\text{grad}}\ \Phi \ .$$

The electromagnetic energy is negligible compared to the elastic energy.

For the second type of wave, the elastic energy is much smaller than the electromagnetic energy.

So, it appears that, even in a strongly piezoelectric material, the interaction between the three elastic waves and the two electromagnetic waves is weak, as their velocities are very different ($v/V \simeq 10^4$ to 10^5). Therefore, the two types of propagation can be considered independently. We shall be concerned with elastic wave propagation, assuming that the electric field can be considered as static with respect to electromagnetic propagation phenomena (quasi static approximation). Our analysis first treats bulk waves (unbounded medium), then surface waves (semi-infinite medium).

6.2.1. Unbounded Medium. Bulk Waves

Our aim is to establish an eigenvalue equation for plane waves, similar to the Christoffel equation (§ 5.1.1); then to discuss the effect of the electromechanical coupling in a few piezoelectric materials, using the slowness curves.

6.2.1.1. Propagation equations. Christoffel tensor

Equation 6.13 for the stress

$$T_{ij} = c^E_{ijk\ell} S_{k\ell} - e_{kij} E_k$$

is transformed by inserting expressions for the strain $S_{k\ell} = 1/2(\partial u_k/\partial x_\ell + \partial u_\ell/\partial x_k)$ and the electric field $E_k = -\partial\Phi/\partial x_k$:

$$T_{ij} = c^E_{ijk\ell} \frac{\partial u_\ell}{\partial x_k} + e_{kij} \frac{\partial \Phi}{\partial x_k} \; .$$

The fundamental equation of dynamics 5.1

$$\rho \frac{\partial^2 u_i}{\partial t^2} = \frac{\partial T_{ij}}{\partial x_j}$$

becomes

$$\boxed{\rho \frac{\partial^2 u_i}{\partial t^2} = c^E_{ijk\ell} \frac{\partial^2 u_\ell}{\partial x_j \partial x_k} + e_{kij} \frac{\partial^2 \Phi}{\partial x_j \partial x_k}} \qquad (6.48)$$

Furthermore, the electric displacement

$$D_j = e_{jk\ell} S_{k\ell} + \varepsilon^S_{jk} E_k = e_{jk\ell} \frac{\partial u_\ell}{\partial x_k} - \varepsilon^S_{jk} \frac{\partial \Phi}{\partial x_k} \qquad (6.49)$$

must obey the Poisson equation $\partial D_j/\partial x_j = 0$ for an insulator:

$$\boxed{e_{jk\ell} \frac{\partial^2 u_\ell}{\partial x_j \partial x_k} - \varepsilon^S_{jk} \frac{\partial^2 \Phi}{\partial x_j \partial x_k} = 0} \qquad (6.50)$$

The propagation equation for the displacement u_i is obtained by eliminating the electric potential Φ between 6.48 and 6.50. In the case of a plane wave propagating in the n_j direction, u_i and Φ are given by

$$u_i = {}^o u_i F\left(t - \frac{n_j x_j}{V}\right) \qquad \Phi = \Phi_0 F\left(t - \frac{n_j x_j}{V}\right) \qquad (6.51)$$

and the electric field is longitudinal:

$$E_j = -\frac{\partial\Phi}{\partial x_j} = \frac{n_j}{V}\Phi_0 F'$$

where F' is the derivative of F. Inserting

$$\frac{\partial^2 u_i}{\partial t^2} = {}^o u_i F'', \qquad \frac{\partial^2 u_\ell}{\partial x_j \partial x_k} = \frac{n_j n_k}{V^2}\,{}^o u_\ell F'', \qquad \frac{\partial^2\Phi}{\partial x_j \partial x_k} = \frac{n_j n_k}{V^2}\Phi_0 F''$$

into 6.48 and 6.50, together with

$$\boxed{\Gamma_{i\ell} = c^E_{ijk\ell} n_j n_k, \quad \gamma_i = e_{kij} n_j n_k, \quad \varepsilon = \varepsilon^S_{jk} n_j n_k} \qquad (6.52)$$

yields

$$\begin{aligned} \rho V^2\,{}^o u_i &= \Gamma_{i\ell}\,{}^o u_\ell + \gamma_i \Phi_0 \\ \gamma_\ell\,{}^o u_\ell - \varepsilon\Phi_0 &= 0 \end{aligned} \qquad (6.53)$$

Eliminating the electric potential Φ_0, we get

$$\rho V^2\,{}^o u_i = \left(\Gamma_{i\ell} + \frac{\gamma_i\gamma_\ell}{\varepsilon}\right){}^o u_\ell \qquad (6.54)$$

The polarizations ${}^o u_i$ of plane elastic waves are obtained, as in non piezoelectric materials, by looking for the eigenvectors of a second rank tensor:

$$\overline{\Gamma}_{i\ell} = \Gamma_{i\ell} + \frac{\gamma_i\gamma_\ell}{\varepsilon} \qquad (6.55)$$

The eigenvalues $\bar{\gamma} = \rho V^2$ give the phase velocities for the corresponding direction. The polarizations of the three plane waves are always mutually orthogonal since the tensor $\overline{\Gamma}_{i\ell}$ is symmetric. In order to show how the additional piezoelectric term depends on the propagation direction, it is useful to complete the expansion of equation 5.8 by that of γ_ℓ components:

$$\begin{aligned} \gamma_\ell = {} & e_{11\ell}n_1^2 + e_{22\ell}n_2^2 + e_{33\ell}n_3^2 + (e_{12\ell}+e_{21\ell})n_1n_2 \\ & + (e_{13\ell}+e_{31\ell})n_1n_3 + (e_{23\ell}+e_{32\ell})n_2n_3 \end{aligned}$$

or for each value of the index ℓ

$$\begin{aligned} \gamma_1 = {} & e_{11}n_1^2 + e_{26}n_2^2 + e_{35}n_3^2 + (e_{16}+e_{21})n_1n_2 \\ & + (e_{15}+e_{31})n_1n_3 + (e_{25}+e_{36})n_2n_3 \\ \gamma_2 = {} & e_{16}n_1^2 + e_{22}n_2^2 + e_{34}n_3^2 + (e_{12}+e_{26})n_1n_2 \\ & + (e_{14}+e_{36})n_1n_3 + (e_{24}+e_{32})n_2n_3 \\ \gamma_3 = {} & e_{15}n_1^2 + e_{24}n_2^2 + e_{33}n_3^2 + (e_{14}+e_{25})n_1n_2 \\ & + (e_{13}+e_{35})n_1n_3 + (e_{23}+e_{34})n_2n_3 \end{aligned} \qquad (6.56)$$

The effect of piezoelectricity on the propagation velocity can be expressed by a change in the stiffness coefficients: the Christoffel tensor can be written, exactly as for a non piezoelectric material, as

$$\overline{\Gamma}_{i\ell} = \overline{c}_{ijk\ell} n_j n_k$$

where

$$\overline{c}_{ijk\ell} = c^E_{ijk\ell} + \frac{(e_{pij} n_p)(e_{qk\ell} n_q)}{\varepsilon^S_{jk} n_j n_k} \qquad (6.57)$$

The terms $\overline{c}_{ijk\ell}$, known as "stiffened" constants, are not true elastic constants since they are defined for plane waves only, and they depend on the propagation direction. So, it is better to use equation 6.55 and the expansions of 6.56 to deal with the following examples.

6.2.1.2. Examples of slowness curves in piezoelectric materials.

a) Propagation in the (001) plane of a cubic piezoelectric crystal ($Bi_{12}GeO_{20}$). Table 6.34 for the coefficients $e_{i\alpha}$ is the same for the two cubic classes $\overline{4}3m$ and 23 which are compatible with piezoelectricity. Since the only non vanishing components are

$$e_{14} = e_{25} = e_{36}$$

the expansions 6.56 reduce to

$$\gamma_1 = 2e_{14}n_2n_3 \qquad \gamma_2 = 2e_{14}n_1n_3 \qquad \gamma_3 = 2e_{14}n_1n_2.$$

In the (001) plane, $n_1 = \cos\phi$, $n_2 = \sin\phi$, $n_3 = 0$:

$$\gamma_1 = \gamma_2 = 0 \quad \text{and} \quad \gamma_3 = e_{14}\sin 2\phi.$$

the components $\overline{\Gamma}_{13}$ and $\overline{\Gamma}_{23}$ vanish, as in § 5.1.6.2a for a nonpiezoelectric cubic crystal. Furthermore,

$$\overline{\Gamma}_{11} = \Gamma_{11}, \qquad \overline{\Gamma}_{22} = \Gamma_{22}, \qquad \overline{\Gamma}_{12} = \Gamma_{12}$$

and only $\overline{\Gamma}_{33}$ is modified

$$\overline{\Gamma}_{33} = \Gamma_{33} + \frac{\gamma_3^2}{\varepsilon^S_{11}} = c^E_{44} + \frac{e^2_{14}}{\varepsilon^S_{11}}\sin^2 2\phi$$

The expression 5.33 for the velocities V_1 and V_2 of quasi-longitudinal and quasi-shear waves polarized in the (001) plane only involves $\overline{\Gamma}_{11}$, $\overline{\Gamma}_{22}$, $\overline{\Gamma}_{12}$. It remains unchanged, so that V_1 and V_2 are still given by equations 5.34 and 5.35, taking the stiffnesses $c^E_{ijk\ell}$ at constant electric field. These modes are not affected by piezoelectricity since the corresponding strains S_{11}, S_{22}, S_{12}, do not generate any longitudinal electric field (there is no e_{ijk} without any index of 3). The waves are not coupled to any electric field. On the other hand, the velocity of the shear wave polarized along [001] is modified. It now depends on direction of propagation; thus

$$V_3 = \sqrt{\frac{\bar{\Gamma}_{33}}{\rho}} = \sqrt{\frac{c_{44}}{\rho}}\left(1 + \frac{e_{14}^2}{\varepsilon_{11}^S c_{44}^E}\sin^2 2\phi\right)^{1/2}$$

The magnitude of this dependence is directly linked to the magnitude of the electromechanical coupling constant defined by

$$K = \frac{e_{14}}{\sqrt{\varepsilon_{11}^S c_{44}^E + e_{14}^2}} \qquad V_3 = \sqrt{\frac{c_{44}}{\rho}} \cdot \sqrt{1 + \frac{K^2 \sin^2 2\phi}{1 - K^2}}$$

In the case of bismuth germanium oxide ($Bi_{12}GeO_{20}$, class 23, constants taken from tables 4.7 and 6.5), the slowness curves are plotted in fig. 6.6, with K = .32; the maximum relative variation of V_3 (direction [110]) is equal to

$$\frac{\Delta V_3}{V_3} \simeq \frac{K^2}{2} \simeq 5\%.$$

b) <u>Propagation in the YZ plane of lithium niobate</u>. Lithium niobate is a ferroelectric crystal (§ 2.5.2d) belonging to class 3m of the trigonal system. The piezoelectric tensor 6.43 and equations 6.56 lead, if the propagation is contained in the YZ symmetry plane ($n_1 = 0$, $n_2 = \sin\theta$, $n_3 = \cos\theta$), to

$$\gamma_1 = 0$$

$$\gamma_2 = e_{22}\sin^2\theta + (e_{15} + e_{31})\sin\theta\cos\theta$$

$$\gamma_3 = e_{15}\sin^2\theta + e_{33}\cos^2\theta$$

$$\varepsilon = \varepsilon_{11}\sin^2\theta + \varepsilon_{33}\cos^2\theta.$$

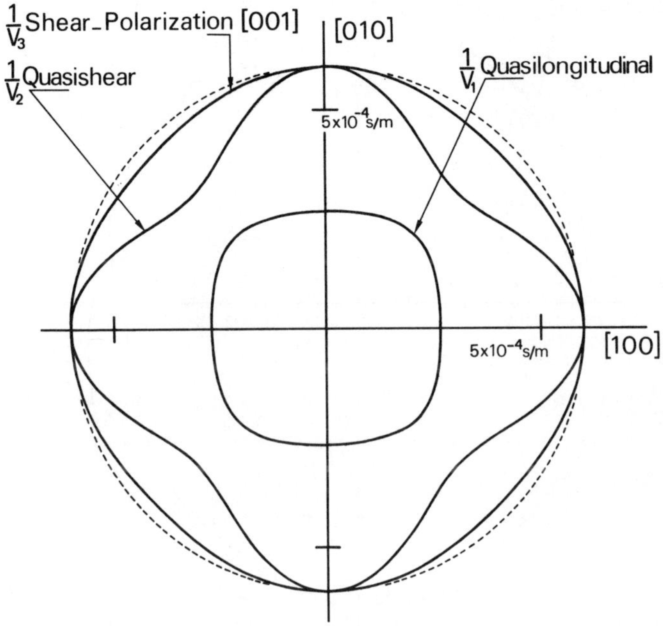

Fig. 6.6. Cross-section of the slowness surface of bismuth-germanium oxide (class 23) in the (001) plane. Only the shear wave is piezoelectrically active, as shown by the deviation from the dotted circle, obtained when ignoring piezo-electricity.

We have seen (§ 5.1.6.2d) that, for this cross section of the slowness surface, $\Gamma_{12} = \Gamma_{13} = 0$, the other components being given by equation 5.54. Therefore, the only nonvanishing components of the Christoffel tensor are

$$\overline{\Gamma}_{11} = \Gamma_{11} \qquad \overline{\Gamma}_{22} = \Gamma_{22} + \frac{\gamma_2^2}{\varepsilon}$$

$$\overline{\Gamma}_{23} = \Gamma_{23} + \frac{\gamma_2\gamma_3}{\varepsilon} \qquad \overline{\Gamma}_{33} = \Gamma_{33} + \frac{\gamma_3^2}{\varepsilon}$$

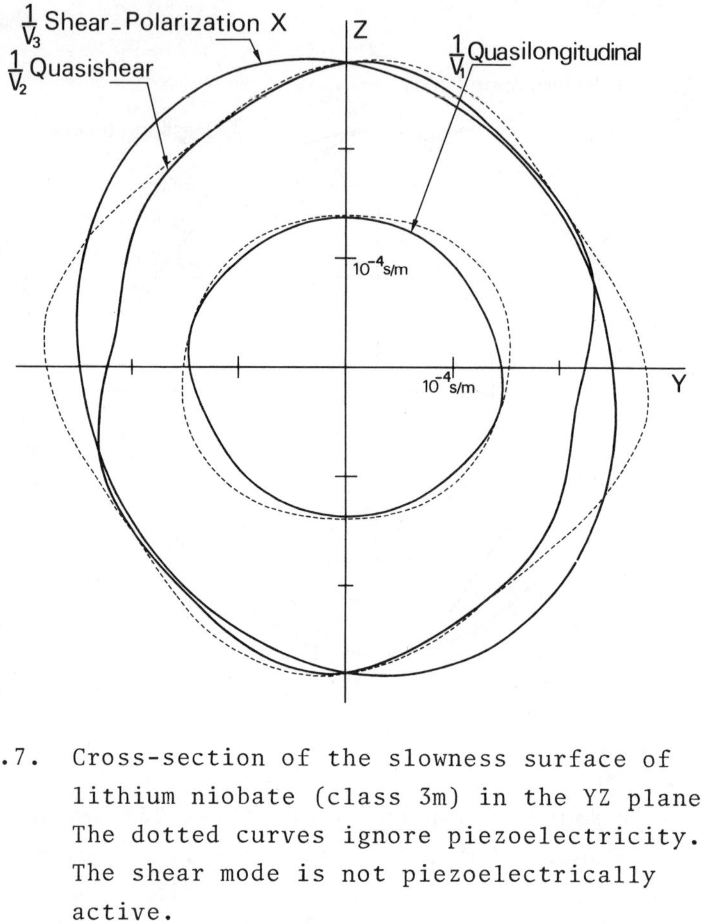

Fig. 6.7. Cross-section of the slowness surface of lithium niobate (class 3m) in the YZ plane. The dotted curves ignore piezoelectricity. The shear mode is not piezoelectrically active.

and, for the shear wave with polarization orthogonal to the YZ plane, the velocity $V_3 = \sqrt{\Gamma_{11}/\rho}$ is unaffected by piezoelectricity. This is not true for the quasi longitudinal wave and for the quasi shear wave polarized in the YZ plane, with velocities V_1 and V_2 such that

$$2\rho(V_{\substack{1\\2}})^2 = \overline{\Gamma}_{22} + \overline{\Gamma}_{33} \pm \sqrt{(\overline{\Gamma}_{22} - \overline{\Gamma}_{33})^2 + 4(\overline{\Gamma}_{23})^2} \quad (6.58)$$

The comparison can be made in fig. 6.7 between curves (solid line), plotted with the coefficients of tables 4.7 and 6.5 and the curves (dotted line), which do not take account of piezoelectricity. The importance of piezoelectric effects in this crystal is plainly visible.

c) Propagation in the azimuthal plane of zinc oxide. Zinc oxide belongs to class 6mm; according to the table 6.39 and equations 6.56:

$$\gamma_1 = (e_{15} + e_{31})n_1n_3$$

$$\gamma_2 = (e_{15} + e_{31})n_2n_3 \qquad (6.59)$$

$$\gamma_3 = e_{15}(n_1^2 + n_2^2) + e_{33}n_3^2$$

$$\varepsilon = \varepsilon_{11}(n_1^2 + n_2^2) + \varepsilon_{33}n_3^2$$

In the azimuthal plane YZ ($n_1 = 0$, $n_2 = \sin\theta$, $n_3 = \cos\theta$),

$$\gamma_1 = 0 \qquad \gamma_2 = \frac{e_{15} + e_{31}}{2}\sin 2\theta$$

$$\gamma_3 = e_{15}\sin^2\theta + e_{33}\cos^2\theta$$

$$\varepsilon = \varepsilon_{11}\sin^2\theta + \varepsilon_{33}\cos^2\theta.$$

Taking account of the results of § 5.1.6.2c:

$$\overline{\Gamma}_{11} = \Gamma_{11} \qquad \overline{\Gamma}_{12} = \Gamma_{12} = 0 \qquad \Gamma_{13} = \overline{\Gamma}_{13} = 0$$

$$\overline{\Gamma}_{22} = \Gamma_{22} + \frac{\gamma_2^2}{\varepsilon} \qquad \overline{\Gamma}_{23} = \Gamma_{23} + \frac{\gamma_2\gamma_3}{\varepsilon} \qquad \overline{\Gamma}_{33} = \Gamma_{33} + \frac{\gamma_3^2}{\varepsilon}$$

Therefore a shear wave, with polarization orthogonal to the YZ plane, velocity V_3 given by equation 5.46, can propagate as well as two waves, one quasi longitudinal and one quasi shear, whose velocities are given by 6.58. These results are depicted in fig. 6.8.

6.2.1.3. Propagation along directions linked to symmetry elements

In order to calculate the coupling constants in major directions of crystals, let us focus on the propagation along a rotation axis, in a plane perpendicular to a tetrad or 6-fold axis, and in a symmetry plane.

a) Along a symmetry axis parallel to the x_3 direction ($n_1 = n_2 = 0$, $n_3 = 1$), the expansions 6.56 give

$$\gamma_1 = e_{35} \qquad \gamma_2 = e_{34} \qquad \gamma_3 = e_{33}.$$

A look at the $e_{i\alpha}$ matrices will show that, for crystals which have a direct n-fold axis or an inverse n-fold axis with $n \neq 2$ along x_3, e_{34} and e_{35} are equal to zero. From the results of § 5.1.3, we know that

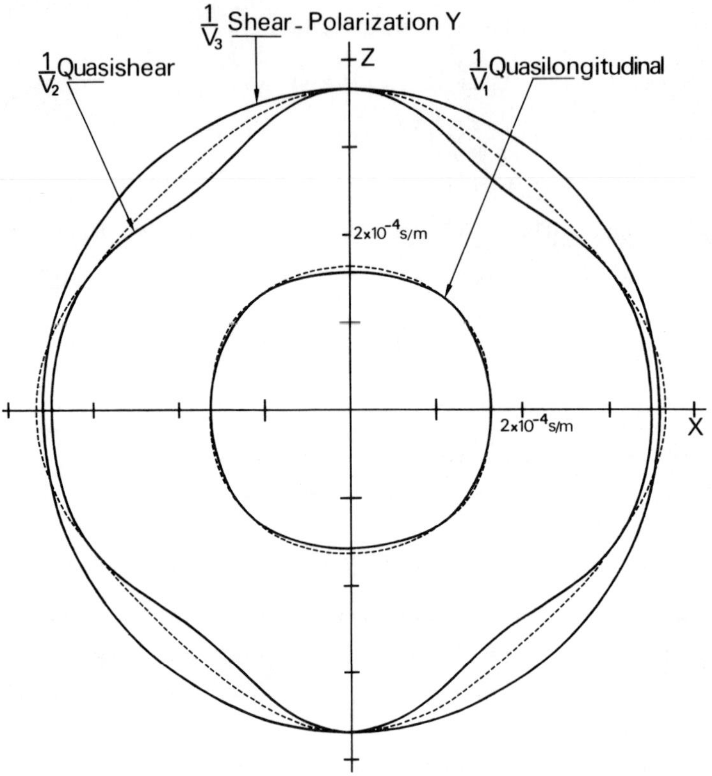

Fig. 6.8. Cross-section of the slowness surface of zinc oxide (class 6mm) by a meridian plane. The dotted curves ignore piezoelectricity. The shear mode is not piezoelectrically active.

$\overline{\Gamma}_{13}$ and $\overline{\Gamma}_{23}$ vanish and only the $\overline{\Gamma}_{33}$ component of the Christoffel tensor is affected by piezoelectricity:

$$\overline{\Gamma}_{33} = c^E_{33} + \frac{e^2_{33}}{\varepsilon^S_{33}} .$$

Therefore, both shear waves propagating along a symmetry axis (different from an $\overline{A}_2$ axis) suffer no piezoelectric coupling. For the longitudinal wave, the electromechanical coupling constant is equal to

$$K = \frac{e_{33}}{\sqrt{\varepsilon_{33}^S c_{33}^E + e_{33}^2}} \qquad A_n // x_3.$$

This constant may also vanish if e_{33} is equal to zero for other symmetry requirements, as in quartz (class 32).

b) <u>In a plane orthogonal to a direct tetrad or 6-fold axis</u>, $n_3 = 0$ and equation 6.56 yields

$$\gamma_1 = e_{11}n_1^2 + e_{26}n_2^2 + (e_{16} + e_{21})n_1n_2$$

$$\gamma_2 = e_{16}n_1^2 + e_{22}n_2^2 + (e_{12} + e_{26})n_1n_2$$

$$\gamma_3 = e_{15}n_1^2 + e_{24}n_2^2 + (e_{14} + e_{25})n_1n_2$$

According to table 6.37, γ_1 and γ_2 vanish since

$$e_{11} = e_{26} = e_{16} = e_{21} = e_{22} = 0$$

and, taking the results of § 5.1.3b into account (especially $\Gamma_{13} = \Gamma_{23} = 0$), one sees that, for all orientations in the plane, there exists a shear wave polarized along the x_3-axis. Furthermore, since $e_{25} = -e_{14}$ and $e_{24} = e_{15}$, the component γ_3 is a constant:

$$\gamma_3 = e_{15}(n_1^2 + n_2^2) = e_{15}.$$

Therefore, the quasi-longitudinal wave and the quasi-shear wave, polarized in the x_1x_2 plane, are not piezoelectrically coupled. The velocity V_3 of the shear wave (T) polarized along x_3 does not depend on the propagation direction in the x_1x_2 plane:

$$V_3 = \sqrt{\frac{c_{44}^E}{\rho} + \frac{e_{15}^2}{\varepsilon_{11}^S \rho}}$$

the coupling constant for this mode is

$$K = \frac{e_{15}}{\sqrt{\varepsilon_{11}^S c_{44}^E + e_{15}^2}} .$$

As in a non piezoelectric solid, the x_1x_2 plane is isotropic for this shear wave. This is no longer true if the axis is not direct ($\bar{A}_4$) see exercise (6.4).

c) In a symmetry plane orthogonal to the x_3 axis ($n_3 = 0$), the γ_3 component vanishes for $e_{15} = e_{24} = e_{14} = e_{25} = 0$ (see table 6.30). A shear wave with polarization orthogonal to the mirror ($\overline{\Gamma}_{13} = \overline{\Gamma}_{23} = 0$) can propagate; it is not piezoelectric since its velocity $V_3 = \sqrt{\overline{\Gamma}_{33}/\rho} = \sqrt{\Gamma_{33}/\rho}$ is not modified.

These results are gathered in table 6.9, which provides the coupling constants for most frequently used piezoelectric crystals.

6.2.2. Surface Waves

When an elastic wave propagates along the surface of a semi-infinite medium, both the propagation equations in the unbounded medium and the boundary conditions at the surface must be satisfied. It has been shown (§ 5.3) that for a non piezoelectric solid, the Rayleigh waves (complex waves with elliptic displacement) are the only solution. Looking for solutions in a piezoelectric medium is somewhat more difficult, firstly because propagation is now described by a set of two equations (6.48 and 6.50), and secondly because, besides the condition of vanishing stress at the free surface, continuity conditions must also be fulfilled for the potential and for the normal component of the electric displacement at the uncharged surface. As well as examining the influence of piezoelectricity on Rayleigh wave propagation, the analysis of the problem will provide evidence for the existence of a linearly polarized surface wave (under certain symmetry conditions), known as a Bleustein-Gulyaev wave.

The general method for searching surface wave modes [4], which is described first, is an extension of the procedure of § 5.3. In the next section we predict how the various types of wave can propagate, depending on the location of the sagittal plane with respect to symmetry elements. Then we focus on Bleustein-Gulyaev wave propagation in a transversely isotropic medium. Finally, from a more general point of view, the influence of electric boundary conditions on surface wave propagation is discussed and some results are given for surface waves in piezoelectric crystals in current use.

Propagation direction	Mode	Coupling coefficient K	$Bi_{12}GeO_{20}$ (23)	ZnO (6mm)	CdS (6mm)	Ceramic PZT4 (6mm)	$LiNbO_3$ (3m)	$LiTaO_3$ (3m)	Quartz (32)
along an axis A_2, A_n or $\bar{A}_n$ (n>2) contained by Ox_3	L	$\frac{e_{33}}{\sqrt{\varepsilon_{33}^S c_{33}^E + e_{33}^2}}$	0	0.27	0.155	0.51	0.16	0.18	0
	T_1, T_2	0	0	0	0	0	0	0	0
in a plane x_1x_2 $\perp$ A_4 or A_6	T_{ox_3}	$\frac{e_{15}}{\sqrt{\varepsilon_{11}^S c_{44}^E + e_{15}^2}}$		0.32	0.19	0.70			
	QL,QT	0		0	0	0			
in a mirror	T	0		meridian plane	meridian plane	meridian plane	YZ plane	YZ plane	
$A_2//Ox_1$	L	$\frac{e_{11}}{\sqrt{\varepsilon_{11}^S c_{11}^E + e_{11}^2}}$	0						0.093
	T_1,T_2	0	0						0

Fig. 6.9. Expressions and numerical values of the coupling constant K in the symmetry directions of some materials. Crossed out boxes are not involved.

6.2.2.1. General procedure for finding surface wave solutions

In the reference frame of fig. 6.10 (the x_1-axis is along the propagation direction, the x_2 axis is orthogonal to the free surface), let us look for a displacement of the form

$$u_k = {}^{o}u_k \; e^{-\chi x_2} \; e^{i\omega(t - x_1/V)} \qquad \text{with} \qquad \mathrm{Re}[\chi] > 0$$

the associated electric potential being

$$\Phi = \Phi_0 \; e^{-\chi x_2} \; e^{i\omega(t - x_1/V)}.$$

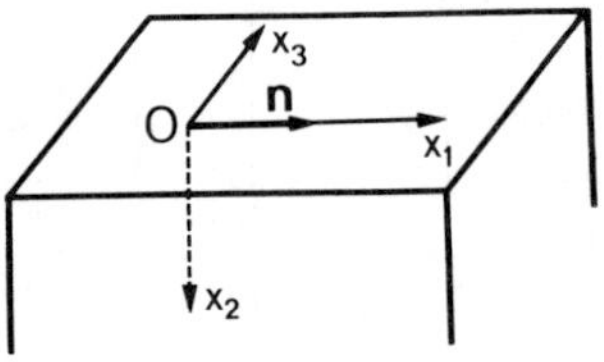

Fig. 6.10.

The magnitudes of u_k and Φ decrease exponentially in the bulk of the material. As in § 5.3, but permutating n_2 and n_3, we get

$$n_1 = 1, \quad n_3 = 0, \quad \chi = \frac{i\omega}{V} n_2 \qquad \text{with} \quad \mathrm{Im}[n_2] < 0$$

The components of displacement and potential can be written as

$$u_k = {}^{o}u_k \; e^{i\omega(t - n_\ell x_\ell / V)} \quad \text{and} \quad \Phi = \Phi_0 \; e^{i\omega(t - n_\ell x_\ell / V)}. \tag{6.60}$$

This can be inserted into the propagation equations 6.48 and 6.50, and leads, in matrix notation, to

$$\begin{pmatrix} \Gamma_{11} - \rho V^2 & \Gamma_{12} & \Gamma_{13} & \gamma_1 \\ \Gamma_{12} & \Gamma_{22} - \rho V^2 & \Gamma_{23} & \gamma_2 \\ \Gamma_{13} & \Gamma_{23} & \Gamma_{33} - \rho V^2 & \gamma_3 \\ \gamma_1 & \gamma_2 & \gamma_3 & -\varepsilon \end{pmatrix} \begin{pmatrix} {}^{o}u_1 \\ {}^{o}u_2 \\ {}^{o}u_3 \\ \Phi_0 \end{pmatrix} = 0 \tag{6.61}$$

which is the equivalent of the set of equations 6.53 in § 6.2.1.1.

The velocity is considered as a parameter which remains to be calculated; the compatibility condition of the above four homogeneous equations turns out to be a real equation of degree 8 in $n_2 = q$, from which we must select the four solutions whose imaginary part is negative. For each of these values, $n_2^{(r)} = q_r$ with r = 1, 2, 3, 4, equations 6.61 provide both the electric displacement ${}^o u_i^{(r)}$ and the potential $\Phi_0^{(r)}$. The general solution is a linear combination of these four waves, propagating at the same velocity V:

$$u_i = \sum_{r=1}^{4} A_r \, {}^o u_i^{(r)} \, e^{i\omega(t - n_\ell^{(r)} x_\ell / V)}$$

$$= \left(\sum_{r=1}^{4} A_r \, {}^o u_i^{(r)} e^{-\chi_r x_2} \right) e^{i\omega(t - x_1/V)} \qquad (6.62)$$

$$\Phi = \sum_{r=1}^{4} A_r \Phi_0^{(r)} \, e^{i\omega(t - n_\ell^{(r)} x_\ell / V)}$$

$$= \left(\sum_{r=1}^{4} A_r \Phi_0^{(r)} \, e^{-\chi_r x_2} \right) e^{i\omega(t - x_1/V)} \qquad (6.63)$$

where $\chi_r = i\frac{\omega}{V} q_r$. The velocity V and the coefficients A_r are determined through the mechanical and electric boundary conditions at the free surface $x_2 = 0$ (§ 6.2.2.4). The equation for V^2 has in fact only one acceptable solution, i.e. such that the χ_r has a positive real part.

The wave, whose displacement has three components, is accompanied by an electric field in the sagittal plane:

$$E_1 = -\frac{\partial \Phi}{\partial x_1} = i\frac{\omega}{V}\Phi \,, \qquad E_2 = -\frac{\partial \Phi}{\partial x_2} \,, \qquad E_3 = 0 \,.$$

This is a complex piezoelectric Rayleigh wave, which we shall denote by $\overline{R}_3$. On the other hand, for some selected sagittal planes, two velocities are acceptable. The first (V_R) corresponds to a simple non piezoelectric Rayleigh wave (polarized in the sagittal plane), denoted by R_2, the second (V_B) to a wave with its polarization orthogonal to the sagittal plane, coupled to an electric field, and called a Bleustein-Gulyaev wave (B). Both waves propagate independently of each other, with different velocities.

Between the previous extreme cases, there is the possibility of a piezoelectric two-component Rayleigh wave ($\overline{R}_2$). It will be recalled that, in the non-piezoelectric case, two types of Rayleigh waves can propagate: a three component wave (R_3) and a two component wave (R_2).

In a piezoelectric solid, the mode R_3 cannot exist because the three components of the displacement generate all the possible strains, and hence an electric field. Likewise, the Bleustein-Gulyaev wave is always associated with a R_2 mode, because the third component of the displacement (which is perpendicular to the sagittal plane) has to be coupled to an electric field if the other two are not.

The five cases indicated above occur for special forms of the equations 6.61, which result from the cancellation (because of symmetry) of some components $\Gamma_{i\ell}$ or γ_i.

6.2.2.2. Symmetry requirements

For a solution of 6.61 to correspond with a displacement orthogonal to the sagittal plane ($u_1 = u_2 = 0,\ u_3 \neq 0$), it is both necessary and sufficient that

$$\Gamma_{13} = \Gamma_{23} = 0 \quad \text{and} \quad \gamma_1 = \gamma_2 = 0 \qquad (6.64)$$

or

$$\Gamma_{13} = \Gamma_{23} = 0 \quad \text{and} \quad \gamma_3 = 0 \Rightarrow \Phi_0 = 0 \qquad (6.65)$$

From the expansion ($n_1 = 1,\ n_2 = q,\ n_3 = 0$)

$$\Gamma_{i\ell} = c^E_{ijk\ell} n_j n_k = c^E_{i11\ell} + c^E_{i22\ell} q^2 + (c^E_{i12\ell} + c^E_{i21\ell}) q \qquad (6.66)$$

the condition $\Gamma_{13} = \Gamma_{23} = 0$ implies that all constants with a single index of 3 are zero. The reduction of the elastic number of constants of monoclinic crystals (table 4.43) shows that it suffices that the sagittal plane is along an elastic symmetry plane (mirror, or plane perpendicular to a direct or inverse A_{2p} axis). From

$$\gamma_i = e_{kij} n_j n_k = e_{1i1} + e_{2i2} q^2 + (e_{1i2} + e_{2i1}) q \qquad (6.67)$$

it is clear that $\gamma_1 = \gamma_2 = 0$ causes all piezoelectric constants without an index of 3 to vanish. This condition is fulfilled if the x_1x_2 sagittal plane is perpendicular to a direct diad axis (§ 6.1.4c).

Equations 6.64 are simultaneously satisfied when <u>the sagittal plane is orthogonal to a direct axis of even order</u> (thus containing a diad axis). The system 6.61 splits into two independent systems; the first

$$\begin{pmatrix} \Gamma_{11} - \rho V^2 & \Gamma_{12} \\ \Gamma_{12} & \Gamma_{22} - \rho V^2 \end{pmatrix} \begin{pmatrix} {}^o u_1 \\ {}^o u_2 \end{pmatrix} = 0 \qquad (6.68)$$

corresponds to a simple <u>non piezoelectric Rayleigh wave (R_2)</u>, the second

$$\begin{pmatrix} \Gamma_{33} - \rho V^2 & \gamma_3 \\ \gamma_3 & -\varepsilon \end{pmatrix} \begin{pmatrix} {}^o u_3 \\ \Phi_0 \end{pmatrix} = 0 \qquad (6.69)$$

to a <u>shear-horizontal piezoelectric wave</u> (polarization ${}^o u_3$) called a Bleustein-Gulyaev wave (B).

Taking (6.65) into account, the set 6.61 leads to

$$(\Gamma_{33} - \rho V^2)\,{}^o u_3 = 0$$

and

$$\begin{pmatrix} \Gamma_{11} - \rho V^2 & \Gamma_{12} & \gamma_1 \\ \Gamma_{12} & \Gamma_{22} - \rho V^2 & \gamma_2 \\ \gamma_1 & \gamma_2 & -\varepsilon \end{pmatrix} \begin{pmatrix} {}^o u_1 \\ {}^o u_2 \\ \Phi_0 \end{pmatrix} = 0 \qquad (6.70)$$

Besides the non piezoelectric shear horizontal wave, there is an $\overline{R}_2$ type solution. The cancellation of γ_3 implies (because of 6.67)

$$e_{15} = e_{24} = 0 \quad \text{and} \quad e_{14} + e_{25} = 0. \qquad (6.71)$$

On the other hand, all elastic constants with only one index of 3 vanish ($\Gamma_{13} = \Gamma_{23} = 0$), so that the Rayleigh wave strains S_{11}, S_{12}, S_{22} do not generate any stress T_{32}. As a consequence, the boundary condition $T_{32} = 0$ or

$$T_{32} = -e_{i23}E_i = 0$$

together with $E_3 = 0$ and $e_{223} = e_{24} = 0$ (equation 6.71), implies $e_{14} = e_{25} = 0$. It was mentioned in § 6.1.4, that the moduli e_{14}, e_{15}, e_{24}, e_{25} vanish only if the x_1x_2 plane is a mirror (table 6.30). Finally, a two-component piezoelectric Rayleigh wave exists if and only if the sagittal plane is a symmetry plane.

The following table summarizes the results:

sagittal plane $\perp A_2$	$\Leftrightarrow$	modes B + R_2
sagittal plane // M	$\Leftrightarrow$	mode $\overline{R}_2$

Applying these two theorems for each symmetry class results in table 6.11.

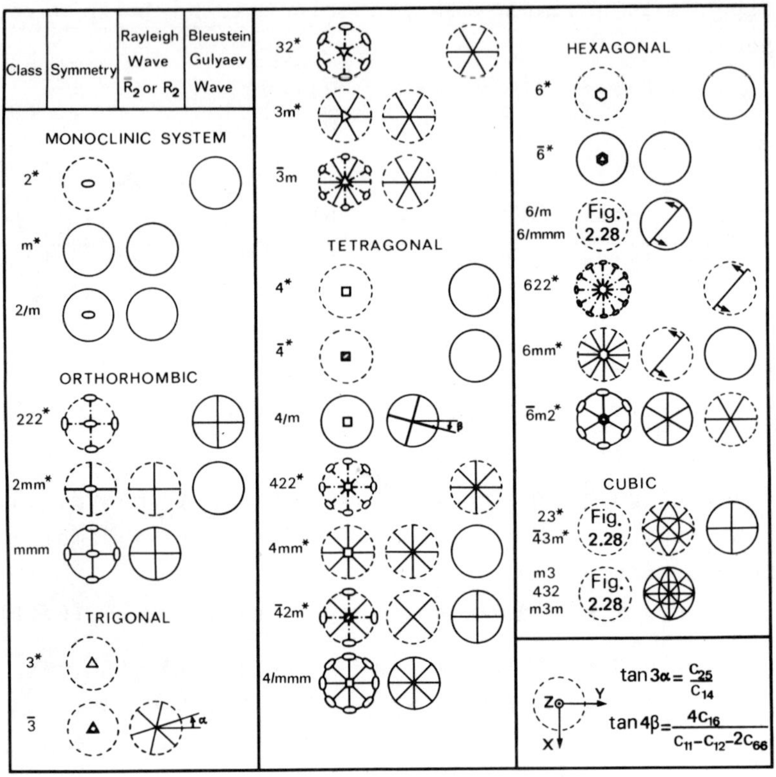

Fig. 6.11. Sagittal planes for the occurrence of Bleustein-Gulyaev waves, and two component ($\bar{R}_2$ or R_2) Rayleigh waves, depending on whether the crystal is piezoelectric or not. Piezoelectric classes are marked by a star. α and β are the angles through which the reference frame must be tilted for the stiffness moduli c_{25} and c_{16} (exercise 5.6 and 5.5) to cancel. A similar table for Love waves can be found in reference [5].

6.2.2.3. Bleustein-Gulyaev wave

A Bleustein-Gulyaev wave [6, 7] propagates when the sagittal plane is orthogonal to an A_6 axis (including an A_2 axis). For a crystal or a piezoelectric ceramic having a 6mm symmetry, the orientation of the x_1 and x_2-axis is arbitrary in the plane orthogonal to the 6-fold axis. With $n_1 = 1$, $n_2 = q$, $n_3 = 0$, equations 5.44 and 6.59 yield

$$\Gamma_{33} = c_{44}^{E}(1 + q^2) \qquad \gamma_3 = e_{15}(1 + q^2) \qquad \varepsilon = \varepsilon_{11}(1 + q^2).$$

The compatibility condition of the system 6.69:

$$(1 + q^2)\begin{vmatrix} c_{44}^{E}(1 + q^2) - \rho V^2 & e_{15} \\ e_{15}(1 + q^2) & -\varepsilon_{11} \end{vmatrix} = 0$$

is satisfied for two values q_1 and q_2 whose imaginary part is negative: $q_1 = -i$ and q_2 such that

$$c_{44}^{E}(1 + q_2^2) - \rho V^2 + \frac{e_{15}^2}{\varepsilon_{11}}(1 + q_2^2) = 0$$

or since $c_{44}^{D} = c_{44}^{E} + e_{15}^2/\varepsilon_{11}$ (exercise 6.2):

$$q_2^2 = \frac{\rho V^2}{c_{44}^{D}} - 1 .$$

q_2 is imaginary if the velocity V of the Bleustein-Gulyaev wave is less than that of the shear bulk wave of identical polarization: $V_T = \sqrt{c_{44}^{D}/\rho}$:

$$q_2 = -i\sqrt{1 - \frac{V^2}{V_T^2}} = -i\beta \qquad (6.72)$$

For $q = q_1$, the first equation of the set 6.69 gives ${}^{o}u_3^{(1)} = 0$; for $q = q_2$, the second equation provides $\Phi_0^{(2)} = (e_{15}/\varepsilon_{11})\,{}^{o}u_3^{(2)}$. If one defines $B = {}^{o}u_3^{(2)}$, the general solution (6.62) and (6.63) reduces to*

$$\begin{cases} u_3 = B\, e^{-\beta(\omega/V)x_2}\, e^{i\omega(t-x_1/V)} & (6.73a) \\ \Phi = \left(\Phi_0^{(1)}\, e^{-(\omega/V)x_2} + \dfrac{e_{15}}{\varepsilon_{11}} B\, e^{-\beta(\omega/V)x_2}\right) e^{i\omega(t-x_1/V)} & (6.73b) \end{cases}$$

The potential $\Phi_0^{(1)}$ and the velocity V are determined by the continuity equations at the free surface $x_2 = 0$. Further on we shall need the normal component of the electrical displacement

$$D_2 = -\varepsilon_{11}\frac{\partial\Phi}{\partial x_2} + e_{15}\frac{\partial u_3}{\partial x_2}$$

or

$$D_2 = \frac{\omega}{V}\varepsilon_{11}\Phi_0^{(1)}\, e^{-(\omega/V)x_2}\, e^{i\omega(t-x_1/V)}. \qquad (6.74)$$

---

* Here, the coefficients A_1 and A_2 are useless: they are included in $\Phi_0^{(1)}$ and ${}^{o}u_3^{(2)} = B$.

The mechanical boundary condition $T_{i2} = 0$ has no bearing except on T_{32}, for, on the one hand, the Bleustein-Gulyaev wave strains S_{32} and S_{31} and, on the other hand, the electric field, with components E_1 and E_2, only generate T_{31} and T_{32} stresses (see table 6.39):

$$T_{32} = c_{44}^{E} \frac{\partial u_3}{\partial x_2} + e_{15} \frac{\partial \Phi}{\partial x_2}$$

Inserting equations 6.73, the condition $T_{32} = 0$ at $x_2 = 0$, implies

$$\beta\left(c_{44}^{E} + \frac{e_{15}^2}{\varepsilon_{11}^{S}}\right)B + e_{15}\Phi_0^{(1)} = 0$$

or

$$\Phi_0^{(1)} = -\beta \frac{c_{44}^{D}}{e_{15}} B = -\frac{\beta}{K^2}\frac{e_{15}}{\varepsilon_{11}} B \qquad (6.75)$$

where

$$K = \frac{e_{15}}{\sqrt{\varepsilon_{11}^{S} c_{44}^{D}}} = \frac{e_{15}}{\sqrt{\varepsilon_{11}^{S} c_{44}^{E} + e_{15}^2}}$$

is the electro-mechanical coupling constant for the bulk shear wave. The Bleustein Gulyaev wave which satisfies the mechanical boundary conditions on the free surface is thus given by

$$\begin{cases} u_3 = B\, e^{-\beta(\omega/V)x_2}\, e^{i\omega(t-x_1/V)} & (6.76a) \\ \Phi = \dfrac{e_{15}}{\varepsilon_{11}} B\left(e^{-\beta(\omega/V)x_2} - \dfrac{\beta}{K^2}\, e^{-(\omega/V)x_2}\right) e^{i\omega(t-x_1/V)} & (6.76b) \end{cases}$$

The decay factor $\beta = \sqrt{1-V^2/V_T^2}$, and accordingly the velocity V, are determined by the electrical boundary conditions. Here we discuss two typical cases:

a) <u>The surface is coated with a thin metal film,</u> whose thickness is much less than the wavelength, at zero electric potential. The surface is still mechanically free. Since $\Phi = 0$ at $x_2 = 0$:

$$\beta = K^2 \qquad (6.77)$$

and the Bleustein-Gulyaev wave velocity

$$V_B = V_T\sqrt{1 - K^4} \qquad (6.78)$$

is very close to the shear wave velocity V_T. The decay in the amplitudes and potential is given by

$$\begin{cases} U_3 = B\, e^{-\beta 2\pi\, x_2/\lambda_B} \\ \Phi_m = \dfrac{e_{15}}{\varepsilon_{11}} B(e^{-\beta 2\pi\, x_2/\lambda_B} - e^{-2\pi\, x_2/\lambda_B}) \end{cases}$$

This decay is greater for the higher coupling constants. The curves of Fig. 6.12 are plotted for cadmium sulphide:

$K = .19 \implies \beta = 0.036 \qquad V_B = .9994\, V_T = 1{,}788$ m/s.

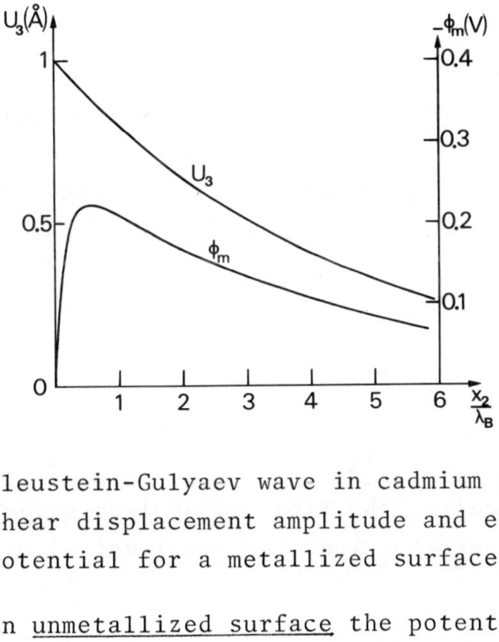

Fig. 6.12. Bleustein-Gulyaev wave in cadmium sulphide. Shear displacement amplitude and electric potential for a metallized surface.

b) For an <u>unmetallized surface</u> the potential Φ and the normal component of the displacement should be continuous. In a vacuum, the electric potential obeys the Laplace equation:

$$\frac{\partial^2 \Phi'}{\partial x_1^2} + \frac{\partial^2 \Phi'}{\partial x_2^2} = 0$$

and must have the same t and x_1 dependence as Φ, if the potential is to be continuous at all times and at every point of the surface. The solution, which vanishes at $x_2 = -\infty$,

$$\Phi' = \Phi_0'\, e^{(\omega/V)x_2}\, e^{i\omega(t - x_1/V)} \tag{6.79}$$

is equal to Φ (e.g. 6.76b) for $x_2 = 0$ if

$$\Phi_0' = \frac{e_{15}}{\varepsilon_{11}} B(1 - \frac{\beta}{K^2}). \tag{6.80}$$

The continuity of the normal component of the electrical displacement requires that (making use of equation 6.74)

$$\varepsilon_{11}\Phi_0^{(1)} = -\varepsilon_0\Phi_0'.$$

This leads, after replacing $\Phi_0^{(1)}$ and Φ_0' by the expressions 6.75 and 6.80, to

$$\frac{\beta}{K^2} e_{15}B = e_{15}B\left(1 - \frac{\beta}{K^2}\right)\frac{\varepsilon_0}{\varepsilon_{11}} .$$

The decay factor

$$\beta = \frac{K^2}{1 + \varepsilon_{11}/\varepsilon_0} \qquad (6.81)$$

is reduced by a factor nearly equal to the relative dielectric constant. A Bleustein-Gulyaev wave penetrates deeper into the material when the surface is unmetallised. In the case of CdS

$$\frac{\varepsilon_{11}}{\varepsilon_0} = 9 \Longrightarrow \beta = .0036.$$

The velocity is also changed, but remains quite close to the bulk shear-wave velocity:

$$V_B = V_T\sqrt{1 - \frac{K^4}{(1 + \varepsilon_{11}/\varepsilon_0)^2}} \qquad (6.82)$$

In both cases, the Bleustein-Gulyaev wave penetration depth is much greater than the wavelength (10 to 100 λ), which accounts for a velocity nearly equal to the bulk shear wave velocity.

6.2.2.4. Effect of the electric boundary conditions

Let us go back to the general case and discuss the effect of electric boundary conditions on the propagation of surface waves, and more specifically on their velocity. The mechanical stress at the free surface $x_2 = 0$ must vanish:

$$T_{i2} = c_{i2k\ell}\frac{\partial u_k}{\partial x_\ell} + e_{ki2}\frac{\partial \Phi}{\partial x_k} = 0 \quad \text{for} \quad x_2 = 0 \text{ and } i = 1,2,3$$

Substituting the expressions 6.62 and 6.63 for the displacement and potential of the surface wave,

$$T_{i2}(x_2=0) = \sum_{r=1}^{4} A_r[c_{i2k\ell}\ {}^{o}u_k^{(r)}\ (-\frac{i\omega}{V}\ n_\ell^{(r)})$$

$$+ e_{ki2}\Phi_0^{(r)}\ (-\frac{i\omega}{V}\ n_k^{(r)})]\ e^{i\omega(t-x_1/V)}$$

leads to a system of three equations (i = 1, 2, 3) and four variables A_r (r = 1, 2, 3, 4), which is written

$$\sum_{r=1}^{4} A_r T_{i2}^{(r)} = 0 \qquad (6.83)$$

where

$$T_{i2}^{(r)} = c_{i2k\ell}\ {}^{o}u_k^{(r)} n_\ell^{(r)} + e_{ki2}\Phi_0^{(r)} n_k^{(r)}. \qquad (6.84)$$

The electrical boundary conditions require the continuity of the potential and of the normal component of the electrical displacement.

$$D_2 = e_{2k\ell}\frac{\partial u_k}{\partial x_\ell} - \varepsilon_{2\ell}\frac{\partial\Phi}{\partial x_\ell}\ .$$

At the surface, $x_2 = 0$

$$D_2(x_2 = 0) = \sum_{r=1}^{4} A_r(-\frac{i\omega}{V})(e_{2k\ell}\ {}^{o}u_k^{(r)} n_\ell^{(r)}$$

$$- \varepsilon_{2\ell}\Phi_0^{(r)} n_\ell^{(r)})\ e^{i\omega(t-x_1/V)}$$

or equivalently

$$D_2(0) = \sum_{r=1}^{4} A_r D_2^{(r)}\ e^{i\omega(t-x_1/V)} \qquad (6.85)$$

where

$$D_2^{(r)} = \frac{i\omega}{V}\ (\varepsilon_{2\ell}\Phi_0^{(r)} n_\ell^{(r)} - e_{2k\ell}\ {}^{o}u_k^{(r)} n_\ell^{(r)}).$$

The electric potential at the surface is readily derived from 6.63:

$$\Phi(0) = \sum_{r=1}^{4} A_r\Phi_0^{(r)}\ e^{i\omega(t-x_1/V)}. \qquad (6.86)$$

The system of three equations (6.83) allows the calculation of three of the coefficients A_r in terms of the fourth. The potential Φ and the component D_2 of the displacement which are linear combinations of all the A_r coefficients, are in fact proportional to only one of these coefficients. The ratio of these two quantities, does not depend on the A_r's and defines a surface impedance Z_s [8]:

$$\frac{\Phi(0)}{D_2(0)} = i\frac{V^2}{\omega} Z_s \; . \qquad (6.87)$$

Taking 6.85 and 6.86 into account, this is put into the form

$$\sum_{r=1}^{4} A_r[\Phi_0^{(r)} - i\frac{V^2}{\omega} Z_s D_2^{(r)}] = 0$$

We add this to the three equations (6.83), to build a linear homogeneous system. The compatibility condition

$$\begin{vmatrix} T_{12}^{(1)} & T_{12}^{(2)} & T_{12}^{(3)} & T_{12}^{(4)} \\ T_{22}^{(1)} & T_{22}^{(2)} & T_{22}^{(3)} & T_{22}^{(4)} \\ T_{32}^{(1)} & T_{32}^{(2)} & T_{32}^{(3)} & T_{32}^{(4)} \\ \Phi_0^{(1)} - \frac{iV^2}{\omega} Z_s D_2^{(1)} & \Phi_0^{(2)} - \frac{iV^2}{\omega} Z_s D_2^{(2)} & \Phi_0^{(3)} - \frac{iV^2}{\omega} Z_s D_2^{(3)} & \Phi_0^{(4)} - \frac{iV^2}{\omega} Z_s D_2^{(4)} \end{vmatrix} = 0$$

is expanded as the sum of two determinants:

$$\Delta_1(V) + VZ_s\Delta_2(V) = 0$$

where Δ_1 and Δ_2 depend on the velocity V. The adjacent medium is also characterized by an impedance Z_o. The continuity of Φ and D_2 at the surface implies $Z_s = Z_o$ and the surface wave velocity is determined by the equation

$$\Delta_1(V) + VZ_o\Delta_2(V) = 0 \; . \qquad (6.88)$$

As for Bleustein-Gulyaev waves, two important cases should be considered:

a) The surface is plated with a very thin metal film at zero potential. The impedance Z_o vanishes and the velocity V_o is given by

$$\Delta_1(V_o) = 0. \qquad (6.89)$$

b) The adjacent medium is the vacuum. The potential Φ' which must satisfy the Laplace equation, i.e. vanish at

$x_2 = -\infty$, and be continuous at $x_2 = 0$, is given by 6.79. Therefore

$$D_2' = -\varepsilon_0 \frac{\partial \Phi'}{\partial x_2} = -\varepsilon_0 \frac{\omega}{V} \Phi'$$

and the impedance

$$Z_o = \frac{\omega}{iV^2} \frac{\Phi'(0)}{D_2'(0)}$$

is equal to

$$Z_o = \frac{i}{\varepsilon_0 V} .$$

With these boundary conditions, the surface wave velocity V_s is a solution of the equation

$$\Delta_1(V_s) + \frac{i}{\varepsilon_0} \Delta_2(V_s) = 0. \qquad (6.90)$$

It is customary [9] to define a coupling constant for surface waves by

$$\frac{K_s^2}{2} = \frac{V_s - V_o}{V_s} . \qquad (6.91)$$

It is intuitively reasonable (but not easy to demonstrate) that this coefficient, which measures the relative effect of a metal film on the velocity V, adequately characterizes the material's ability to transform an electric signal into an elastic surface wave through a transducer consisting of metal strips deposited on the surface (fig. 7.19 in § 7.2).

Once the velocity V is known, the decay factors $n_2^{(r)}$ and the partial amplitudes ${}^o u_i^{(r)}$ and $\Phi_0^{(r)}$ are, in principle, derived from equations 6.61. The system 6.83 provides the set of the A_r's, except for a proportionality factor which is linked to the elastic power transported by the surface wave. Next, from 6.62 and 6.63, the displacement components u_i and the potential Φ are derived. In fact, the algebra is so involved that it is impossible to get an exact analytical solution, but for the Bleustein-Gulyaev waves which have been considered earlier. The detailed calculations can be performed only with the aid of a computer.

6.2.2.5. Results for some specific materials

As an illustration of the previous discussion, we give a few curves, for lithium niobate, bismuth-germanium oxide, and quartz. In fig. 6.13, the dependence of V_o (metallized surface) and V_S (unmetallized surface) on the Rayleigh wave propagation direction in the XZ plane (Y cut) of lithium niobate is depicted. The maximum difference between the two velocities occurs when the propagation is along the Z axis: $V_o = 3{,}404$ m/s, $V_s = 3{,}488$ m/s. The coupling constant is then quite high: $K_R^2 = 4.8\ 10^{-2}$.

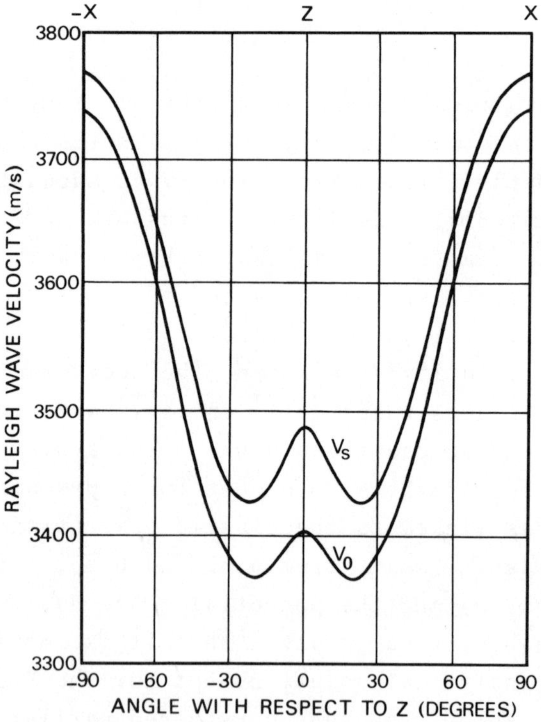

Fig. 6.13. Rayleigh wave in lithium niobate (Y cut). Velocities V_o (metallized surface) and V_s (unmetallized surface) vs propagation direction in the XZ plane (fig. 3 of reference [10]).

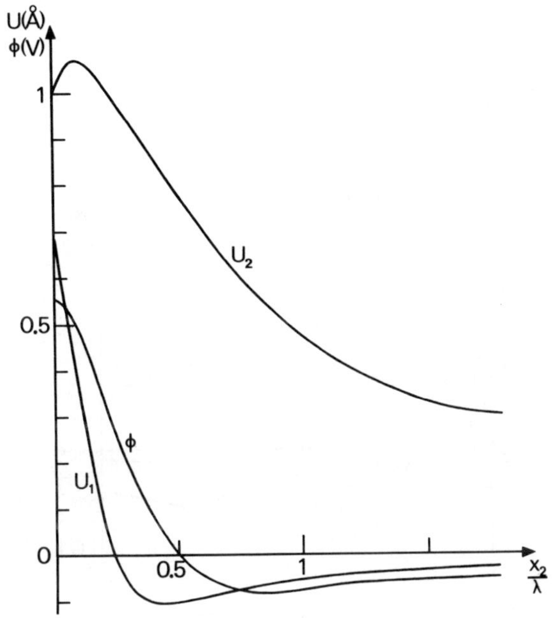

Fig. 6.14. Rayleigh waves in $LiNbO_3$ (Y cut, Z propagation). Displacement and potential vs depth, from an unmetallized surface (after fig. 1 of reference [11]).

Since the sagittal plane is a mirror, the displacement has only two components ($\overline{R}_2$ type). The penetration depth dependence of the displacement amplitude and of the potential is sketched in fig. 6.14 and 6.15 for both major electric boundary conditions. In this direction, the Rayleigh wave is pure, since the Poynting vector, already parallel to the surface, must be contained in the sagittal plane (YZ symmetry plane).

The curves in fig. 6.16 refer to bismuth-germanium oxide ($Bi_{12}GeO_{20}$, class 23), for the [001] propagation direction in the (110) plane. The sagittal plane $(1\overline{1}0)$ is not a mirror; however, an $\overline{R}_2$ type mode exists (table 6.11) because $(1\overline{1}0)$ is a symmetry plane for the elastic and piezoelectric properties. The tensors $c_{ijk\ell}$ and e_{ijk}

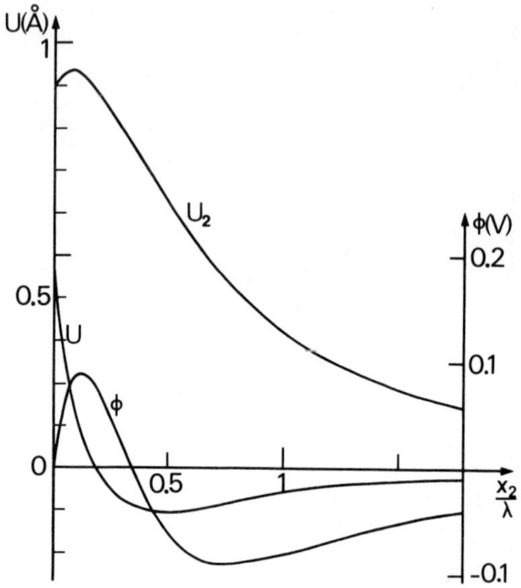

Fig. 6.15. Rayleigh waves in $LiNbO_3$ (Y cut, Z propagation). Displacement and potential vs depth for a metallized surface (after fig. 2 of reference [11]).

for class 23 have the same vanishing components as in class $\bar{4}3m$, where the plane $(1\bar{1}0)$ is a mirror. The comparison of the electric potential values in fig. 6.14 and 6.16 shows that this material is noticeably less piezoelectric than lithium niobate.

Quartz (class 32) has no mirror plane and is not suitable for $\bar{R}_2$ type Rayleigh wave propagation, as shown in Fig. 6.11. However pure mode directions for $\bar{R}_3$ waves do exist: e.g. the diad X axis in the XZ plane (Y cut).

The extremity of the displacement vector describes an ellipse, the plane of which is obtained from the sagittal plane by a 43° rotation about the X-axis. The velocity is 3,154 m/s. Fig. 6.17 shows the velocity of Rayleigh waves propagating in the X direction, when the normal to the surface (in the YZ plane) and the Y-axis make an angle β. The ST cut corresponding to $\beta = 42.5^{\circ}$ is noteworthy since it has zero temperature coefficient at $25^{\circ}C$ [14].

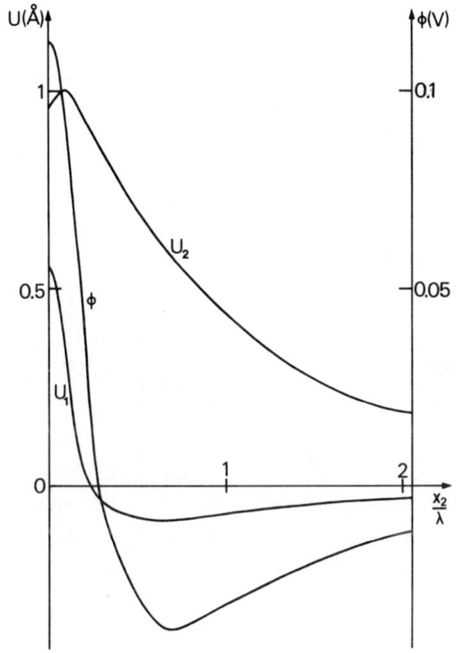

Fig. 6.16. Rayleigh waves in $Bi_{12}GeO_{20}$ ([110] cut, propagation along [001]). Displacement and potential vs depth for an unmetallized surface (after fig. 1 of reference [12]).

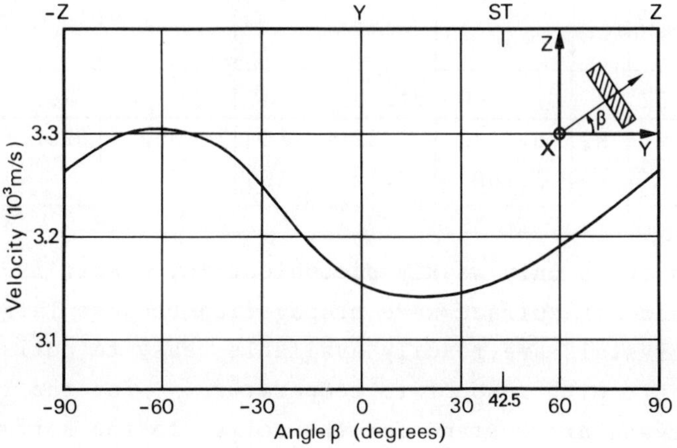

Fig. 6.17. Rayleigh wave velocity in the X direction vs rotation angle β of the cut in a quartz crystal (Fig. 3 of reference [13]).

Table 6.18. Interesting cuts and directions for pure mode propagation of surface waves, for some piezoelectric materials.

Cut and direction of propagation	Type of wave	Velocity V_s (m/s)	Coupling coefficient $K_s^2 x10^2$	Ref.	Velocity V_s (m/s)	Coupling coefficient $K_s^2 x10^2$	Ref.
Cubic system		$Bi_{12}GeO_{20}$			AsGa		
(100)[010]	R_2	1,624	0	[15]	2,720	0	[9]
(100)[011]	$\overline{R}_2$	1,683	1.4	-	2,865	0.07	-
(110)[001]	$\overline{R}_2$	1,625	0.8	-	2,820	0.02	-
(110)[1$\overline{1}$0]	B	1,755	1.0	-	3,354	0.02	-
(110)[1$\overline{1}$0]	R_2	2,160	0	-	2,400	0	-
(111)[1$\overline{1}$0]		1,708	1.7	-			
Class 6mm		ZnO			CdS		
any Z	$\overline{R}_2$	2,702	0.88	[9]	1,731	0.47	[9]
XY	B	2,838	0.35	-	1,800	0.12	-
XY	R_2	2,623	0	-	1,730	0	-
XZ	$\overline{R}_2$	2,680	1.1	-	1,718	0.52	-
Class 3m		$LiNbO_3$			$LiTaO_3$		
YZ	$\overline{R}_2$	3,488	4.8	[10]	3,230	0.66	[18]
ZY	$\overline{R}_2$	3,903	2.2	[10]	3,329	1.18	[18]
[Y+131.5°]X	$\overline{R}_3$	4,000	5.54	[16]			
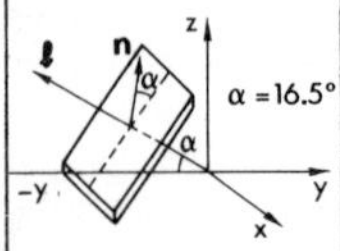	$\overline{R}_3$	3,503	5.36	[17]			
Class 32		Quartz α			Se		
YX	$\overline{R}_3$	3,154	0.22	[13]	810	1.7	[19]
ST(Y+42.5°)X	$\overline{R}_3$	3,160	0.16	[13]			

Although it is only weakly piezoelectric, quartz is very often used for surface wave propagation because large single crystals are readily available, easy to polish, and very stable with respect to temperature. For the $\underline{Y}\underrightarrow{X}$ cut (the bar and arrow stand for the normal to the surface and for the propagation direction), the coupling constant is $K_R^2 = .22.10^{-2}$.

Table 6.18 indicates cuts of particular interest and their major features for some materials.

REFERENCES

[1] G. Quentin and J.M. Thuillier. Solid State Commun. 2, 115 (1964).
[2] J. Bouat and J.M. Thuillier. Phys. Letters, 37 A (1971).
[3] G. Arlt and P. Quadflieg. Phys. Status Solidi, 25, 323 (1968).
[4] G.W. Farnell. Properties of elastic surface waves. In Physical Acoustics (W.P. Mason and R.N. Thurston, eds), vol. 6, chap. 3, paragr. VI. New York: Academic Press (1970).
[5] C. Maerfeld and C. Lardat. C.R. Acad. Sc. Paris, 270, 1187 (1970).
[6] J.L. Bleustein. Appl. Phys. Lett., 13, 412 (1968).
[7] Yu. V. Gulyaev. Soviet Phys. JETP Lett., 9, 63 (1969).
[8] K.A. Ingebrigtsen. J. Appl. Phys., 40, 2681 (1969).
[9] J.J. Campbell and W.R. Jones. J. Appl. Phys., 41, 2796 (1970).
[10] J.J. Campbell and W.R. Jones. IEEE Trans. Son. Ultrason., SU-15, 209 (1968).
[11] M. Moriamez, E. Bridoux, J.M. Desrumaux, J.M. Rouvaen and M. Delannoy. Revue Phys. Appl., 6, 333 (1971).
[12] E. Bridoux, J.M. Rouvaen, G. Coussot and E. Dieulesaint. Appl. Phys. Lett., 19, 523 (1971).
[13] G.A. Coquin and H.F. Tiersten. J. Acoust. Soc. Am., 41, 921 (1967).
[14] M.B. Schulz, B.J. Matsinger and M.G. Holland. J. Appl. Phys., 41, 2755 (1970).
[15] R.G. Pratt, G. Simpson, W.A. Crossley. Electron. Lett., 8, 127 (1972).
[16] A.J. Slobodnik Jr. and E.D. Conway. Electron. Lett., 6, 171 (1970).
[17] A.J. Slobodnik Jr. and T.L. Szabo. Electron. Lett., 7, 257 (1971).

[18] A.J. Slobodnik Jr. and E.D. Conway. Microwave acoustics handbook, vol. 1, AFCRL Report No. 70-0164, Air Force Cambridge Research Laboratories, Bedford, Mass., U.S.A. (1970).

[19] E. Dieulesaint, D. Royer, J. Barbot and J.C. Thuillier. IEEE Ultrasonics Symposium Proc., p. 383 (1973).

[20] C. Maerfeld and P. Tournois. Appl. Phys. Lett., 19, 117 (1971).

BIBLIOGRAPHY

Physical Mechanism of piezoelectricity

T. Ogawa. A linear chain model for piezoelectricity in zincblende and wurtzite type crystals. Japan J. Appl. Phys., 10, 72 (1971).

G. Arlt and P. Quadflieg. Reference [3].

Tensor formulation of piezoelectricity. Reduction due to crystal symmetry

J.F. Nye. Physical Properties of Crystals, chap. VII. Oxford (1964).

S. Bhagavantam. Crystal symmetry and Physical properties. London and New York: Academic Press (1966).

D.A. Berlincourt, D.R. Curran and H. Jaffe. Piezoelectric and piezomagnetic materials and their function in transducers. In: Physical Acoustics (W.P. Mason and R.N. Thurston, eds) vol. 1 A, chap 3, New York: Academic Press (1964).

Elastic waves in an unbounded piezoelectric solid

M. Cotte. Propagation d'ondes élastiques dans un milieu piézoélectrique. C.R. Acad. Sc. Paris, 445 (1944).

J.J. Kyame. Wave propagation in piezoelectric crystals. J. Acoust. Soc. Am., 21, 159 (1949).

M. Cotte. La mécanique des milieux piézoélectriques. Actes du Colloque International de Mécanique, 5, Poitiers (1950).

B.A. Auld. Acoustic fields and waves in solids, vol. I, chap. 8. New York: Wiley-Interscience (1973).

Surface waves

G.W. Farnell. Reference [4].

C.C. Tseng and R.M. White. Propagation of piezoelectric and elastic surface waves on the basal plane of hexagonal piezoelectric crystals. J. Appl. Phys., 38, 4274 (1967).

C.C. Tseng. Piezoelectric surface waves in cubic crystals. J. Appl. Phys., 41, 2270 (1970).

C.A.A.J. Greebe, P.A. Van Dalen, T.J.B. Swanenburg, J. Wolter. Electric coupling properties of acoustic and electric surface waves. Physics Reports, 1 C, 235 (1971).

EXERCISES

6.1. From the Curie principles, find the crystal classes for which a hydrostatic pressure generates an electric polarization.

Solution. A hydrostatic pressure, which is exerted equally in all directions, has spherical symmetry. Therefore, the crystal must correspond to a subgroup of the characteristic symmetry $A_\infty \infty M$ of a polarization. This is true for the ten polar classes: 1, 2, m, 2mm, 3, 3m, 4, 4mm, 6, 6mm, which also exhibit pyroelectricity.

6.2. Express the coefficients $c^D_{\alpha\beta}$ in terms of the stiffness coefficients $c^E_{\alpha\beta}$ and of $e_{i\alpha}$.

Solution. Equation 6.15b, expressed in terms of β^S_{ki}, yields

$$\beta^S_{ki} D_i = E_k + \beta^S_{ki} e_{i\beta} S_\beta$$

or

$$E_i = \beta^S_{ik} D_k - \beta^S_{ik} e_{k\beta} S_\beta .$$

Inserting this into 6.15a, we get

$$T_\alpha = (c^E_{\alpha\beta} + e_{i\alpha}\beta^S_{ik} e_{k\beta}) S_\beta - e_{i\alpha}\beta^S_{ik} D_k$$

Hence

$$c^D_{\alpha\beta} = c^E_{\alpha\beta} + e_{i\alpha}\beta^S_{ik}e_{k\beta}$$

6.3. Show that the piezoelectric tensor for classes 4mm and 6mm is invariant under any rotation about the x_3-axis.

Solution. The new components e'_{ijk} are calculated through the rotation matrix α, given by 3.12:

$$e'_{ijk} = \alpha^p_i\alpha^q_j\alpha^r_k e_{pqr} \; .$$

Since the only non vanishing components (see table 6.39) are

$$e_{113} = e_{223} = e_{15}, \; e_{333} = e_{33} \quad \text{and } e_{311} = e_{322} = e_{31}$$

and from $\alpha^3_k = \delta_{k3}$:

$$e'_{ijk} = (\alpha^1_i\alpha^1_j + \alpha^2_i\alpha^2_j)\delta_{k3}e_{15} + \delta_{i3}\delta_{j3}\delta_{k3}e_{33} + (\alpha^1_j\alpha^1_k + \alpha^2_j\alpha^2_k)\delta_{i3}e_{31}.$$

Applying this equation to the constants of table 6.37 for classes 4 and 6 yields

$e'_{123} = (\alpha^1_1\alpha^1_2 + \alpha^2_1\alpha^2_2)e_{15} = 0 \Rightarrow$ <u>$e'_{14} = 0$</u>

$e'_{113} = (\alpha^1_1\alpha^1_1 + \alpha^2_1\alpha^2_1)e_{15} = e_{15} \Rightarrow$ <u>$e'_{15} = e_{15}$</u>

$e'_{333} = e_{333} \Rightarrow$ <u>$e'_{33} = e_{33}$</u> and

$e'_{311} = (\alpha^1_1\alpha^1_1 + \alpha^2_1\alpha^2_1)e_{31} = e_{31} \Rightarrow$ <u>$e'_{31} = e_{31}$</u> .

The underlined equations show that the new tensor is the same as the old one.

6.4. Calculate the velocity V_3 of the shear wave polarized along an $\bar{A}_4$ axis.

Solution. Making use of 6.46, and expansions 6.56, $n_1 = \cos\phi$, $n_2 = \sin\phi$, $n_3 = 0$:

$\gamma_1 = \gamma_2 = 0 \quad \gamma_3 = e_{15}\cos 2\phi + e_{14}\sin 2\phi.$

According to the results of § 5.1.3b:

$$\rho V_3^2 = \overline{\Gamma}_{33} = \Gamma_{33} + \frac{\gamma_3^2}{\varepsilon} = c_{44}^E + \frac{(e_{15}\cos 2\phi + e_{14}\sin 2\phi)^2}{\varepsilon_{11}^S}.$$

6.5. For class 622, table 6.11 indicates the existence of a Bleustein-Gulyaev wave for any sagittal plane containing A_6, even if it is not orthogonal to a diad axis. Prove this statement.

<u>Solution</u>. It is sufficient to show that the e_{ijk} tensor is invariant under any rotation about the x_3-axis//A_6. For this class (table 6.38), only $e_{123} = e_{14}$ and $e_{213} = -e_{14}$ are non-zero $\Rightarrow$ $e'_{ijk} = (\alpha_i^1\alpha_j^2\alpha_k^3 - \alpha_i^2\alpha_j^1\alpha_k^3)e_{14}$. Among the constants of table 6.37

$e'_{31} = 0$ for $k \neq 3$, $e'_{113} = e'_{15} = 0$ and

$e'_{333} = e'_{33} = 0$ for $i = j$, $k = 3$

$e'_{14} = e'_{123} = (\cos^2\phi + \sin^2\phi)e_{14} \Longrightarrow e'_{14} = e_{14}$.

6.6. Two identical piezoelectric crystals, of symmetry 6mm, are rigidly bounded; the A_6 axes are in opposite directions (fig. 6.19). Show that a surface wave, polarized along the A_6 axes, propagates along the boundary $x_2 = 0$ (from reference [20]).

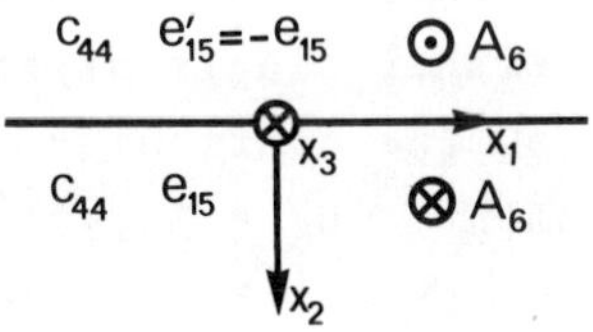

Fig. 6.19.

<u>Solution</u>. The propagation equations are the same in both crystals, and identical to those of § 6.2.2.3, the results of which can be used with $c'_{44} = c_{44}$, $e'_{15} = -e_{15}$ for $x_2 < 0$ because of the A_6 axis reversal. The displacement u_3 and the potential Φ are given by 6.73a and b

for $x_2 > 0$. For $x_2 < 0$, the sign of e_{15} must be changed as well as the signs in the exponents, in order that the wave vanish at $x_2 \to -\infty$:

$$\begin{cases} u'_3 = B\, e^{\beta\frac{\omega}{V}x_2}\, e^{i\omega(t-x_1/V)} \\ \Phi' = (\Phi'^{(1)}_0\, e^{\frac{\omega}{V}x_2} - \frac{e_{15}}{\varepsilon_{11}} B\, e^{\beta\frac{\omega}{V}x_2})\, e^{i\omega(t-x_1/V)} . \end{cases}$$

The coefficient B is the same because of the continuity of displacement at $x_2 = 0$ and, from 6.72: $\beta = \sqrt{1 - V^2/V_T^2}$. We now write the continuity equations at $x_2 = 0$. It has already been established that the continuity of mechanical stress must be checked only for T_{32}:

$$-\beta c^D_{44}B - e_{15}\Phi^{(1)}_0 = \beta c^D_{44}B - e_{15}\Phi'^{(1)}_0$$

hence

$$\Phi'^{(1)}_0 - \Phi^{(1)}_0 = \frac{2\beta c^D_{44}}{e_{15}} B$$

The continuity of potential ensures

$$\Phi'^{(1)}_0 - \Phi^{(1)}_0 = \frac{2e_{15}}{\varepsilon_{11}} B .$$

Both equations are compatible only if

$$\beta = \frac{e^2_{15}}{\varepsilon^S_{11} c^D_{44}} = K^2 ,$$

which leads to the same propagation velocity $V = V_T\sqrt{1-K^4}$ as for the B-G wave on the metallized surface of either single crystal. This is accounted for by the continuity of normal electric displacement: $D_2 = \frac{\omega}{V}\varepsilon_{11}\Phi^{(1)}_0$ according to 6.74 and $D'_2 = -\frac{\omega}{V}\varepsilon_{11}\Phi'^{(1)}_0$ because of the sign reversal in the exponents for $x_2 < 0$. With

$$\Phi'^{(1)}_0 = -\Phi^{(1)}_0 = \frac{e_{15}}{\varepsilon_{11}} B ,$$

the potentials Φ and Φ' vanish at the boundary:

$$\Phi = -\frac{e_{15}}{\varepsilon_{11}} B(e^{-\frac{\omega}{V}x_2} - e^{-\beta\frac{\omega}{V}x_2})e^{i\omega(t-x_1/V)}$$

$$\Phi' = \frac{e_{15}}{\varepsilon_{11}} B(e^{\frac{\omega}{V}x_2} - e^{\beta\frac{\omega}{V}x_2})e^{i\omega(t-x_1/V)} .$$

Chapter 7

Generation and Detection of Elastic Waves

In chapter 5 (dynamic elasticity), and also in section II of chapter 6, we dealt with the propagation of the major types of elastic waves without discussing their generation. In chapter 6, devoted to piezoelectricity, we described the mutual dependence of elastic and electric properties in crystals which have no centre of symmetry. We now discuss the piezoelectric generation and detection firstly of bulk shear or longitudinal waves and, secondly of surface waves, and more specifically of Rayleigh waves. The bulk wave transducer is analyzed through a conventional model which is valid for a piezoelectric single crystal rigidly bonded to the propagation medium, and also for a piezoelectric film deposited on the medium which is the case for high frequencies ($f > 1000$ MHz). As far as surface waves are concerned we have described the most frequently used, the interdigital transducer. A small paragraph is devoted to the technology of each type of transducer.

7.1. BULK WAVE PIEZOELECTRIC TRANSDUCER

As indicated in fig. 7.1, a transducer consists essentially of a piezoelectric material bearing two metallic electrodes. The electric field of the signal applied to the electrodes induces a vibration of the piezoelectric material, whose thickness is a fraction of the elastic wavelength. The elastic wave beam, the size of which is controlled by the outer electrode, crosses the inner electrode, bonded to the propagation medium. The crystallographic orientation of the piezoelectric solid is selected so that the desired mode be excited preferentially. For example, in the case of zinc oxide, the 6-fold axis should be parallel to the electric field if longitudinal waves are to be launched. The propagation medium is so oriented with respect to the transducer that the elastic mode is a pure mode, for which the energy propagates along the axis of the structure.

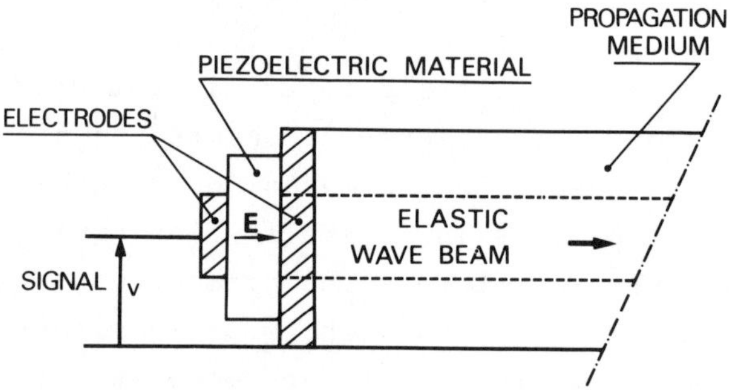

Fig. 7.1. Structure of a bulk wave transducer. The piezoelectric material is excited by the electric field $\vec{E}$ of the signal applied between the electrodes.

This transducer can be analyzed in two ways. The first way is to calculate directly the elastic power, by expressing the fact that all propagation equation solutions must satisfy the boundary conditions. The second way is to represent all parts of the transducer by equivalent circuits, connecting these with each other and then to utilize Kirchhoff equations.

7.1.1. Direct Calculation of the Elastic Power

Let us calculate the elastic power provided to the propagation medium in the practical case where only one (longitudinal or shear) wave is generated and can propagate. This restriction imposes conditions on the elastic and piezoelectric constants.

7.1.1.1. Conditions of validity of a one dimensional model

The reference frame $x_1'x_2'x_3'$ is defined in fig. 7.2; the $x_1x_2x_3$ frame specifically refers to the crystallographic axes. Because the thickness of the piezoelectric plate where acoustic waves are produced is much less than all other characteristic lengths, the electric field is along the x_3' axis and the planes $x_3' = C^{te}$ are equiphase planes. The propagation direction (the x_3' axis) is conserved at each boundary: $n_1' = n_2' = 0$, $n_3' = 1$.

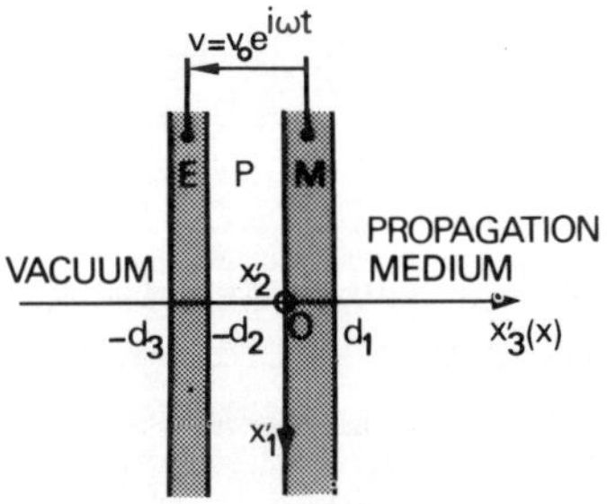

Fig. 7.2. Analytic model.

A longitudinal wave can propagate only if the x_3' axis is a principal axis of the Christoffel tensor: $\Gamma_{13}' = \Gamma_{23}' = 0$. With $n_1' = n_2' = 0$, $n_3' = 1$, expansion of 5.8 yields

$$c_{34}' = c_{35}' = 0 \qquad (7.1)$$

In the piezoelectric medium: $\overline{\Gamma}_{i\ell}' = \Gamma_{i\ell}' + \gamma_i'\gamma_\ell'/\varepsilon$ (§ 6.2.1.1). Moreover it is necessary that $\gamma_1' = \gamma_2' = 0$ or, according to (6.56)

$$e_{34}' = e_{35}' = 0 \qquad (7.2)$$

The non zero modulus e_{33}' allows an excitation of this wave by an electric field parallel to the x_3' axis. Likewise, the conditions for shear wave (with x_1' polarization) generation and propagation are

$$c'_{45} = c'_{35} = 0, \quad e'_{33} = e'_{34} = 0 \quad \text{and } e'_{35} \neq 0. \tag{7.3}$$

If these conditions are satisfied, equation 6.15a becomes (without any subscripts)

$$T = c^E \frac{\partial u}{\partial x} - eE \tag{7.4}$$

where $x = x'_3$, $E = E'_3$ and the stress T represents T'_{33} or T'_{31}, the displacement u represents u'_3 or u'_1, the stiffness c^E represents c'_{33} or c'_{55}, the piezoelectric modulus e, e'_{33} or e'_{35}, according to whether the modes are longitudinal or shear.

7.1.1.2. Mathematical formulation

The expressions* for the particle displacement and stress in the various parts of the transducer are

$$u_S = a_S\, e^{-ik_S x} + b_S\, e^{ik_S x} \tag{7.5}$$

$$T_S = -ik_S c^E_S (a_S\, e^{-ik_S x} - b_S\, e^{ik_S x}) - eE\delta_{SP} \tag{7.6}$$

when the subscript S represents M (earth electrode), P (piezoelectric solid) or E (external electrode) (fig. 7.2). The Kronecker symbol δ_{SP} indicates that the eE term exists in the piezoelectric solid only. In the propagation medium, the wave is progressive

$$u = a\, e^{-ikx} \tag{7.7}$$

$$T = -ikca\, e^{-ikx}. \tag{7.8}$$

The seven quantities a, a_M, b_M, a_P, b_P, a_E and b_E, are determined by the seven boundary conditions which express the continuity of particle displacement and stress between the inner electrode and the propagation medium:

$$u_M(d_1) = u(d_1) \tag{7.9a}$$

$$T_M(d_1) = T(d_1) \tag{7.9b}$$

between the piezoelectric solid and the earth electrode

$$u_P(0) = u_M(0) \tag{7.10a}$$

$$T_P(0) = T_M(0) \tag{7.10b}$$

---

* Since equations are linear, we have ignored the factor $e^{i\omega t}$.

between the external electrode and the piezoelectric solid

$$u_E(-d_2) = u_p(-d_2) \qquad (7.11a)$$

$$T_E(-d_2) = T_p(-d_2) \qquad (7.11b)$$

and at the external electrode free surface

$$T_E(-d_3) = 0. \qquad (7.12)$$

Solving this system directly provides the amplitudes a, a_s and b_s in terms of the electric field values at the extremities (0 and $-d_2$) of the piezoelectric material. In fact, the only measurable quantity is the potential difference $v = v_0 e^{i\omega t}$ applied to the electrodes:

$$v_0 = \int_o^{-d_2} -E\,dx.$$

Furthermore, since x_3' is the only position coordinate, the Poisson equation for insulators $\partial D_i'/\partial x_i' = 0$ leads to $\partial D_3'/\partial x_3' = 0$. The electrical displacement component $D_3' = D$ is constant and is given by the simplified form of equation 6.15b:

$$D = \varepsilon^S E + e\frac{\partial u_p}{\partial x} \qquad \text{where} \qquad \varepsilon^S = \varepsilon_{33}' \qquad (7.13)$$

Integrating from 0 to $-d_2$ yields:

$$-Dd_2 = -\varepsilon^S v_0 + e[u_p(-d_2) - u_p(0)]. \qquad (7.14)$$

After eliminating E from equations 7.4 and 7.13, the piezoelectric material stiffness at constant D comes out:

$$c^D = c^E + \frac{e^2}{\varepsilon^S}$$

$$T_p = c_p^D \frac{\partial u_p}{\partial x} - \frac{e}{\varepsilon^S} D. \qquad (7.14a)$$

The electrical displacement, as derived from 7.14 and 7.5, is

$$D = \frac{\varepsilon^S v_0}{d_2} - \frac{e}{d_2}[a_p(e^{ik_p d_2} - 1) + b_p(e^{-ik_p d_2} - 1)]$$

Inserting this into 7.14a for the stress T_p, leads to:

$$T_p = -ik_p c_p^D\{a_p[e^{-ik_p x} + \frac{ie^2}{\varepsilon^S c_p^D k_p d_2}(e^{ik_p d_2} - 1)] - b_p[c.c.]\} - e\frac{v_0}{d_2}$$

where c.c. denotes complex conjugate. The factor $e^2/\varepsilon^S c_p^D k_p d_2$ involves the square of the coupling constant (see § 6.2.1.2)

$$K^2 = \frac{e^2}{\varepsilon^S c_p^E + e^2} = \frac{e^2}{\varepsilon^S c_p^D}$$

The piezoelectric slab thickness d_2 is of the order of half a wavelength: $k_p d_2 \simeq \pi$. The ratio

$$\frac{e^2}{\varepsilon^S c_p^D k_p d_2} = \frac{K^2}{k_p d_2} \simeq \frac{K^2}{\pi}$$

is less than 1 for a moderately piezoelectric material ($K < .3 \Rightarrow K^2 < .1$) and the stress T_p is readily expressed in terms of the voltage v_o:

$$T_p = - ik_p c_p^D(a_p\ e^{-ik_p x} - b_p\ e^{ik_p x}) - e\,\frac{v_o}{d_2} \qquad (7.15)$$

Knowing the amplitude a as a function of v_o, the useful acoustic power may be calculated (exercise 5.1)

$$\mathscr{P} = \frac{1}{2}\, Z\omega^2 |a|^2 A \qquad (7.16)$$

where Z is the elastic impedance per unit area of the propagation medium and A is the beam cross-section. Although a numerical solution is possible with the help of a computer in the most general case, even taking possible sublayers into account, algebraic expressions on the other hand soon become complicated and the large number of parameters makes the results difficult to interpret. Therefore we apply the method to two simple structures. In the first (fig. 7.3) the electrode thicknesses are very small compared to the wavelength and do not affect elastic wave propagation. The corresponding results - frequency response for various values of the piezoelectric crystal vs propagation medium elastic impedance ratio - can be applied to low frequency transducers (below 100 MHz). The second structure, takes account of the inner electrode thickness; it corresponds to high frequency transducers where a thin layer of piezoelectric material is deposited on the crystal. For a given combination of piezoelectric material and crystal, we analyze the effect of the inner electrode nature and thickness on the transducer frequency response.

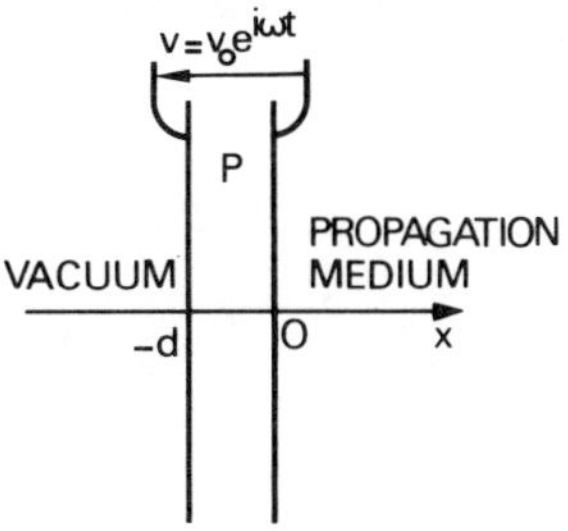

Fig. 7.3. Simplified model, valid when the electrode thickness is much less than that of the piezoelectric crystal.

7.1.1.3. Structure consisting of piezoelectric crystal and propagation medium

Using the notation of fig. 7.3, the boundary conditions are

$$\begin{cases} u_P(0) - u(0) = 0 \\ T_P(0) - T(0) = 0 \\ T_P(-d) = 0 \end{cases}$$

Their expansion, together with $\phi = k_P d$ and with the expression 7.15 for T_P, leads to*

$$\begin{cases} a_P + b_P - a = 0 \\ k_P c_P a_P - k_P c_P b_P - kca = i\,\dfrac{ev_o}{d} \\ k_P c_P a_P\, e^{i\phi} - k_P c_P b_P\, e^{-i\phi} = i\,\dfrac{ev_o}{d} \end{cases}$$

or bringing out the elastic impedances of the two media

$$Z_P = \frac{k_P c_P}{\omega}, \qquad Z = \frac{kc}{\omega}$$

$$\begin{cases} a_P + b_P = a \\ \dfrac{Z_P}{Z}(a_P - b_P) = a + i\,\dfrac{ev_o}{\omega d Z} \\ \dfrac{Z_P}{Z}(a_P - b_P)\cos\phi + i\,\dfrac{Z_P}{Z}(a_P + b_P)\sin\phi = i\,\dfrac{ev_o}{\omega d Z} \end{cases}$$

---

* One must bear in mind that c_P denotes the piezoelectric material stiffness at constant electric displacement D.

The last equation, equivalent to

$$(a + i \frac{ev_o}{\omega dZ}) \cos \phi + i \frac{Z_P}{Z} a \sin \phi = i \frac{ev_o}{\omega dZ}$$

provides the unknown quantity a:

$$a = i \frac{ev_o}{\omega dZ} \cdot \frac{1 - \cos \phi}{\cos \phi + i \frac{Z_P}{Z} \sin \phi} = 2i \frac{ev_o}{\omega dZm_o}$$

where

$$m_o = (\cos \phi + i \frac{Z_P}{Z} \sin \phi)/\sin^2 \frac{\phi}{2} .$$

According to equation 7.16, the elastic power launched into the delay line

$$\mathscr{P} = \frac{2e^2v_o^2}{\varepsilon^S dZ|m_o|^2} \frac{\varepsilon^S A}{d}$$

is proportional to the square of the electromechanical coupling constant $K^2 = e^2/\varepsilon^S c_P$ and to the capacitance $C_o = \varepsilon^S A/d$ of the rigidly bonded transducer:

$$\mathscr{P} = \frac{2K^2C_o c_P}{dZM_o} v_o^2$$

where the form factor M_o is defined by

$$M_o = |m_o|^2 = \frac{\cos^2 \phi + (\frac{Z_P}{Z})^2 \sin^2 \phi}{\sin^4 \frac{\phi}{2}} \qquad (7.17)$$

Let us now introduce the piezoelectric crystal resonance frequency f_P, for which the thickness d is equal to a half wavelength:

$$f_P = \frac{V_P}{2d} .$$

Since $c_P = Z_P V_P$, the elastic power is

$$\mathscr{P} = 4K^2C_o f_P \frac{Z_P}{ZM_o} v_o^2. \qquad (7.18)$$

In this expression, only the form factor $M_o(\phi)$ is frequency dependent, through the angle

$$\phi = k_P d = \pi \frac{f}{f_P}$$

From 7.17

$$M_o\left(\frac{f}{f_P}\right) = \frac{\left(\cos \pi \frac{f}{f_P}\right)^2 + \left(\frac{Z_P}{Z} \sin \pi \frac{f}{f_P}\right)^2}{\left(\sin \frac{\pi}{2} \frac{f}{f_P}\right)^4} \qquad (7.19)$$

For a constant voltage amplitude v_o, the form factor determines the transducer behaviour with respect to the electric signal frequency: the radiation resistance R_A, defined by

$$\mathcal{P} = \frac{v_o^2}{2R_A}$$

is proportional to M_o:

$$R_A = \frac{1}{8K^2C_o f_P} \frac{Z}{Z_P} M_o. \qquad (7.20)$$

The influence of the elastic impedance ratio Z_P/Z on the transducer frequency response is depicted by the curves of Fig. 7.4, which show the variation of the quantity

$$10 \log \left(M_o \frac{Z}{Z_P}\right) = 10 \log \left(8K^2C_o f_P R_A\right)$$

vs the normalized frequency f/f_P, for various values of Z_P/Z.

The shape of these curves can be interpreted as follows: when the propagation medium elastic impedance is small compared to that of the piezoelectric crystal, this crystal, the two faces of which are nearly free, vibrates at half wavelength (see the curve $Z_P/Z = 2$). As a result, the radiation resistance is a minimum at $f = f_P$, and consequently, the elastic power is a maximum. Conversely, when Z is large compared to Z_P ($Z_P/Z = 1/2$ or $1/4$), the transducer, one of whose faces is free and the other almost rigidly bonded, vibrates at $\lambda_P/4$ or $3\lambda_P/4$. The radiation resistance minimum occurs in the vicinity of $f = .5\ f_P$ or $f = 1.5\ f_P$. The critical value $Z_P/Z = 1/\sqrt{2}$, beyond which there is only one minimum left, leads to a flat frequency response around the central frequency f_P.

7.1.1.4. Structure consisting of piezoelectric crystal, electrode and propagation medium

The boundary conditions for this structure are

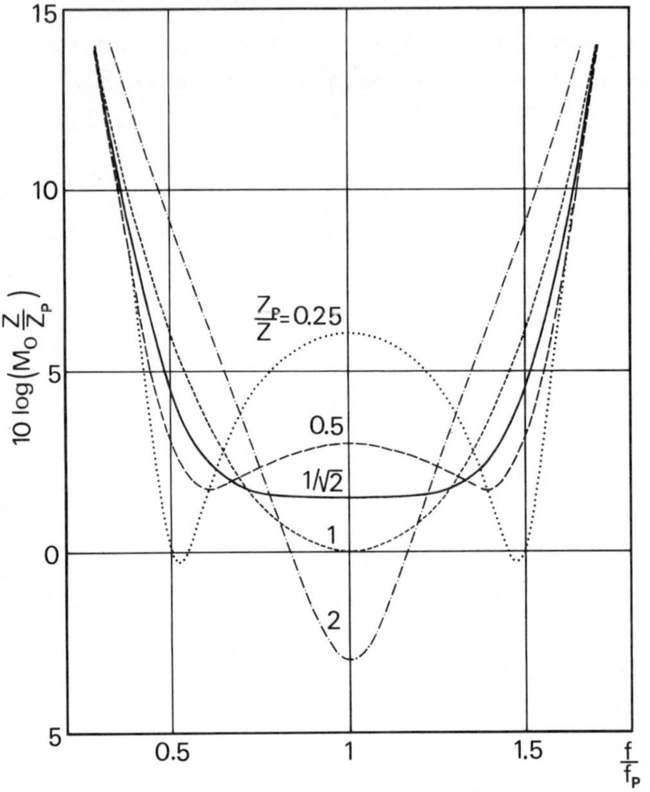

Fig. 7.4. Piezoelectric crystal - propagation medium structure. If the elastic impedance Z of the medium is smaller than the crystal impedance Z_P, the latter vibrates at $\lambda/2$ (ex. $Z_P/Z = 2$). In the opposite case, it vibrates at $\lambda/4$ and $3\lambda/4$ ($Z_P/Z = .25$).

expressed by equations 7.9, 7.10 and by the relation $T_P(-d_2) = 0$. Using equation 7.15 for T_P, and defining

$$\phi_M = k_M d_1 \qquad \phi_P = k_P d_2$$

yields the following expansion

$$\begin{cases} a_M\, e^{-i\phi_M} + b_M\, e^{i\phi_M} - a\, e^{-ikd_1} = 0 \\ k_M c_M a_M\, e^{-i\phi_M} - k_M c_M b_M\, e^{i\phi_M} - kca\, e^{-ikd_1} = 0 \\ a_P + b_P - a_M - b_M = 0 \\ k_P c_P a_P - k_P c_P b_P - k_M c_M a_M + k_M c_M b_M = \dfrac{iev_o}{d_2} \\ k_P c_P a_P\, e^{i\phi_P} - k_P c_P b_P\, e^{-i\phi_P} = \dfrac{iev_o}{d_2} \end{cases}$$

or, in matrix form, and bringing out the elastic impedances of the three media:

$$Z_P = \frac{k_P c_P}{\omega} \qquad Z_M = \frac{k_M c_M}{\omega} \qquad Z = \frac{kc}{\omega}$$

$$\begin{pmatrix} 0 & 0 & e^{-i\phi_M} & e^{i\phi_M} & -1 \\ 0 & 0 & e^{-i\phi_M} & -e^{i\phi_M} & -\dfrac{Z}{Z_M} \\ 1 & 1 & -1 & -1 & 0 \\ 1 & -1 & -\dfrac{Z_M}{Z_P} & \dfrac{Z_M}{Z_P} & 0 \\ e^{i\phi_P} & -e^{-i\phi_P} & 0 & 0 & 0 \end{pmatrix} \begin{pmatrix} a_P \\ b_P \\ a_M \\ b_M \\ a\, e^{-ikd_1} \end{pmatrix} = \frac{iev_o}{\omega Z_P d_2} \begin{pmatrix} 0 \\ 0 \\ 0 \\ 1 \\ 1 \end{pmatrix}$$

The unknown quantity $a\, e^{-ikd_1}$ is given by Kramer's theorem

$$a\, e^{-ikd_1} = \frac{iev_o}{\omega Z_P d_2} \frac{\Delta_1}{\Delta} = \frac{2iev_o}{\omega Z_P d_2 m}$$

where

$$m = 2\,\frac{\Delta}{\Delta_1}\ .$$

Expanding both determinants Δ_1 and Δ leads to:

$$m = \frac{\cos\phi_P \cos\phi_M - \dfrac{Z_P}{Z_M} \sin\phi_P \sin\phi_M}{\sin^2 \dfrac{\phi_P}{2}}$$

$$+\, i\, \frac{\dfrac{Z_M}{Z} \cos\phi_P \sin\phi_M + \dfrac{Z_P}{Z} \sin\phi_P \cos\phi_M}{\sin^2 \dfrac{\phi_P}{2}} \qquad (7.21)$$

Introducing the halfwave frequencies f_M and f_P respectively for the inner electrode and the piezoelectric layer:

$$f_M = \frac{V_M}{2d_1} \qquad f_P = \frac{V_P}{2d_2}$$

the elastic power provided to the line is derived from equation 7.18, where M_o has been replaced by the form factor $M = |m|^2$:

$$\mathscr{P} = 4\,\frac{Z_P}{Z}\,\frac{K^2}{M}\,C_o f_P v_o^2. \qquad (7.22)$$

The radiation resistance

$$R_A = \frac{ZM}{8Z_P K^2 C_o f_P} \qquad (7.23)$$

is proportional to the form factor M, which depends on frequency through the angles

$$\phi_P = k_P d_2 = \frac{\pi f}{f_P} \qquad \text{and} \qquad \phi_M = \frac{\pi f}{f_M}$$

according to 7.21:

$$M\left(\frac{f}{f_P},\frac{f}{f_M}\right) = \frac{\left(\cos\frac{\pi f}{f_P}\cos\frac{\pi f}{f_M} - \frac{Z_P}{Z_M}\sin\frac{\pi f}{f_M}\sin\frac{\pi f}{f_P}\right)^2}{\left(\sin\frac{\pi f}{2f_P}\right)^4} + \frac{\left(\frac{Z_M}{Z}\cos\frac{\pi f}{f_P}\sin\frac{\pi f}{f_M} + \frac{Z_P}{Z}\sin\frac{\pi f}{f_P}\cos\frac{\pi f}{f_M}\right)^2}{\left(\sin\frac{\pi f}{2f_P}\right)^4} \qquad (7.24)$$

It is straightforward to check that the above expression reverts to M_o when the electrode thickness vanishes ($f_M \to \infty$):

$$M\left(\frac{f}{f_P},0\right) = M_o\left(\frac{f}{f_P}\right).$$

Form factor variation. Examining the symmetry properties of the function $M(f/f_P, f/f_M)$ shows that it is sufficient to evaluate it only for values of f/f_P and f/f_M ranging between 0 and 1. In fig. 7.5 and 7.7, the variation of 10 log M is sketched in the form of cross sections (at constant M) of the corresponding surface. These contour lines have been plotted [1] for two typical

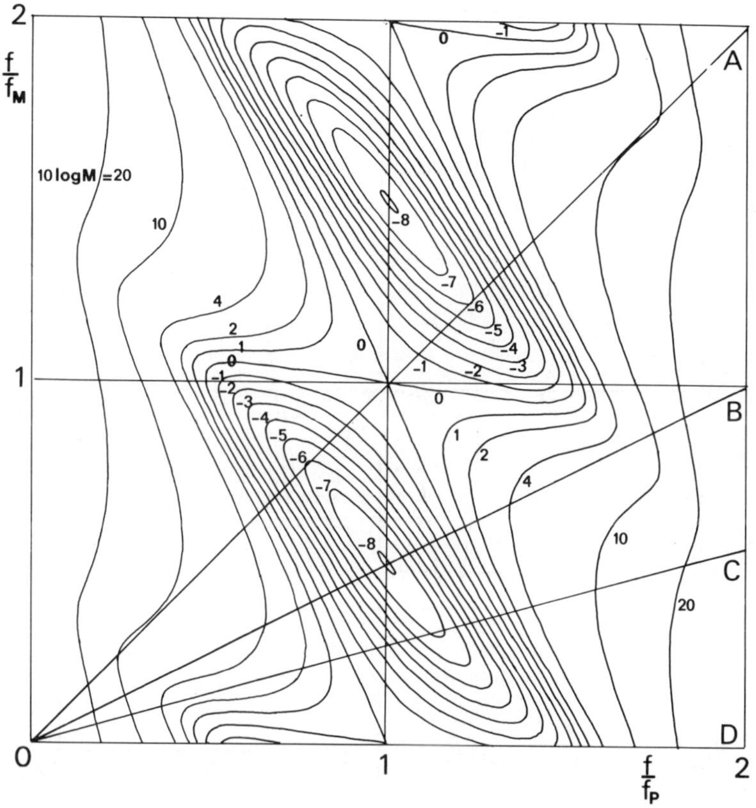

Fig. 7.5. Contour lines of the form factor M, of a CdS-Al-Al_2O_3 structure, for the longitudinal mode. A transducer such that $f_P/f_M = .25$ is shown by a vertical plane, whose intersection is OC.

structures from the point of view of elastic impedances: they comprise either an aluminium electrode or a gold electrode between a cadmium sulphide layer and a sapphire propagation medium. The 6-fold axis of cadmium sulphide is parallel to the electric field so that only longitudinal waves are excited; they propagate along the triad axis in sapphire: thus the one dimensional model is valid.

Since the ratio $\dfrac{f/f_M}{f/f_P} = \dfrac{f_P}{f_M}$ has a constant value,

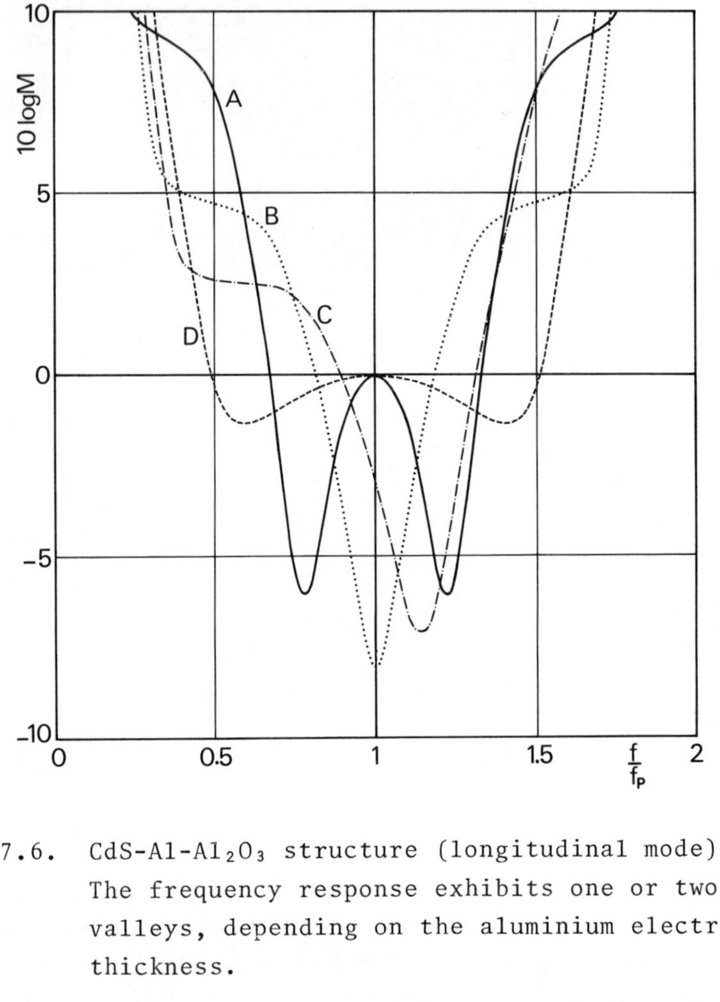

Fig. 7.6. CdS-Al-Al_2O_3 structure (longitudinal mode). The frequency response exhibits one or two valleys, depending on the aluminium electrode thickness.

determined only by the layer thicknesses and the velocities, the behaviour of a given transducer is represented by a vertical cross section of the surface $M(f/f_P, f/f_M)$, whose plane contains the M axis. The lines A, B, C, D in fig. 7.5 and 7.7 are the intersections of such planes. The frequency response is plotted in fig. 7.6 for the CdS-Al-Al_2O structure and in fig. 7.8 for the CdS-Au-Al_2O_3 structure. The following argument can account for the shape of these curves: the elastic impedance of cadmium sulphide (Z_P = 21.5 10^6 kg/m²/s) is definitely lower than that of

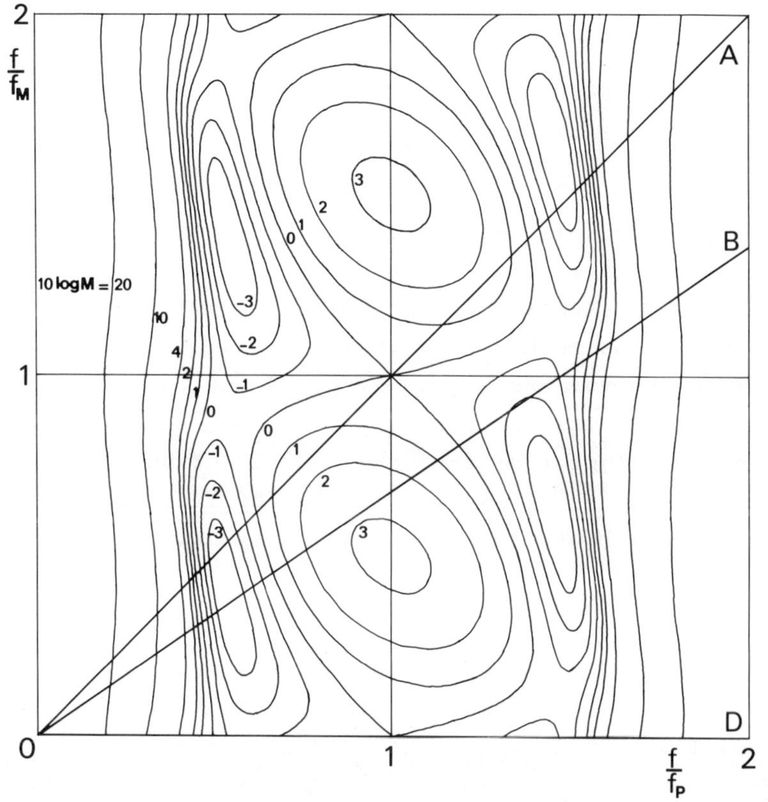

Fig. 7.7. Contour lines of the form factor, of a CdS-Au-Al_2O_3 structure, for the longitudinal mode. A transducer of given f_P/f_M is shown by a vertical plane, whose intersection is analogous to OB.

sapphire (Z = 44.5 10^6 kg/m²/s). Therefore, for a vanishing metal thickness (curve D) the transducer, one of whose faces is free and the other bonded, vibrates at quarter wavelength (f/f_P = .5) or $3\lambda/4$ (f/f_P = 1.5). The form factor M exhibits two valleys around these two points, each corresponding to a maximum elastic power. Adding a gold film of impedance* Z_M = 64.10^6 kg/m²/s can only enhance the valleys, whatever its thickness (curves A and B in fig. 7.8). On the other hand, an aluminium layer, of acoustic impedance* $Z_M = 17.5\ 10^6\ kg/m^2/s$, inserted between the cadmium sulphide crystal and the sapphire crystal with a suitable thickness (equal to $\lambda/4$) acts as an impedance transformer. The transformed impedance at the inner face of the transducer ($x = 0$) is equal to (see exercise 7.1):

$$Z_T(x = 0) = \frac{Z_M^2}{Z} = 6.9\ 10^6\ kg/m^2/s$$

so that this face is nearly free. The transducer vibrates at halfwave length ($f = f_P$) and the frequency response now presents only one valley (curves B and C in fig. 7.6). The form factor evolution is observed, when going from two valleys to only one, by the rotating intersections of the characteristic plane about the origin.

The form factor of the ZnO-Pt-Al_2O_3 structure depicted in fig. 7.9, is quite similar to that of the CdS-Au-Al_2O_3 structure, which has an analogous series of elastic impedances:

$Z_{ZnO} = 35\ 10^6\ kg/m^2/s$ $\quad Z_{Pt} = 87\ 10^6$ $\quad Z_{Al_2O_3} = 44.5\ 10^6$.

So far, we have not taken the outer electrode into account; it usually consists of an aluminium or silver layer. Provided its thickness is much less than that of the piezoelectric layer (less than 10%) it only shifts the curves of form factor variation towards lower frequencies.

Order of magnitude. For the last structure (ZnO-Pt-Al_2O_3) at a frequency $f_P = 1$ GHz, the zinc oxide layer thickness ($V_P = 6{,}330$ m/s) is

$$d_2 = \frac{V_P}{2f_P} = 3.2\ \mu m$$

and the static capacitance of a transducer, for an area $A = 1\ mm^2$ is equal to

$$C_0 = \frac{\varepsilon_{33}^S A}{d_2} = 26\ pF.$$

* The metal layer deposited on the sapphire crystal is not a single crystal: however the triad axis of sapphire enhances crystal growth in the [111] direction. The elastic impedance value is between the [100] and the [111] values.

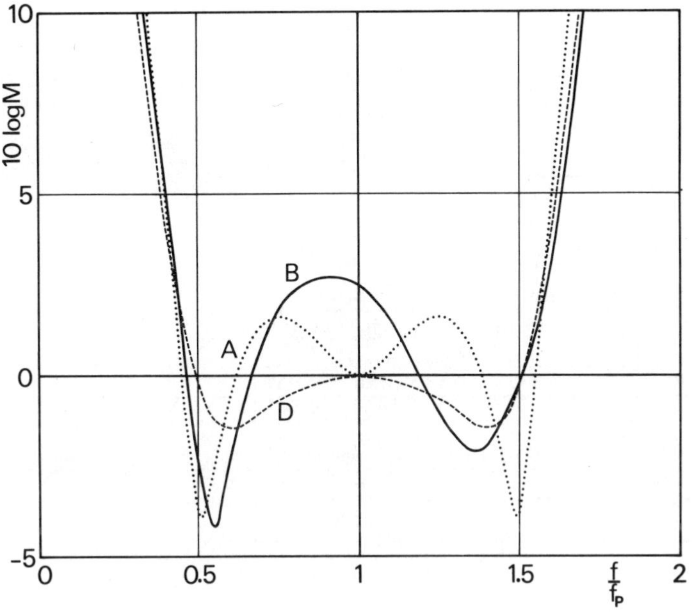

Fig. 7.8. Structure CdS-Au-Al_2O_3 (longitudinal mode). The frequency response always exhibits two valleys.

Given the value of the longitudinal mode coupling constant:

$$K^2 = \frac{e_{33}^2}{\varepsilon_{33}^S c_{33}^E + e_{33}^2} = 0.073$$

the radiation resistance is (in ohms)

$$R_A = \frac{ZM}{8Z_P K^2 C_o f_P} \simeq 84\ M.$$

A voltage of amplitude v_o = 1 V, at a frequency such that M = 1, generates an elastic power

$$P = \frac{v_o^2}{2R_A} \simeq 6\ mW.$$

7.1.1.5. Electrical matching. Conversion loss

If the piezoelectric material is assumed to be an insulator and a perfect dielectric, then the only electric current is the displacement current, whose density per unit area is

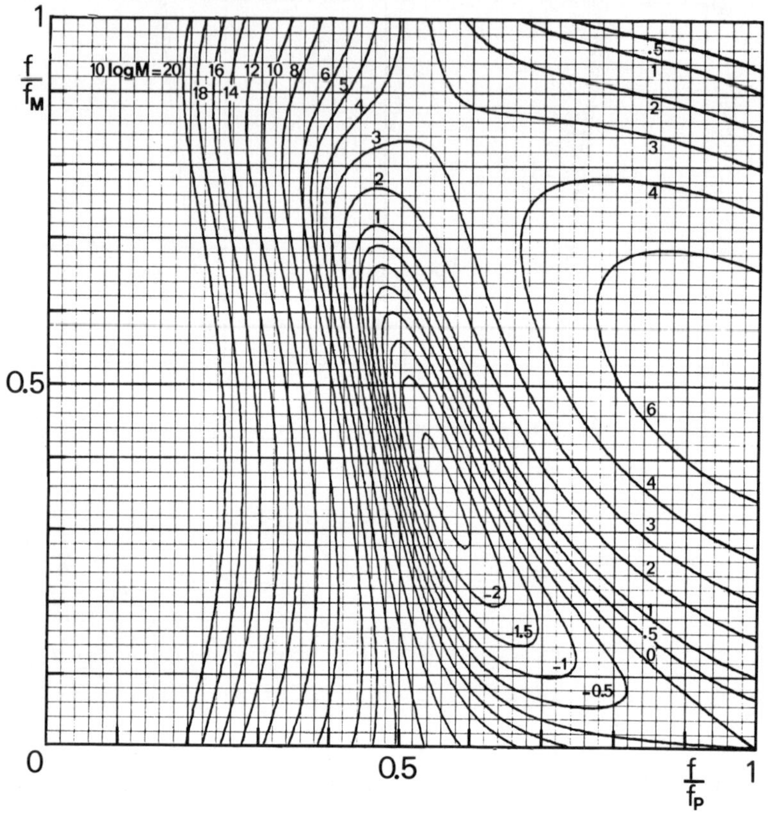

Fig. 7.9. Contour lines of the ZnO-Pt-Al_2O_3 structure for the longitudinal mode, f/f_P and f/f_M ranging from 0 to 1. Note the similarity with fig. 7.7.

$$j = \frac{\partial D}{\partial t} = i\omega D.$$

Given the expression 7.14 for the voltage:

$$v_o = \frac{d_2 D}{\varepsilon^S} - \frac{e}{\varepsilon^S}\,[u_p(0) - u_p(-\,d_2)]$$

the electric impedance of the transducer splits into two terms:

$$Z_e = \frac{v_o}{jA} = \frac{d_2}{i\omega\varepsilon^S A} - \frac{e}{i\omega\varepsilon^S DA}\,[u_p(0) - u_p(-\,d_2)]$$

The first is due to the static capacitance $C_o = \varepsilon^S A/d_2$. The second, for a moderately piezoelectric material ($K^2 << \pi$), leads to the radiation resistance R_A which represents the electric to mechanical energy transformation.

In fact, this equivalent circuit must be completed by a resistance R due to electric power dissipation (other than by the generation of elastic waves) (fig. 7.10). This loss is important in high frequency transducers for the following reasons:

- piezoelectric materials (CdS, ZnO) are semiconductors, not insulators;
- dielectric losses are high at frequencies above 1 GHz;
- the metal layers are not perfectly conducting;
- elastic waves are damped in non-single crystal layers. It is known from experiment that the loss-equivalent resistance R is smaller than the radiation resistance R_A.

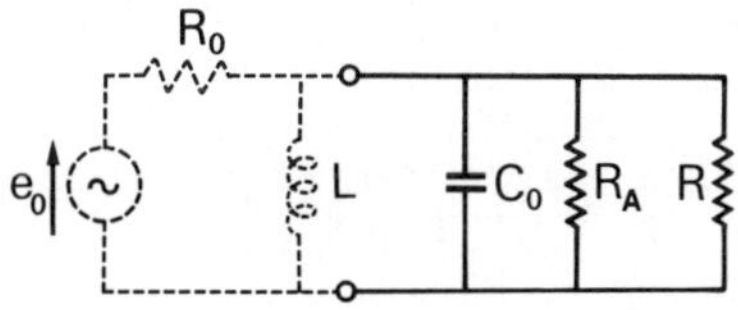

Fig. 7.10. Equivalent circuit of a transducer. C_o is the static capacitance, R_A the radiation resistance (which depends on frequency). The resistance R includes both electric and elastic losses in the layers.

A generator of emf e_o, of internal resistance R_o, provides a maximum electrical power to the transducer when impedance matching is achieved. The conversion loss $\mathcal{T}$ of the transducer is by definition the ratio of this maximum available electrical power from the source

$$\mathcal{P}_d = \frac{e_o^2}{8R_o} = \frac{V_o^2}{2R_o}$$

to the elastic power launched into the crystal $\mathcal{P} = \frac{V_o^2}{2R_A}$:

$$\mathcal{T} = \frac{\mathcal{P}_d}{\mathcal{P}} = \frac{R_A}{R_o}$$

or, since the impedances are matched:

$$R_o = \frac{R\,R_A}{R+R_A} \quad \Rightarrow \quad \mathcal{T} = 1 + \frac{R_A}{R}$$

This definition makes sense only if matching is efficient over the whole transducer bandwidth, i.e., since R_A is frequency dependent, if R is much smaller than R_A. The capacitance C_o is tuned with an inductance L, at the centre frequency f_P (fig. 7.10). Therefore the reciprocal Q-factor of the transducer

$$\frac{1}{Q} = \frac{1}{R_A C_o \omega_P}\left(1 + \frac{R_A}{R}\right) \qquad (7.25)$$

must be larger than the form factor bandwidth, which is about 0.5 (fig. 7.4, 6, 8). Since the term

$$\frac{1}{R_A C_o \omega_P} = \frac{4Z_P}{\pi ZM} K^2 \simeq K^2 \qquad (7.26)$$

is less than 1, this condition is satisfied because of $R \ll R_A$. The generator is then matched to the resistance R and the conversion loss, which is large, is proportional to the form factor

$$\mathcal{T} \simeq \frac{R_A}{R} = \frac{ZM}{8Z_P K^2 R C_o f_P} \qquad (7.27)$$

7.1.2. Equivalent Circuit

With the model of fig. 7.2, is associated an electromechanical network [2], obtained by putting together the circuits equivalent to the various parts of the transducer; the forces and particle velocities correspond to voltages and electric currents (exercise 1.3).

a) An infinite medium of cross section A, where a single progressive wave propagates, is represented by its acoustic impedance $\mathcal{Z} = (ck/\omega)A$, i.e. the ratio of force ($F = -AT$) to particle velocity ($\dot{u} = \partial u/\partial t$):

$$F = -Ac\frac{\partial u}{\partial x} = \frac{Ack}{\omega}\frac{\partial u}{\partial t} = \mathcal{Z}\dot{u}.$$

b) The equivalent circuit for a slab of finite thickness d and cross section A, bounded by the planes $x = x_1$ and $x = x_2$, is derived from the relationships between the forces F_1, F_2 and the particle velocities $\dot{u}_1$, $\dot{u}_2$ at the two sides:

$$\dot{u}_1 = i\omega(a\, e^{-ikx_1} + b\, e^{ikx_1})$$

$$\dot{u}_2 = i\omega(a\, e^{-ikx_2} + b\, e^{ikx_2})$$

Extracting a and b from this system:

$$i\omega a = \frac{\dot{u}_1\, e^{ikx_2} - \dot{u}_2\, e^{ikx_1}}{2i \sin kd}$$

$$i\omega b = \frac{\dot{u}_2\, e^{-ikx_1} - \dot{u}_1\, e^{-ikx_2}}{2i \sin kd}$$

then inserting them into the expression for the force (analogous to 7.6):

$$F = ickA(a\, e^{-ikx} - b\, e^{ikx}) = \mathscr{Z}(i\omega a\, e^{-ikx} - i\omega b\, e^{ikx})$$

yields, for $x = x_1$

$$F_1 = \mathscr{Z}\left(\frac{\dot{u}_1}{i \tan kd} - \frac{\dot{u}_2}{i \sin kd}\right)$$

and for $x = x_2$

$$F_2 = \mathscr{Z}\left(\frac{\dot{u}_1}{i \sin kd} - \frac{\dot{u}_2}{i \tan kd}\right)$$

or, recalling the identity

$$\frac{1}{\tan kd} = \frac{1}{\sin kd} - \tan\frac{kd}{2}$$

$$\begin{cases} F_1 = \dfrac{\mathscr{Z}}{i \sin kd}(\dot{u}_1 - \dot{u}_2) + \left(i\mathscr{Z} \tan\dfrac{kd}{2}\right)\dot{u}_1 & (7.28a) \\ F_2 = \dfrac{\mathscr{Z}}{i \sin kd}(\dot{u}_1 - \dot{u}_2) - \left(i\mathscr{Z} \tan\dfrac{kd}{2}\right)\dot{u}_2 & (7.28b) \end{cases}$$

These relations can also be found by applying Kirchhoff laws to the circuit in Fig. 7.11, which is the equivalent circuit for the slab being considered.

c) When the material is piezoelectric, the stress T is given by equation 7.14a which, denoting $h = e/\varepsilon^S$, reads

$$T = c^D \frac{\partial u}{\partial x} - hD.$$

The previous results can be applied to T + hD instead of T. Furthermore, since the normal component of the

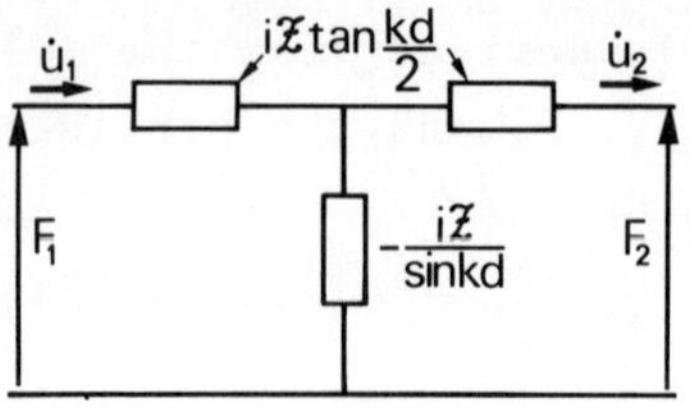

Fig. 7.11. Mechanical equivalent circuit for a non-piezoelectric slab of thickness d and of elastic impedance $\mathcal{Z}$.

electrical displacement D is constant, it suffices to add the term hDA to the right hand sides of equations 7.28 in order to obtain the forces $F_{\frac{1}{2}} = - AT_{\frac{1}{2}}$.

The displacement current $I = i\omega DA$ (through a cross section of area A) is calculated from the voltage v between metallized faces (fig. 7.12):

$$v = \int_{x_1}^{x_2} E\,dx = \frac{Dd}{\varepsilon^S} - h(u_2 - u_1) \qquad (7.29)$$

or, introducing the particle velocities $\dot{u}_1$ and $\dot{u}_2$

$$v = \frac{d}{i\omega\varepsilon^S A} I - \frac{h}{i\omega} (\dot{u}_2 - \dot{u}_1)$$

or

$$I = i\omega C_o v - hC_o(\dot{u}_1 - \dot{u}_2). \qquad (7.30)$$

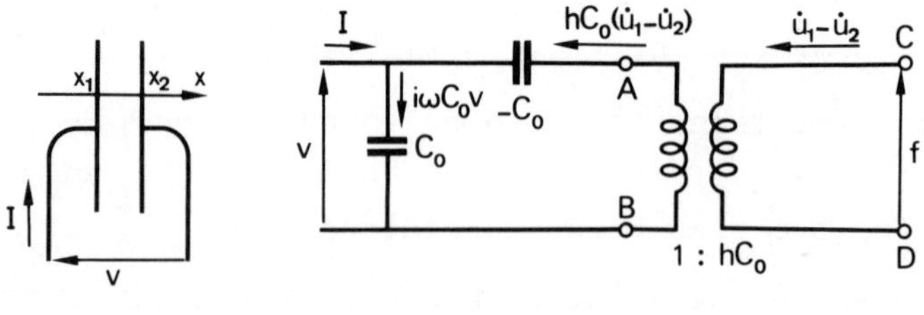

Fig. 7.12. Electromechanical transformer taking piezoelectricity into account.

It will be recalled that $C_o = \varepsilon^S A/d$ is the capacitance of the rigidly bonded transducer. The extra term $hDA = hI/i\omega$ can be obtained from an electromechanical transformer in whose secondary coil "flows" a velocity $(\dot{u}_1 - \dot{u}_2)$ and in whose primary coil flows the current $hC_o(\dot{u}_1 - \dot{u}_2) = i\omega C_o v - I$; the transformer ratio should be $N = hC_o$ (fig. 7.12). In fact, due to the capacitance $-C_o$ in the primary circuit, the force between terminals C and D is

$$f = hC_o(v_A - v_B) = hC_o[v - \frac{hC_o(\dot{u}_1 - \dot{u}_2)}{i\omega C_o}]$$

from 7.30: $f = hI/i\omega$.

The equivalent circuit depicted in fig. 7.13 has one electrical input and two mechanical outputs.

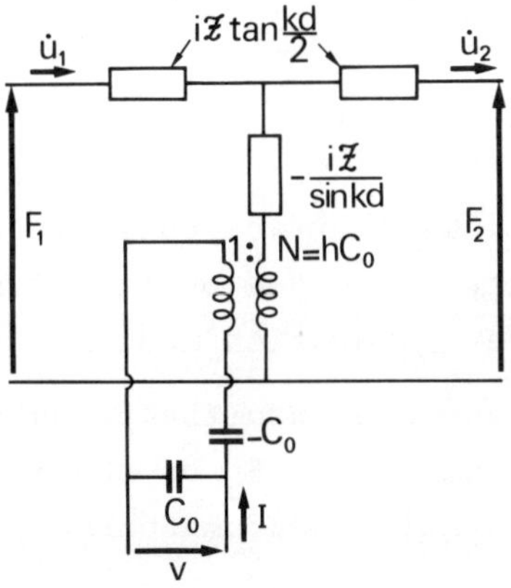

Fig. 7.13. Electromechanical equivalent circuit for a piezoelectric slab of thickness d and of elastic impedance $\mathscr{Z}$. It includes one electric port and two mechanical ports.

The equivalent circuit for the whole transducer (Fig. 7.14) is obtained by cascading the circuits corresponding to the various elements that constitute the transducer. The terminals for the free face of the outer electrode are short-circuited by the very weak acoustic impedance of air.

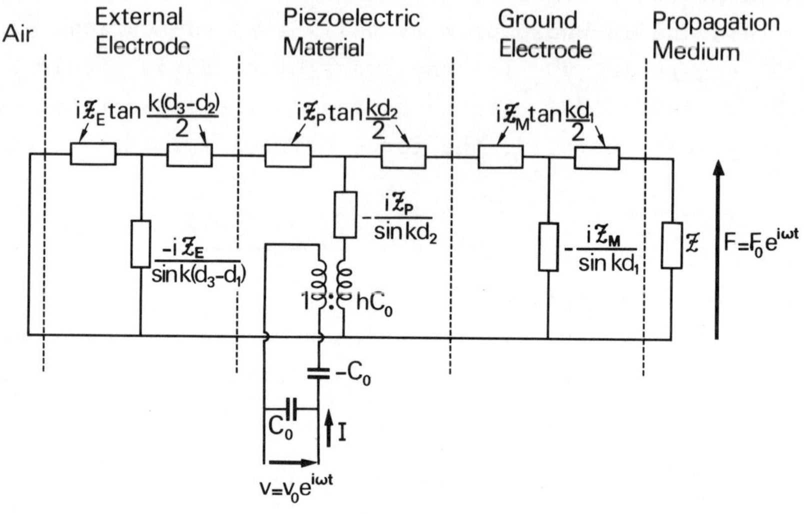

Fig. 7.14. Overall equivalent circuit of the transducer of Fig. 7.1. The left hand end, in contact with air is short circuited. The other extremity is loaded by the elastic impedance of the propagation medium.

The elastic power launched into the propagating medium is calculated from the amplitude F_0 of the force which is applied to the propagation medium boundary:

$$\mathcal{P} = \frac{F_0^2}{2\mathcal{Z}} \; .$$

d) As an example, let us now deal with a <u>resonator</u>, in the form of a thin piezoelectric plate metallized on both faces. The electrode thickness is assumed to be negligible; so the equivalent circuit is obtained by short circuiting both mechanical ports in the circuit of fig. 7.13. The secondary coil of the transformer is closed by the mechanical impedance

$$\mathcal{Z}_2 = \frac{i\mathcal{Z}}{2}\left(\tan\frac{kd}{2} - \frac{2}{\sin kd}\right) = \frac{-i\mathcal{Z}}{2\tan\frac{kd}{2}}$$

which, transformed at the primary circuit, is reduced by a factor $N^2 = h^2C_o^2$. The resonator electric impedance is

$$Z_e = \frac{V}{I} = \frac{1}{iC_o\omega}\left(\frac{\frac{\mathcal{Z}_2}{N^2} - \frac{1}{iC_o\omega}}{\frac{\mathcal{Z}_2}{N^2}}\right)$$

or

$$Z_e = \frac{1}{iC_o\omega}\ (1 - 2\ \frac{h^2C_o}{\omega\mathcal{Z}}\ \tan\frac{kd}{2}).$$

Since $\mathcal{Z} = \frac{c}{V}\ A,\ C_o = \frac{\varepsilon^S A}{d},\ h = \frac{e}{\varepsilon^S}$:

$$\frac{h^2C_o}{\omega\mathcal{Z}} = \frac{e^2}{\varepsilon^S c}\ \frac{V}{\omega d} = K^2\ \frac{V}{\omega d}$$

where the electromechanical coupling constant K is involved:

$$Z_e = \frac{1}{iC_o\omega}\ [1 - K^2\ \frac{\tan(\omega d/2V)}{\omega d/2V}]. \qquad (7.31)$$

Fig. 7.15 is a plot of the frequency dependence of the modulus of the admittance of the resonator $Y = 1/Z_e$. This quantity vanishes at odd multiples of the antiresonant frequency

$$f_a = \frac{V}{2d} \qquad (7.32)$$

and is infinite ($Z_e = 0$) for resonant frequencies $f_r^{(n)}$ such that

$$K^2\ \tan\ (\frac{\pi f_r^{(n)}}{V}\ d) = \frac{\pi f_r^{(n)}}{V}\ d$$

or, introducing f_a:

$$K^2\ \tan\ (\frac{\pi}{2}\ \frac{f_r^{(n)}}{f_a}) = \frac{\pi}{2}\ \frac{f_r^{(n)}}{f_a}. \qquad (7.33)$$

Measuring the antiresonant frequency f_a and the first resonance frequency $f_r = f_r^{(1)}$ leads to the propagation velocity $V = 2f_a d$ and the electromechanical coupling coefficient of the excited mode:

$$K^2 = \frac{\pi}{2}\ \frac{f_r}{f_a}\ \tan\ (\frac{\pi}{2}\ .\ \frac{f_a - f_r}{f_a}). \qquad (7.34)$$

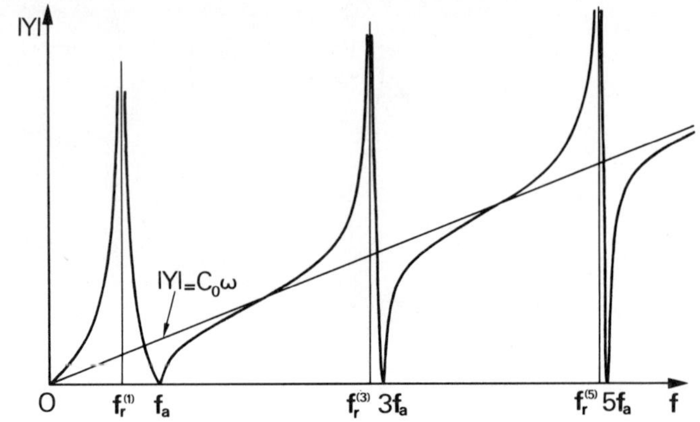

Fig. 7.15. Frequency dependence of the admittance modulus for a piezoelectric resonator. The line $|Y| = C_0\omega$ shows the admittance of the static capacitor C_0. In fact, at the resonance frequencies, the admittance is limited to a finite value by propagation and reflexion losses. This is also true for the impedance at the antiresonance frequencies.

But for high coupling constant materials ($K > .3$), the difference $(f_a - f_r)/f_a$ is quite small:

$$K^2 \simeq \frac{\pi^2}{4} \cdot \frac{f_a - f_r}{f_a} . \tag{7.35}$$

For the AT cut of quartz (shear mode) the coupling constant is 0.07 and the relative separation of resonant and antiresonant frequencies is only 0.2%. This separation varies as $1/n^2$ for n-th order vibrational modes (called "partial" modes) (see exercise 7.3). It is possible to simplify the resonator equivalent circuit in the vicinity of resonance (exercise 7.4).

7.1.3. Electromechanical Coupling Constant

Throughout the study of elastic wave propagation in piezoelectric materials (§ 6.2.1.2) the factor $K = e_{i\alpha}/\sqrt{\varepsilon^S_{jk}c^E_{\beta\gamma} + e^2_{i\alpha}}$ naturally emerged. We called it the electromechanical coupling constant. The relative influence of piezoelectricity on the velocity is of the order of $K^2/2$. Again this constant appeared twice in the present chapter: the elastic power supplied by a transducer (equation 7.22) and the separation between the resonant and antiresonant frequencies are both proportional to the square of the coupling constant, at least when it is less than 1. It is a figure of merit which provides information about the coupling between electric and mechanical properties; it also indicates the ability of piezoelectric materials to convert electrical energy into mechanical energy (or vice versa). Its expression is simple only for directions related to symmetry elements (table 6.9).

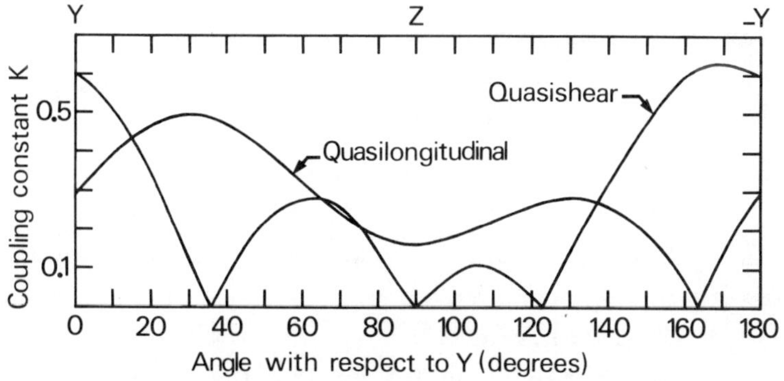

Fig. 7.16. Variation of electromechanical coupling constants K_L and K_T, with direction in the YZ plane of $LiNbO_3$ (fig. 13 of reference 3).

Equation 7.34 may be considered as a definition of the coupling constant in the direction which is orthogonal to the resonator faces. It is measured experimentally from the resonant and antiresonant frequencies of the excited vibrational modes. Its variation within the YZ symmetry plane of $LiNbO_3$ is plotted in fig. 7.16. The two curves refer to the quasi-longitudinal mode and to the quasi shear mode which is polarized in the YZ plane; no coupling takes place with the third mode. The directions for which one coupling constant vanishes are most interesting because only one mode is generated. The most remarkable ones are the +36° direction (with respect to the Y axis) which exhibits a very high quasi longitudinal coupling constant (K_L = .49) and the +163° direction for which the quasi shear mode coupling constant is even larger (K_T = .62). The corresponding cuts are used to build transducers for frequencies up to a few hundreds of megahertz. Note that for low loss transducers ($R >> R_A$), a high coupling constant has the advantage of allowing matching over a wide frequency range, since the electric Q-factor (according to formulae 7.25 and 7.26) is of the order of $1/K^2$.

In the definition of equation 7.34, $f_a = V/2d$ and, for $K^2 << 1$, $f_r \simeq V_o/2d$, where V_o is the velocity calculated while neglecting piezoelectric phenomena. This explains why the shape of the curves in fig. 7.16 can be derived by examining fig. 6.7.

7.1.4. Technology

The piezoelectric material thickness is always a fraction of the wavelength λ. Therefore, as far as technology is concerned, there are two types of transducer, according to whether the value of λ allows or forbids the handling of single crystals.

If λ/2 is greater than 0.1 mm (typically f = 25 MHz if V = 5,000 m/s) a piezoelectric single crystal (or ceramic) plate, after being metallized on both faces, can be directly bonded to the propagation medium with a rather thick binder such as an epoxy resin. If λ/2 ranges between 10 and 100 μm, the method is to solder a thick (≥ .1 mm) single crystal by indium metal diffusion [4], then to make it thinner (by conventional polishing) down to the desired value. It has been possible to reach a thickness of a few microns, in the laboratory by finishing with an ion bombardment.

Indium soldering involves evaporation in a vacuum on to one face of each piece, of a thin chromium film which serves as an adherent interface (150 to 300 Å), then evaporation of a gold layer (1500 Å) and finally evaporation of an indium layer (2500 Å). Once the indium is deposited the vacuum must not be broken so as to prevent oxidation. The two crystals are brought into contact and subjected by means of a hydraulic jack to a pressure of 200 kg/cm^2 for a period of about twenty minutes.

If $\lambda/2$ is of the order of 1 μm the piezoelectric material is vacuum-deposited on the substrate which has previously been coated by the inner electrode metal layer. The useful fraction of the piezoelectric film (Ø ≃ 1 mm) is determined by the size of the outer electrode, also deposited in a vacuum. It is difficult to control the crystal orientation of the piezoelectric layer, which is not a single crystal, and whose growth conditions depend on the state of the underlying metal layer which depends in turn on the cut of the substrate. There is a large difference from the former technique where the orientation of the single crystal transducer is not influenced by that of the substrate. This accordingly makes the choice of elastic mode quite free (§ 7.1.3).

Thin layers of cadmium sulphide and zinc oxide have been extensively studied, the substrate being a sapphire single crystal, which may be a parallelepiped (20 x 6 x 6 mm^3) or a cylinder (Ø ≃ 3 mm). In order to launch longitudinal waves along the triad rotation axis A_3 of sapphire, the 6-fold axis of symmetry A_6 of CdS or ZnO layers must be parallel to A_3. CdS is often associated with a gold electrode, ZnO with a platinum electrode. Titanium or chromium sublayers improve the metal adherence. Chromium, which has nearly the same elastic impedance as sapphire, (respectively 41.2 and 44.5 10^6) has almost no influence on the frequency response, whatever its thickness. For all these materials, the thickness is checked (during deposition) by the frequency variation of a quartz oscillator located in the vicinity of the substrate.

Cadmium sulphide can be deposited by direct evaporation [5, 6] or by the three temperature method [7] from two distinct sources (S and Cd or S and CdS). The thickness of the transparent CdS layer is measured, during its growth, by the interference of two optical beams, one of which is reflected by the gold layer and the other one by the CdS layer. Precautions must be taken: using liquid nitrogen traps, argon rinsing, and prolonged heating to diminish the residual gas pressure (oil, water, nitrogen). Typical temperatures are 220°C for the substrate, 100°C for the sulphur crucible and 750°C for the cadmium sulphide source. The layer growth rate is about 350 Å/mn. The crystallographic orientation of the CdS layer depends on the substrate inclination with respect to the direction of the vapour beam [8]. It is therefore theoretically possible to choose an orientation that excites mostly longitudinal or shear waves (A_6 axis at 40° or 90°). When made under good conditions by this method, transducers generally have losses of above 10 dB; the frequency response is as predicted by the analysis.

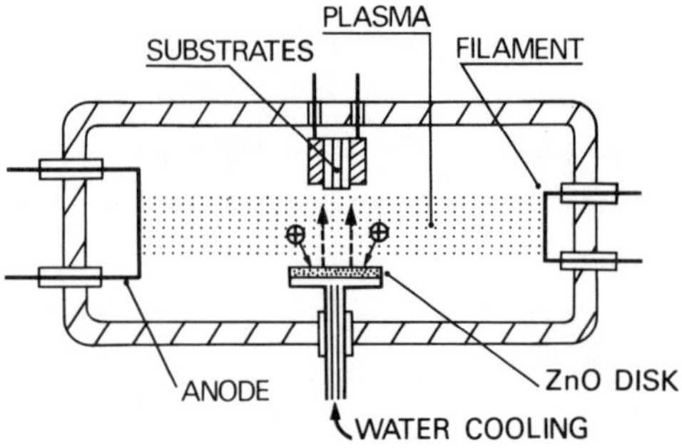

Fig. 7.17. Zinc oxide deposition by cathodic sputtering. The zinc oxide which constitutes the target is bombarded by ions extracted from the plasma, and deposited on the sapphire substrate at a rate of 200 Å/mn.

In our opinion, zinc oxide provides better and, above all, more reproducible results when it is deposited by continuous ion sputtering [9], for instance, using a triode layout. This method consists of bombarding a ZnO sintered powder disc, biased at a negative potential of around 2,000 volts, with ions extracted from a plasma column of argon and oxygen (fig. 7.17). On the substrate, located in front of the target, the ZnO layer grows at a rate typically of 200 Å/mn if the cathode current intensity is 50 mA. The A_6 axis of the deposited ZnO layer is always perpendicular to the surface: only longitudinal waves can be launched. The outer electrode is usually made of aluminium.

Fig. 7.18 illustrates an example of a good result. It refers to a $ZnO-Pt-Al_2O_3$ transducer. The zinc oxide layer thickness is 2.6 μm, the platinum layer thickness .4 μm. The frequency response curve, plotted from .5 to 4.5 GHz, very clearly shows the λ/4, 3λ/4, 5λ/4 and 7λ/4 resonances.

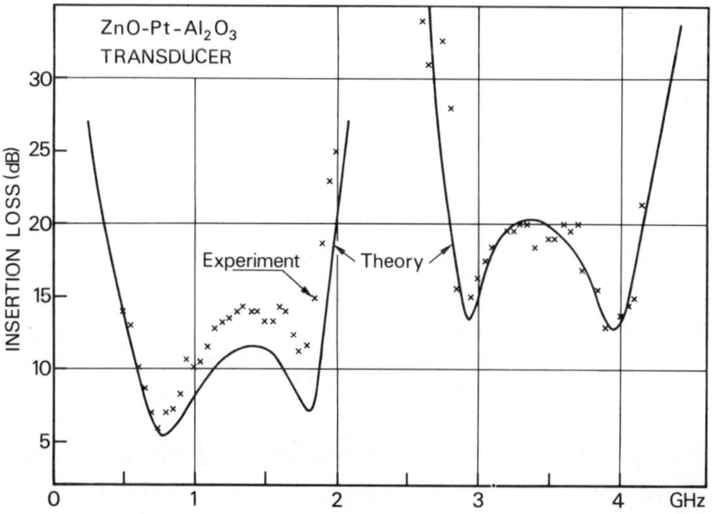

Fig. 7.18. Theoretical and experimental frequency response of a transducer made of a thin zinc oxide layer.

7.2. INTERDIGITAL TRANSDUCERS FOR SURFACE (RAYLEIGH) WAVES

The phrase "surface wave" refers, in principle, to any deformation that, when propagating, affects only a small depth of the material (of the order of the wavelength); normally included are Rayleigh waves, Bleustein-Gulyaev waves and Love waves (see §§ 5.3, 6.2.2.3 and 5.2.4).

It will be recalled that the last two types of waves are shear waves, retained near the surface by the material piezoelectricity (Bleustein-Gulyaev waves) or the medium inhomogeneity (thin layer on a substrate, Love waves). Rayleigh waves, which we are going to deal with, are complex waves that can propagate along the surface of any medium. In simple cases, they consist of a longitudinal displacement and a $\pi/2$-phase shifted shear displacement, which vanish completely at a depth of the order of two wavelengths. However, this complexity is compensated for by a major advantage: Rayleigh waves can be generated (or detected) on piezoelectric substrates by easily constructed interdigital transducers,* which can also be employed for other purposes than electromechanical conversion. Before studying the properties of this transducer, which account for the wide range of Rayleigh wave applications in electronics, it should be noted that these waves can be more or less adequately generated from bulk waves, by grazing refraction or by scattering at an array of equally spaced obstacles (grooves, metal strips). Reference [10] contains a detailed description of the various procedures involved in this technology.

7.2.1. Principle of Operation

The interdigital transducer is composed of (fig. 7.19) two metal comb shaped electrodes, deposited on a piezoelectric substrate. The voltage applied to the electrodes produces an electric field, which creates compressional and dilatational strains in the vicinity of the surface giving rise to various types of elastic waves (see § 7.2.3). As far as the Rayleigh waves are concerned which are generated

---

* Such transducers are also employed for Bleustein-Gulyaev or Love wave generation, but to date these waves have been less widely used because Bleustein-Gulyaev waves require very restrictive symmetry conditions for the materials and Love waves necessitate a homogeneous thin film of constant thickness, often difficult to achieve in practice.

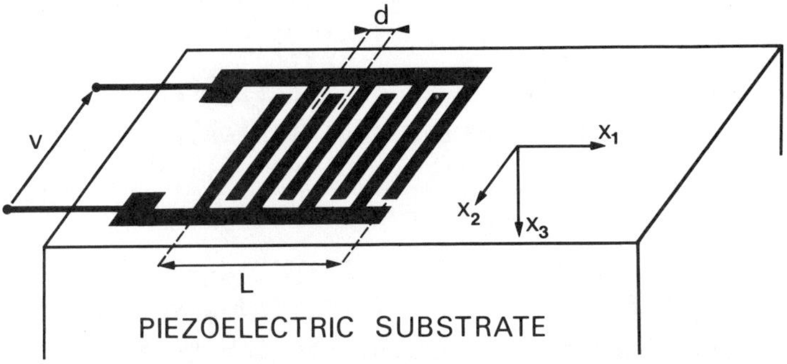

Fig. 7.19. Interdigital transducer. The voltage applied to the comb shaped electrodes generates stresses at the material surface.

in the direction orthogonal to the comb fingers, the transducer behaves as a sequence of ultrasonic sources. For an applied sinusoidal voltage, all vibrations interfere constructively only if the distance d between two neighbouring fingers is equal to half the elastic wavelength. Any stress created at time t by a finger pair, for a given polarity of the applied voltage, travels at the Rayleigh wave velocity V_R, along a distance $\lambda/2$ during the half period T/2. At time t + T/2, this stress has reached the next pair of fingers precisely when the voltage, which has changed sign, generates an in-phase stress; the elastic excitation due to the second pair is added to that of the first pair. The frequency $f_o = V_R/2d$ that corresponds to this cumulative effect is called the synchronous frequency or the resonance frequency. If the frequency departs from this value, the interference between the elastic signals due to various finger pairs is no longer constructive and the overall vibration is weaker. Thus, the bandwidth of a transducer is narrower when there are more fingers.

The frequency response of an N-finger transducer can be derived from its impulse response. A pulse whose duration is short compared to the elastic wave transit time between two fingers excites the (N - 1) ultrasonic sources simultaneously when applied to the electrodes.

Since the electric field changes sign at every interdigital spacing, the elastic signal is periodic. The spatial period is equal to twice the distance d between neighbouring finger axes. The time duration Θ of the signal is equal to the active length of the transducer [$L = (N - 1)d$] divided by the Rayleigh wave velocity:

$$\Theta = \frac{(N - 1)d}{V_R} = \frac{(N - 1)}{2f_o} . \qquad (7.36)$$

If this impulse-response is intuitively regarded as a sine signal of frequency $f_o = V_R/2d$ and duration Θ, then the frequency response, i.e. the Fourier transform of the impulse response, is a (sin x)/x curve (see equation 1.27) with

$$x = \pi\Theta(f - f_o) = (N - 1)\,\frac{\pi}{2}\cdot\frac{f - f_o}{f_o} .$$

The 3dB bandwidth ($x = \pm .885\ \pi/2$) is thus proportional to the reciprocal of the number of intervals:

$$\frac{\Delta f}{f_o} \simeq \frac{1.77}{N - 1} . \qquad (7.37)$$

In practice to determine the transducer impulse response, one must transform the generated elastic wave packet into an electric signal through a pick up transducer, e.g. of the interdigital type. The electric field associated with the elastic wave, when passing under the electrodes, induces a time varying voltage. The shape of this signal depends on the number of receiver fingers. Let us examine two typical cases:

a) <u>Single finger pair receiver</u> (fig. 7.20a). Since the detection is localized on a narrow band parallel to the wave fronts, the electric signal faithfully reproduces the elastic signal as it passes under the receiver.

b) <u>Receiver identical to the transmitter</u> (fig. 7.20b). The voltage amplitude increases from the time $\tau = \ell/V_R$ when the elastic wave packet front reaches the first interdigital receiver interval; reaches a maximum at time $\tau + \Theta$ when all ultrasonic receivers are simultaneously excited, then decreases. The overall impulse response, which has a triangular envelope and duration 2Θ, is the

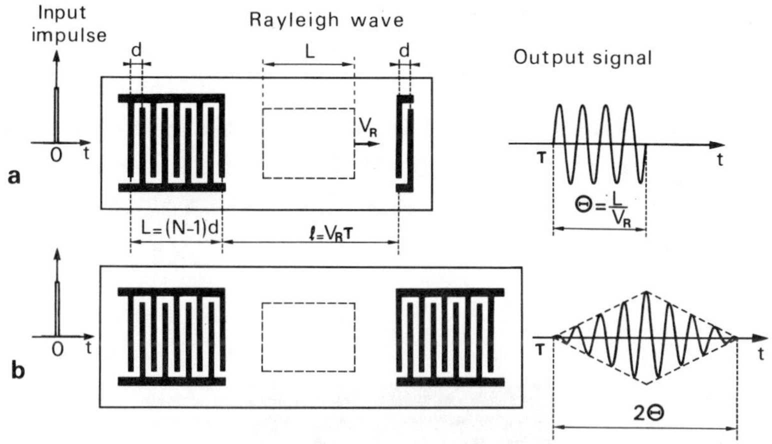

Fig. 7.20. Impulse response of a delay line consisting of two interdigital transducers: a) the receiver has only two fingers; b) the receiver is identical to the emitter.

autocorrelation of the single transducer impulse response. The transfer function of the two transducers is (exercise 1.9) a $(\sin x/x)^2$ curve, whose 3 dB bandwidth is

$$\frac{\Delta f}{f_o} \simeq \frac{1.27}{N-1} . \tag{7.38}$$

Fig. 7.21 shows the impulse response of a delay line consisting of two identical transducers each with 60 fingers, a centre frequency $f_o = 17$ MHz, and a quartz substrate (Y cut, X propagation). The diamond-shaped envelope, together with its duration $2\Theta = 3.5$ μs, confirms the fact that the single transducer transfer function is a $(\sin x)/x$ curve ($\Theta = L/V_R = (N-1)/2f_o = 1.74$ μs).

This intuitive method is not completely rigorous because the elastic signal corresponding to a short pulse does have a periodicity and a duration controlled (as described above) by the transmitting transducer, but it is not necessarily a sinusoidal signal. Accordingly, the transfer function consists of $(\sin x)/x$ curves of central frequencies f_o, $3f_o$, $5f_o$, ... Furthermore, it has been assumed that the transducer launches Rayleigh waves; this is so if the strains of the Rayleigh wave are indeed coupled to the electric field by the moduli of the piezoelectric material. A more detailed analysis of elastic wave generation by a comb shaped electrode transducer requires, first of all, a knowledge of the electric field distribution. It must be pointed out that this analysis will involve rather long calculations. Readers who are ready to accept that the interdigital transducer does launch Rayleigh waves may wish to omit the following sections and continue at § 7.2.4.

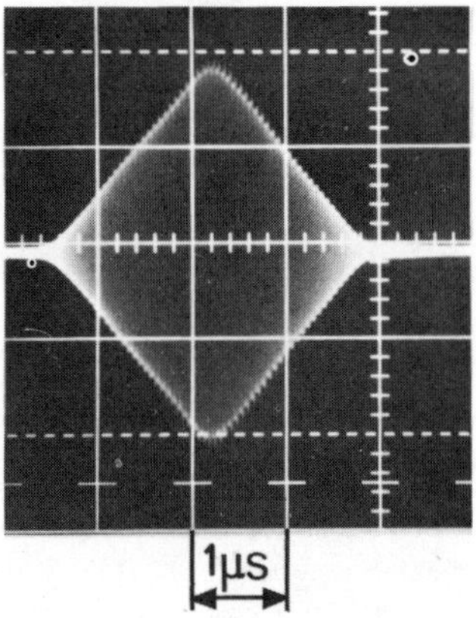

Fig. 7.21. Impulse response of a line consisting of two identical transducers of 60 fingers and centre frequency 17 MHz.

7.2.2. Electric Field Distribution

Fig. 7.22 indicates the transducer geometry and the coordinate system. The fingers are sufficiently long for no quantity to be dependent on x_2. The thickness of the metal layer is small compared to the distance d. To simplify the calculations, we assume the medium to be isotropic within its free surface plane i.e. it is either a class 6mm crystal or a piezoelectric ceramic whose axis of symmetry is directed along the x_3 axis, towards the core of the material. The applied voltage is a sine function: $v = v_o\, e^{i\omega t}$. Within the framework of the quasistatic approximation, as mentioned in § 6.2, the electric field is derived from a potential

$$\Phi(x_i, t) = \Phi(x_i)\, e^{i\omega t}.$$

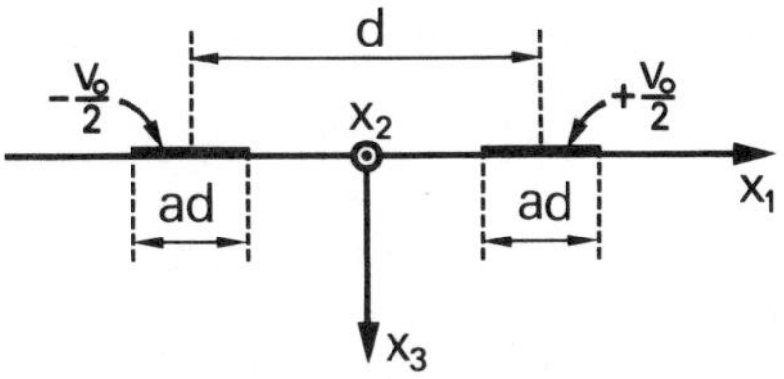

Fig. 7.22. Elementary cell of the transducer.

The electric displacement obeys the Poisson equation $\partial D_i/\partial x_i = 0$ for an insulator. When the electromechanical coupling constant is weak, the electric displacement due to the generated elastic waves can be neglected. Equation 6.15b can now be written as

$$D_i = \varepsilon^S_{ij} E_j = -\varepsilon^S_{ij} \frac{\partial \Phi}{\partial x_j}$$

and the potential Φ obeys the Laplace equation

$$\varepsilon^S_{ij} \frac{\partial^2 \Phi}{\partial x_i \partial x_j} = 0$$

Since the material is transversally isotropic and $\partial/\partial x_2 = 0$, this is expanded as

$$\varepsilon^S_{11} \frac{\partial^2 \Phi}{\partial x_1^2} + \varepsilon^S_{33} \frac{\partial^2 \Phi}{\partial x_3^2} = 0. \qquad (7.39)$$

The alternation of electrodes implies (for an infinite transducer)

$$\Phi(x_1 + d, x_3) = -\Phi(x_1, x_3) \qquad \forall\ x_1, x_3 \qquad (7.40)$$

and

$$\Phi(x_1 + 2d, x_3) = \Phi(x_1, x_3) \qquad \forall\ x_1, x_3.$$

Let us seek a solution of the form

$$\Phi(x_1, x_3) = f(x_1) g(x_3)$$

for which equation 7.39 becomes

$$\frac{f''}{f} = -\frac{\varepsilon^S_{33}}{\varepsilon^S_{11}} \frac{g''}{g} = -\chi^2$$

χ being a real constant (since f should be a periodic

function of x_1). The solutions are

$$f = F \sin \chi(x_1 + \delta) \tag{7.41}$$

and

$$g = G\, e^{-r\chi x_3} \quad \text{with} \quad \chi > 0 \quad \text{and} \quad r = \left(\frac{\varepsilon_{11}^S}{\varepsilon_{33}^S}\right)^{1/2} \tag{7.42}$$

(because the potential must decay inside the material). The symmetry condition 7.40 reads

$$f(x_1 + d) + f(x_1) = 0$$

or

$$2F \sin \left[\chi(x_1 + \delta) + \frac{\chi d}{2}\right] \cos \frac{\chi d}{2} = 0$$

thus setting the allowed values for χ:

$$\chi_m = (2m + 1)\frac{\pi}{d} \qquad m = 0, 1, 2\ldots \tag{7.43}$$

The general solution is a linear combination of odd harmonics of the spatial period 2d:

$$\Phi(x_1, x_3) = -\sum_{m=0}^{\infty} \frac{F_m}{\chi_m} e^{-r\chi_m x_3} \sin \chi_m(x_1 + \delta_m) \quad x_3 > 0 \tag{7.44}$$

(taking $G_m = -1/\chi_m$).

The electric potential in the vacuum ($x_3 < 0$) where $\varepsilon_{11} = \varepsilon_{33} = \varepsilon_0$, which must vanish at $x_3 = -\infty$, can be written in a similar form with $r = -1$, and the same coefficients F_m and δ_m (because of the continuity of Φ at $x_3 = 0$). These constants are determined by the electric boundary conditions at the free surface. The tangential component of the electric field

$$E_1 = -\frac{\partial \Phi}{\partial x_1} = \sum_{m=0}^{\infty} F_m\, e^{-r\chi_m x_3} \cos \chi_m(x_1 + \delta_m) \tag{7.45}$$

should be zero at the conductive electrodes:

$$\sum_{m=0}^{\infty} F_m \cos\left[(2m + 1)\frac{\pi}{d}(x_1 + \delta_m)\right] = 0$$

$$\text{for } \frac{d}{2}(1 - a) < |x_1| < \frac{d}{2}. \tag{7.46}$$

As the coefficients F_m and δ_m are the same in the crystal and in the vacuum, the continuity of the normal component of the electrical displacement between the fingers imposes

$$D_3 = -\varepsilon_{33}^S \frac{\partial\Phi}{\partial x_3} = -(\varepsilon_{11}^S\varepsilon_{33}^S)^{1/2} \sum_{m=0}^{\infty} F_m \sin\chi_m(x_1 + \delta_m) \qquad x_3 = 0_+$$

and

$$D_3 = \varepsilon_0 \sum_{m=0}^{\infty} F_m \sin\chi_m(x_1 + \delta_m) \qquad x_3 = 0_-$$

The only possibility is $D_3 = 0$

$$\sum_{m=0}^{\infty} F_m \sin\left[(2m+1)\frac{\pi}{d}(x_1 + \delta_m)\right] = 0 \quad \text{for} \quad |x_1| < \frac{d}{2}(1 - a). \tag{7.47}$$

Equations 7.46 and 7.47 are satisfied if δ_m vanishes and if F_m is equal to the Legendre polynomial P_m [11]:

$$F_m = HP_m[\cos\pi(1 - a)] = HP_m(-\cos\pi a). \tag{7.48}$$

In the expression for the potential (7.44)

$$\Phi(x_1, x_3) = -\frac{Hd}{2\pi}\sum_{m=0}^{\infty}\frac{P_m(-\cos\pi a)}{m + 1/2} e^{-r\chi_m x_3} \sin\chi_m x_1 \tag{7.49}$$

the constant H depends on the applied voltage v_0 across the electrodes

$$v_0 = 2\Phi\left[x_1 = \frac{d}{2}(1 - a), 0\right]$$

or,

$$v_0 = -H\frac{d}{\pi}\sum_{m=0}^{\infty}\frac{P_m(-\cos\pi a)}{m + 1/2}\sin\left[(m + \tfrac{1}{2})\pi(1 - a)\right].$$

Introducing the variable $s = \sin\pi a/2$, the sum is then the negative of the complete elliptic integral $K(s') = K'(s)$ of the complementary modulus $s' = \sqrt{1 - s^2}$ [11]:

$$v_0 = \frac{Hd}{\pi}K'(s) \tag{7.50}$$

Hence the spatial harmonic amplitudes for the electric field, as derived from 7.48:

$$\boxed{F_m = \frac{\pi v_0}{d}\frac{P_m(2s^2 - 1)}{K'(s)} = \frac{v_0}{d}\mathscr{F}_m}. \tag{7.51}$$

Because of the Legendre Polynomials, the coefficients $\mathscr{F}_m$ strongly depend - through the parameter $s = \sin\pi a/2$ - on the ratio a of the electrode width to the half period d. Fig. 7.23 shows that all harmonic amplitudes may vanish, except the first. In particular, since $P_{2p+1}(0) = 0$, there is no 3rd, 7th, ... (4p+3)-th harmonic when the

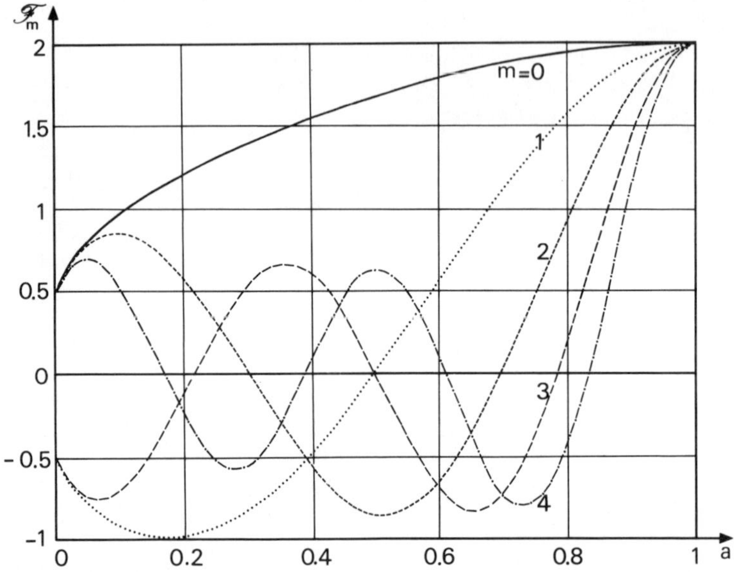

Fig. 7.23. Spatial distribution of the electric field. Variation of the amplitude of the harmonic terms $\mathscr{F}_m$ (m = 0...4) of the electric field, versus the relative finger width a of the electrodes (Fig. 2 in ref. [11]).

finger width is equal to the interfinger gap (a = 1/2). This is confirmed by experiment for low coupling constant materials, as illustrated by the impulse response of a transducer for which a = 1/2, fabricated on quartz (Fig. 7.24). This is no longer true for strongly piezoelectric materials such as lithium niobate or ceramics.

When the transducer length L is finite, the electric potential is no longer periodic, it is defined by its spatial Fourier transform:

$$\hat{\Phi}(k, x_3) = \int_{-\infty}^{+\infty} \Phi(x_1, x_3) e^{ikx_1} \, dx_1 \qquad (7.52)$$

such that

$$\Phi(x_1, x_3) = \frac{1}{2\pi} \int_{-\infty}^{+\infty} \hat{\Phi}(k, x_3) \, e^{-ikx_1} \, dk. \quad (7.53)$$

The Laplace equation 7.39 leads to

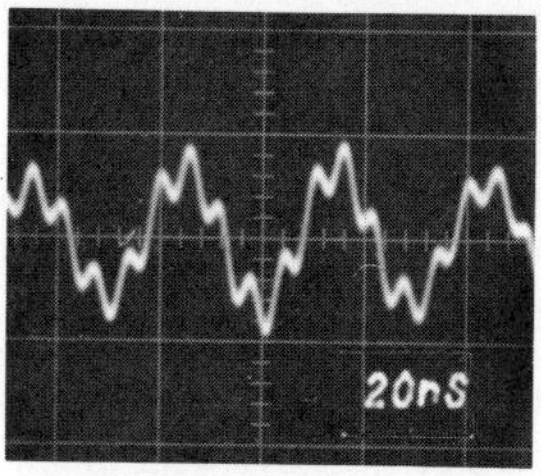

Fig. 7.24. Impulse response of an interdigital transducer such that a = 1/2. The 5th harmonic does exist, but there is no 3rd harmonic. The substrate is α quartz ($\underline{Y}$ $\vec{X}$).

$$-\varepsilon_{11}^{S} k^2 \hat{\Phi} + \varepsilon_{33}^{S} \frac{\partial^2 \hat{\Phi}}{\partial x_3^2} = 0$$

or

$$\hat{\Phi}(k, x_3) = \hat{\Phi}_0(k) e^{-rkx_3} \quad \text{where} \quad rk > 0 \qquad (7.54)$$

In order to calculate $\hat{\Phi}_0(k) = \hat{\Phi}(k, 0)$ we assume that the electric field is zero outside the transducer ($|x_1| > L/2$) and equal for $|x_1| < L/2$, to that created by an infinite transducer of the same dimensions. Once more, these assumptions lead us to discard the electric field accompanying the elastic wave and to suppose a high number of fingers. According to the differentiation theorem (§ 1.3.2), the Fourier transform of the component $E_1 = -\partial\Phi/\partial x_1$ is

$$\hat{E}_1(k, x_3) = \int_{-\infty}^{+\infty} E_1(x_1, x_3) e^{ikx_1}\, dx_1 = ik\hat{\Phi}(k, x_3).$$

This equation, together with equation 7.45 for the electric field ($\delta_m = 0$):

$$E_1(x_1, 0) = \sum_{m=0}^{\infty} F_m \cos \chi_m x_1$$

leads to

$$\hat{\Phi}_0(k) = -\frac{i}{k} \sum_{m=0}^{\infty} F_m \int_{-L/2}^{L/2} \cos \chi_m x_1\, e^{ikx_1}\, dx_1$$

or

$$\hat{\Phi}_0(k) = -\frac{i}{k}\sum_{m=0}^{\infty} F_m\left[\frac{\sin(k-\chi_m)L/2}{k-\chi_m} + \frac{\sin(k+\chi_m)L/2}{k+\chi_m}\right]. \tag{7.55}$$

7.2.3. Emitted Elastic Waves

Before solving the equation of motion it is interesting to show from a qualitative point of view, that interdigital transducers are able to generate Rayleigh waves [12]. Taking into account the piezoelectric matrix of a transversally isotropic medium, the horizontal and normal components (E_1 and E_3) of the electric field create the following stresses:

$$(T_1, T_2, T_3, T_4, T_5, T_6) = (E_1, 0, E_3)\begin{pmatrix} 0 & 0 & 0 & 0 & e_{15} & 0 \\ 0 & 0 & 0 & e_{15} & 0 & 0 \\ e_{31} & e_{31} & e_{33} & 0 & 0 & 0 \end{pmatrix}$$

The two stresses $T_{11} = e_{31}E_3$ and $T_{13} = e_{15}E_1$ can produce respectively a longitudinal and a shear wave propagating along the x_1-axis. The harmonics of a given order of the components E_1 and E_3 exhibit a spatial $\pi/2$ phase shift in the x_1 direction; the same applies to the displacement components u_1 and u_3, which can constitute the longitudinal and shear components of a Rayleigh wave emitted in the x_1 direction. The stress components $T_{22} = e_{31}E_3$ and $T_{33} = e_{33}E_3$ give rise to spurious phenomena.

7.2.3.1. Solution of propagation equation

In order to obtain the different waves emitted, it is necessary to solve the propagation equation 6.48, with the potential distribution implied by the transducer geometry:

$$\rho\frac{\partial^2 u_i}{\partial t^2} = c^E_{ijk\ell}\frac{\partial^2 u_\ell}{\partial x_j \partial x_k} + e_{kij}\frac{\partial^2 \Phi}{\partial x_j \partial x_k}$$

This can be rewritten for a transversally isotropic medium and if $\partial/\partial x_2 = 0$:

$$\begin{cases} \left(c_{11}\dfrac{\partial^2}{\partial x_1^2} + c_{44}\dfrac{\partial^2}{\partial x_3^2} + \rho\omega^2\right)u_1 + (c_{13} + c_{44})\dfrac{\partial^2 u_3}{\partial x_1 \partial x_3} \\ \qquad\qquad = -(e_{15} + e_{31})\dfrac{\partial^2 \Phi}{\partial x_1 \partial x_3} \\ \left(c_{66}\dfrac{\partial^2}{\partial x_1^2} + c_{44}\dfrac{\partial^2}{\partial x_3^2} + \rho\omega^2\right)u_2 = 0 \\ (c_{13} + c_{44})\dfrac{\partial^2 u_1}{\partial x_1 \partial x_3} + \left(c_{44}\dfrac{\partial^2}{\partial x_1^2} + c_{33}\dfrac{\partial^2}{\partial x_3^2} + \rho\omega^2\right)u_3 \\ \qquad\qquad = -e_{15}\dfrac{\partial^2 \Phi}{\partial x_1^2} - e_{33}\dfrac{\partial^2 \Phi}{\partial x_3^2} \end{cases} \qquad (7.56a,b,c)$$

Since u_2 is related neither to the electric field, nor to the other components u_1 and u_3, the particle displacement takes place within the sagittal plane x_1x_3. In view of the elimination of x_1 from the other two coupled differential equations, let us use the Fourier transforms $\hat{u}_i(k, x_3)$ of the displacements [13]. These are defined by

$$u_i(x_1, x_3) = \frac{1}{2\pi}\int_{-\infty}^{+\infty} \hat{u}_i(k, x_3)e^{-ikx_1}\, dk \qquad (7.57)$$

and likewise for the potential in equation 7.53. Introducing the phase velocity $V = \omega/k$, eq. 7.56a and c, together with $d\hat{\Phi}/dx_3 = -rk\hat{\Phi}$, lead to

$$\begin{cases} \left[(\rho V^2 - c_{11})k^2 + c_{44}\dfrac{d^2}{dx_3^2}\right]\hat{u}_1 - ik(c_{13} + c_{44})\dfrac{d\hat{u}_3}{dx_3} \\ \qquad\qquad = -irk^2(e_{15} + e_{31})\hat{\Phi} \\ -ik(c_{13} + c_{44})\dfrac{d\hat{u}_1}{dx_3} + \left[(\rho V^2 - c_{44})k^2 + c_{33}\dfrac{d^2}{dx_3^2}\right]\hat{u}_3 \\ \qquad\qquad = k^2(e_{15} - r^2 e_{33})\hat{\Phi} \end{cases} \qquad (7.58a,b)$$

The general solution of such a differential system is obtained by adding one solution of the whole system to the general solution of the homogeneous associated system. Let us seek a solution (for the whole system) of the form

$$\hat{u}_i = b_i\hat{\Phi} \qquad (7.59)$$

After dividing by k^2, b_1 and b_3 are determined by

$$\begin{cases} (\rho V^2 - c_{11} + r^2 c_{44})b_1 + ir(c_{13} + c_{44})b_3 = - ir(e_{15} + e_{31}) & (7.60a) \\ ir(c_{13} + c_{44})b_1 + (\rho V^2 - c_{44} + r^2 c_{33})b_3 = (e_{15} - r^2 e_{33}). & (7.60b) \end{cases}$$

The general solution of the homogeneous equation is

$$\hat{u}_i = {}^o\hat{u}_i \; e^{-iqkx_3} \qquad (7.61)$$

where the values of q are determined by the compatibility condition of the homogeneous system which is obtained on inserting (7.61) into the homogeneous system associated with (7.58):

$$\begin{cases} (c_{11} - \rho V^2 + c_{44}q^2)\,{}^o\hat{u}_1 + q(c_{13} + c_{44})\,{}^o\hat{u}_3 = 0 & (7.62a) \\ q(c_{13} + c_{44})\,{}^o\hat{u}_1 + (c_{44} - \rho V^2 + c_{33}q^2)\,{}^o\hat{u}_3 = 0 & (7.62b) \end{cases}$$

or:

$$c_{33}c_{44}q^4 + q^2[c_{33}(c_{11} - \rho V^2) + c_{44}(c_{44} - \rho V^2) - (c_{13} + c_{44})^2] + (c_{11} - \rho V^2)(c_{44} - \rho V^2) = 0. \quad (7.63$$

Note the similarity between 7.63 and equation 5.108 which gives the Rayleigh wave decaying factors. The difference is due to the fact that the medium is hexagonal instead of cubic. In the same way, for ${}^o\hat{u}_1 = 1$, the amplitude $p = {}^o\hat{u}_3$ is given by an expression analogous to 5.109:

$$p = {}^o\hat{u}_3 = - \frac{c_{11} - \rho V^2 + c_{44}q^2}{(c_{13} + c_{44})q} \qquad (7.64)$$

Multiplying both sides of equation 7.63 by k^4, in order to isolate the product qk:

$$c_{33}c_{44}(qk)^4 + (qk)^2[c_{33}c_{11}(k^2 - k_L^2) + c_{44}^2(k^2 - k_T^2) - (c_{13} + c_{44})^2k^2] + c_{11}c_{44}(k^2 - k_L^2)(k^2 - k_T^2) = 0 \quad (7.65)$$

$k_L = \omega/V_L$ and $k_T = \omega/V_T$ are the longitudinal and shear wave numbers; the velocities are respectively $V_L = \sqrt{c_{11}/\rho}$ and $V_T = \sqrt{c_{44}/\rho}$ (for propagation along x_1). If the displacement $\hat{u}_i$ is to be finite, the corresponding roots (q_1k) and (q_3k) with negative imaginary parts are the only possible solutions. The corresponding amplitudes are ${}^{o}\hat{u}_1^{(1)} = p_1$ and ${}^{o}\hat{u}_3^{(3)} = p_3$. Taking 7.59 into account, the general solution is obtained:

$$\begin{cases} \hat{u}_1 = A_1\, e^{-iq_1kx_3} + A_3\, e^{-iq_3kx_3} + b_1\hat{\Phi}_0\, e^{-rkx_3} & (7.66a) \\ \hat{u}_3 = p_1A_1\, e^{-iq_1kx_3} + p_3A_3\, e^{-iq_3kx_3} + b_3\hat{\Phi}_0\, e^{-rkx_3} & (7.66b) \end{cases}$$

The coefficients A_1 and A_3 are determined by the boundary conditions at the free surface $x_3 = 0$:

$$T_{i3} = c_{i3k\ell}\frac{\partial u_\ell}{\partial x_k} + e_{ki3}\frac{\partial \Phi}{\partial x_k} = 0 \quad \text{for} \quad x_3 = 0 \quad \forall x_1 \qquad (7.67)$$

This condition is automatically verified in the case of

$$T_{23} = c_{2323}\left(\frac{\partial u_3}{\partial x_2} + \frac{\partial u_2}{\partial x_3}\right) + e_{223}\frac{\partial \Phi}{\partial x_2}$$

because $\partial/\partial x_2 = 0$ and $u_2 = 0$. The other two equations ($i = 1$ and $i = 3$)

$$\begin{cases} c_{1313}\left(\dfrac{\partial u_1}{\partial x_3} + \dfrac{\partial u_3}{\partial x_1}\right) + e_{113}\dfrac{\partial \Phi}{\partial x_1} = 0 & (7.68a) \\ c_{3333}\dfrac{\partial u_3}{\partial x_3} + c_{3311}\dfrac{\partial u_1}{\partial x_1} + e_{333}\dfrac{\partial \Phi}{\partial x_3} = 0 & (7.68b) \end{cases}$$

become, when introducing the Fourier transforms $\hat{u}_1$, $\hat{u}_3$ and $\hat{\Phi}$:

$$\begin{cases} c_{44}\left(\dfrac{d\hat{u}_1}{dx_3} - ik\hat{u}_3\right) = ike_{15}\hat{\Phi} \\ c_{33}\dfrac{d\hat{u}_3}{dx_3} - ikc_{13}\hat{u}_1 = -e_{33}\dfrac{d\hat{\Phi}}{dx_3} \end{cases}$$

and, inserting the expressions for $\hat{u}_1$, $\hat{u}_3$ and $\hat{\Phi}$, (at $x_3 = 0$):

$$\begin{cases} c_{44}(q_1 + p_1)A_1 + c_{44}(q_3 + p_3)A_3 \\ \qquad = -(e_{15} + c_{44}b_3 - irc_{44}b_1)\hat{\Phi}_0 & (7.69a,\ b) \\ (c_{13} + c_{33}p_1q_1)A_1 + (c_{13} + c_{33}p_3q_3)A_3 \\ \qquad = (ire_{33} + irc_{33}b_3 - c_{13}b_1)\hat{\Phi}_0 \end{cases}$$

The unknown quantities A_1 and A_3 are proportional to $\hat{\Phi}_0$:

$$A_1 = \frac{B_1}{\Delta}\hat{\Phi}_0 \quad \text{and} \quad A_3 = \frac{B_3}{\Delta}\hat{\Phi}_0$$

where Δ is the determinant of the coefficients of A:

$$\Delta = (p_1 + q_1)(c_{13} + c_{33}p_3q_3) - (p_3 + q_3)(c_{13} + c_{33}p_1q_1). \qquad (7.71)$$

On replacing p_i and q_i by their expressions in terms of V, the solution of the equation $\Delta(V) = 0$, which is analogous to 5.116, is the Rayleigh wave velocity V_R. As a function of wavenumber $\Delta(k)$ vanishes for $k = k_R = \omega/V_R$. In order to obtain the displacements $u_i(x_1, x_3)$, we make use of the Fourier transform 7.57:

$$u_1(x_1, x_3) = \frac{1}{2\pi}\int_{-\infty}^{+\infty} \left(\frac{B_1}{\Delta} e^{-iq_1kx_3} + \frac{B_3}{\Delta} e^{-iq_3kx_3}\right)\hat{\Phi}_0 e^{-ikx_1} dk + \frac{1}{2\pi}\int_{-\infty}^{+\infty} \hat{\Phi}_0 b_1 e^{-rkx_3} e^{-ikx_1} dk \qquad (7.72a)$$

$$u_3(x_1, x_3) = \frac{1}{2\pi}\int_{-\infty}^{+\infty} \left(p_1 \frac{B_1}{\Delta} e^{-iq_1kx_3} + p_3 \frac{B_3}{\Delta} e^{-iq_3kx_3}\right)\hat{\Phi}_0 e^{-ikx_1} dk + \frac{1}{2\pi}\int_{-\infty}^{+\infty} b_3 e^{-rkx_3} \hat{\Phi}_0 e^{-ikx_1} dk \qquad (7.72b)$$

The poles of the integrand, i.e. the values $k = \pm k_R$ which make the determinant Δ equal to zero, are located on the path of integration because all attenuation has been neglected. The pole with positive real part which corresponds to propagation in the positive x_1 direction, must have a negative imaginary part ($k'_R = k_R - i\eta$, with $\eta > 0$) since the propagation factor

$$e^{-ik_Rx_1} e^{-\eta x_1}$$

must tend to zero as x_1 tends to infinity. Similarly, the attenuation of the propagation in the negative x_1 direction requires that the other pole be of the form $k''_R = -k_R + i\eta$. We need not take the attenuation into account provided that the path of integration is modified, and leaves the poles just below ($+k_R$) or above ($-k_R$) (fig. 7.25).

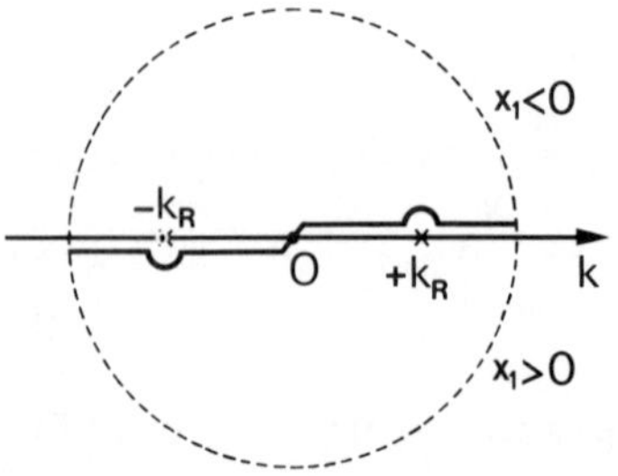

Fig. 7.25. Path of integration for a medium with no attenuation.

7.2.3.2. Bulk waves

In order to distinguish between the emitted bulk and surface waves, we can evaluate the integrals far from the transducer ($R \to \infty$) in a given direction θ (fig. 7.26). With $x_1 = R \sin\theta$, $x_3 = R\cos\theta$, the first two terms of the displacements (formulae 7.72a and b) come from integrals such as

$$I = \int_{-\infty}^{+\infty} e(k) \exp\left[-\, iRk(q\cos\theta + \sin\theta)\right]dk \quad (7.73)$$

where

$$e(k) = \frac{1}{2\pi}\,\frac{B(k)\hat{\Phi}_0(k)}{\Delta(k)} \quad (7.74)$$

The contribution of the last term vanishes as $R \to +\infty$ since

$$e^{-rkx_3} = e^{-rRk\cos\theta} \to 0 \qquad (rk > 0)$$

If $\phi(k)$ is defined as

$$\phi(k) = k[q(k)\cos\theta + \sin\theta] \quad (7.75)$$

the integral I becomes

$$I(\theta) = \int_{-\infty}^{+\infty} e(k)\, e^{-iR\phi(k)}\, dk. \quad (7.76)$$

For large values of R, the exponential oscillates rapidly where $\phi(k)$ has a strong dependence on k. The limiting value of the integral, for large distances, is in § 1.3.2; k represents time t, θ represents the angular frequency ω and $\alpha = -R\phi$. Equation 1.24 leads to

$$I(\theta) = \sqrt{\frac{2\pi}{R|\phi''(k_0)|}}\; e^{-i\pi/4}\, e(k_0)\, e^{-iR\phi(k_0)} \quad (7.77)$$

k_0 being that value of k which makes the phase ϕ stationary:

$$\left(\frac{d\phi}{dk}\right)_{k_0} = 0. \quad (7.78)$$

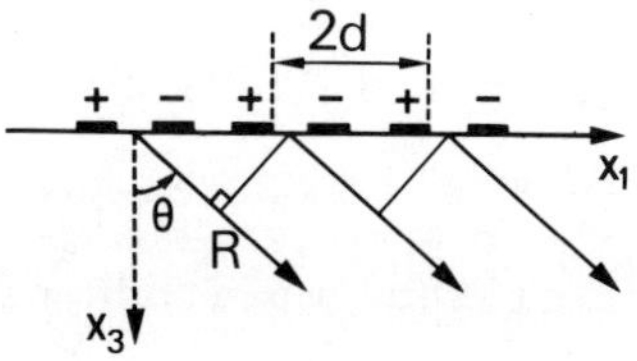

Fig. 7.26. Far from the transducer, bulk waves are emitted preferentially in directions corresponding to constructive interference.

The $R^{-\frac{1}{2}}$ decay of the integral $I(\theta)$ reflects the conservation of elastic energy. Far from the transducer, the generated waves are almost plane waves; their phase velocity can only be the quasi longitudinal wave velocity (V_1) or the quasi shear wave velocity (V_2). These velocities depend on the propagation direction

$$V_{\frac{1}{2}} = f(n_1, n_2, n_3)$$

i.e. on the angle θ, through $n_1 = \sin\theta$, $n_2 = 0$, $n_3 = \cos\theta$:

$$V_{\frac{1}{2}}(\theta) = f(\sin\theta, 0, \cos\theta).$$

The explicit expressions (5.47 and 5.48) were established in § 5.1.6.2c. The parameter $V = \omega/k$, introduced in the previous paragraph, is equal to

$$V(q) = f(1, 0, q)$$

because for a plane wave

$$n_1 = \frac{i}{k}\frac{\partial}{\partial x_1} = 1 \qquad n_2 = \frac{i}{k}\frac{\partial}{\partial x_2} = 0 \quad \text{and} \quad n_3 = q$$

and since f is a homogeneous function of degree 1 (§ 5.1.5.2):

$$f(\sin\theta, 0, q\sin\theta) = V(q)\sin\theta.$$

the left hand side of this equation represents $V_{\frac{1}{2}}(\theta)$ for $q = q_o = \text{cotan}\,\theta$:

$$V_{\frac{1}{2}}(\theta) = V(q_o)\sin\theta.$$

The identity relating the phase of I and the phase of a plane wave with velocity V_1 or V_2:

$$\phi(k_o) = \frac{\omega}{V_{\frac{1}{2}}(\theta)} = \frac{\omega}{V(q_o)\sin\theta} \qquad (7.79)$$

also reads (because of 7.75)

$$k_o[q(k_o)\cos\theta + \sin\theta] = \frac{k(q_o)}{\sin\theta}$$

It is satisfied if

$$q(k_o) = q_o = \frac{\cos\theta}{\sin\theta} \Rightarrow k_o = k(q_o) = \frac{\omega}{V(q_o)}$$

or equivalently (according to 7.79) for the two values

$$k_{01} = \frac{\omega}{V_1(\theta)}\sin\theta \qquad k_{02} = \frac{\omega}{V_2(\theta)}\sin\theta.$$

The θ-dependence of the amplitude of the corresponding two waves

$$U(\theta) = \frac{1}{\sqrt{2\pi R|\phi''(k_o)|}}\frac{B(k_o)}{\Delta(k_o)}\hat{\Phi}_0(k_o) \qquad (7.80)$$

provides the radiation pattern of the transducer. If the transducer has many fingers (L >> d), the function $\Phi_0(k)$, as given by (7.55), takes appreciable values only if k is close to $\pm\chi_m$. The bulk waves are chiefly launched in the θ_m directions:

$$\frac{\omega}{V_{\frac{1}{2}}(\theta_m)} \sin\theta_m = \pm (2m+1)\frac{\pi}{d}$$

satisfying the constructive interference condition (fig. 7.26)

$$2d \sin\theta_m = \pm (2m+1)\lambda_{\frac{1}{2}}(\theta_m)$$

which is achieved if the frequency is larger than $V(\pi/2)/2d$ where $V(\pi/2)$ is the longitudinal (or shear) wave velocity in the x_1 direction.

7.2.3.3. Rayleigh waves

The contribution of the poles $-k_R$ and $+k_R$ in e(k) is calculated by the method of residues:

$$I_R = 2\pi i \text{ Res } [e(k)]_{k_R} \; e^{-iR\phi(k_R)} \qquad (7.81)$$

and must be added to I. A glance at equation (7.75) for $\phi(k)$ shows that the factor

$$e^{-iR\phi(k)} = e^{-iRk_R q(k_R)\cos\theta} \; e^{-iRk_R \sin\theta}$$

tends to zero as R tends to infinity because the acceptable roots qk of equation 7.65 have a negative imaginary part. This contribution of the poles, which vanishes except for $\theta = \pi/2$ ($\cos\theta = 0$), corresponds to a Rayleigh wave propagating along the free surface. The displacements u_1 and u_3 are calculated by the method of residues by closing the path of integration with a semicircle of infinite radius in the lower half plane when $x_1 > 0$ and in the upper half plane when $x_1 < 0$, so that the factor exp $-ikx_1$ vanishes on this semicircle (fig. 7.25). The pole $+k_R$ corresponds to a wave propagating in the positive x_1 direction, the pole $-k_R$ to a wave emitted in the negative x_1 direction. As b_1 and b_3 are analytic functions of k, the third part of integrals 7.72a and b is zero. The other two integrals are derived from I_R with the residue:

$$\mathrm{Res}[e(k)]_{k_R} = \frac{1}{2\pi}\,\frac{B(k_R)\hat{\Phi}_0(k_R)}{[\frac{d\Delta}{dk}]_{k_R}} = -\frac{1}{2\pi}\,\frac{B(k_R)}{V_R(\frac{d\Delta}{dV})_{V_R}}\,k_R\hat{\Phi}_0(k_R).$$

The components of the particle displacement are:

$$u_1^R(x_1,x_3,t) = -\frac{ik_R\hat{\Phi}_0(k_R)}{V_R[\frac{d\Delta}{dV}]}(B_1e^{-iq_1k_Rx_3}+B_3e^{-iq_3k_Rx_3})e^{i\omega(t-x_1/V_R)} \tag{7.82a}$$

$$u_3^R(x_1, x_3,t) = -\frac{ik_R\hat{\Phi}_0(k_R)}{V_R[\frac{d\Delta}{dV}]}(p_1B_1e^{-iq_1k_Rx_3}+p_3B_3e^{-iq_3k_Rx_3})e^{i\omega(t-x_1/V_R)} \tag{7.82b}$$

The index R indicates that all quantities in the right hand side have been calculated for $V = V_R$ or $k = k_R = \omega/V_R$. Since they are functions of V^2 or k^2, Rayleigh waves are launched with the same amplitude in both directions. The amplitude depends on the frequency through the term $H = ik_R\hat{\Phi}_0(k_R)$ or, from 7.55:

$$H(\omega) = \sum_{m=0}^{\infty} F_m\left[\frac{\sin(\frac{\omega}{V_R} - \chi_m)\frac{L}{2}}{\frac{\omega}{V_R} - \chi_m} + \frac{\sin(\frac{\omega}{V_R} + \chi_m)\frac{L}{2}}{\frac{\omega}{V_R} + \chi_m}\right]$$

Inserting the value of χ_m (equation 7.43):

$$\frac{\omega}{V_R} - \chi_m = \frac{2\pi}{V_R}[f - (2m + 1)f_0]$$

leads to ($F_m = (v_0/d)\mathcal{F}_m$):

$$H(f) = v_0\frac{V_R}{d}\sum_{m=0}^{\infty}\mathcal{F}_m\frac{\sin\frac{\pi L}{V_R}[f-(2m+1)f_0]}{2\pi[f-(2m+1)f_0]} + \mathcal{F}_m\frac{\sin\frac{\pi L}{V_R}[f+(2m+1)f_0]}{2\pi[f+(2m+1)f_0]} \tag{7.83}$$

The transducer frequency response consists of (sin x)/x curves, each centred around odd harmonics of the synchronous frequency f_0; the bandwidth is constant and the amplitudes are proportional to the coefficients $\mathcal{F}_m$. From equation 1.27, it appears that the impulse response has duration $\Theta = L/V_R$ and amplitude $A = V_Rv_o/d$:

$$h(t) = V_R \frac{v_o}{d} \Pi\left(\frac{t}{\Theta}\right) \sum_{m=0}^{\infty} \mathscr{F}_m \cos(2m+1) f_o t. \quad (7.84)$$

Note that, for a given applied voltage of amplitude v_o, the impulse response is proportional to the reciprocal of the interdigital spacing d^{-1}, i.e. to the resonance frequency f_o (as the electric field $E \propto v_o/d$).

The relative finger width a of the transducer also modifies the amplitude of the emitted Rayleigh wave. It has been shown [14, 15] that the amplitude is a maximum for $a = 1/2$, i.e. for a finger width equal to the interval in between ($\lambda_o/4$).

7.2.4. The Method of Discrete Sources

The preceding two paragraphs pointed out that, even in the simplest case of a transversally isotropic, weakly piezoelectric medium, with regularly spaced fingers of constant length, analyzing how an interdigital transducer operates is a difficult task. However, the shape of the frequency response is independent of the material. It is therefore wise to have a general method available for calculating the frequency response of a transducer whose fingers have a variable spacing and different lengths. In the discrete source method or "delta" method [16, 17], each finger pair is considered as an infinitely narrow ultrasonic source (or receiver) located for example midway between two fingers, as shown in fig. 7.27. The amplitude A assigned to each source is proportional to the overlap length of the two fingers, with + or - sign depending on the direction of the electric field.

With these assumptions, the impulse response of a transducer is a series of Dirac functions:

$$h(t) = \sum_n s_n A_n \delta(t - t_n) \quad (7.85)$$

The sign s_n is equal to $(-1)^n$ when the electric field direction is reversed from one interdigital interval to the next. The frequency response

$$H(f) = \int_{-\infty}^{+\infty} h(t)\, e^{-i2\pi ft}\, dt$$

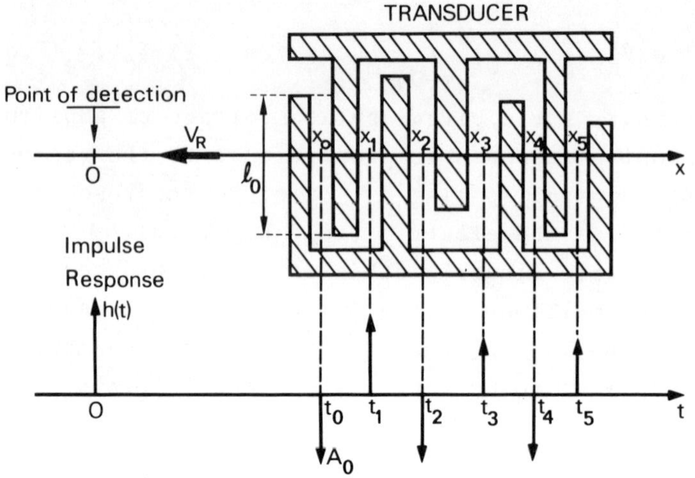

Fig. 7.27. Method of discrete sources. The transducer is regarded as a discrete sequence of sources, each one located midway between two fingers, and of amplitude proportional to the overlap length.

is equal to

$$\boxed{H(f) = \sum_n s_n A_n \, e^{-i2\pi f t_n}} \tag{7.86}$$

In order to check the validity of this method of discrete sources let us apply it to a transducer which has N equally spaced fingers of equal length for which

$$s_n = (-1)^n \quad A_n = A_o = C^{te} \quad t_n = t_o + n\frac{d}{V_R} = t_o + \frac{n}{2f_o} \,.$$

The frequency response is

$$H(f) = A_o \, e^{-i2\pi f t_o} \sum_{n=0}^{N-2} (-1)^n \exp\left(-\,in\pi\frac{f}{f_o}\right)$$

or, since $(-1)^n = e^{in\pi}$:

$$H(f) = A_o \, e^{-i2\pi f t_o} \sum_{n=0}^{N-2} \exp\left(-\,in\pi\frac{f-f_o}{f_o}\right)$$

The summation leads to

$$H(f) = A_o \, e^{-i2\pi f t_o} \frac{1 - \exp\left[-i(N-1)\pi\frac{f-f_o}{f_o}\right]}{1 - \exp\left(-i\pi\frac{f-f_o}{f_o}\right)}$$

or

$$H(f) = A_o \exp\left[-i\left(2\pi f t_o + (N-2)\frac{\pi}{2}\cdot\frac{f-f_0}{f_0}\right)\right] \frac{\sin\left[(N-1)\frac{\pi}{2}\cdot\frac{f-f_0}{f_0}\right]}{\sin\left(\frac{\pi}{2}\cdot\frac{f-f_0}{f_0}\right)}$$

If the phase shift τ due to the average time delay is ignored

$$\tau = t_o + \frac{(N-2)}{4f_0} = t_o + \frac{N-2}{2V_R}d$$

the transfer function reads

$$H(f) = A_o \frac{\sin(N-1)\pi\frac{f-f_0}{2f_0}}{\sin\pi\frac{f-f_0}{2f_0}} \qquad (7.87)$$

At frequencies around $f_m = (2m+1)f_0$ $(m = 0, \pm 1, \pm 2 \ldots)$ such that

$$\frac{f-f_0}{2f_0} = \frac{f-f_m}{2f_0} + m$$

and for which both the numerator and the denominator vanish in (7.87), H(f) is equivalent to

$$H_m(f) = A_o \frac{\sin(N-1)\pi\frac{f-f_m}{2f_0}}{\pi\frac{f-f_m}{2f_0}} = A_o \frac{V_R}{d}\cdot\frac{\sin\pi\frac{L}{V_R}(f-f_m)}{\pi(f-f_m)} \qquad (7.88)$$

The only difference from the analysis of the previous paragraph (equation 7.83) arises from the fact that all odd harmonics have the same amplitude whereas previously they were proportional to the coefficients $\mathcal{F}_m$.

The method of discrete sources is an adequate tool for designing a transducer that has a given impulse response (or a given transfer function). For example, let

$$h(t) = e(t)\cos\phi(t) \qquad (7.89)$$

be the desired impulse response, whose amplitude is modulated by the positive envelope function e(t), and whose frequency is modulated by the phase law (fig. 7.28)

$$f = \frac{1}{2\pi}\frac{d\phi}{dt}. \qquad (7.90)$$

In order to set the positions of the discrete source $x_n = V_R t_n$, it is always possible to define a function g(t) such that

$$g(t_n) = n \qquad n \text{ integer.}$$

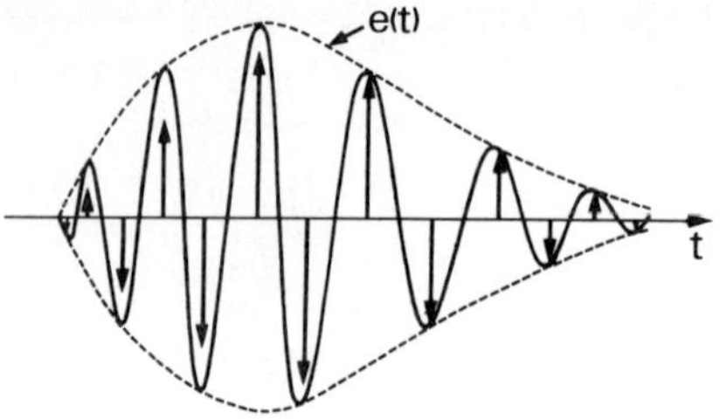

Fig. 7.28. Synthesis of an impulse response. The response is represented by samples whose phases are equidistant and whose amplitudes are proportional to the value of the envelope and to the reciprocal of the instantaneous frequency.

Assigning an amplitude $h(t_n)$ to each source, leads to a sampling of the impulse response:

$$h_e(t) = \sum_n h(t_n)\delta[g(t) - n] = h(t) \sum_n \delta[g(t) - n] \quad (7.91)$$

or, using the results of exercise 1.11:

$$h_e(t) = \sum_n h(t_n) \frac{\delta(t-t_n)}{|g'(t_n)|} = \sum_n \frac{e(t_n)}{|g'(t_n)|} \cos\phi(t_n)\delta(t-t_n). \quad (7.92)$$

It is difficult to achieve this distribution of sources whose signs are (a priori) either positive or negative with an interdigital transducer. It is better to choose the instants t_n such that

$$\boxed{\phi(t_n) = n\pi} \quad \Rightarrow \quad \cos\phi(t_n) = (-1)^n \quad (7.93)$$

so as to make the signs of the sources alternate (fig. 7.28). From $g(t) = \phi(t)/\pi$, expression 7.92 becomes

$$h_e(t) = \sum_n (-1)^n \frac{\pi e(t_n)}{\phi'(t_n)} \delta(t - t_n). \quad (7.94)$$

In fact, this series of delta functions represents more than the function $h(t) = e(t)\cos\phi(t)$, as could be predicted from fig. 7.29. Applying equation 1.40 yields:

$$\sum_{n=-\infty}^{+\infty} \delta(x - n) = \sum_{m=-\infty}^{+\infty} e^{-i2\pi mx} = 1 + 2\sum_{m=1}^{\infty} \cos(2\pi mx)$$

(which is proved in exercise (1.10)). Inserting this result into 7.91 where $x = g = \phi/\pi$, leads to

$$h_e(t) = h(t) + 2e(t) \sum_{m=1}^{\infty} \cos\phi \cos(2m\phi)$$

or, after the product has been turned into a sum:

$$h_e(t) = 2h(t) + 2e(t) \sum_{m=1}^{\infty} \cos(2m + 1)\phi.$$

The sampled impulse response

$$\frac{h_e(t)}{2} = h(t) + e(t) \sum_{m=1}^{\infty} \cos(2m + 1)\phi \qquad (7.95)$$

represents, besides the desired response $h(t) = e(t) \cos\phi(t)$, functions with the same envelope but whose instantaneous frequency is an odd harmonic of the frequency of the fundamental mode $h(t)$.

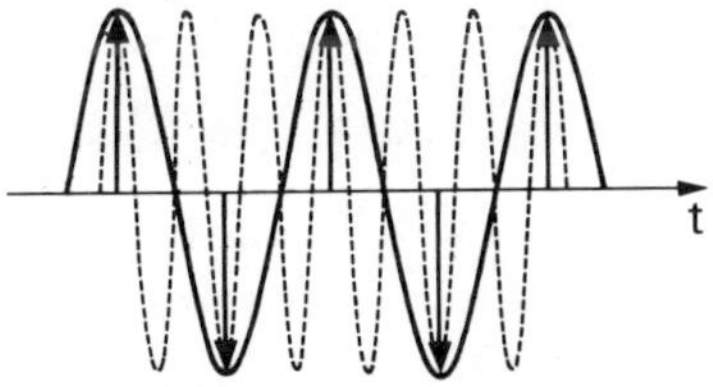

Fig. 7.29. A sequence of delta functions represents not only the sine wave at the centre frequency f_o, but also all the harmonic sine waves at frequencies $(2m + 1)f_o$.

The overlap finger length ℓ_n is proportional to the source amplitude A_n, itself derived from relation 7.94:

$$\boxed{A_n = \frac{\pi e(t_n)}{2\phi'(t_n)}} \; . \qquad (7.96)$$

If the interval between two fingers and the finger width are chosen to be equal to a quarter of the instantaneous wavelength, the distance between the axes of the two fingers that constitute the n-th source is

$$d_n = \frac{\lambda(t_n)}{2} = \frac{V_R}{2f(t_n)} = \frac{\pi V_R}{\phi'(t_n)} \; . \qquad (7.97)$$

The desired impulse response is obtained on eliminating the harmonics by means of a band pass filter, consisting of the receiving transducer and the matching circuits.

In the case of dispersive delay lines, described in § 9.3.5.1, the number of transducer fingers may be very high. For example a pulse of duration $\Theta = 10$ μs and central carrier frequency $f_o = 60$ MHz, needs $2f_o\Theta = 1{,}200$ fingers. Reducing this number renders mask design easier and cuts down the spurious interactions which arise when a wave emitted at one extremity of the transducer passes under the other fingers. This is achieved by using one of the harmonic responses, as shown in fig. 7.30. The distance between the axes of the source fingers, still at times t_n, is $\lambda(t_n)/2Q$. The Q-th harmonic $(Q = 2m + 1)$ is selected by a filter centred at the frequency Qf_o. The impulse response

$$h_Q(t) = e(t) \cos [Q\phi(t)]$$

which has a frequency Q times higher is thus achieved with the same number of fingers. Similarly, the fundamental signal is obtained with Q times fewer samples, located at times t_n such that

$$\phi(t_n) = Qn\pi \tag{7.98}$$

so that the number of fingers is divided by the sampling rate Q.

Note. Wave front distortion. In the above one-dimensional analysis, the wave emitted by the transducer is assumed to be a plane wave, since the wave front is parallel to the fingers. An investigation of the beam, in the direction orthogonal to the propagation direction, reveals that the wave front is distorted when the finger length varies, especially if the electromechanical coupling constant of the substrate is high (this is the case for $LiNbO_3$). This distortion comes from inhomogeneity of the transducer: the Rayleigh wave velocity at the surface of a piezoelectric medium is affected by the existence of a

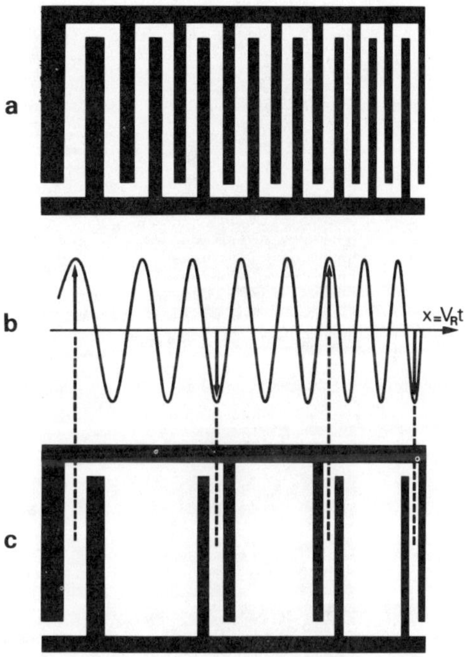

Fig. 7.30. Reduction of the number of transducer fingers by sampling. The desired impulse response (b) is obtained by placing

- either a source at each time t_n, such that $\phi(t_n) = n\pi$ which leads to transducer (a);
- or a source at each time t_n such that $\phi(t_n) = Qn\pi$ (here Q = 5), which leads to transducer (c).

metal layer* (§ 6.2.2.4). If the finger length is not constant, the waves originating from various points of a

* If the material is not piezoelectric, the Rayleigh wave velocity is also affected by the metal mass. This purely mechanical effect, known as "mass loading", is negligible for a thin film of a light metal such as aluminium.

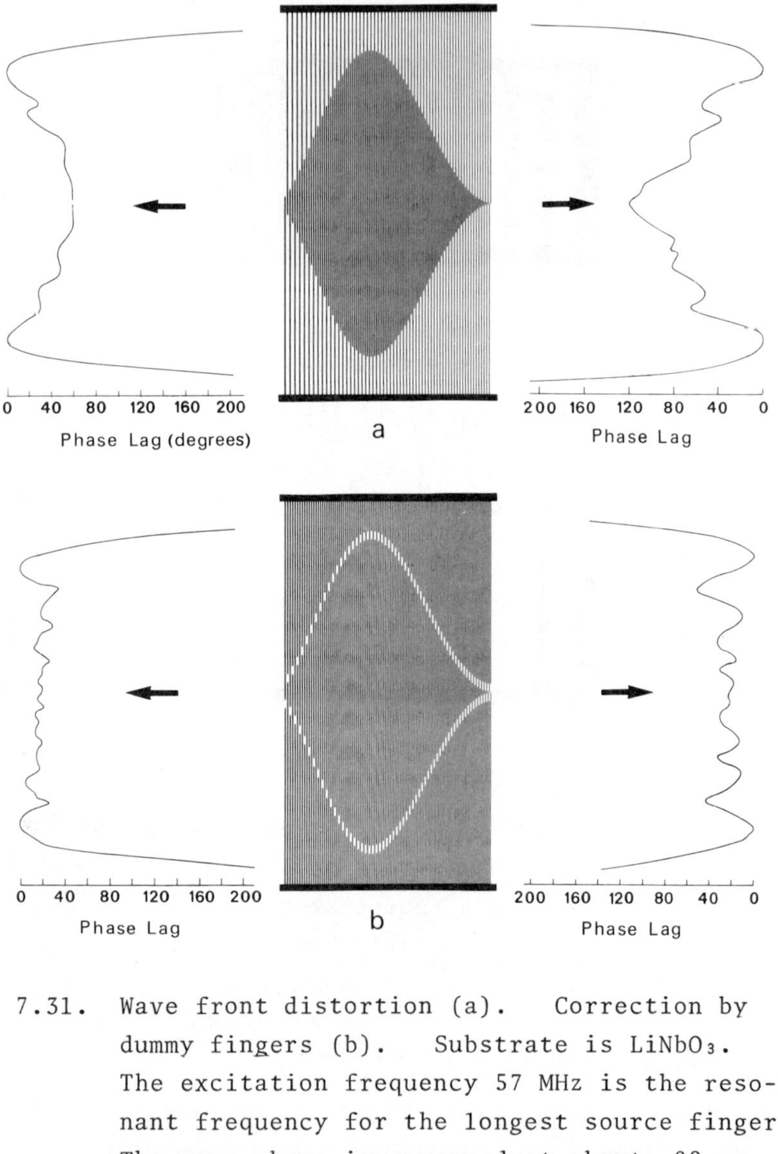

Fig. 7.31. Wave front distortion (a). Correction by dummy fingers (b). Substrate is $LiNbO_3$. The excitation frequency 57 MHz is the resonant frequency for the longest source fingers. The wave phase is measured at about .89 cm from the centre of the transducer (Fig. 2 of reference [18], by courtesy of R.H. TANCRELL).

linear source travel along different lengths of metallized path and undergo different phase shifts (lens effect). An example [18] is given in fig. 7.31a. The experimental measurements have been performed using a capacitative probe, consisting of a tungsten needle in contact with the surface [19]. The phase lag for the waves emitted from the central zone and of maximum acoustic path reaches 120° with respect to those emitted from the sides. The wavefront distortion for waves emitted in the opposite direction is not the same because the transducer is not symmetric. These significant phase shifts, which give rise to unwanted interferences on a straight finger receiver transducer, are very much damped when dummy fingers are added (fig. 7.31b). These fingers are inactive, since they are at the same potential as their neighbours; their role is to make the surface metal distribution more uniform.*

7.2.5. Equivalent Circuit

A two electrode interdigital transducer is equivalent, from the electrical point of view, near the synchronous

---

* Another similar effect exists: the variation of surface impedance (mechanical and electric), due to the metal fingers, gives rise to reflections when an elastic wave arrives under the transducer. The analysis of such effects is far from simple. The higher the coupling constant of the substrate, the more important the effects. They depend on wave frequency as well as on the load of the transducer. Some patterns have been devised to avoid these effects, such as splitting a finger into two parts. The waves reflected on the similar edges of ordinary neighbouring fingers ($\lambda/4$ wide with a $\lambda/4$ spacing), add up since the path difference induces a 2π phase shift. In principle, these reflections cancel when each finger consists of two sub-fingers ($\lambda/8$ with $\lambda/8$ spacing). The distance between the axes of neighbouring fingers remaining equal to $\lambda/2$, the interference is destructive (see "Reflection of a surface wave from three types of ID transducers" by A.J. de Vries, R.L. Miller and T.J. Wojcik, 1972, IEEE Ultrasonics Symposium Proc., p. 353, and "Applications of double electrodes in acoustic surface wave device design" by T.W. Bristol, W.R. Jones, P.B. Snow and W.R. Smith, 1972, IEEE Ultrasonics symposium Proc., p. 343.)

frequency, to a capacitance and a resistance. The resistance splits into two components: the ohmic electrode resistance and the radiation resistance. The capacitance depends on the number of fingers, the spacing and on the dielectric constant of the substrate [11, 20].

In order to calculate the impedance of an interdigital transducer, it is possible [21] to separate it into sections of length 2d; then, to each section is assigned an equivalent circuit, derived from Mason's circuit for a bulk wave transducer (§ 7.1.2). This decomposition is simplified if one electric field component plays a dominant role. If the horizontal component is the major one ("in-line field" approximation), the basic circuit is that of fig. 7.13. If the normal component is the major one ("crossed field" approximation), the basic circuit is deduced from the previous one by short circuiting the capacitor $-C_o$ (exercise 7.2). The inversion of the electric field from one section to the other requires that the association of two basic circuits be a series connection for the mechanical outputs and a shunt connection for the electric inputs.

The expression which is often used for the radiation conductance can be found (except for a numerical factor) from the following argument. The power $\mathcal{P}_s$ emitted by a source in the form of Rayleigh waves is proportional to the frequency f_o and to the mean dielectric energy stored in the substrate between two adjacent fingers:

$$\mathcal{P}_s = \eta C_s V_o^2 f_o$$

where η is a proportionality factor and C_s the capacitance per interdigital spacing. At the synchronous frequency $f_o = V_R/2d$, the (N - 1) sources that constitute the transducer radiate in phase and the overall power is $(N-1)^2\mathcal{P}_s$:

$$\mathcal{P} = (N-1)^2 \eta C_s V_o^2 f_o .$$

The radiation conductance is defined by

$$G_o = \frac{2\mathcal{P}}{V_o^2} = 2(N-1)^2 \eta C_s f_o . \qquad (7.99)$$

If the frequency departs from the synchronism the sources are no longer in phase and the conductance G is equal to G_o times the square of expression 7.87:

$$G = G_o\left(\frac{\sin x}{x}\right)^2 \quad \text{with} \quad x = (N - 1)\,\frac{\pi}{2}\cdot\frac{f - f_o}{f_o} \,. \qquad (7.100)$$

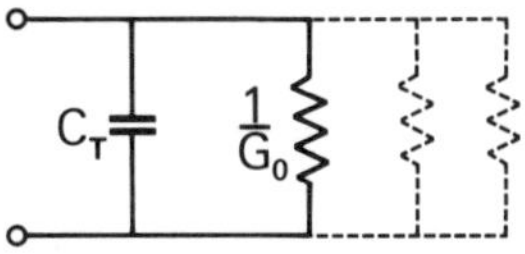

Fig. 7.32. Equivalent circuit of an interdigital transducer.

Equations 7.99 and 7.100 can be identified with those of reference [21] by taking $\eta = 2K_R^2$:

$$G_o = 4(N - 1)K_R^2 C_T f_o \qquad (7.101)$$

where $C_T = (N - 1)C_s$ is the overall transducer capacitance. In fact, since the transducer also launches bulk waves (§ 7.2.3) the electrical equivalent circuit (fig. 7.32) comprises several conductances. In order to ensure generator matching, the capacitance C_T is tuned with an inductance, chosen in such a way that resonance occurs at the synchronous frequency f_o.

7.2.6. Technology

The finger width is generally equal to $\lambda_o/4$. If the frequency is less than 300 MHz, the wavelength is greater than 10 μm for most materials used in practice and the transducers are made by photolithographic techniques, which are well known in micro-electronics. At higher frequencies, the finger width is of the order of optical wavelengths - at 1,000 MHz, $\lambda \simeq 3$ μm - and the effects of light diffraction become prohibitive. One must then resort to electronic engraving (or "masking") to fabricate the fingers. The associated wavelengths for electrons are very small indeed: $\lambda < 1$ Å for 10 keV electrons.

To our knowledge, the surface wave delay lines now used in systems (see chapter 9: applications) have all been implemented using conventional integrated circuit technology [22]. This consists, except for a few variations [23], in the vacuum deposition of an homogeneous and uniform metal layer (for instance a 6,000 Å aluminium layer) on a plane polished substrate (a slab shaped quartz or lithium niobate single crystal, a few centimetres long, one centimetre wide and two millimetres thick). The metal is then coated with a polymer, known as photoresist, spread out by a spinner. The glass plate bearing the transducer pattern is applied to the photoresist layer, then illuminated by ultraviolet light. The resin undergoes photochemical decomposition. The exposed part of the resin, and then the metal which is no longer protected, are removed by chemical etching. The photographic mask comes from a large scale pattern drawn on a coordinatograph ($1 \times 1\ m^2$). If possible, both transducers are engraved simultaneously.

Electron lithography is a laboratory technique [24] which requires an apparatus similar to a scanning microscope and which consists of "insolating" a polymer (electroresist) by means of an electron beam ($\emptyset < 500$ Å). The sequence of operations in the so-called liftoff technique [30] may be as follows [25]: (i) deposition of a thin aluminium film (200 Å) on the insulating substrate in order to prevent the surface from being electrostatically charged; (ii) spreading of the electroresist (4000 Å thick methyl polymetacrylate film); (iii) exposure to the electron beam at 20 keV, which automatically reproduces the transducer shape; (iv) selective chemical etching of the exposed resin; (v) evaporation of a second aluminium layer (1500 Å) in the engraved grooves and on the remaining electroresist; (vi) dissolution of the polymer and its coating of aluminium; the metal fingers, which are going to be the transducer fingers, remain in relief on the very first aluminium film; (vii) ion bombardment of the whole device in order to eliminate this aluminium layer between the fingers: the finger height is diminished by this amount. Transducers operating at frequencies of about 3,500 MHz [27] have been made by this method. Each finger is about 0.15 μm wide. Nevertheless, the exploitation of delay lines with such transducers gives rise to problems, because of the high propagation loss of Rayleigh waves at these frequencies (> 10 dB/μs for $LiNbO_3$). It seems difficult to go beyond centre frequencies of 1,500 MHz. We should mention that a different type of transducer, the grating transducer, has also been employed [28]. It consists of a series of parallel and electrically isolated metal strips, the first and the last being the electrodes. For a given line width, it operates at a frequency which is twice that of the interdigital transducer. It is therefore easier to build; however losses are heavier.

Once the transducers have been fabricated, the delay lines, whatever their operating frequency, are placed in sealed cases since the presence of dust or vapour condensation perturbs Rayleigh wave propagation.

REFERENCES

[1] J. Trotel, E. Dieulesant and B. Autin. Electron. Letters, 4, 156 (1968).
[2] W.P. Mason. Electromechanical transducers and wave filters, pp. 201-399. Princeton N.J.: Van Nostrand (1948).
[3] Reference [20] of chapter 4.
[4] E.K. Sittig, A.W. Warner and H.D. Cook. Ultrasonics, 7, 108 (1969).
[5] N.F. Foster. IEEE Trans. Son. Ultrason. SU-11, 63 (1964).
[6] J. de Klerk and E.F. Kelly. Appl. Phys. Lett., 5, 2 (1964).
[7] N.F. Foster, G.A. Coquin, G.A. Rozgonyi and F.A. Vannata. IEEE Trans. Son. Ultrason. SU-15, 28 (1968).
[8] N.F. Foster. Proc. IEEE, 53, 1400 (1965).

[9] G.A. Rozgonyi and W.J. Polito. Appl. Phys. Lett., 8, 220 (1966).

[10] R.M. White. Proc. IEEE, 58, 1238 (1970).

[11] H. Engan. IEEE Trans. Electron. Devices, ED-16, 1014 (1969).
- ELAB Report TE-91, Electronics Res. Lab., Norwegian Institute of Technology, Trondheim, Norway (1967).

[12] J. de Klerk. Ultrasonics, 9, 35 (1971).

[13] S.G. Joshi and R.M. White. J. Acoust. Soc. Am., 46, 17 (1969).

[14] G.A. Coquin and H.F. Tiersten. J. Acoust. Soc. Am., 41, 921 (1967).

[15] M. Redwood and R.F. Milsom. Electron. Letters, 6, 437 (1970).

[16] R.H. Tancrell and M.G. Holland. Proc. IEEE, 59, 393 (1971).

[17] C. Atzeni and L. Masotti. Acoustic surface waves and acousto-optic devices (T. Kallard Ed.), Vol. 4, p. 69-80. New York: Optosonic Press (1971).

[18] R.H. Tancrell and R.C. Williamson. Appl. Phys. Lett., 19, 456 (1971).

[19] B.A. Richardson and G.S. Kino. Appl. Phys. Lett., 16, 82 (1970).

[20] G.W. Farnell, I.A. Cermak, P. Silvester and S.K. Wong. IEEE Trans. Son. Ultrason. SU-17, 188 (1970).

[21] W.R. Smith, H.M. Gerard, J.H. Collins, T.M. Reeder and H.J. Shaw. IEEE Trans. Microwave Theory Tech. MTT-17, 856 (1969).

[22] R. Lyon-Caen. Circuits logiques intégrés, chap. 2. Paris, Masson et Cie (1968).

[23] H.I. Smith, F.J. Bachner and N. Efremow. J. Electrochemical society, 118, 821 (1971).

[24] A.N. Broers, E.G. Lean and M. Hatzakis. Appl. Phys. Lett., 15, 98 (1969).

[25] P. Hartemann, C. Arnodo and R. Gaudry. J. Physique, C. 6, 33, 266 (1972).

[26] O. Cahen, R. Sigelle, J. Trotel. 141st National Meeting, The Electrochemical Society, Houston, U.S.A. (1972).

[27] E.G. Lean and A.N. Broers. The Microwave Journal, 13, 97 (1970).

[28] A.J. Bahr, R.E. Lee and A.F. Podell. Proc. IEEE, 60, 443 (1972).

[29] A.P. Van den Huevel. Appl. Phys. Lett., 21, 280 (1972).

[30] H.I. Smith. Proc. IEEE, 62, 1361 (1974).

BIBLIOGRAPHY

Bulk waves transducers. Resonators

C.F. Brockelsby, J.S. Palfreeman, R.W. Gibson. Ultrasonic Delay Lines, chap. 3, London, Iliffe L.T.D. (1963).

Y. Kikuchi. Ultrasonic Transducers, Corona Pub. Co. Tokyo (1969).

T.R. Meeker. Thickness mode piezoelectric transducers. Ultrasonics, 10, 26 (1972).

E.K. Sittig. Design and technology of piezoelectric transducers for frequencies above 100 MHz. In Physical Acoustics (W.P. Mason and R.N. Thurston, Eds.) Vol. 9, p. 221-275. New York: Academic Press (1972).

T.M. Reeder and D.K. Winslow. Characteristics of microwave acoustic transducers for volume wave excitation. IEEE Trans. MTT-17, 927 (1969).

H.F. Tiersten. Linear Piezoelectric Plate Vibrations, New York: Plenum Press (1969).

W.P. Mason. Use of Piezoelectric Crystals and Mechanical Resonators in Filters and Oscillators. In Physical Acoustics (W.P. Mason and R.N. Thurston Eds.) Vol. 1 A, p. 335-416. New York: Academic Press (1964).

Interdigital transducers

G.F. Miller and H. Pursey. The field and radiation impedance of mechanical radiators on the free surface of a semi-infinite isotropic solid. Proc. Roy. Soc., A 223, 521 (1954).

R.M. White and F.W. Voltmer. Direct piezoelectric coupling to surface elastic waves, Appl. Phys. Lett., 7, 314 (1965).

C.C. Tseng. Frequency response of an interdigital transducer for excitation of surface elastic waves, IEEE Trans. Electron Devices ED-15, 586 (1968).

J.H. Collins, H.M. Gerard, T.M. Reeder and H.J. Shaw. Unidirectional surface wave transducer, Proc. IEEE (Letters), 57, 833 (1969).

M. Redwood and R.F. Milsom. Piezoelectric coupling coefficient of interdigital Rayleigh-wave transducers, Electron. Lett., 6, 437 (1970).

J. de Klerk. Surface wave transducers, Ultrasonics, 9, 35 (1971).

R.F. Milsom and M. Redwood. Piezoelectric generation of surface waves by interdigital array, Proc. IEEE, 118, 831 (1971).

T. Krairojananan and M. Redwood. Equivalent electrical circuits of interdigital transducers for piezoelectric generation and detection of Rayleigh waves, Proc. IEEE, 118, 305 (1971).

R.V. Schmidt. Excitation of shear elastic waves by an interdigital transducer operated at its surface-wave center frequency, J. Appl. Phys., 43, 2498 (1972).

W.R. Smith, H.M. Gerard and W.R. Jones. Analysis and design of dispersive interdigital surface-wave transducers, IEEE Trans. Microwave Theory Tech., MTT-20, 458 (1972).

F.S. Hickernell and J.W. Brewer. Surface-elastic-wave properties of dc-triode-sputtered zinc oxide films, Appl. Phys. Lett., 21, 389 (1972).

G.S. Kino and R.S. Wagers. Theory of interdigital couplers on non piezoelectric substrates, J. Appl. Phys., 44, 1480 (1973).

A.K. Ganguly and M.O. Vassell. Frequency response of acoustic surface wave filters, J. Appl. Phys., 44, 1072 (1973).

Other types of transducers for Rayleigh waves

R.M. Arzt, E. Salzmann and K. Dransfeld. Elastic surface waves in quartz at 316 MHz, Appl. Phys. Lett., 10, 165 (1967). Grating transducer.

R.F. Humphryes, E.A. Ash. Acoustic bulk-surface-wave transducer, Electron. Lett., 5, 175 (1969).

M.L. Dakss and E.G. Lean. Excitation of surface waves with a grating-coupled Gunn oscillator, Appl. Phys. Lett., 18, 137 (1971).

L.T. Claiborne, E.J. Staples, J.L. Harris. MOSFET ultrasonic surface wave detectors for programmable matched filters, Appl. Phys. Lett., 19, 58 (1971).

M. Bruun, S. Ludvik and C.F. Quate. Field effect transistors on epitaxial GaAs as transducers for acoustic surface waves, Appl. Phys. Lett., 18, 118 (1971).

E.W. Greeneich and R.S. Muller. Acoustic-wave detection via a piezoelectric field-effect transducer, Appl. Phys. Lett., 20, 156 (1972).

EXERCISES

7.1. Calculate the elastic impedance at x = 0 of the line located on the side of positive x (fig. 7.2). Consider the case $d_1 = \lambda_M/2$, $d_1 = \lambda_M/4$.

Solution. For $0 < x < d_1$:

$$u = A\, e^{i(\omega t-kx)} + B\, e^{i(\omega t+kx)} \quad \text{and} \quad T = c_M \frac{\partial u}{\partial x}$$

The elastic impedance

$$Z_T = \frac{-T}{\dot{u}} = Z_M \frac{A\, e^{-ikx} - B\, e^{ikx}}{A\, e^{-ikx} + B\, e^{ikx}}$$

is equal to Z for $x = d_1$:

$$B = A \frac{Z_M - Z}{Z_M + Z} e^{-2ikd_1}$$

Hence

$$Z_T(x = 0) = Z_M \frac{Z + iZ_M \tan kd_1}{Z_M + iZ \tan kd_1}$$

- For $d_1 = \lambda_M/2$ $(kd_1 = \pi)$: $Z_T(0) = Z$.

- For $d_1 = \lambda_M/4$ $(kd_1 = \pi/2)$: $Z_T(0) = Z_M^2/Z$.

7.2. Consider an infinite piezoelectric bar, of width L and thickness $\ell \ll L$, cut and metallized so that the electric field launches a longitudinal wave along the x_3-axis (fig. 7.33). What is the velocity of the wave? By a procedure similar to that of § 7.1.2.c, find the equivalent circuit for a slab of length d.

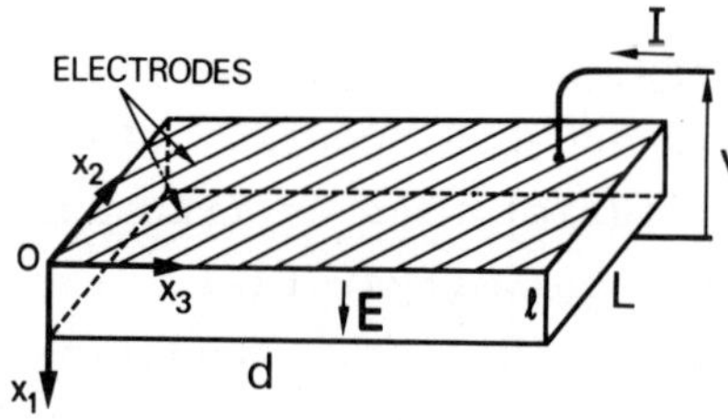

Fig. 7.33.

Solution. The strain components $(S_{33} = \partial u_3/\partial x_3)$ and the electric field $\vec{E}(E_1, 0, 0)$ yield

$$T_{33} = c_{33}^E \frac{\partial u_3}{\partial x_3} - e_{13}E_1 \qquad (7.102)$$

$$D_1 = \varepsilon_{11}^S E_1 + e_{13}\frac{\partial u_3}{\partial x_3} \qquad (7.103)$$

and the equation of propagation

$$\rho\frac{\partial^2 u_3}{\partial t^2} = \frac{\partial T_{33}}{\partial x_3} = c_{33}^E\frac{\partial^2 u_3}{\partial x_3^2} - e_{13}\frac{\partial E_1}{\partial x_3}$$

shows that the wave velocity is $V = \sqrt{c_{33}^E/\rho}$, since E_1 does not depend on x_3 (planes $x_1 = 0$ and $x_1 = \ell$ are equipotential surfaces). We shall hereafter use the simplified notation:

$x_3 = x$, $u_3 = u$, $E_1 = E$, $D_1 = D$, $c_{33}^E = c$,

$$\varepsilon_{11}^S = \varepsilon,\ e_{13} = e = h\varepsilon,\ T_{33} = T = -\frac{F}{\ell L}\ .$$

Equation 7.102 shows that the forces on sides $x = 0$ and $x = d$ are obtained by adding the constant term

$$eE/L = hmC_o v \qquad (7.104)$$

to equations 7.28, with $m = \ell/d$, $C_o = \varepsilon Ld/\ell$ (capacitance), $v = E\ell$ (voltage).

The current is given by $I = dQ/dt = i\omega Q$ where Q is the charge of the electrode $x_1 = 0$:

$$Q = \int D\, ds = L \int_0^d D\, dx = Ld\varepsilon E - Le(u_1 - u_2).$$

Introducing the velocities $\dot{u}_1 = i\omega u_1$ (at $x = 0$) and $\dot{u}_2 = i\omega u_2$ (at $x = d$):

$$I = i\omega C_o v - hmC_o(\dot{u}_1 - \dot{u}_2) \qquad (7.105)$$

and the equivalent circuit, as derived from 7.104 and 7.105, no longer includes the capacitance $-C_o$, the transformer ratio is $N = hmC_o$.

7.3. How does the relative separation $\varepsilon_n = [f_a^{(n)} - f_r^{(n)}]/f_a^{(n)}$ between the resonant and antiresonant frequencies depend on the order $n = 2p + 1$?

Solution. As $f_a^{(n)} = nf_a$: $f_r^{(n)} = nf_a(1 - \varepsilon_n)$ and the left hand side of 7.33 becomes

$$K^2 \tan \frac{\pi}{2} \frac{f_r^{(n)}}{f_a} = \frac{K^2}{\tan(\frac{\pi}{2} n\varepsilon_n)} \simeq \frac{K^2}{\frac{\pi}{2} n\varepsilon_n} \quad \text{for } \varepsilon_n << 1$$

Hence

$$\varepsilon_n \simeq \frac{4K^2}{\pi^2 n^2}$$

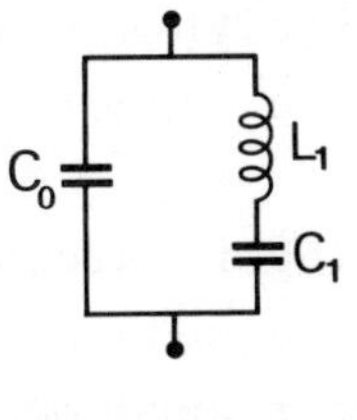

Fig. 7.34.

7.4. Show that the equivalent circuit for a bar shaped resonator (exercise 7.2) can be equated with the circuit of Fig. 7.34 in the vicinity of resonance. Calculate the values of L_1 and C_1.

Solution. The impedance of the transformer secondary circuit: $\mathcal{Z}_2 = -i\mathcal{Z}/2 \tan(\omega d/2V)$ is equivalent (near resonance $\omega_r = \pi V/d$) to

$$\mathcal{Z}_2 = -\frac{i\mathcal{Z}}{2\tan\frac{\pi}{2}(1+\varepsilon)} \simeq \frac{i\pi\mathcal{Z}}{4}\varepsilon \quad \text{where } \varepsilon = \frac{\omega - \omega_r}{\omega_r}$$

This can be equated to the impedance of a $\mathcal{L}\mathcal{C}$ circuit around the resonant frequency $\omega_o = \omega_r$:

$$\mathcal{Z}' = i\left(\omega\mathcal{L} - \frac{1}{\mathcal{C}\omega}\right) \simeq 2i\mathcal{L}\omega_r\varepsilon$$

where

$$\mathcal{L} = \frac{\pi\mathcal{Z}}{8\omega_r} \quad \text{and} \quad \mathcal{C} = \frac{8}{\pi\mathcal{Z}\omega_r}.$$

Now, in the primary circuit of the transformer (the ratio being $N = hmC_o = eL$), the capacitance is

$$C_1 = N^2\mathcal{C} = \frac{8e^2L^2d}{\pi^2\rho V^2\ell L} = \frac{8}{\pi^2}\frac{e^2}{\varepsilon^S c_E}\frac{\varepsilon^S Ld}{\ell} = \frac{8}{\pi^2}\frac{K^2}{1-K^2}C_o$$

$$= \frac{8}{\pi^2}K^2C_o^T$$

where $C_o^T = \varepsilon^T/\varepsilon^S\, C_o$ is the capacitance at constant (zero) mechanical stress; and

$$L_1 = \frac{\mathcal{L}}{N^2} = \frac{\rho\ell Ld}{8N^2} = \frac{\text{mass}}{8N^2}.$$

Orders of magnitude: $\ell = .5$ mm, $L = 4$ mm, $d = 10$ mm, $\varepsilon^T = 4 \times 10^{-11}$ F/m, $e = .1$ C/m^2, $\rho = 3 \times 10^3$ kg m^{-3}, $K^2 = 10^{-3} \Rightarrow C_o = 3.2$ pF, $C_1 = 2.6 \times 10^{-3}$ pF, $L_1 = 47$ H.

7.5. Assuming that Rayleigh waves are solely and uniformly generated under the fingers, draw the impulse response

(i) - of a three-finger interdigital transducer;

(ii) - of a delay line made of a three finger input and a four finger output transducer.

Solution. (i) Fig. 7.35a.

(ii) correlation of a and b $\Rightarrow$ c.

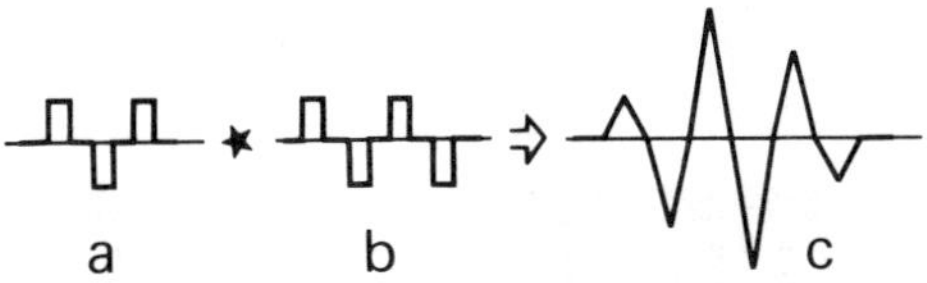

Fig. 7.35.

7.6. Two identical transducers T_1 and T_2 are separated by a distance ℓ; they emit waves at frequency f_o in both directions along the same axis. How can ℓ and the phase shift ϕ between the voltages applied to T_1 and T_2 be chosen for the pair T_1 and T_2 to emit in only one direction?

<u>Solution</u>. In one direction, the electric (ϕ) and spatial (ψ) phase shifts add up. In the other direction, they subtract from each other. The transducer pair is unidirectional if

$$\phi + \psi = \pi \quad \text{and} \quad \phi - \psi = 0 \Rightarrow \phi = \psi = \frac{\pi}{2} \Rightarrow$$

$$\ell = \frac{\lambda_o}{4} = \frac{V}{4f_o} .$$

7.7. A delay line consists of an interdigital input transducer with N slanted fingers (N >> 1) and an upright output transducer (fig. 7.36). What is the frequency response of the line if the source finger inclination θ_n varies linearly with $x_n = n\lambda_o/2$, ranging from $-\theta_o$ to $+\theta_o$ (θ_o is small) (after reference [29])?

<u>Solution</u>. Each source is decomposed into elements dy; the signal from the n-th source is then

$$A_n = A_o \, e^{-i\,\omega\ell/V} \int_{-b/2}^{b/2} e^{i\,\omega x/V} \frac{dy}{b}$$

where $x = x_n + \theta_n y$ and A_o is the amplitude of the wave emitted by an upright source of length b:

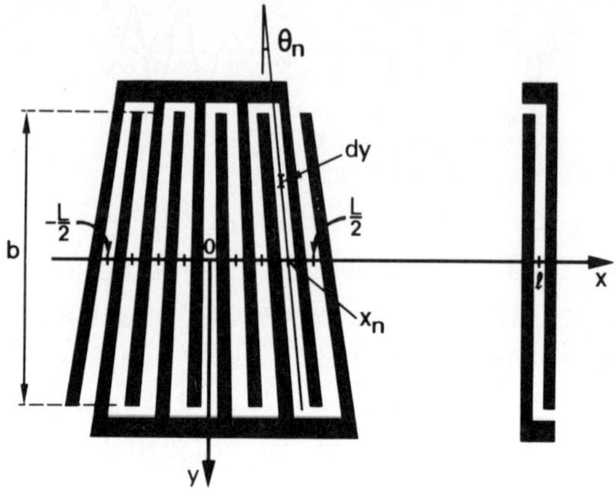

Fig. 7.36.

$$A_n = A_o\, e^{-(i\omega/V)(\ell - x_n)}\, \frac{\sin(\frac{b\theta_n}{2V}\,\omega)}{\frac{b\theta_n}{2V}\,\omega}$$

For large values of N, the source distribution is nearly continuous and the frequency response $H(\omega) = \sum_n (-1)^n A_n = \sum_n A_n \cos(2\pi f_o x_n/V)$ is calculated from the integral

$$H(\omega) = \int_{-L/2}^{+L/2} A(x) \cos\left(\frac{2\pi f_o}{V}\, x\right) \frac{dx}{L} \quad \text{with}$$

$$A(x) = A_o\, e^{i\,\omega x/V}\, \frac{\sin(\frac{b\theta(x)\omega}{2V})}{\frac{b\theta(x)\omega}{2V}}$$

(omitting the average delay $\tau = \ell/V$). For a linear variation $\theta = 2\theta_o x/L$:

$$H(\omega) = \int_{-L/2}^{+L/2} \frac{A_o}{L}\, \frac{\sin(\frac{b\theta_o\omega}{LV}\, x)}{\frac{b\theta_o\omega}{LV}\, x} \cos\left(\frac{2\pi f_o}{V}\, x\right) e^{i\,\omega x/V}\, dx.$$

The integration can be extended to $\pm\, \infty$ ($N \gg 1$) and the frequency is considered to be constant within the factor $(\sin\alpha)/\alpha$ (narrow bandwidth:

$\omega \simeq \omega_o$). So the frequency response is the Fourier transform of a $(\sin \alpha)/\alpha$ signal with a carrier:

$$H(\omega) = \frac{A_o V}{L} \int_{-\infty}^{+\infty} \frac{\sin \pi B t}{\pi B t} \cos 2\pi f_o t \; e^{i2\pi f t} \, dt$$

with $t = \frac{x}{V}$ and $B = \frac{2b\theta_o f_o}{L}$.

$H(\omega)$ is thus a square signal, of centre frequency $\pm f_o$ (§ 1.3.2):

$$H(\omega) = \frac{A_o \lambda_o}{4b\theta_o} [\Pi(\frac{f - f_o}{B}) + \Pi(\frac{f + f_o}{B})]$$

where the fractional bandwidth is

$$\frac{B}{f_o} = \frac{2b\theta_o}{L} = 4 \frac{b\theta_o}{(N - 1)\lambda_o} .$$

7.8. For a simple transducer, matched with an inductance, what is the number of fingers N_o for which the fractional bandwidth is maximum? Apply to $LiNbO_3$ (YZ), SiO_2 (YX), using values from table 6.18.

Solution. From equation 7.37, the Q-factor due to the cumulative effect is $Q_a = (N-1)/1.77$. The electrical Q-factor arising from the matching is (equation 7.101):

$$Q_e = \frac{C_T \omega_o}{G_o} = \frac{\pi}{2(N - 1)K_R^2} .$$

For large values of N, $\Delta f/f_o = 1/Q_a$, for small values of N, $\Delta f/f_o = 1/Q_e$; $\Delta f/f_o$ is maximum when $Q_a = Q_e \Rightarrow N_o - 1 = 1.67/K_R$ and $(\Delta f/f_o)\max = 1.06\ K_R$.

$LiNbO_3$: $N_o = 9$, $(\Delta f/f_o)\max = .23$

SiO_2 : $N_o = 37$, $(\Delta f/f_o)\max = .05$.

Chapter 8

Interaction Between Light Waves and Elastic Waves

The interaction between elastic waves and light was investigated by Léon Brillouin [1] as early as 1922. The first experiments were performed in 1932 by René Lucas and Pierre Biquard [2] in France and by P. Debye and F.W. Sears [3] in the U.S.A.

Evidence for the diffraction of light by ultrasonic waves was provided by these experiments; since then this effect has been used mostly for measurements of elastic wave propagation velocity in various materials. The upper limit for the ultrasonic frequency was of the order of thirty megahertz.

In the last few years, there has been a new interest in the subject, firstly because lasers are now readily available and are adequate sources for light beams of well defined shape and high energy density and secondly because of the technological improvement in transducers, which convert electric energy into elastic energy in the gigahertz domain (§ 7.1) and which were primarily designed for delay lines.

The appearance of these new elements (lasers, high frequency transducers) provides access, as far as acousto-optic interactions are concerned, to a wide and interesting field of applications, linked to the possibility of rapidly and significantly modifying the intensity, the direction, and even the frequency of a light beam. This diffraction of light by elastic waves also remains a valuable tool for measuring the characteristic quantities (homogeneity, attenuation, radiation pattern) of an elastic wave beam propagating in a solid; it then plays the role of a probe which penetrates into the solid.

After a review of the conditions for the propagation of light in a crystal, this chapter aims to define the acousto-optic tensor, and (without any extensive theoretical discussion) to describe the major results in light wave-elastic wave interaction. First, however it seems useful to sketch the various circumstances under which the interaction, known as the Brillouin effect, Debye-Sears effect, Lucas-Biquard effect, Raman-Nath effect, or Bragg effect, occurs.

8.1. MAJOR CASES OF INTERACTION

Three major cases should be distinguished:

- The first is depicted in fig. 8.1; it refers to a light beam of diameter less than the elastic wavelength. The slow variation of the optical refractive index only bends the light rays (mirage effect).

- The second is shown in fig. 8.2. The width of the light beam is large compared to the elastic wavelength. The periodic index variation generates beams of different intensities and directions. This can be interpreted by assuming that, because of the light ray curvature, several foci are formed and act as sources [2]. The interference effects between waves emitted from various sources account for the division into several distinct beams. The phenomenon can also be analyzed by looking at the phase shifts of the light wave due to the progressive index variation wave [4]. We shall adopt this latter point of view. The reader interested in more general theories, based directly on Maxwell's equations, may refer to the papers in the bibliography.

- In the third case (fig. 8.3) the angle between the incident light beam and the elastic beam together with the elastic beam width are chosen in such a way that a single light beam emerges; the other beams cancel due to interference effects.

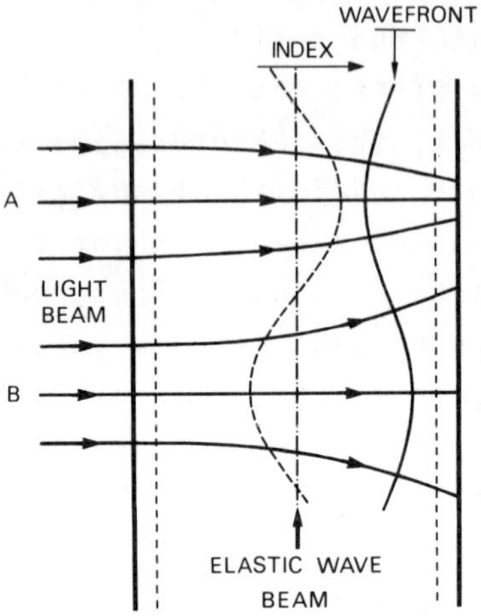

Fig. 8.1. Light ray curvature due to an elastic wave. Rays of the light beam A, propagating in a region of increased index, converge whereas rays of the light beam B, propagating in a region of decreased index, diverge.

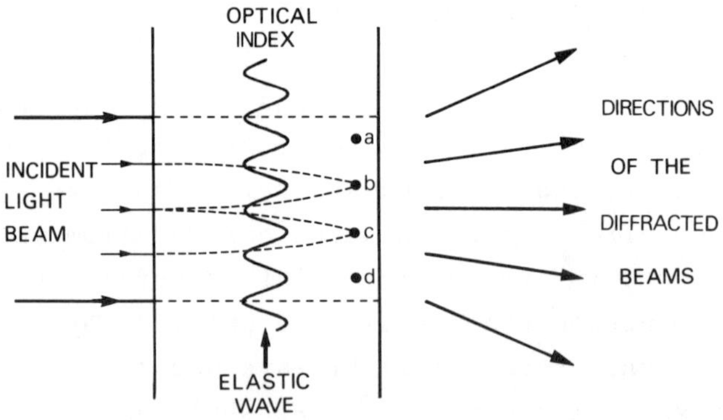

Fig. 8.2. Grating effect. When the light beam width is large compared to the elastic wavelength, several sources a, b,... are formed by light ray curvature [2]. Interference between waves emitted from these sources account for the overall emission along distinct directions. This is also interpreted by considering the light wave phase shift produced by the progressive index variation wave (§ 8.4).

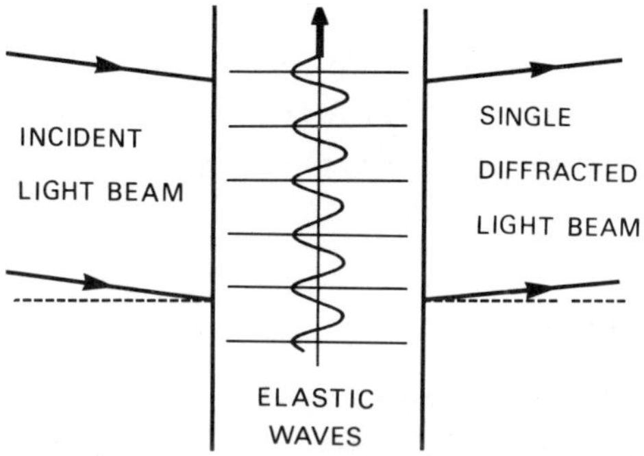

Fig. 8.3. Bragg effect. For the correct inclination of the incident beam, and provided the light rays travel across several elastic wave planes, it is possible to obtain a single diffracted beam, the others (in fig. 8.2) being destroyed by interference.

Generally, the term "Brillouin effect" refers to any coupling between electromagnetic waves and elastic (coherent or incoherent) waves. Depending upon the country in question, one of the three pairs (Debye-Sears, Lucas-Biquard, Raman-Nath) is referred to to describe the second case. The third case is often called "Bragg effect" because it reminds one of X-ray diffraction by lattice planes.

8.2. LIGHT WAVE PROPAGATION IN CRYSTALS

The transmission of light through crystals at rest involves such phenomena as double refraction, which are related to the natural crystal anisotropy. For example a calcite slab ($CaCO_3$) splits an incident light beam into two beams polarized in two orthogonal directions. The anisotropy, which causes this double refraction, may be enhanced, or even created in isotropic media or in cubic system crystals where it does not occur spontaneously, by applying external "forces" such as an electric field (electro-optic effect) or mechanical strain (elasto-optic or acousto-optic effect, which we are presently studying).

In the case of light waves the behaviour of crystals is usefully interpreted with the help of the index ellipsoid - or of the index surface - to which we are led by Maxwell's equations (the equations for electromagnetic wave propagation). We summarize next the main points to be borne in mind.

8.2.1. Index Ellipsoid

In an isotropic medium, whatever the propagation direction and its polarization, a light wave propagates with velocity v:

$$v = \frac{1}{\sqrt{\varepsilon\mu}} = \frac{1}{\sqrt{\varepsilon_0\mu_0}} \cdot \frac{1}{\sqrt{\varepsilon_r\mu_r}} = \frac{v_0}{\sqrt{\varepsilon_r\mu_r}} = \frac{v_0}{n}$$

where ε_0, μ_0 is the dielectric and magnetic constants in a vacuum and ε, μ, $\varepsilon_r = \varepsilon/\varepsilon_0$, $\mu_r = \mu/\mu_0$ are the absolute and relative constants of the medium, $n = \sqrt{\varepsilon_r\mu_r}$ is the refractive index of the medium. For a non magnetic material, $\mu_r = 1 \Rightarrow n = \sqrt{\varepsilon_r}$. In this chapter we shall only use the relative dielectric constant. In order to avoid complicated notations, we shall omit the index r. The index ellipsoid, with equation

$$\frac{x_1^2 + x_2^2 + x_3^2}{\varepsilon} = \frac{x_i^2}{n^2} = 1 \quad \text{or} \quad Bx_i^2 = 1 \quad \text{where} \quad B = \frac{1}{\varepsilon}$$

is a sphere.

The dielectric properties of an anisotropic medium are described by the symmetric tensor ε_{ij} or by the inverse tensor B_{ij}, called the impermittivity tensor

$$\varepsilon_{ij}B_{jk} = \delta_{ik} \tag{8.1}$$

to which there corresponds an index ellipsoid of equation

$$B_{ij}x_ix_j = 1 \tag{8.2}$$

Taking the equation $D_i = \varepsilon_0\varepsilon_{ij}E_j$ into account, the Maxwell equations are solved for and, for a given direction, lead to the possible occurrence of two plane waves, polarized in two orthogonal directions, and propagating with different velocities. The polarizations and velocities of these two waves can be derived from the index ellipsoid:

The directions of the electric displacement vectors of the two waves $\vec{D}^{(1)}$ and $\vec{D}^{(2)}$ are the axes of the ellipse obtained from the cross section of the index ellipsoid, for that crystal, with the equatorial plane P which is perpendicular to the wave vector $\vec{K}$ (fig. 8.4). The indices, which correspond to the velocities, are equal to the lengths of the semi-axes of the ellipse.

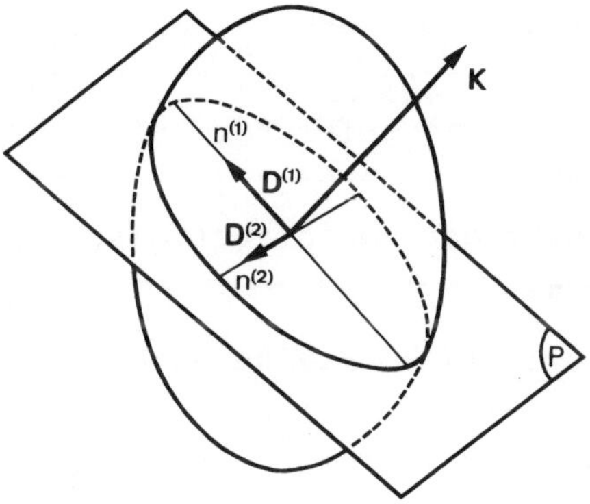

Fig. 8.4. Index ellipsoid. Two waves, polarized along $\vec{D}^{(1)}$ and $\vec{D}^{(2)}$ (mutually orthogonal), can propagate in the $\vec{K}$ direction, with velocities $v_1 = v_o/n^{(1)}$ and $v_2 = v_o/n^{(2)}$.

The index ellipsoid obviously possesses the symmetry of the crystal whose optical properties are described. Therefore, crystals should be separated into three groups. This result is also deduced from analysis of the symmetric tensor B_{ij}, after its components have been reduced by crystal symmetry (§ 3.4.3).

Biaxial crystals. These have no axis of order larger than 2, and therefore belong to the triclinic, monoclinic or orthorhombic systems. The three principal axes of an ordinary ellipsoid are diad axes. If the crystal has a diad axis (monoclinic system) or three diad axes (orthorhombic system), they must coincide with one of the axes or with the three axes of the ellipsoid. For triclinic crystals, there is no particular requirement. In the principal axis reference frame, the equation of the index ellipsoid is

$$\frac{x_1^2}{n_1^2} + \frac{x_2^2}{n_2^2} + \frac{x_3^2}{n_3^2} = 1 \tag{8.3}$$

where n_1, n_2, n_3* - the semi-axis lengths - are the principal indices. An ordinary ellipsoid has two circular equatorial cross sections which are symmetric with respect to two of the principal axes and which contain the third axis. The directions orthogonal to these particular cross sections are the optical axes. Only one index remains (the radius of the circle) and no special polarization is enforced; along these directions, all waves have the same velocity, i.e. there is no double refraction. For these two directions, crystals behave as isotropic materials.

Uniaxial crystals. Having an A_n axis of symmetry with $n > 2$, these belong to the trigonal, tetragonal, or hexagonal system. One of the principal axes coincides with A_n and is an axis of revolution of the index ellipsoid. The plane wave which is orthogonal to A_n describes a circular cross section, thus the polarization is not specified and the index (the radius of the circle) is unique, so that the crystal behaves as an isotropic material for waves propagating along A_n, which is the optical axis. The index ellipsoid of uniaxial crystals is therefore determined by two numbers, n_o and n_e, where n_o is the ordinary index equal to the radius of the circular cross section, and n_e is the extraordinary index equal to the semi-axis in the A_n direction. The equation for the index ellipsoid is

$$\frac{x_1^2 + x_2^2}{n_o^2} + \frac{x_3^2}{n_e^2} = 1 \tag{8.4}$$

---

* In the preceding chapters, K, n_1, n_2, n_3 represented the coupling constant, and components of the propagation direction. As these quantities are no longer used in this chapter, no confusion is possible.

The ellipsoid is prolate (along the optical axis) if $n_e > n_o$; the crystal is positive uniaxial (quartz). If $n_e < n_o$, it is oblate, the crystal is negative uniaxial. The aim of exercise 8.1 is to show that the index of one vibration does not depend on the propagation direction (n_o), the index of the other ranges from n_o to n_e.

Optically isotropic crystals. These have several axes of order greater than 2 and therefore belong to the cubic system ($4A_3$). The index ellipsoid must have several axes of revolution, thus it is a sphere, defined by a single index. In these crystals, any light wave propagates as in glass or in a liquid, with the same velocity and maintaining the original polarization, whatever the direction.

We must add two comments:

- The range between the principal indices is always small, whatever the crystal, unlike what is currently sketched in figures (for the sake of clarity). The index ellipsoid is very close to a sphere. For quartz, $n_o = 1.5442$ and $n_e = 1.5533$.

- Dielectric constants are frequency dependent; every fixed frequency domain corresponds to one specific index ellipsoid. In a piezoelectric material, the dielectric constants which determine the index ellipsoid are not the same as the constants involved in the expression for the elastic wave velocities.

8.2.2. The Index Surface

It follows from the above discussion that the index ellipsoid is obtained by making the radius proportional to the optical index along every direction of vibration. Another useful surface is the index surface: a length proportional to the index is drawn along the wave vector direction $\vec{K}$. Since the index is proportional to the reciprocal of the velocity, this characteristic surface is quite similar to the slowness surface in the theory of elasticity. It has two sheets, since two orthogonal vibrations can propagate along any given direction. The shape of the index surface is shown in fig. 8.5 for a positive uniaxial crystal.

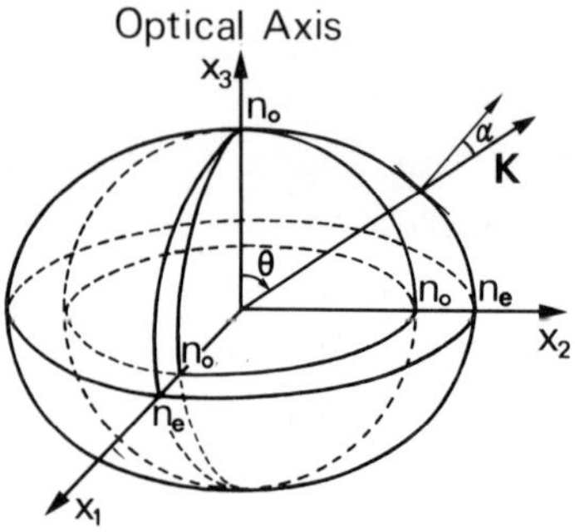

Fig. 8.5. Index surface of a positive uniaxial crystal ($n_e > n_o$). It is constructed by drawing the two lengths $n^{(1)} = n_o = C^{te}$ and $n^{(2)} = n_e$ on the wave vector $\vec{K}$. The extraordinary ray makes an angle α with $\vec{K}$, which vanishes only for $\theta = 0$ and $\theta = \pi/2$.

One sheet is a sphere since one of the vibrations - the ordinary one - always propagates at velocity $v_{or} = v_o/n_o$, whatever the direction. The other sheet is an ellipsoid, tangential to the sphere on the optical axis, with semi-axis n_e. The energy propagation direction is orthogonal to the index surface, at an angle α with the wave vector of the elliptic sheet; of course α depends on the propagation direction and vanishes only for the principal axes. The major aim of the index surface is to provide a simple construction of the refracted wave when the incident wave is known (see the wave vector diagram in § 8.4.2.5).

8.3. THE ACOUSTO-OPTIC TENSOR

Thus the optical properties of a crystal at rest are described by the index ellipsoid from the equation

$$B_{ij}x_ix_j = 1.$$

A strain in the crystal will generate a change in the index ellipsoid and, accordingly, a variation ΔB_{ij} of the tensor B_{ij}. Two major cases must be discussed. In the first case, the medium undergoes a strain which does not imply any rotation of the index ellipsoid other than a rigid rotation. However, in the second case, the strain causes a local rotation of the index ellipsoid, this rotation varying from one point to the other. In addition, if the crystal is piezoelectric, another term stemming from the electro-optic interaction has to be taken into account.

8.3.1. Pockels' Theory

Pockels' theory is valid provided the following requirements are satisfied: either the crystal is optically isotropic and non piezoelectric (cubic classes m3m, m3, 432) or the strain is homogeneous and static or due to a longitudinal wave. Then the variation ΔB_{ij} can be directly derived from the strain $S_{k\ell}$ by the equation

$$\boxed{\Delta B_{ij} = p_{ijk\ell}S_{k\ell}} \qquad (8.5)$$

The dimensionless quantities $p_{ijk\ell}$ are the components of an acousto-optic tensor, of rank 4.

The variation ΔB_{ij} can also be expressed in terms of the stress tensor $T_{k\ell}$:

$$\Delta B_{ij} = \pi_{ijk\ell}T_{k\ell} \qquad (8.6)$$

The tensor $\pi_{ijk\ell}$ is the piezo-optic tensor. Taking Hooke's law

$$T_{k\ell} = c_{k\ell mn}S_{mn}$$

into account, one obtains

$$p_{ijmn} = \pi_{ijk\ell}c_{k\ell mn}. \qquad (8.7)$$

Since the tensors B_{ij} and $S_{k\ell}$ are symmetric, the subscripts of the acousto-optic tensor components can be contracted, by denoting

$$B_{ij} = B_\alpha$$

and, as in § 4.3

$$S_\beta = \begin{cases} S_{k\ell} & \text{if } \beta \le 3 \\ 2S_{k\ell} & \text{if } \beta > 3 \end{cases}$$

the equation 8.5 reads, in matrix notation

$$\Delta B_\alpha = p_{\alpha\beta} S_\beta \tag{8.8}$$

with

$$p_{\alpha\beta} = p_{ijk\ell} \qquad \forall\ \alpha,\ \beta = 1,\ 2,\ \ldots,\ 6.$$

However, $p_{\alpha\beta}$ is generally different from $p_{\beta\alpha}$, in contrast to the elastic constants for which the existence of a thermodynamic potential requires the α-β symmetry (§ 4.4). In the triclinic system, the acousto-optic tensor has thus 36 independent components. Because of crystal symmetry, the number of components can be reduced by the same method as used in § 4.5. The results are displayed in fig. 8.6 and are also valid for the piezo-optic tensor $\pi_{ijk\ell}$. In order to obtain the relationship between the components $\pi_{\alpha\beta}$, as defined by $\Delta B_\alpha = \pi_{\alpha\beta} T_\beta$, it is necessary to recall

$$\pi_{\alpha\beta} = \pi_{ijk\ell} \text{ if } \beta \le 3 \quad \text{and} \quad \pi_{\alpha\beta} = 2\pi_{ijk\ell} \text{ if } \beta > 3. \tag{8.9}$$

The acousto-optic coefficients for commonly used solids show little variation, ranging from 0.2 to 0.3, as indicated in table 8.7. Therefore, with $S \simeq 10^{-5}$, the variations ΔB_{ij} are small, of the order of 10^{-6}.

It may be useful to derive the variations ΔB_{jk} and $\Delta\varepsilon_{i\ell}$ of the tensors B_{ij} and $\varepsilon_{i\ell}$. Since such variations are small, they are obtained by differentiating the equation (8.1) defining the tensor B_{jk}:

$$\Delta\varepsilon_{ij} B_{jk} + \varepsilon_{ij} \Delta B_{jk} = 0.$$

After a contraction with $\varepsilon_{k\ell}$, and using $B_{jk}\varepsilon_{k\ell} = \delta_{j\ell}$:

$$\Delta\varepsilon_{i\ell} = -\ \varepsilon_{ij} \Delta B_{jk} \varepsilon_{k\ell}$$

or, in terms of the strain tensor

$$\boxed{\Delta\varepsilon_{i\ell} = -\ \varepsilon_{ij} p_{jkmn} \varepsilon_{k\ell} S_{mn}} \tag{8.10}$$

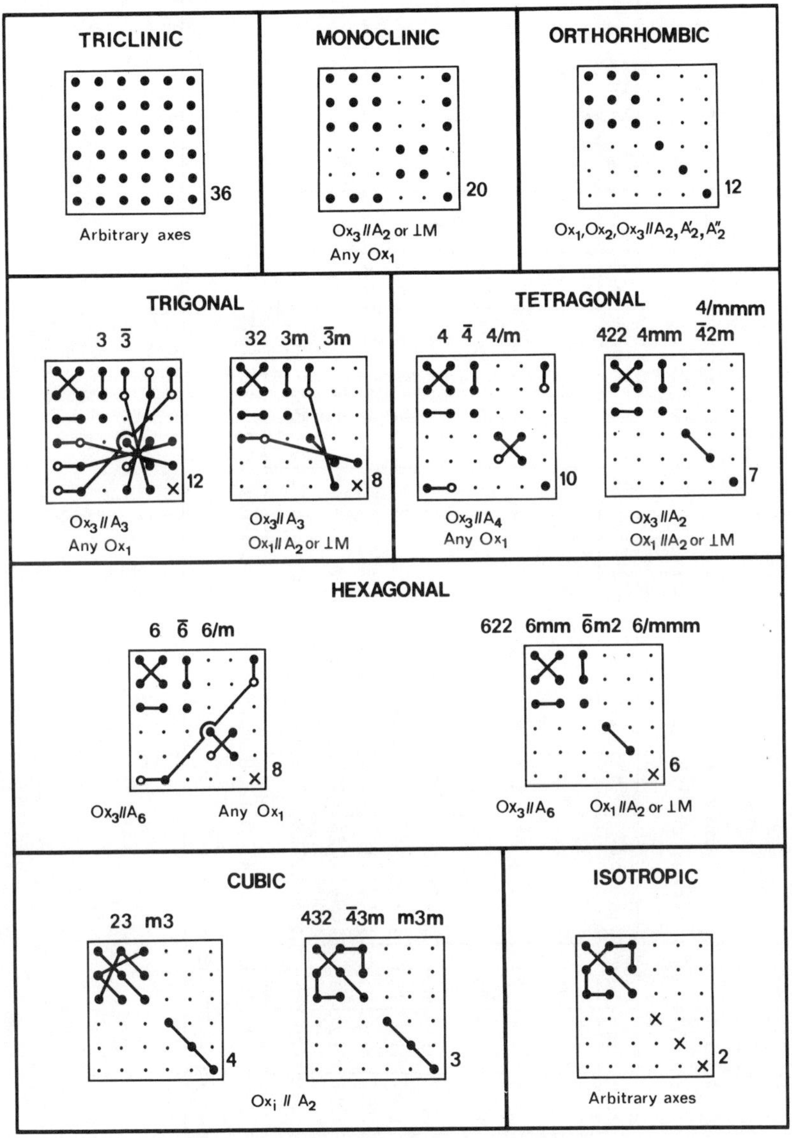

Fig. 8.6. Table for the acousto-optic matrices $p_{\alpha\beta}$ (Pockels' theory).

● ○ non zero component

●—● equal components

• zero component

●—○ opposite components

x component equal to $(p_{11} - p_{12})/2$.

Table 8.7. Acousto-optic constants of some materials, measured by diffraction of a He-Ne laser (0.6328 μm), except for the case of quartz (Λ_o = 0.589 μm).

Materials	p_{11}	p_{12}	p_{21}	p_{22}	p_{13}	p_{31}	p_{33}	p_{23}	p_{32}	p_{14}	p_{41}	p_{44}	p_{45}	p_{55}	p_{16}	p_{61}	p_{66}	Ref.
Silica (isotropic)	.121	.270	p_{12}	p_{11}	p_{12}	p_{12}	p_{11}	p_{12}	p_{12}	0	0	[-.075]	0	p_{44}	0	0	p_{44}	[5]
α-HIO_3(222)	.406	.277	.279	.343	.304	.503	.334	.305	.310	0	0	-	0	-	0	0	.092	[6]
$PbMoO_4$(4/m)	.24	.24	p_{12}	p_{11}	.255	.175	.300	p_{13}	p_{31}	0	0	.067	-.01	p_{44}	.017	.013	.05	[14] chap.4
TiO_2(4/mmm)	.011	.172	p_{12}	p_{11}	.168	.096	.058	p_{13}	p_{31}	0	0	-	0	p_{44}	0	0	-	[7]
TeO_2(422)	.007	.187	p_{12}	p_{11}	.340	.090	.240	p_{13}	p_{31}	0	0	-.17	0	p_{44}	0	0	-.046	[8]
$LiNbO_3$(3m)	.036	.072	p_{12}	p_{11}	.092	.178	.088	p_{13}	p_{31}	.07 (a)	.155	-	0	p_{44}	0	0	[-.018]	[7]
$LiTaO_3$(3m)	.080	.080	p_{12}	p_{11}	.094	.086	.150	p_{13}	p_{31}	.031	.024	.022	0	p_{44}	0	0	[.00]	[7]
α-Quartz (32)	.138	.250	p_{12}	p_{11}	.259	.258	.098	p_{13}	p_{31}	-.029	-.042	-.068	0	p_{44}	0	0	[-.056]	[9]

(a) Value determined by J. Reintjes and M.B. Schulz. J. Appl. Phys., 39, 5254 (1968).

Values between brackets are equal to: $(p_{11} - p_{12})/2$.

As an example, let us calculate the variations $\Delta\varepsilon_{i\ell}$ and Δn for a strain S_{11} in a cubic crystal of the class m3m. The dielectric tensor of a cubic crystal is

$$\varepsilon_{ij} = \varepsilon\delta_{ij}$$

so that

$$\Delta\varepsilon_{i\ell} = - \varepsilon^2 p_{i\ell mn} S_{mn}$$

or, for a strain S_{11}

$$\Delta\varepsilon_{i\ell} = - \varepsilon^2 p_{i\ell 11} S_{11} \; .$$

From table 8.6, it appears that the only non-zero acousto-optic coefficients, for the class m3m, are

$p_{1111} = p_{11}$ $\quad p_{2211} = p_{12}$ and $p_{3311} = p_{12}$.

The dielectric constant variations are:

$\Delta\varepsilon_{11} = - \varepsilon^2 p_{11} S_{11}$ $\quad \Delta\varepsilon_{22} = \Delta\varepsilon_{33} = - \varepsilon^2 p_{12} S_{11}$

and

$$\Delta\varepsilon_{i\ell} = 0 \quad \text{if} \quad i \neq \ell .$$

Since the tensor ε_{ij} remains diagonal, the reference system consists of the principal axes of the index ellipsoid of the distorted crystal. The principal indices are

$$n_1 = \sqrt{\varepsilon + \Delta\varepsilon_{11}} \quad n_2 = n_3 = \sqrt{\varepsilon + \Delta\varepsilon_{22}} \; .$$

The variations $\Delta\varepsilon_{ii}$ are small and $n = \sqrt{\varepsilon}$, so that

$$n_i \simeq n\left(1 + \frac{\Delta\varepsilon_{ii}}{2\varepsilon}\right) = n + \Delta n_i$$

with

$$\Delta n_i = \frac{\Delta\varepsilon_{ii}}{2n}$$

or

$$\Delta n_1 = - \frac{n^3}{2} p_{11} S_{11} \tag{8.11}$$

$$\Delta n_2 = \Delta n_3 = - \frac{n^3}{2} p_{12} S_{11} \; . \tag{8.12}$$

The crystal becomes uniaxial, with the x_1 axis as the optical axis. It is worth noting that double refraction occurs for any light wave (whatever the polarization) entering the crystal along x_2 or x_3; however, the polarization is preserved if it is parallel to a principal axis. The wave only suffers a phase lag (or lead) if the index increases (decreases) along this axis. We shall hereafter (§ 8.4) refer only to this simple case.

8.3.2. Theory of Nelson and Lax

A more general formulation of the photoelastic tensor is required when local rotations occur as happens when shear elastic waves propagate through the crystal [10]. In equation 8.5 the six-component symmetric tensor $S_{k\ell}$ has to be replaced by the nine-component displacement gradient $\partial u_k/\partial x_\ell$

$$\Delta B_{ij} = P_{ijk\ell} \frac{\partial u_k}{\partial x_\ell} \qquad (8.13)$$

This new acousto-optic tensor $P_{ijk\ell}$ is no longer symmetric with respect to the last two indices k and ℓ. Let $P_{ij(k\ell)}$ be its symmetric part and $P_{ij[k\ell]}$ its antisymmetric part. As

$$\frac{\partial u_k}{\partial x_\ell} = S_{k\ell} + R_{k\ell} \qquad (8.14)$$

where the antisymmetric tensor

$$R_{k\ell} = \frac{1}{2}\left(\frac{\partial u_k}{\partial x_\ell} - \frac{\partial u_\ell}{\partial x_k}\right) = - R_{\ell k}$$

accounts for the local rotation, it follows that

$$\Delta B_{ij} = (P_{ij(k\ell)} + P_{ij[k\ell]})(S_{k\ell} + R_{k\ell})$$

or

$$\boxed{\Delta B_{ij} = P_{ij(k\ell)}S_{k\ell} + P_{ij[k\ell]}R_{k\ell}} \qquad (8.15)$$

since the contracted product of a symmetric tensor and an antisymmetric tensor vanishes.

The comparison of equation 8.5 with equation 8.15 shows that the symmetric part $P_{ij(k\ell)}$ is identical to the tensor $p_{ijk\ell}$. The antisymmetric part $P_{ij[k\ell]}$ which refers solely to a local rotation of the index ellipsoid can be expressed in terms of the components B_{ij}. In fact, when an infinitesimal volume experiences a local rotation, its coordinates x_i become x_i' and the components B_{ij} become B_{ij}' so that*

* This transformation of components is a generalization of those stated in paragraph 3.2, when the new coordinates are no longer linear functions of the previous ones.

$$B'_{ij} = \frac{\partial x'_i}{\partial x_k}\frac{\partial x'_j}{\partial x_\ell} B_{k\ell}$$

The displacement u_i is defined by $x'_i = x_i + u_i$, hence

$$\frac{\partial x'_i}{\partial x_j} = \delta_{ij} + \frac{\partial u_i}{\partial x_j}$$

Setting $S_{ij} = 0$ in equation 8.14 gives

$$\frac{\partial x'_i}{\partial x_j} = \delta_{ij} + R_{ij}$$

Thus, the variation $\Delta B_{ij} = B'_{ij} - B_{ij}$ can be expanded as

$$\Delta B_{ij} = (\delta_{ik} + R_{ik})(\delta_{j\ell} + R_{j\ell})B_{k\ell} - B_{ij}$$

If the second order terms are discarded,

$$\Delta B_{ij} = R_{ik}B_{kj} + R_{j\ell}B_{i\ell}$$

or

$$\Delta B_{ij} = (B_{i\ell}\delta_{jk} - B_{kj}\delta_{i\ell})R_{k\ell}$$

Comparing with equation 8.15 and setting $S_{k\ell} = 0$, $P_{ij[k\ell]}$ turns out to be the antisymmetric part of the above bracketed tensor. Then

$$\boxed{P_{ij[k\ell]} = \frac{1}{2}(B_{i\ell}\delta_{jk} + B_{\ell j}\delta_{ik} - B_{ik}\delta_{j\ell} - B_{kj}\delta_{i\ell})} \qquad (8.16)$$

Except for triclinic and monoclinic systems, the principal axes coincide with the crystallographic axes. Then, in this reference frame,

$$B_{i\ell} = \left(\frac{1}{n_i^2}\right)\delta_{i\ell}$$

where $n_i = n_1, n_2, n_3$ are the principal optical indices. Equation 8.16 becomes:

$$P_{ij[k\ell]} = \frac{1}{2}\left(\frac{1}{n_i^2} - \frac{1}{n_j^2}\right)(\delta_{i\ell}\delta_{jk} - \delta_{ik}\delta_{j\ell}) \qquad (8.17)$$

and the only non vanishing components are

$$\begin{cases} P_{12[12]} = -P_{12[21]} = -\frac{1}{2}\left(\frac{1}{n_1^2} - \frac{1}{n_2^2}\right) \\ P_{23[23]} = -P_{23[32]} = -\frac{1}{2}\left(\frac{1}{n_2^2} - \frac{1}{n_3^2}\right) \\ P_{13[13]} = -P_{13[31]} = -\frac{1}{2}\left(\frac{1}{n_1^2} - \frac{1}{n_3^2}\right) \end{cases} \qquad (8.18)$$

With respect to Pockels' photoelastic tensor, the only modifications affect the last three terms (p_{44}, p_{55}, p_{66}) of the main diagonal of the matrices in Fig. 8.6. For isotropic media and cubic crystals, there is no change at all ($P_{ijk\ell} = p_{ijk\ell}$) since $n_1 = n_2 = n_3$. For uniaxial crystals (hexagonal, tetragonal, trigonal systems)

$$P_{12[12]} = 0, \quad P_{23[23]} = P_{13[13]}$$

since $n_1 = n_2$. The equation of Pockels' theory (see table 8.6) $p_{44} = p_{55}$ i.e. $P_{23(23)} = P_{13(13)}$, splits into two equations

$$\begin{cases} P_{2323} = P_{23(23)} + P_{23[23]} = P_{1313} \\ P_{2332} = P_{23(23)} - P_{23[23]} = P_{1331} \end{cases} \tag{8.19}$$

In the case of orthorhombic crystals the three antisymmetric components $P_{12[12]}$, $P_{23[23]}$, $P_{13[13]}$ exist. In each matrix, a circle can conventionally stand for one of these extra components. For instance, the following matrix.

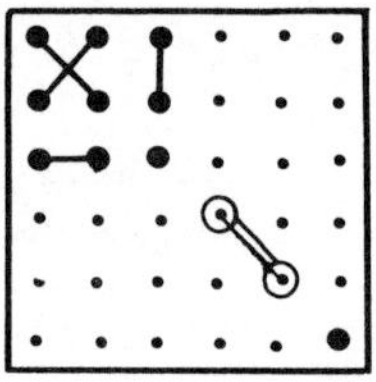

corresponds to the symmetry class 422. The lines joining the dots or the open circles account for equation 8.19. The photoelastic tensors of the monoclinic and triclinic classes comprise respectively 7 and 18 extra components.

This theory has been confirmed by fairly recent experiments carried out by Nelson et al. with strongly birefringent non piezoelectric crystals: rutile (4/mmm) and calcite ($\bar{3}$m):

- for rutile [11]

$$P_{2323} = P_{1313} = 9.10^{-4}, \; P_{2332} = P_{1331} = -255.10^{-4}$$

- for calcite [12]

$$P_{2332} = P_{1313} = -11.10^{-3}, \; P_{2332} = P_{1331} = -105.10^{-3}$$

Equation 8.10 which expresses the variations of the permittivities can readily be generalized:

$$\Delta\varepsilon_{i\ell} = -\varepsilon_{ij}P_{jkmn}\,\varepsilon_{k\ell}\,\frac{\partial u_m}{\partial x_n} \tag{8.20}$$

8.3.3. Piezoelectric Crystals. Electro-optic Effect

A modification of the index ellipsoid may also occur when an electric field is applied to a crystal. This electro-optic effect is expressed by the equation

$$\Delta B_{ij} = r_{ijp}E_p$$

Like the piezoelectric tensor, the third rank electro-optic tensor r_{ijp} vanishes for centrosymmetric crystals.

In a piezoelectric crystal, the propagation of elastic waves is generally accompanied by a longitudinal electric field. Along a direction parallel to the unit vector $\vec{n}$, this field can be deduced from equation 6.49

$$E_p = -n_p\,\frac{n_q e_{qk\ell}}{\varepsilon_{pq}n_p n_q}\,\frac{\partial u_\ell}{\partial x_k}$$

There exists an indirect photoelastic effect

$$\Delta B_{ij} = -\left(\frac{r_{ijp}n_p n_q e_{qk\ell}}{\varepsilon_{pq}n_p n_q}\right)\frac{\partial u_\ell}{\partial x_k}$$

This effect is characterized by the above bracketed term which depends on the direction of propagation

$$P^{\text{ind.}}_{ijk\ell}(\vec{n}) = -\frac{r_{ijp}n_p n_q e_{qk\ell}}{\varepsilon_{pq}n_p n_q} \tag{8.21}$$

The contribution of this electro-optic term may be significant or even predominant in a certain number of cases. For example, a wave propagating in the [111] direction in a α-iodic acid (class 222) generates an indirect photoelastic effect with $P^{\text{ind.}}_{3212} = 0.013$ whereas the direct component is zero [13].

8.4. DIFFRACTION OF OPTICAL WAVES BY A BEAM OF ELASTIC WAVES

As a result of the above investigation, the dielectric tensor and the optical index are both functions of the strain which the medium undergoes. Therefore, any elastic wave propagating in the medium is accompanied by an index variation wave travelling at the same velocity. In order to analyze its action on optical waves, we first focus on the simplest case where the polarization is maintained, namely an optically isotropic medium, an elastic longitudinal sine wave, a monochromatic light beam, with polarization parallel or perpendicular to the elastic wave vector, and normal or Bragg angle incidence.

8.4.1. Normal Incidence

The elastic wave, propagating in the x_1 direction, involves a succession of compressions and dilatations, and thus gives rise to increases or decreases of the optical index with respect to the value n_o at rest

$$n = n_o + \Delta n \sin \omega(t - \frac{x_1}{V}) \tag{8.22}$$

where ω and V denote the angular frequency and the velocity of the elastic wave.

The optical waves in the medium travel slower (faster) when the optical index is larger (smaller). The expression for the light wave propagating in the x_2 direction is

$$a = A \cos (\Omega t - \frac{2\pi}{\Lambda_o} n x_2) \tag{8.23}$$

The origin for the phase angle is taken at the origin of the acoustic beam, $x_2 = 0$ (fig. 8.8); Λ_o is the wavelength of the light in a vacuum.

Disregarding any attenuation, the expression for the light wave after passing through the elastic beam (of thickness e) is

$$a = A \cos [\Omega t + \Phi_0 + \Delta\Phi \sin \omega(t - \frac{x_1}{V})] \tag{8.24}$$

where $\Phi_0 = - (2\pi n_o/\Lambda_o)e = - (2\pi/\Lambda)e$ is the phase shift occurring for the mean index n_o and $\Delta\Phi = - (2\pi e/\Lambda_0)\Delta n$.

At every point of the elastic beam boundary ($x_2 = e$, x_1 fixed) this is a phase modulated wave, whose modulation is dependent only on time. Let us look for its spectrum. Denoting $t - x_1/V$ by τ, we obtain

$$\frac{a}{A} = \cos (\Delta\Phi \sin \omega\tau)\cos(\Omega t+\Phi_0)- \sin(\Delta\Phi \sin \omega\tau)\sin(\Omega t+\Phi_0). \tag{8.25}$$

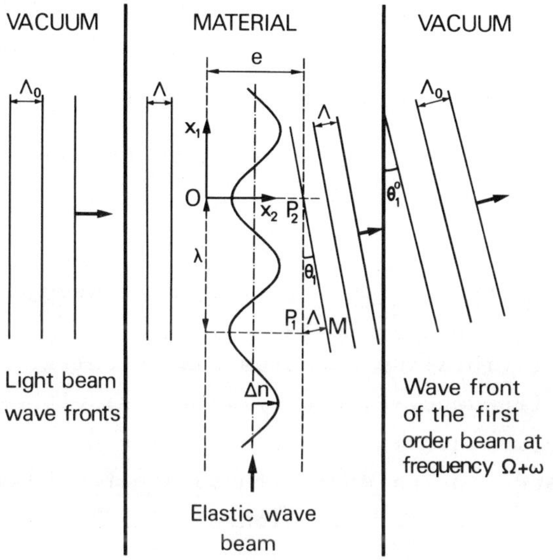

Fig. 8.8. The 2π phase shift (due to the index variation Δn) between P_1 and P_2, separated by one wavelength λ, determines the inclination (θ_1) of the wavefront for the first order beam, at frequency $\Omega + \omega$: $\sin \Theta_1 = \Lambda/\lambda$. In a vacuum, on emerging from the material, the angle of deflection is θ_1^o such that $\sin \theta_1^o = \Lambda_0/\lambda$.

The expressions $\cos(\Delta\Phi \sin \omega\tau)$ and $\sin(\Delta\Phi \sin \omega\tau)$ are expanded through Bessel functions:

$$\cos(\Delta\Phi \sin \omega\tau) = J_o(\Delta\Phi) + 2 \sum_{p=1}^{\infty} J_{2p}(\Delta\Phi) \cos 2p\omega\tau$$

$$\sin(\Delta\Phi \sin \omega\tau) = 2 \sum_{p=1}^{\infty} J_{2p-1}(\Delta\Phi) \sin (2p - 1)\omega\tau.$$

Inserting this into (8.25), yields

$$\frac{a}{A} = J_o(\Delta\Phi)\cos(\Omega t+\Phi_0) + \sum_{p=1}^{\infty} J_{2p}(\Delta\Phi) 2 \cos(\Omega t+\Phi_0)\cos 2p\omega\tau$$

$$- \sum_{p=1}^{\infty} J_{2p-1}(\Delta\Phi) 2 \sin(\Omega t+\Phi_0)\sin(2p-1)\omega\tau$$

The products of sine or cosines are converted into sums, so that

$$\frac{a}{A} = J_o(\Delta\Phi)\ \cos\ (\Omega t + \Phi_0)$$

$$+ \sum_{p=1}^{\infty} J_{2p}(\Delta\Phi)[\cos(\Omega t+2p\omega\tau+\Phi_0)+ \cos(\Omega t-2p\omega\tau+\Phi_0)]$$

$$+ \sum_{p=1}^{\infty} J_{2p-1}(\Delta\Phi)[\cos(\Omega t+(2p-1)\omega\tau+\Phi_0)- \cos(\Omega t-(2p-1)\omega\tau+\Phi_0)]$$

or, more concisely

$$\frac{a}{A} = J_o(\Delta\Phi)\cos(\Omega t+\Phi_0)$$

$$+ \sum_{N=1}^{\infty} J_N(\Delta\Phi)[\cos(\Omega t+N\omega\tau+\Phi_0)+(-1)^N\cos(\Omega t-N\omega\tau+\Phi_0)]. \quad (8.26)$$

The light vibration spectrum has a central peak (carrier) at frequency Ω, and symmetric sidelobes at frequencies $\Omega \pm N\omega$ since $\tau = t - x_1/V$, where N is an integer. The carrier and side wave amplitudes are the Bessel functions J_o, J_1, J_2 of the phase variation $\Delta\Phi$ (fig. 8.9).

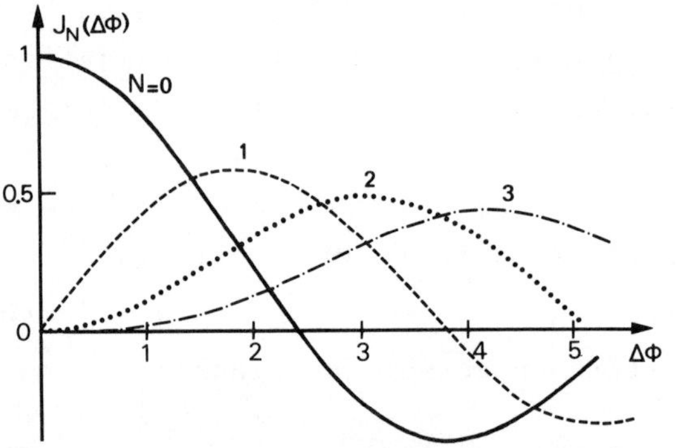

Fig. 8.9. Bessel functions J_o, J_1, J_2 ... representing the amplitudes of the carrier wave and of the spectral components, in terms of the phase excursion $\Delta\Phi$.

At a given moment, the carrier phase angle $\Phi_0 = -2\pi e/\Lambda$ does not depend on x_1; that is to say the wave planes are parallel to the incident wave planes. This is no longer true for the side spectral components, such as

$$J_N(\Delta\Phi) \cos [(\Omega + N\omega)t + \Phi_0 - \frac{N\omega x_1}{V}]$$

whose phase (Φ_N) depends on x_1, i.e. on the point where the ray crosses the boundary:

$$\Phi_N = \Phi_0 - \frac{N\omega x_1}{V} = \Phi_0 - 2\pi \frac{N x_1}{\lambda}$$

where λ is the acoustic wavelength. Interference of waves of same order N emitted at the boundary ($x_2 = e$) is constructive only in the direction θ_N (fig. 8.8). In this direction, for waves emerging at P_1 and P_2, the phase lag $-(2\pi/\Lambda)P_1M$ due to the path difference P_1M cancels out the initial phase lead $\Delta\Phi_N = (2\pi N/\Lambda)P_2P_1$:

$$-\frac{2\pi}{\Lambda} P_1M + 2\pi \frac{N}{\lambda} P_2P_1 = 0$$

or, since $P_1M = P_2P_1 \sin \theta_N$:

$$\sin \theta_N = N \frac{\Lambda}{\lambda} . \qquad (8.27)$$

The spectral component of order N, at frequency $\Omega + N\omega$ is deflected by an amount θ_N (given by 8.27). As for the $(\Omega - N\omega)$ component, it undergoes a symmetric deflection (fig. 8.10).

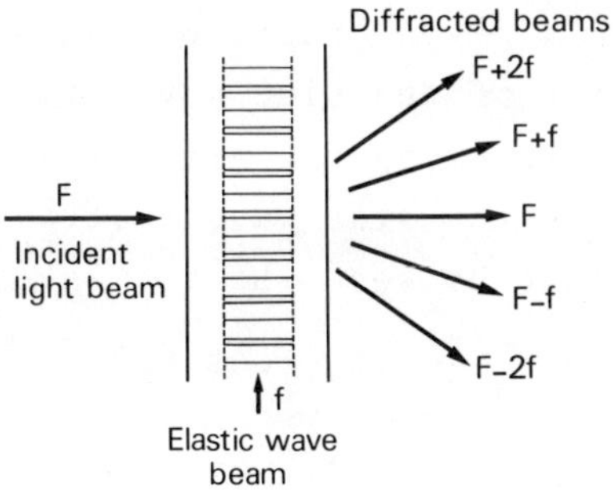

Fig. 8.10. Light diffraction by an elastic wave beam. Normal incidence. The light beam is split into several beams, which have symmetric inclinations with respect to the incident beam.

Finally, the sinusoidal variation of the index (due to the elastic wave) has an effect similar to that of a phase grating on a light wave. The light beam which penetrates into the crystal and is parallel to the elastic wave planes is separated into several beams, at symmetric angles θ_N with respect to the incident direction:

$$\boxed{\sin \theta_N = \pm N \frac{\Lambda}{\lambda}} .$$

In the case of silica, with longitudinal elastic waves at 200 MHz and for the red light of a He-Ne laser:

$$V_L = 5.96\ 10^3 \mathrm{m/s} \qquad \Rightarrow \lambda = 30\ \mu\mathrm{m}$$

$$\Lambda_0 = 0.633\ \mu\mathrm{m} \quad \text{and} \quad n = 1.46 \Rightarrow \Lambda = 0.433\ \mu\mathrm{m}$$

$$\sin \theta_1 \simeq \theta_1 = 1.45\ 10^{-2}\ \mathrm{rd} = 0^\circ 50' .$$

<u>Critical thickness</u>. The above argument is valid only if the elastic beam thickness e is below a critical value e_c. In fact, the side waves are generated all along the carrier path within the ultrasonic beam, and not only at the boundary. Let us mentally divide the elastic beam into slabs parallel to its propagation direction x_1, and repeat the above analysis for each slab, thus the frequencies $\Omega + N\omega$ and the propagation directions θ_N of the side waves are the same for the slabs at x_2 or $x_2 + \ell$ (fig. 8.11). For a given order, we add up both contributions. The front of the side wave emitted at time $t - \ell/v$ and point x_2, along θ_N, arrives at Q_1 at time t, whereas the front of the side wave emitted at $x_2 + \ell$ is at Q_2. The phase shift between the two waves is

$$\eta = \frac{2\pi\ell}{\Lambda}(1 - \cos \theta_N)$$

or, θ_N being small

$$\cos \theta_N \simeq 1 - \frac{1}{2}\left(\frac{N\Lambda}{\lambda}\right)^2 \Rightarrow \eta \simeq \frac{\pi\ell\Lambda}{\lambda^2} N^2 .$$

The phase shift η is equal to π for a path:

$$\ell_N = \frac{\lambda^2}{\Lambda} \cdot \frac{1}{N^2}$$

and for this distance, destructive interference occurs between the two waves. If the ultrasonic beam thickness is larger than ℓ_N, the effect caused by one slab is cancelled by the effect of another slab at distance ℓ_N.

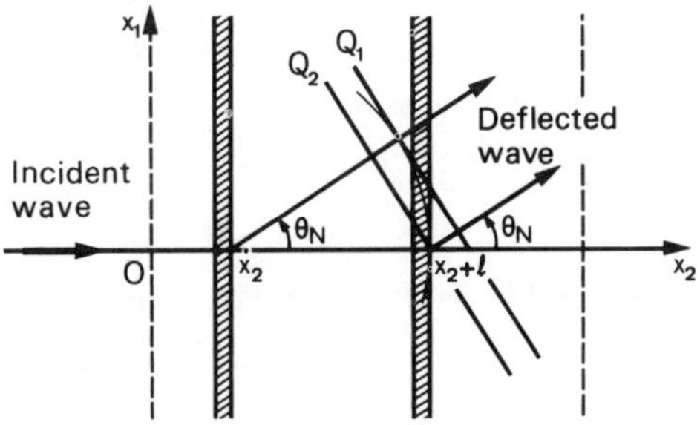

Fig. 8.11. Normal incidence. Critical thickness. The front of the optical side wave of order N, emitted at time $t - \ell/v$ and at x_2, propagates along θ_N and arrives at Q_1 at time t. The front of the side wave of order N emitted at $x + \ell$ at time t is at Q_2. Both waves exhibit a phase shift

$$\eta = \frac{2\pi\ell}{\Lambda}(1 - \cos\theta_N) \simeq \frac{\pi\ell\Lambda}{\lambda^2} N^2 .$$

Ideally the acoustic beam thickness should not exceed the first order critical value

$$e_c = \ell_1 = \frac{\lambda^2}{\Lambda} . \qquad (8.28)$$

The higher the frequency, the more difficult the condition is to satisfy. As an example, for silica and $\Lambda_0 = .633$ µm ($\Lambda = .433$ µm):

$$\text{at } 100 \text{ MHz } e_c = 8.4 \text{ mm,}$$
$$\text{at } 300 \text{ MHz } e_c = .93 \text{ mm.}$$

8.4.2. Bragg Angle Incidence

In practice, the use of Bragg angle incidence is more important, because there is only one diffracted beam. It is obtained on rotating the incident beam through an angle α, so that the interference for the first order angular frequency $\Omega + \omega$ is constructive (fig. 8.12).

8.4.2.1. Angle of incidence

We first look for the diffraction angle θ corresponding to an angle of incidence α. For two points P_1 and P_2 in a slab parallel to the x_1 axis, the phase lag due to the optical path difference cancels the phase lead due to the elastic wave:

$$-\frac{2\pi}{\Lambda}(HP_1 + P_1K) + 2\pi\frac{P_1P_2}{\lambda} = 0$$

or

$$\sin\alpha + \sin(\theta - \alpha) = \frac{\Lambda}{\lambda}. \qquad (8.29)$$

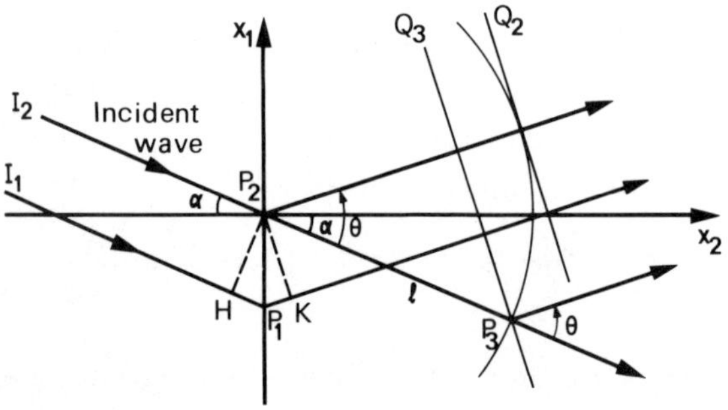

Fig. 8.12. Oblique incidence. In order that the interference between first order side waves be constructive, the effects produced by the incident rays I_1P_1 and I_2P_2 must add up: $\sin\alpha + \sin(\theta - \alpha) = \Lambda/\lambda$. This is also true for the effects created by each ray along its path

$$1 - \cos\theta = \frac{\Lambda}{\lambda}\sin\alpha \quad\Rightarrow\quad \alpha = \frac{\theta}{2}.$$

Let us now express the fact that the side waves emitted from P_2 and P_3 on an incident ray are in phase, whatever their distance ℓ. At a given time t, the phase shift η between the wave planes Q_2 and Q_3 involves an oblique propagation term $(2\pi\ell/\Lambda)(1 - \cos\theta)$ and an elastic wave term $(-2\pi\ell/\lambda)\sin\alpha$:

$$\eta = 2\pi\ell\left(\frac{1 - \cos\theta}{\Lambda} - \frac{\sin\alpha}{\lambda}\right).$$

The phase shift vanishes for all ℓ, provided the equation

$$1 - \cos\theta = \frac{\Lambda}{\lambda}\sin\alpha$$

is satisfied; taking equation 8.29 into account, it is equivalent to

$$\frac{\Lambda}{\lambda} = 2\sin\frac{\theta}{2}\cos\left(\frac{\theta}{2} - \alpha\right) ,$$

if

$$\sin\frac{\theta}{2} = \cos\left(\frac{\theta}{2} - \alpha\right)\sin\alpha .$$

This equation is satisfied for $\alpha = \theta/2$. The angle of incidence α is given by

$$\boxed{\sin\alpha = \frac{\Lambda}{2\lambda} = \frac{\Lambda f}{2V}} \qquad (8.30)$$

and the light beam is deflected by the amount 2α. The elastic wave fronts seem to act as mirrors for the incident beam (fig. 8.13) and the condition (8.30), which reminds one of the selective X-ray diffraction by lattice planes, is called the Bragg condition. The Bragg angle incidence enhances the first order beam only. However, the carrier generates side waves of all orders all along its path. The effect is cumulative for the first order, and destructive for the others. It is stronger if the incident beam crosses more wave planes, which requires (for small α):

$$e\alpha \gg \lambda \quad \Rightarrow \quad e \gg \frac{2\lambda^2}{\Lambda} = 2e_c .$$

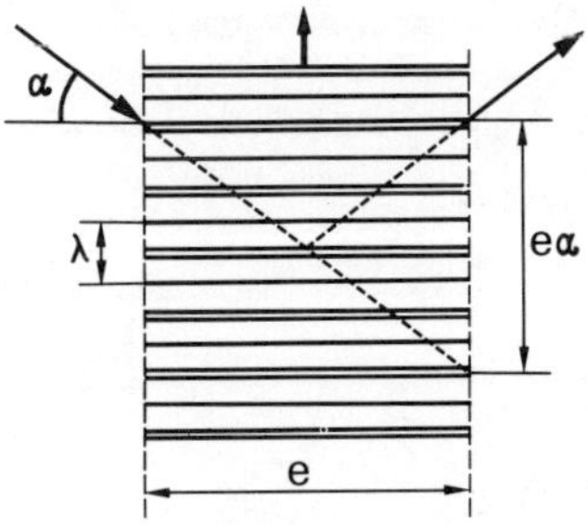

Fig. 8.13. Bragg condition. A light ray must travel across several elastic wave fronts.

Notice that if the incident light beam is symmetric to that of Fig. 8.13 with respect to elastic wave planes, the first order beam at angular frequency $\Omega - \omega$ is enhanced.

8.4.2.2. Intensity of the deflected beam. Figure of merit

In order to compare the acousto-optic properties of different materials, it is useful to introduce a figure of merit which represents the fraction of the light beam which is deflected at Bragg incidence. Let A_o and $A(x_2)$ be respectively the carrier wave amplitude on entering and after passing a distance x_2 through the acoustic beam. The approximation

$$J_N(x) \simeq \frac{x^N}{2^N N!} \qquad \text{for} \qquad x \ll 1$$

allows one to reduce equation 8.26 for the overall wave at point $x_2 + dx_2$ (keeping only first order terms) to

$$a(x_2+dx_2,t) = A(x_2)\cos(\Omega t+\delta\Phi_0) + \frac{A(x_2)}{2}\, d\Phi[\cos(\Omega t+\omega\tau+\delta\Phi_0) - \cos(\Omega t-\omega\tau+\delta\Phi_0)] \qquad (8.31)$$

where

$$d\Phi = -\frac{2\pi}{\Lambda_0}\Delta n\, dx_2 = \frac{\Delta\Phi}{e}dx_2 \quad \text{and} \quad \delta\Phi_0 = -\frac{2\pi}{\Lambda}dx_2$$

$d\Phi$ is the amplitude of the phase shift introduced by the slab of thickness dx_2. For the component which satisfies the Bragg condition, say for the $(\Omega + \omega)$ component, the amplitudes of all waves, emitted at various points of the acoustic beam, should be added, because of the constructive interference. The increase dB in the amplitude $B(x_2)$ of the side wave $\Omega + \omega$, due to the slab x_2, is, according to (8.31)

$$dB = \frac{A}{2}d\Phi = A\frac{\Delta\Phi}{2e}dx_2 \; . \qquad (8.32)$$

The carrier wave simultaneously decreases in amplitude. Energy conservation

$$A^2 + B^2 = A_o^2$$

implies

$$A\,dA + B\,dB = 0$$

or, making use of equation 8.32

$$\frac{dB}{dx_2} = A\frac{\Delta\Phi}{2e} \quad \text{and} \quad \frac{dA}{dx_2} = -B\frac{\Delta\Phi}{2e} \; .$$

The amplitudes A and B satisfy the same differential equation

$$\frac{d^2f}{dx_2^2} + \left(\frac{\Delta\Phi}{2e}\right)^2 f = 0$$

the solutions of which must obey the boundary conditions $B = 0$, $A = A_o$ at $x_2 = 0$. In addition energy must be conserved. These solutions are

$$B = A_o \sin \frac{|\Delta\Phi|}{2e} x_2$$

$$A = A_o \cos \frac{|\Delta\Phi|}{2e} x_2 \ .$$

On emergence from the elastic beam ($x_2 = e$), the relative intensity of the diffracted beam is

$$\frac{i}{i_o} = \left(\frac{B}{A_o}\right)^2 = \sin^2 \frac{|\Delta\Phi|}{2} \qquad (8.33)$$

Let us now introduce the mean elastic power per unit area:

$$P = \left(\frac{1}{2}\, cS^2\right)V = \frac{1}{2}\, \rho V^3 S^2$$

into the equation $|\Delta\Phi| = (2\pi e/\Lambda_0)|\Delta n|$. The index variation is linked to the strain $S = \sqrt{2P/\rho V^3}$ by 8.11:

$$|\Delta n| = \frac{n^3}{2}\, pS = \sqrt{\frac{p^2 n^6}{2\rho V^3}\, P} \qquad (8.34)$$

Defining the figure of merit of the material [14] by

$$\boxed{M = \frac{p^2 n^6}{\rho V^3}} \qquad (8.35)$$

yields

$$|\Delta\Phi| = \frac{2\pi e}{\Lambda_0}\sqrt{\frac{MP}{2}} \ .$$

The relative intensity of the diffracted beam, given by

$$\frac{i}{i_o} = \sin^2 \left(\pi\, \frac{e}{\Lambda_0}\sqrt{\frac{MP}{2}}\right) \qquad (8.36)$$

may, in principle, reach unity (total deflection of the incident beam). The elastic power density required to deflect a given fraction of the incident light is proportional to the reciprocal of the figure of merit M of the material. As we have already pointed out, the photoelastic coefficient, for the best crystal cuts ranges between .2 and .3. The relevant parameters (they are not in fact independent [15]) are the refractive index n and the velocity V which are involved in the expression for M, with exponents of 6 and 3. The refractive index hardly goes above 2.5 for transparent materials in the visible spectrum. The elastic longitudinal wave velocity ranges between 1,000 and 10,000 m/s. Generally, a low velocity in solids corresponds to a high attenuation coefficient for elastic waves. We must note that, despite their high figures of merit, liquids cannot be used at frequencies above 50 MHz. For such frequencies, the attenuation in water is over 5 dB/cm and it increases as the square of the frequency. Table 8.14 shows the attenuation coefficient at 500 MHz for some materials, together with the figure of merit referred to that of fused silica:

$$M_o = 1.51\ 10^{-15}\ s^3 kg^{-1}$$

($n = 1.46$, $p - 0.27$, $\rho = 2.2\ 10^3$ kg m^{-3}, $V = 5{,}960$ m/s)

α-iodic acid has a high figure of merit (55) but suffers the disadvantage of being highly hygroscopic. The very high figure of merit (515) of paratellurite (TeO_2) is induced by the very low shear wave velocity along direction [110]. Unfortunately its attenuation factor is prohibitive beyond 100 MHz (6 dB/cm). By now, lead molybdate ($PbMoO_4$) seems to be the most interesting material. TiO_2 and $LiNbO_3$ are adequate in the high frequency domain because of their low attenuation coefficient. Let us calculate the elastic power $\mathscr{P} = Pe^2$ (provided by a square transducer of size e) required to deflect one half of a He-Ne laser beam (Λ_0 = .6328 μm). According to equation 8.36

$$\mathscr{P} = \frac{\Lambda_0^2}{8M}$$

a power of 33 watts is needed for silica, and only 1.4 watt for lead molybdate.

8.4.2.3. Interaction bandwidth

Although the Bragg angle condition relating the incidence angle α_0 and the elastic wave frequency f_o is quite restrictive, it is possible to obtain a non-zero bandwidth from the spontaneous divergence of the elastic beam. We have already mentioned (§ 1.3.6) that the radiation pattern far from the source (transducer) of width $e = a\lambda$ is

$$A(\Delta\alpha) = A_o \frac{\sin(\pi es/\lambda)}{\pi es/\lambda} \quad \text{where} \quad s = \sin\Delta\alpha \simeq \Delta\alpha$$

Table 8.14. Interaction conditions and figures of merit for some materials.

Material	Elastic wave			Incident light wave (Λ_0 = 0.6328 μm)				Appropriate p_{ij}	Figure of merit
	propag.-polar.	Velocity m/s	Attenuation dB/cm (500 MHz)	Useful domain (μm)	propag.-polar. (a)	indices n_o	n_e		$M = \frac{n^6 p^2}{\rho V^3}$
Silica	arbitrary-L	5,960	3.0	0.2-2.5	arbitrary-⊥	1.46		p_{12}	1 (1.51 10^{-15})
α-HIO_3	[001]-L	2,440	2.5[6]	0.4-1.3	[010]-⊥([100])	n_1=1.98 (b)		p_{31}	55
$PbMoO_4$	[001]-L	3,630	2.5 (c)		[010]-// or ⊥	2.38	2.26	p_{33} or p_{31}	23.7
TiO_2	[100]-L	8,015	0.5 (d)	0.2-0.9	[001]-⊥([010])	2.58	2.9	p_{12}	2.6
TeO_2	[110]-T[1$\bar{1}$0]	616	50 [8]	0.35-5	[001]-arbitrary	2.26	2.41	$\frac{p_{11}-p_{12}}{2}$	515
	[001]-L	4,200	2.3[8]		[010]-[100]			p_{13}	23
$LiNbO_3$	X-L	6,560	0.03 (e)	0.5-4.5	Y-Z	2.29	2.20	p_{31} (*)	1.8
$LiTaO_3$	Z-L	6,180	0.02 (e)		arbitrary-Z	2.175	2.18	p_{33}	0.9
α-quartz	Z-L	6,363	0.6 (b)		Y-Z	1.54	1.55	p_{13}	0.87

(a) The polarization is defined parallel or orthogonal to the elastic and optic wave vector plane.
(b) The other two principal indices of α-HIO_3 (biaxial) are n_2 = 1.96 and n_3 = 1.84.
(c) Ref. [14] of chapter IV.
(d) M. Dutoit, IEEE Trans. Son. Ultrason., SU-20, 279 (1973).
(e) E.G. Spencer and P.V. Lenzo, J. Appl. Phys., 38, 423 (1967).
(f) C.D.W. Wilkinson and D.E. Caddes, J. Acoust. Soc. Am., 40, 498 (1966).
(*) For the beam undergoing no change of polarization.

$\Delta\alpha$ is the angular deviation from the mean beam direction (i.e. from the crystal axis). The angle of incidence of the light beam with respect to the crystal is maintained, the interaction takes place at frequency f for elastic waves inclined at an angle $\Delta\alpha$ (fig. 8.15), such (equation 8.30) that

$$\Delta\alpha = \frac{\Lambda}{2V}(f - f_o).$$

In the weak interaction hypothesis, the intensity of the deflected beam is proportional to the elastic power and

$$\frac{i(f)}{i(f_o)} = \left[\frac{A(\Delta\alpha)}{A_o}\right]^2 = \frac{\sin^2\left[\frac{\pi e\Lambda f(f - f_o)}{2V^2}\right]}{\left[\frac{\pi e\Lambda f(f - f_o)}{2V^2}\right]^2}$$

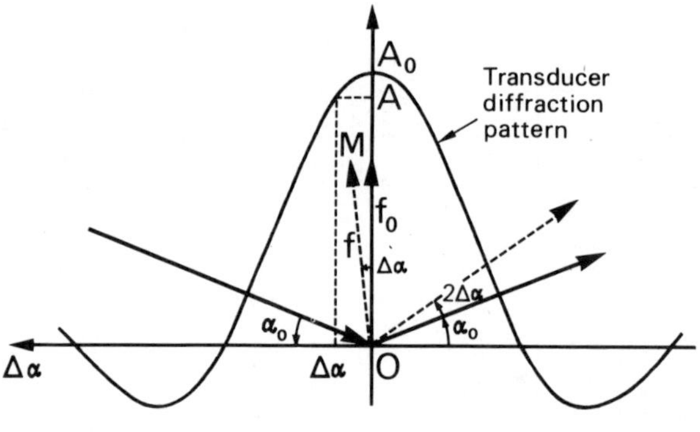

Fig. 8.15. Interaction bandwidth, including effects of the elastic beam divergence. At a frequency f different from f_o, the interaction takes place with those elastic waves for which the propagation direction OM satisfies the Bragg condition $\alpha_0 + \Delta\alpha = \Lambda f/2V$.
The intensity of the deflected beam, proportional to A^2 decreases.

The 3dB bandwidth (i.e. half deflected beam intensity) which is assumed to be narrow ($f \simeq f_o$) is:

$$\Delta f = \frac{1.77\ V^2}{e\Lambda f_o} \qquad (8.37)$$

or Δf = 100 MHz for V = 4,000 m/s, e = 1.5 mm, Λ_0 = .63 μm, n = 2, f_0 = 600 MHz.

Conversely, if the crystal is rotated and the signal frequency kept constant, the variation of the deflected beam intensity follows the transducer radiation pattern (except for a dilatation of angles $\Delta\theta = 2\Delta\alpha$):

$$\frac{i(\theta_0 + \Delta\theta)}{i(\theta_0)} = \frac{\sin^2\left(\frac{\pi e f_0 \Delta\theta}{2V}\right)}{\left(\frac{\pi e f_0 \Delta\theta}{2V}\right)^2}$$

8.4.2.4. Number of resolvable directions

The angular deflection θ of the light beam caused by the acoustic beam at frequency f is twice the Bragg angle:

$$\theta = 2\alpha \simeq \frac{\Lambda}{\lambda} = \frac{\Lambda}{V} f.$$

A change Δf in the frequency causes a change $\Delta\theta$ where

$$\Delta\theta = \frac{\Lambda}{V} \Delta f.$$

Assuming that the angular width of the light beam (whose diameter is D) is defined by the natural diffraction angle $\Delta\beta = \Lambda/D$, the number N_B of directions that can be distinguished in angle $\Delta\theta$ is

$$N_B = \frac{\Delta\theta}{\Delta\beta} = \frac{D}{V} \Delta f$$

or

$$N_B = T_B \Delta f$$

where T_B is the transit time of elastic waves through the light beam. In order to obtain N_B = 300, one must (for example) have Δf = 100 MHz and T_B = 3 μs. If the elastic wave velocity is 4,000 m/s or 4 mm/μs, the light beam diameter should be 12 mm. Generally, it is necessary to enlarge the laser beam with an appropriate device. In addition the crystal length, which is at least equal to $T_B V$, must not give rise to an exceedingly high attenuation of the elastic waves at the operating frequency, which is itself determined by the required bandwidth. For lead molybdate, at 300 MHz, the attenuation is above 1 dB/cm.

To sum up, the three major features of Bragg interaction are the following:

- the intensity of the single diffracted beam depends on the elastic beam intensity;
- the angle by which the light beam is deflected is proportional to the elastic wave frequency;
- the elastic wave frequency is found in the deflected light beam.

The first feature accounts for light beam modulation (§ 9.8), the second for deflection (§§ 9.8 and 9.3.6). The third has been exploited to delay an electric signal continuously (§ 9.2.3).

Let us now discuss a few cases of interaction where there is a change of polarization, especially the colinear interaction, where there is no deflection.

8.4.2.5. Wave vector diagram. Change of polarization

The equation 8.30, which defines the Bragg incidence angle $\alpha = \theta/2$, may be written

$$\frac{1}{\Lambda} \sin \frac{\theta}{2} = \frac{1}{2\lambda}$$

This leads to the triangle of fig. 8.16, that relates the optical incident ($\vec{K}_I$) and diffracted ($\vec{K}_D$) wave vectors to the elastic wave vector $\vec{k}$. This construction explicitly illustrates momentum conservation:

$$\boxed{\vec{K}_D = \vec{K}_I \pm \vec{k}} \qquad (8.38)$$

while the angular frequency equation

$$\Omega_D = \Omega \pm \omega \qquad (8.39)$$

ensures energy conservation.

In the above conditions (optically isotropic material, longitudinal waves, identical polarization for the incident and diffracted beam) the triangle is nearly isosceles ($f \ll F \Rightarrow K_D \simeq K_I$). For anisotropic crystals wave vectors $\vec{K}_I$ and $\vec{K}_D$ do not generally have the same magnitude especially if the incident and diffracted wave polarizations are different, so that the propagation velocities are no longer identical. The triangle is no longer

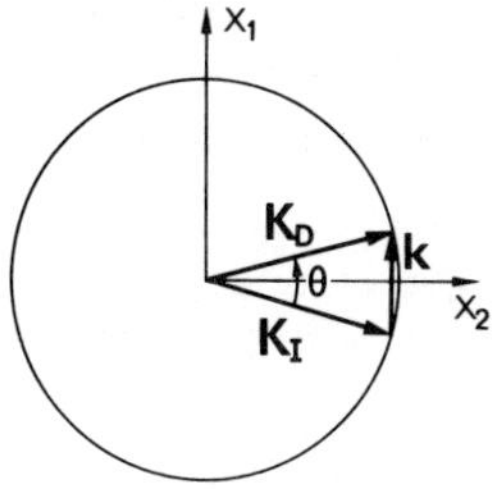

Fig. 8.16. Wave vector diagram. The triangle is nearly isosceles if the material is optically isotropic.

isosceles and the incidence and diffraction angles, α_I and α_D, are no longer equal [16].

As an example, let us consider a trigonal crystal belonging to one of the three classes 3m, 32, $\bar{3}$m. Its acousto-optic tensor is given by table 8.6. A shear elastic wave propagating along the x_1 axis (crystallographic axis X) with x_3 polarization will generate a variation $\Delta\varepsilon_{i\ell}$ as derived from equation 8.20

$$\Delta\varepsilon_{i\ell} = - \varepsilon_{ii}\varepsilon_{\ell\ell}P_{i\ell 31} \frac{\partial u_3}{\partial x_1}$$

The only non vanishing $P_{i\ell 31}$ components are P_{1231} and P_{1331} (see table 8.6). The electro-optic effect (from equation 8.21 where $n_1 = 1$, $n_2 = n_3 = 0$)

$$P^{\text{ind.}}_{i\ell 31} = - \frac{r_{i\ell 1}}{\varepsilon_{11}} e_{131}$$

has no effect for classes 32 and $\bar{3}$m where $e_{15} = 0$. For class 3m only the components P_{1231} and P_{1331} are modified:

$$P^{\text{ind.}}_{1231} = - r_{61}e_{15}/\varepsilon_{11} \quad \text{and} \quad P^{\text{ind.}}_{1331} = -r_{51}e_{15}/\varepsilon_{11}$$

The tensor $\Delta\varepsilon_{i\ell}$ is as follows

$$\Delta\varepsilon_{i\ell} = \begin{bmatrix} 0 & \Delta\varepsilon_{12} & \Delta\varepsilon_{13} \\ \Delta\varepsilon_{12} & 0 & 0 \\ \Delta\varepsilon_{13} & 0 & 0 \end{bmatrix}$$

where

$$\Delta\varepsilon_{12} = \varepsilon_{22}(r_{61}e_{15} - \varepsilon_{11}P_{1231}) \frac{\partial u_3}{\partial x_1}$$

and

$$\Delta\varepsilon_{13} = \varepsilon_{33}(r_{51}e_{15} - \varepsilon_{11}P_{1331}) \frac{\partial u_3}{\partial x_1}$$

The x_3-polarized incident light wave (fig. 8.17), is accompanied by an electric field, with components (0, 0, $E_3 = E \cos \Omega t$), which creates a displacement*

$$D_i = (\varepsilon_{i3} + \Delta\varepsilon_{i3})E_3$$

Since

$$\Delta\varepsilon_{i3} = (\Delta\varepsilon_{i3})_o \sin \omega t$$

the diffracted wave at $\Omega + \omega$ comes from the product $\Delta D_i = \Delta\varepsilon_{i3}E_3$. It is x_1-polarized because only $\Delta\varepsilon_{13}$ is non zero. The acousto-optic interaction changes the light beam polarization. The wave vector diagram now involves both sheets of the index surface, i.e. (in the x_1x_2 plane) two circles, whose radii are n_e and n_o, since the optical axis x_3 is an axis of revolution for this surface. The end of the incident wave vector $\vec{K}_I$ lies on the circle of radius n_e, corresponding to an x_3-polarization, whereas the end of $\vec{K}_D$ for the diffracted wave, which is polarized in the x_1x_2 plane, lies on the circle of radius n_o (fig. 8.17). Note that for a given extraordinary incident ray, there are generally two diffracted rays, corresponding to different acoustic frequencies.

Still within the framework of 32, 3m $\bar{3}$m classes, let us now focus on the effect of a longitudinal elastic wave propagating along the x_1-axis generating a strain S_{11} on a x_3 polarized light beam propagating in the x_1 direction (colinear interaction). According to Pockel's theory, valid for this example, the dielectric tensor variation

$$\Delta\varepsilon_{i\ell} = -\varepsilon_{ii}\varepsilon_{\ell\ell}p_{i\ell 11}S_{11}$$

reduces (using the table for the $p_{\alpha\beta}$ to

$$\Delta\varepsilon_{i\ell} = \begin{bmatrix} \Delta\varepsilon_{11} & 0 & 0 \\ 0 & \Delta\varepsilon_{22} & \Delta\varepsilon_{23} \\ 0 & \Delta\varepsilon_{23} & \Delta\varepsilon_{33} \end{bmatrix}$$

---

* We are only interested in the polarization of the diffracted wave. We take $\varepsilon_0 = 1$ in the formulae for D_i.

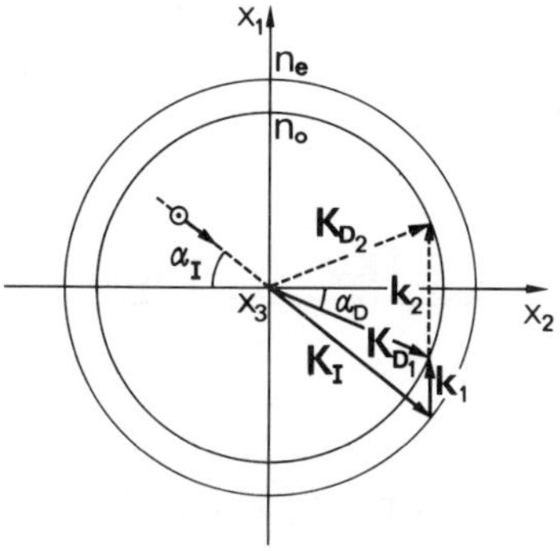

Fig. 8.17. Wave vector diagram. Uniaxial crystal. The interaction of an incident extraordinary ray (index n_e, wave vector $\vec{K}_I$) with a shear elastic wave (x_3 polarization, wave vector $\vec{k}_1$ or $\vec{k}_2$) leads to an ordinary diffracted ray (index n_o, wave vector $\vec{K}_{D_1}$ or $\vec{K}_{D_2}$).

where

$$\Delta\varepsilon_{11} = -(\varepsilon_{11})^2 p_{11} S_1 \qquad \Delta\varepsilon_{22} = -(\varepsilon_{11})^2 p_{12} S_1$$

$$\Delta\varepsilon_{23} = -\varepsilon_{11}\varepsilon_{33} p_{41} S_1 \qquad \Delta\varepsilon_{33} = -(\varepsilon_{33})^2 p_{31} S_1$$

The electro-optic effect, that appears only in class 32 ($p^{ind.}_{i\ell 11} = -r_{i\ell 1} e_{11}/\varepsilon_{11}$), modifies the components p_{11}, p_{12} and p_{41}.

The wave at $\omega + \Omega$ is polarized along

$$\Delta D_i = \Delta\varepsilon_{i3} E_3$$

or

$$\Delta D_1 = 0, \quad \Delta D_2 = \Delta\varepsilon_{23} E_3, \quad \Delta D_3 = \Delta\varepsilon_{33} E_3.$$

For this colinear interaction, the Bragg condition cannot be satisfied with a finite elastic wave vector $\vec{k}$, except if the polarization of the light beam is changed (otherwise $K_D = K_I$ and $k = 0$). The diffracted ray has an ordinary polarization along the x_2-axis $(0, \Delta\varepsilon_{23} E_3, 0)$. The wave vector diagram is shown in fig. 8.18 for a positive uniaxial crystal ($n_e > n_o$). It imposes the condition

$$k = K_I - K_D = \frac{2\pi}{\Lambda_0}(n_e - n_o) = \frac{2\pi}{V} f \qquad (8.40)$$

and therefore the elastic wave frequency must be

$$f = \frac{V}{\Lambda_0} (n_e - n_o). \tag{8.41}$$

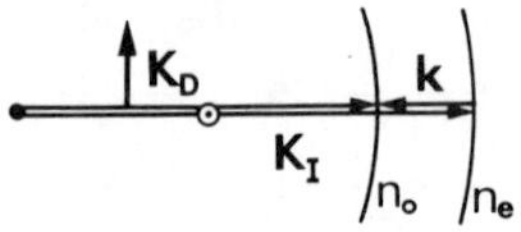

Fig. 8.18. Colinear interaction. In order to obtain an ordinary diffracted ray ($\vec{K}_D$) from an extraordinary ray ($\vec{K}_I$) of wavelength Λ_0, the elastic wave frequency must be $f = V(n_e - n_o)/\Lambda_0$.

Conversely, for a given elastic wave frequency, only light waves satisfying the condition 8.40 can be diffracted. This is the principle of optical filters, tunable by a variation of elastic (stationary or progressive) wave frequency [17, 18]. For instance, if the shear elastic wave frequency ranges between 40 and 68 MHz in a calcium molybdate ($CaMoO_4$) crystal, light waves can be selected with a bandwidth of 8 Å in the 6,700 - 5,100 Å interval.

8.5. INTERACTION OF LIGHT WAVES AND SURFACE ACOUSTIC WAVES

So far, in this chapter, we have dealt with light diffraction by acoustic longitudinal or shear bulk waves. Optical waves can also be diffracted by surface waves especially by Rayleigh waves. As these waves have at least a longitudinal and a shear displacement component, a combination of the previously described effects can generally be expected; for example, only a fraction of the diffracted light has a rotated polarization [19]. In addition, the surface experiences a deformation due to Rayleigh waves and diffracts the reflected light [20]. Two major cases can be considered.

The first, shown in Fig. 8.19, refers to the reflection onto the surface of the crystal of a light beam whose diameter is small with respect to the acoustic wavelength [21].

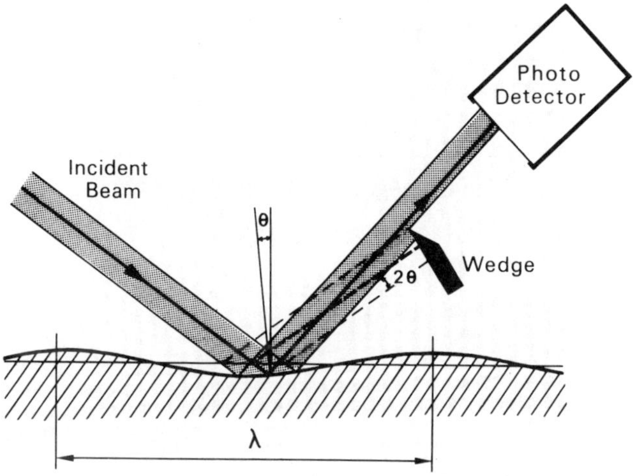

Fig. 8.19. Configuration for detection of surface ripple by narrow beam deflection technique. The reflected pencil of light oscillates at the acoustic frequency.

Then the modulation of permittivity ε_{ij} need not be taken into account. As the strain propagates, the angle of reflection of the light pencil varies at the acoustic frequency because of the rippling of the surface. An obstacle, such as a wedge, is inserted in the path of the reflected beam so that the intensity of the light received by the photodetector varies at the same acoustic frequency. The electric signal from the photodetector is phase-shifted, with respect to the electric signal applied to the transducer which generates the wave, by an amount that depends on the point at which the light pencil impinges. Two points, half a wavelength apart, provide two signals whose phases differ by π. The exploration of the surface can be achieved by moving the light beam. The corrugated surface may be displayed on the screen of an oscilloscope if this motion is synchronized with that of the electron beam whose intensity is a function of the phase at the point where the beam impinges. Figure 8.20 shows the result of an experiment [21] carried out with a steel plate notched by two

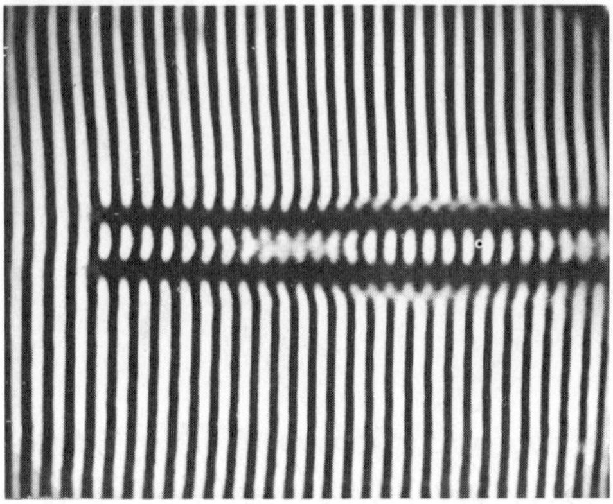

Fig. 8.20. Display, using the light pencil deflection method, of Rayleigh waves (f = 8 MHz) propagating on a steel plate. The two black lines perpendicular to the wavefronts define two grooves (observed length: 15 mm) (Fig. 11.a from reference [21], by courtesy of R. Adler, Zenith, Chicago).

grooves aligned along the Rayleigh wave propagation direction (f = 8 MHz). The break of the wavefronts defines the grooves.

The orientation of the wedge is important. According to whether its edge is perpendicular or parallel to the acoustic wavefronts, the amount of light that reaches the photodetector varies between zero and a maximum while the reflected beam oscillates. Thus waves propagating along various directions are made visible by an appropriate rotation of the wedge. Power densities as low as one μW/cm corresponding to amplitudes smaller than 0.1 Å can be detected.

In the second case, the modulation of the dielectric permittivity is exploited. This effect is significant [22] only if the light wave crosses the acoustic beam sideways. However, this interaction condition demands a high concentration of optical energy near the surface. A good approach consists of guiding the light beam in a thin layer (Fig. 8.21) deposited on the piezoelectric substrate which

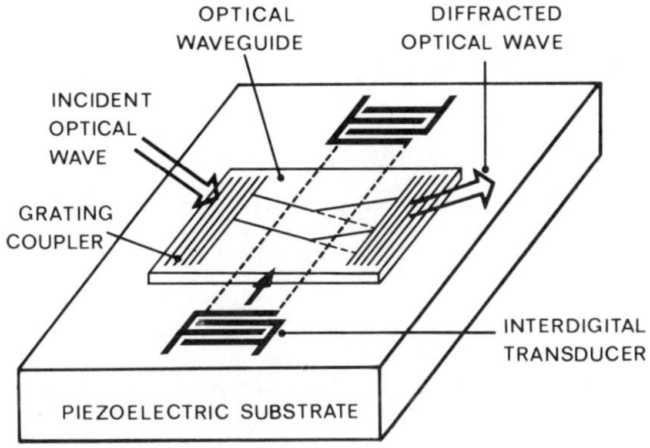

Fig. 8.21. Diffraction of light, guided in a thin layer, by Rayleigh waves.

is the acoustic propagation medium [23]. The plane layer is made of glass with an index higher than that of the substrate. The light waves are fed into and extracted from this layer through two gratings [24]. The guide thickness (≃ 1 μm) is less than the acoustic wavelength (32 μm in quartz at 100 MHz). In another process, the optical waveguide and the acoustic propagation medium are made of the same material, lithium niobate or tantalate. Then the plane waveguide is fabricated by a thermal out-diffusion of lithium [25, 26] or, with better results, by in-diffusion into $LiNbO_3$ of metals such as Ni, Au or Ag [27]. What was explained in previous sections in the context of bulk waves applies to some extent here. At Bragg incidence, only one beam is diffracted. With an electrical drive power of the order of one Watt and with the layer procedure almost all the light is deflected [28, 29]. The structure of the indiffused guide is efficient; only 0.1 Watt electrical drive power is required for 100% deflection of the incident beam [30]. However the bandwidth of these deflectors is limited to a few tens of MHz. More sophisticated transducer configurations are needed [31] to increase this bandwidth.

REFERENCES

[1] L. Brillouin. Ann. de Phys., 17, 88 (1922).

[2] R. Lucas and P. Biquard, J. Physique, 10, 464 (1932).

[3] P. Debye and F.W. Sears, Proc. Nat. Acad. Sc., 18, 409 (1932).

[4] R. Adler, IEEE Spectrum, 4, 42 (1967).

[5] W. Primak and D. Post, J. Appl. Phys., 30, 779 (1959).

[6] D.A. Pinnow and R.W. Dixon, Appl. Phys. Letters, 13, 156 (1968).

[7] R.W. Dixon, J. Appl. Phys. 38, 5149 (1967).

[8] N. Uchida and Y. Ohmachi, J. Appl. Phys. 40, 4692 (1969).

[9] F. Pockels, Ann. Physik. Chem., 37, pp. 269-372 (1889).

[10] D.F. Nelson and M. Lax, Phys. Rev. Lett., 24, 379 (1970).

[11] D.F. Nelson and P.D. Lazay, Phys. Rev. Lett., 25, 1187 (1970).

[12] D.F. Nelson, P.D. Lazay and M. Lax, Phys. Rev., B6, 3109 (1972).

[13] D.F. Nelson and M. Lax, Phys. Rev. B3, 2778 (1971).

[14] T.M. Smith and A. Korpel, IEEE J. Quantum Electronics, QE-1, 283 (1965).

[15] D.A. Pinnow, IEEE J. Quantum Electronics, QE-6, 223 (1970).

[16] R.W. Dixon, IEEE J. Quantum Electronics, QE-3, 85 (1967).

[17] S.E. Harris, S.T.K. Nieh and D.K. Winslow, Appl. Phys. Letters, 15, 325 (1969).

[18] S.E. Harris, S.T.K. Nieh and R.S. Feigelson, Appl. Phys. Letters, 17, 223 (1970).

[19] R.M. Montgomery and E.H. Young Jr., J. Appl. Phys. 42, 2585 (1971).

[20] A. Korpel, L.J. Laub and H.C. Sievering, Appl. Phys. Lett., 10, 295 (1967).

[21] R. Adler, A. Korpel and P. Desmares, IEEE Trans. Son. Ultrason. SU-15, 157 (1968).

[22] E. Salzmann and D. Weismann, J. Appl. Phys., 40, 3408 (1969).

[23] P.K. Tien, R. Ulrich and R.J. Martin, Appl. Phys. Lett., 14, 291 (1969).

[24] M.L. Dakss, L. Kuhn, P.F. Heidrich and B.A. Scott, Appl. Phys. Lett., 17, 265 (1970).

[25] I.P. Kaminow, J.R. Carruthers, Appl. Phys. Lett., 22, 326 (1973).

[26] R.V. Schmidt, I.P. Kaminow, J.R. Carruthers, Appl. Phys. Lett., 23, 417 (1973).

[27] R.V. Schmidt and I.P. Kaminow, Appl. Phys. Lett., 25, 458 (1974).

[28] L. Kuhn, M.L. Dakss, P.F. Heidrich and B.A. Scott, Appl. Phys. Letters, 17, 265 (1970).

[29] P. Hartemann, D. Ostrowsky, M. Reiber and R. Torguet, J. Physique, C6, 33, 216 (1972).

[30] R.V. Schmidt and I.P. Kaminow, IEEE J. Quantum Electronics, QE-11, 57 (1975).

[31] C.S. Tsai, Le T. Nguyen, S.K. Yao and M.A. Alhaider, Appl. Phys. Lett., 26, 140 (1975).

BIBLIOGRAPHY

<u>Crystal optics. Acousto-optic tensor</u>

G. Bruhat and A. Kastler. "Cours de physique générale - Optique", chap. XX, Paris: Masson (1965).

J.F. Nye. "Physical properties of crystals", chap. 13, Oxford: Clarendon Press (1960).

S. Bhagavantam. "Crystal symmetry and Physical properties", chap. 16, London and New York: Academic Press (1966).

<u>Light wave. Elastic bulk wave interaction</u>

References 1 to 4.

C.V. Raman and N.S.N. Nath. "The diffraction of light by high frequency sound: Part I", Proc. Indian Acad. Sci., 2, 406 (1935); "Part II", 2, 413 (1935); "Part III", 3, 75 (1936); "Part IV", 3, 119 (1936); "Part V", 3, 459 (1936). "Generalized Theory", 4, 222 (1936).

A.B. Bhatia and W.S. Noble. "Diffraction of light by ultrasonic waves", Proc. Roy. Soc., 220A, 356 (1953).

J.C. Slater. "Interaction of waves in crystals", Rev. Mod. Phys., 30, 197 (1958).

C.F. Quate, C.D. Wilkinson and D.K. Winslow. "Interaction of light and microwave sound", Proc. IEEE, 53, 1604 (1965).

M.G. Cohen and E.I. Gordon. "Acoustic beam probing using optical techniques", Bell Syst. Tech. J., 44, 693 (1965).

E.I. Gordon. "A Review of acoustooptical deflection and modulation devices", Proc. IEEE, 54, 1391 (1966).

W.R. Klein and Bill D. Cook. "Unified approach to ultrasonic light diffraction", IEEE Trans. Son. Ultrason. SU-14, 123 (1967).

E.G. Spencer, P.V. Lenzo and A. Ballman. "Dielectric materials for electrooptic, elastooptic and ultrasonic device applications", Proc. IEEE, 55, 2074 (1967).

R.W. Dixon. "Acoustooptics interactions and devices", IEEE Trans. Electron. Devices, ED-17, 229, (1970).

R.W. Damon, W.T. Maloney and D.H. McMahon. "Interaction of Light with Ultrasound and Applications". In Physical Acoustics (W.P. Mason and R.N. Thurston, Eds), Vol. 7, chap. 5, New York: Academic Press (1970).

J.M. Bauza, C. Carles and R. Torguet. "Etude théorique et expérimentale de la lumière par de fortes interactions acousto-électriques", Acoustica, vol. 30, no. 4 (janv. 1974).

J. Sapriel. "L'Acousto-optique", Paris, Masson, 1976.

Light wave-elastic surface wave interaction

A.J. Slobodnik, Jr. "Microwave frequency acoustic surface wave propagation losses in $LiNbO_3$", Appl. Phys. Lett., 14, 94 (1969).

E.G. Lean and C.G. Powell. "Optical probing of surface acoustic waves", Proc. IEEE, 58, 1939 (1970).

L. Kuhn, P.F. Heidrich and E.G. Lean. "Optical guided mode conversion by an acoustic surface wave", Appl. Phys. Lett., 19, 428 (1971).

E.G. Lean. "Interaction of light and acoustic surface waves". In Progress in Optics (E. Wolf, Ed.), Vol. 11, chap. 3, Amsterdam: North-Holland (1973).

H. Sasaki, J. Kushibiki and N. Chubachi. "Efficient acousto-optic TE⇄TM mode conversion in ZnO films", Appl. Phys. Lett., 25, 476 (1974).

E. Bridoux, J.M. Rouvaen, M. Moriamez, R. Torguet and P. Hartemann. "Optoacoustic testing of microsound devices", J. Appl. Phys., 45, 5156 (1974).

EXERCISES

8.1. In a uniaxial crystal, how do the indices depend on the direction of propagation defined by its polar angle θ? (fig. 8.22).

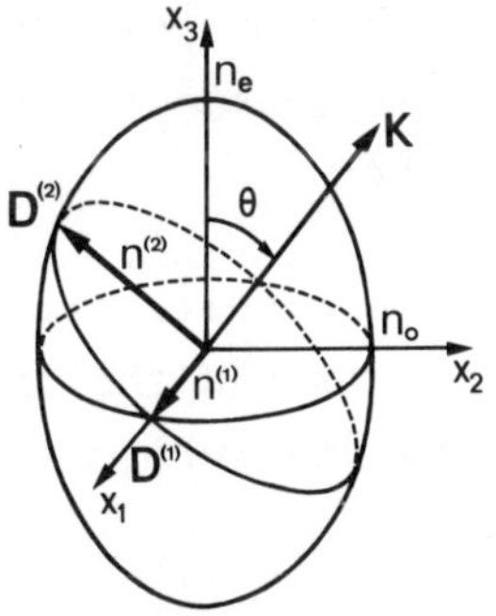

Fig. 8.22.

Solution. For vibration $D^{(1)}: n^{(1)} = n_o \quad \forall\, \theta$

For vibration $D^{(2)}$

$$\frac{x_1^2 + x_2^2}{n_o^2} + \frac{x_3^2}{n_e^2} = 1$$

where $x_1 = 0 \quad x_2 = -n \cos\theta \quad x_3 = n \sin\theta$

Hence

$$n^{(2)} = \frac{n_o n_e}{(n_e^2 \cos^2\theta + n_o^2 \sin^2\theta)^{1/2}} .$$

8.2. A monochromatic light beam penetrates along the x_2-axis into a crystal where elastic waves of the same frequency propagate in every direction of the x_1x_3 plane.

What can be seen on a screen parallel to the x_1x_3 plane at a distance D? What happens in the case of an isotropic solid?

Solution. The first order deflected beam is at an angle $\theta \simeq \Lambda/\lambda = vf/FV$ (formula 8.27). In the direction of propagation $\vec{n}$ of the active elastic wave, it produces a spot at a distance $d = Dvf/FV$, proportional to the reciprocal of the elastic wave velocity in the $\vec{n}$ direction. Since this latter direction can be anywhere in the x_2x_3 plane, the three cross sections of the slowness surface in the x_1x_3 plane, appear on the screen (§ 5.1.6); these are two circles for an isotropic solid. This method, due to Schaefer and Bergmann, provides a measurement of the elastic wave velocity.

8.3. For the interaction conditions of Table 8.14, what are the optical coefficients and the indices involved in the figure of merit of paratellurite (TeO_2)?

Solution.

1. An elastic wave polarized along $[1\bar{1}0]$ and propagating along [110] generates the strains

$$S_{11} = \frac{\partial u_1}{\partial x_1}, \quad S_{22} = \frac{\partial u_2}{\partial x_2} = -S_{11},$$

$$S_{12} = \frac{1}{2}\left(\frac{\partial u_1}{\partial x_2} + \frac{\partial u_2}{\partial x_1}\right) = 0$$

since $\partial/\partial x_1 = \partial/\partial x_2$ and ${}^{o}u_2 = -{}^{o}u_1$. The elastic power density $P = \langle\mathscr{E}\rangle V$, where

$$\langle\mathscr{E}\rangle = \frac{1}{2}\, c_{ijk\ell}S_{ij}S_{k\ell} = \frac{1}{2}[c_{11}+c_{22}-2c_{12}]S_{11}^2$$

$$= (c_{11}-c_{12})S_{11}^2$$

can be put into the form 8.34 where $S = 2S_{11}$:

$$P = 2\rho V^3 S_{11}^2 = \frac{1}{2}\,\rho V^3 S^2.$$

According to equation 8.20:

$$\Delta\varepsilon_{i\ell} = -\,\varepsilon_{ii}\varepsilon_{\ell\ell}\left[(P_{i\ell 11}-P_{i\ell 22})\frac{\partial u_1}{\partial x_1} + (P_{i\ell 21}-P_{i\ell 12})\frac{\partial u_2}{\partial x_1}\right]$$

i.e. $\Delta\varepsilon_{i\ell} = - \varepsilon_{ii}\varepsilon_{\ell\ell}(p_{i\ell 11}-p_{i\ell 22})S_{11}$

because for an uniaxial crystal $p_{i\ell 21} = p_{i\ell 12}$.
It yields

$$\Delta\varepsilon_{12} = \Delta\varepsilon_{13} = \Delta\varepsilon_{23} = 0$$

and

$$\Delta\varepsilon_{33} = - \varepsilon_{33}^2(p_{31} - p_{32})S_{11} = 0$$

for

$$p_{32} = p_{31} .$$

Given the polarization (E_1, E_2, 0) of the incident light wave along the Z-axis (index n_o), the diffracted wave arising from
$\Delta D_i = \Delta\varepsilon_{i1}E_1 + \Delta\varepsilon_{i2}E_2$:

$$\Delta D_1 = \Delta\varepsilon_{11}E_1, \quad \Delta D_2 = - \Delta\varepsilon_{11}E_2, \quad \Delta D_3 = 0$$

has an amplitude proportional to $\Delta\varepsilon_{11}$. This variation is linked to $S = 2S_{11}$:

$$\Delta\varepsilon_{11} = - \varepsilon_{11}^2 \frac{p_{11} - p_{12}}{2} S = - \varepsilon^2 pS$$

through the acousto-optic coefficient

$$p = \frac{p_{11} - p_{12}}{2} .$$

2. For the longitudinal wave along [001], $S = S_{33}$ and

$$\Delta\varepsilon_{i\ell} = - \varepsilon_{ii}\varepsilon_{\ell\ell}p_{i\ell 33}S_{33}$$

or

$$\Delta\varepsilon_{11} = - \varepsilon_{11}^2 p_{13}S = \Delta\varepsilon_{22}, \quad \Delta\varepsilon_{33} = - \varepsilon_{33}^2 p_{33}S$$

and $\Delta\varepsilon_{ij} = 0$ if $i \neq j$.

The light wave polarized along (E_1, 0, 0), whose index is n_o, is diffracted into a wave of polarization ($\Delta D_1 = \Delta\varepsilon_{11}E_1$, 0, 0) with modulus $p = p_{13}$.

8.4. Check that the values for the figure of merit of $PbMoO_4$ are very close to each other, for both conditions of table 8.14 (light polarization parallel or orthogonal to the wave vector plane).

Solution. The strain S_{33} induces the variations

$\Delta\varepsilon_{22} = \Delta\varepsilon_{11} = -\varepsilon_{11}^2 p_{13} S_{33}$ and $\Delta\varepsilon_{33} = -\varepsilon_{33}^2 p_{33} S_{33}$.

a) Parallel polarization $(0, 0, E_3) \Rightarrow$ index n_e; $\Delta D_3 = \Delta\varepsilon_{33} E_{33} \Rightarrow$ acousto-optic constant $p_{33} \Rightarrow M = n_e^6 p_{33}^2/\rho V^3 = 23.9\ M_o$.

b) Orthogonal polarization $(E_1, 0, 0) \Rightarrow$ index n_o; $\Delta D_1 = \Delta\varepsilon_{11} E_1 \Rightarrow$ acousto-optic constant $p_{13} \Rightarrow M = n_o^6 p_{13}^2/\rho V^3 = 23.5\ M_o$.

8.5. Show that the interaction bandwidth is proportional to the reciprocal of the time needed for a wave front to travel across a light ray at the Bragg angle.

Solution.

$$\Delta t = \frac{e\alpha_o}{V} = \frac{e\Lambda}{V^2} f_o \quad \text{or} \quad \Delta f = \frac{1.77}{\Delta t} \quad \text{(equation 8.42).}$$

Chapter 9

Applications to Signal Processing

Since the development of the amplifier tube during World War I, elastic waves have played an important rôle in electronics. Until then the major goal of the study of elastic waves was the analysis of the propagation modes of earthquakes. By recording the arrival times of the disturbances at various points, geophysicists tried to locate the seismic epicentre.

C. Chilowsky and P. Langevin in 1916 suggested that elastic waves could be exploited in the detection of submarine objects by means of echoes. Relying on the work of Jacques and Pierre Curie - who, after their discovery of piezoelectricity in 1880 had designed a quartz balance - Paul Langevin designed piezoelectric transducers (quartz-steel sandwich) to transmit elastic waves in water at about 150 kHz and to detect the waves that had been reflected by obstacles. This device, now known as sonar,* has undergone significant development in both the civil and military areas. It is now established that it played an essential part (by locating enemy submarines) in reducing allied shipping losses, especially in the North Atlantic, during World War II. Sonar is currently used for detecting shoals of fish, mapping the ocean floor, and monitoring the position of prospecting ships.

---

* From SOund NAvigation and Ranging.

Although this book is concerned with elastic wave propagation in solids, we mention this application because of its major historical and practical importance [1].

In about 1920, W.G. Cady utilized the high Q-factor of quartz piezoelectric resonators for signal filtering and for stabilizing the frequency of oscillators. The resonator, as we have seen at § 7.1.2, is a special case of the transducer. Quartz filters, made of several resonators, are now common components, even though they are still evolving (monolithic filters). We recall their major characteristics (§ 9.4) only for comparison with Rayleigh wave filters.

The storing of information in a memory in the form of elastic waves dates from the development of radar and of the electronic computer (1940-1945). The first delay lines with a liquid propagation medium (water, mercury) were studied in England and in the USA; the first delay lines with a solid propagation medium (glass, silica) were built in Germany (see historical summary in ref. 2). These lines, equipped with quartz transducers, could operate at 10 MHz with bandwidths of a few MHz. They could provide delays of about 100 μs. Since then, this delay has been increased by a factor of 10 by the use of some ingenious structures (multiple reflexion polygonal lines [2]).

At the beginning of the sixties, several people at Bell Laboratories [3] revealed the relevance of pulse compression in radar systems (RAdio Detecting And Ranging), and ways of producing dispersion effects using elastic waves were studied. Since that time, several principles have been utilized in the construction of dispersive lines, namely group velocity variation for specific modes in waveguides (Lamb waves); bulk wave, then surface wave (Rayleigh wave) interference effects, with various transducer structures, association of propagation media with different velocities (Love waves).

The aim of this chapter is to present some relatively recent devices which illustrate the application of the principles. Although we may often give orders of magnitude, we do not intend to quote performances. Lines are grouped according to their function: delay, pulse compression, filtering, memory, convolution. As far as possible, the state of the art is specified: some lines are still at the laboratory stage, others have been developed commercially. Finally, we describe a surface wave multistrip coupler, a light modulator and a light deflector based on acousto-optic interaction. It seems useful to discuss first the general structure of an elastic wave delay line.

9.1. GENERAL STRUCTURE OF AN ELASTIC WAVE DELAY LINE

Elastic wave delay lines dealt with in this chapter aim to process electronic signals, i.e. intentionally to change their parameters. Whatever the nature of the process, the most involved line may be represented by the layout of fig. 9.1.

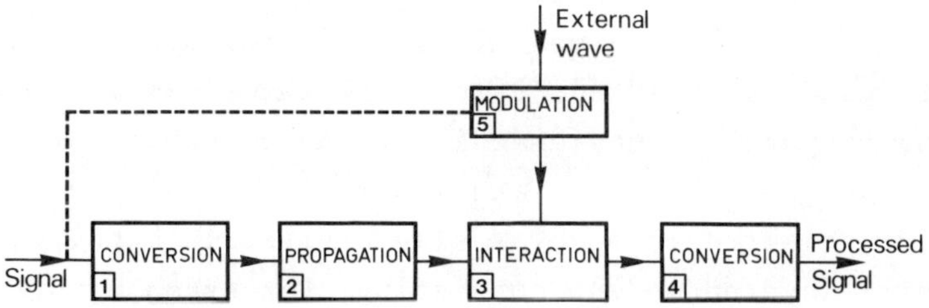

Fig. 9.1. General structure of an elastic wave delay line.

9.1.1. Conversion of the Electric Signal into an Elastic Wave

The electric signal to be processed, phase or amplitude modulated, is converted into an elastic wave input to the line (part 1).

This conversion of energy, the principle of which consists of generating vibrations in a piezoelectric material from the electric field of the signal, can be achieved in such a way that the transducer is equivalent to a single localized source (§ 7.1) or to a set of sources distributed on a surface (§ 7.2). In the latter case, "box" no. 1 can act as more than a mere transducer if the interference effects of the waves emitted by all these sources are exploited. The distribution of ultrasonic sources can be adequately determined so as to produce the desired signal modification.

The transducer, which has to be preceded by a matching circuit, gives rise to losses which depend firstly on the central frequency of the signal, secondly on the bandwidth of the signal, and thirdly on the nature of the elastic waves to be generated. These losses range from a few dB to ten or more dB. The elastic wave frequencies now employed in such lines lie between a few MHz and several GHz.

9.1.2. Elastic Wave Propagation

The elastic waves propagate either in the bulk of the material or along its surface, with a velocity about 10^5 times lower than the electromagnetic wave velocity. Thus they travel only a few millimeters per microsecond, which is used in "box" no. 2 to delay the signal.

The propagation medium which we have supposed to be a solid can be homogeneous and isotropic as is fused silica or anisotropic such as single crystal sapphire, or a multi-layered medium consisting, for example, of a layer on a substrate. In the case of a crystal, the cut must be chosen so that propagation takes place with the desired mode. The velocity is constant for all frequencies, because the wavelength is much larger than the atomic spacing (§ 1.2.2). For the layered material, the wave velocity is frequency dependent, ranging between the layer velocity and the substrate velocity. The delay of the signal depends on frequency. A dispersive effect can also be created in a homogeneous medium provided that the waves are guided (§ 1.1.2.2).

Many variations are possible, for example, the propagation medium may be a bar, a polygonal block (the path of the wave reflected at the polygon sides is a broken line and the delay can be as long as 1,000 μs), a long metal strip (dispersive delay line). Mode conversion can be generated deliberately by a reflexion (§ 5.2.3).

The coefficient of attenuation of elastic waves (fig. 9.3) depends on the medium and polarization of the wave and increases with frequency. Defects become of major importance when their size is of the order of the wavelength. Losses are also greater in a polycrystal than in a single crystal.

9.1.3. Interaction of the Elastic Wave with an External Wave

"Box" 3 symbolizes an interaction between the elastic wave and an external wave which may be a light wave or a space charge wave linked to a movement of electrons.

If the interaction is with an optical wave, it yields a diffracted optical wave which carries the elastic beam characteristics (frequency, power) together with those of the incident light beam (space and time modulation). The diffraction comes from the index variation wave which propagates with any elastic wave (§ 8.4). "Box" 5 symbolizes a modulation, generally related to the waveform of the signal. For example, it can be an amplitude mask, i.e. a spatial replica of the signal code.

If it is an interaction with electrons, drifting in the direction of elastic wave propagation, the coupling provides an amplification or attenuation of the elastic wave depending on whether the carriers travel at above or below the acoustic velocity. Coupling between an elastic wave and charge carriers requires a piezoelectric material in order that an electric field be associated with the elastic wave, and a semiconductor where charge carriers can move [4]. A crystal - cadmium sulphide, zinc oxide - may exhibit both features (semi-conductivity and piezoelectricity): coupling takes place in the bulk of the material. However, more possibilities are available when each feature is associated with a distinct material: coupling between the two waves is induced by bringing the two crystals together. Surface waves are adequate for this second approach, especially when the semiconductor is a layer deposited on the piezoelectric substrate. The elastic wave amplification can reach several tens of dB [5]. However, we are not going into more detail about these amplification phenomena which, to our knowledge, have not yet been exploited outside the laboratory.

9.1.4. Conversion of the Resulting Wave into an Electric Signal

In order to obtain the processed signal at the line output, the wave that results from interaction must be converted into an electric signal. Depending upon the nature of wave, this conversion is carried out by a localized transducer, a transducer on the surface analogous to the input device, or by any other type of transducer such as a photomultiplier or a photodiode if an optical wave carries the information after the interaction. The transducer characteristics and the matching structure must be chosen so as not to alter the transformation of the signal.

9.2. DELAY FUNCTION

In this section, we shall restrict ourselves to the description of fairly recent devices, namely bulk wave delay lines with thin layer transducers working at frequencies greater than one gigahertz, Rayleigh wave delay lines, acousto-optic interaction lines with continuously variable delay. We shall not consider bulk wave delay lines with a more traditional technology (single crystal transducers soldered with indium and thinned* for f < 1000 MHz, or ceramic bonded transducers for f < 50 MHz). Many types of low frequency delay lines have been constructed [7].

9.2.1. Bulk Wave Delay Lines

Before the advent of these delay lines, no device existed capable of providing a microsecond delay in the gigahertz frequency domain. A signal can be delayed by a cable, but a 2 μs delay requires about 400 metres of cable, which implies a large volume and a prohibitive loss (more than 100 dB at 1 GHz). In practice cables are limited to delays of 0.1 μs at frequencies below 1 GHz. Elastic wave delay lines are now able to provide a delay of more than ten microseconds, in a volume of a few cm^3, with reasonable losses up to 5 GHz.

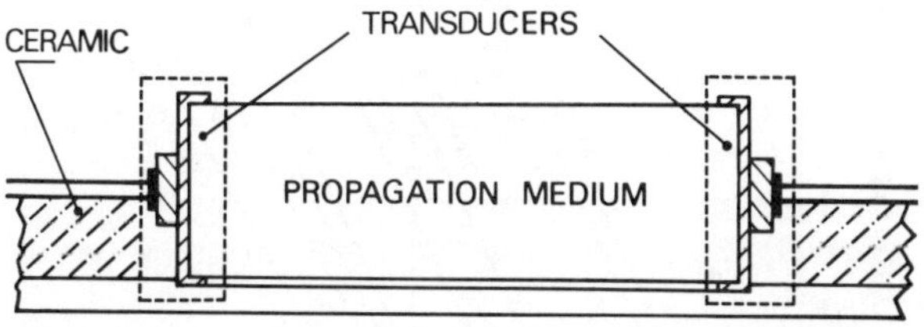

Fig. 9.2. Structure of a bulk wave delay line.

9.2.1.1. Structure

The structure, shown in fig. 9.2, consists of an input transducer, a zinc oxide transducer, for example, launching longitudinal waves, then a sapphire single

---

* This soldering technique by metal diffusion has been improved [6]. The lithium niobate transducer thickness can now be reduced to 1 μm (f ≃ 3 GHz) by ion bombardment. Delay lines made in this way may compete with the lines described here, since single crystal transducers have lower losses than layered transducers and allow for a wider choice of propagation media (§ 7.1.4).

crystal, whose length controls the delay, and finally a zinc oxide output transducer. Both transducers are alike, but the ZnO layers generally have different thicknesses and accordingly different resonant frequencies, in order to enlarge the bandwidth. Sapphire is chosen as the propagation medium because very long (up to 30 cm) single crystals are available at a moderate price and also because of its comparatively low attenuation coefficient (1 dB/μs at 2 GHz), as indicated in Fig. 9.3 where the propagation loss of a few materials is plotted. These losses increase as approximately the square of frequency.

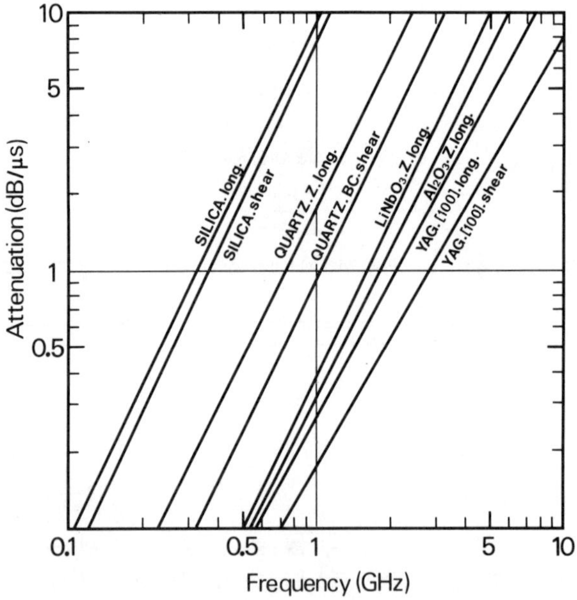

Fig. 9.3. Elastic wave attenuation vs frequency, in a few solids (after fig. 4 of reference 88).

Between the cable that carries the signal and each transducer, a strip line, made on a ceramic, ensures that the generator and transducer are matched (fig. 9.4). The strip and transducer are connected by a gold wire (diameter 50 μm) soldered to the aluminium outer electrode by

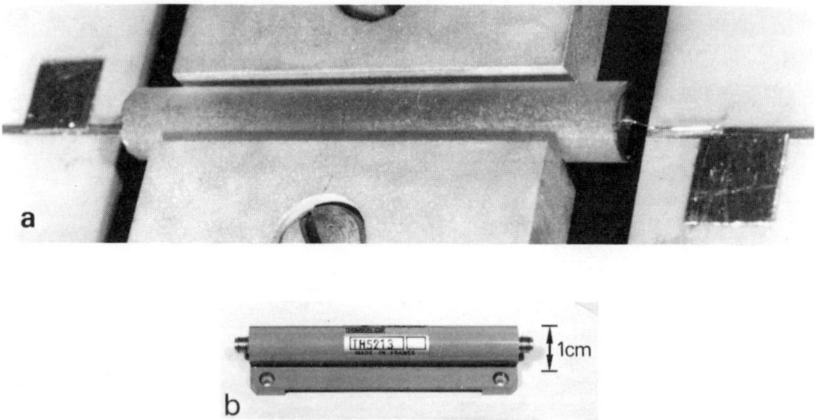

Fig. 9.4. Bulk wave delay line.
a) Sapphire single crystal, of diameter 3 mm, with, at each end, zinc oxide thin film transducer.
b) Coaxial structure. (Thomson CSF).

thermocompression. The diameter of this electrode, which is less than one millimetre, determines the transducer capacitance and the size of its piezoelectrically active part.

9.2.1.2. Examples of characteristics

Since matching at both ends of the line of length ℓ cannot be perfect, the delay is measured with an accuracy of a few parts in 10^4, by determining the resonance frequencies

$$f = n \frac{V}{2\ell}, \quad n \text{ an integer}$$

where, because a stationary wave is superimposed on the progressive wave, the output signal is a maximum. The time delay τ is derived from the number $n_2 - n_1$ of maxima between frequencies f_2 and f_1:

$$\tau = \frac{n_2 - n_1}{2(f_2 - f_1)} \qquad (9.1)$$

Figure 9.5 shows the principal characteristics of a 2 μs delay line [8], of centre frequency 1.5 GHz. The loss is 33 dB for a 700 MHz bandwidth. The standing wave

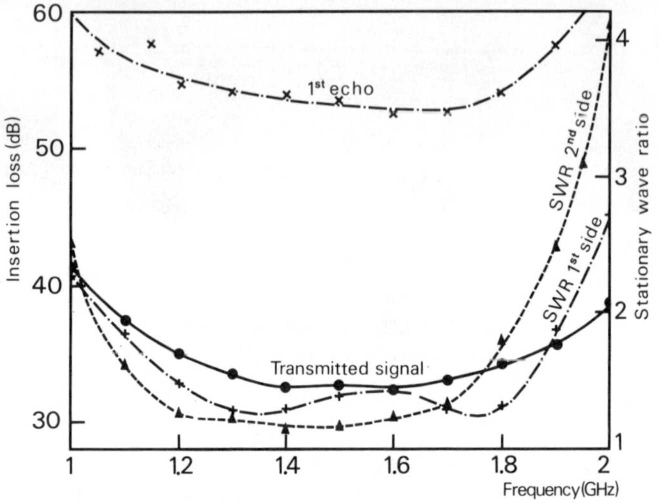

Fig. 9.5. Example of characteristics for a 2 μs delay line with centre frequency 1.5 GHz.

ratio (§ 1.1.2.1) at the input and at the output is less than 2. The spurious signal, the first echo, is at a level 20 dB below the transmitted signal. The signal which is not delayed, i.e. which reaches the output by direct radiation, is at -30 dB with respect to the main signal.

The following table indicates the domain of operation of such lines, but the figures given should not be regarded as limits.

- centre frequency f_o	.5 to 10 GHz
- bandwidth	≤ 50%
- time delay τ	.5 to 10 μs
- temperature dependence of τ	
(-50°C to 120°C): $\frac{1}{\tau}\frac{d\tau}{d\theta}$ (Al_2O_3)	25.10^{-6} $(^oC)^{-1}$
- standing wave ratio	≤ 2
- radiated signal level	- 30 dB
- first echo level	-15 to -30 dB
- insertion loss	20 to 80 dB

The last figures include:

- the transduction loss, which increases with bandwidth;
- the propagation loss, proportional to the delay and to the square of the frequency;
- the diffraction loss (beam spreading), which depends on transducer size, frequency and delay (§ 1.3.6).

The echo delay line comprises only one transducer and a free end, polished so that the elastic wave is reflected; it is a variant of delay line. About ten echoes, separated by a time interval equal to twice the transit time, are easily generated by a very high frequency electric pulse. If the ends of the propagation medium are adequately shaped, so as to focus the beam after each reflexion and to avoid spurious echoes, very large delays can be achieved, but at the expense of bandwidth. For example, delays of 100 μs can be obtained with bandwidths of the order of 5%.

9.2.2. Rayleigh Wave Delay Lines

Rayleigh waves propagate on the surface of solids. This property has been used to design multiple output fixed delay lines, but up to now at lower frequencies (f < 400 MHz) than those achieved for bulk waves. However, delay lines operating beyond one gigahertz have been constructed in the laboratory [9, 10].

The substrate is a piezoelectric single crystal: quartz (Y or ST cut, X propagation) or lithium niobate (Y cut, Z propagation). A one microsecond delay results from a 3.15 mm path in quartz or 3.49 mm in $LiNbO_3$. Fig. 9.6 shows a quartz line with ten outputs; the bandwidth is greater than 6 MHz about the centre frequency of 31 MHz. Each transducer has four pairs of fingers of 26 μm width. The loss is about 35 dB and, between the first and the last receiver, varies only by a few dB.

Delay lines with a single output and a large bandwidth (50%) have been designed as shown in fig. 9.7. The finger spacing varies but the output transducer is derived from

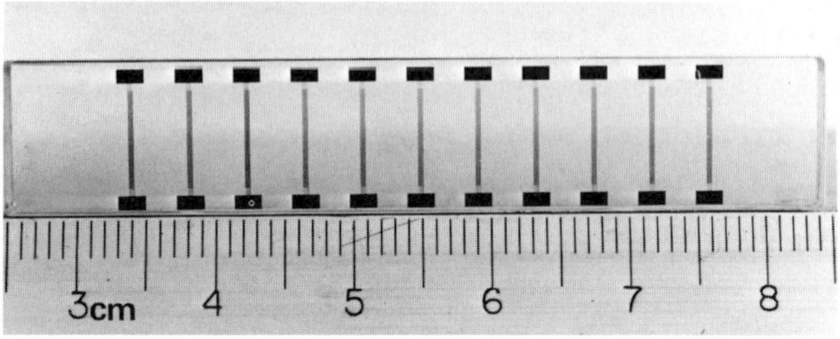

Fig. 9.6. Raylcigh wave delay line. Each transducer has 8 fingers, of width 26 μm.

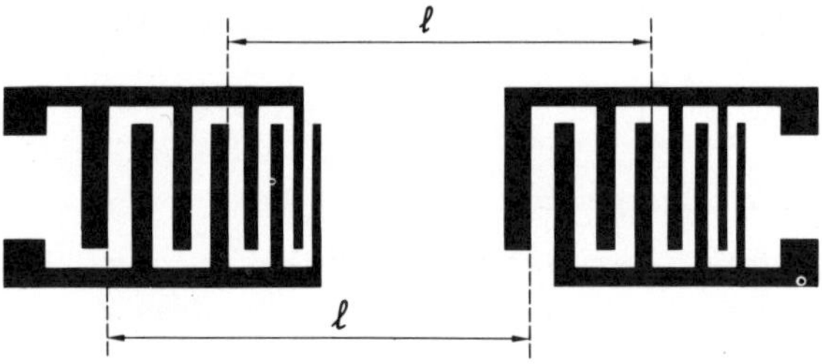

Fig. 9.7. Broadband delay line. Both transducers are unchanged after translation through a distance ℓ. The delay is ℓ/V_R.

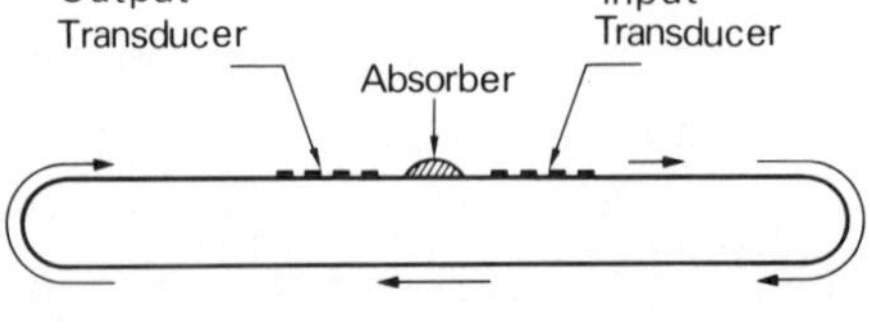

Fig. 9.8. Wraparound delay line. The radius of curvature is large compared to the wavelength.

the input transducer by a translation which defines the delay, the same for all frequencies.

Rayleigh waves can propagate with reasonable loss on curved surfaces if the radius of curvature is large compared to the wavelength [11, 12]. This is exploited in wraparound delay lines (fig. 9.8) and even in helical path delay lines [13, 14].

Large time delays (1 ms) can in principle be obtained with low propagation velocity materials: a loop 10 cm long on a $Bi_{12}GeO_{20}$ crystal ((111) cut, $[1\bar{1}0]$ propagation, velocity 1,708 m/s) creates a delay of 58.5 μs. In order to avoid beam spreading and related loss, the wave can be guided by a strip deposited on the surface [15]. It has also been suggested [16, 17] that acousto-electric amplification by coupling with drifting carries, as mentioned in § 9.1.3, could compensate for the attenuation due to propagation (a few dB per turn at 100 MHz) in lines of several tens of turns.

Multiple output delay lines are an essential part of digital filters: samples of the signal are tapped by the different transducers, then weighted and added up (§ 9.3.5.2).

Rayleigh wave delay lines are also employed in oscillators [18]. They have the advantage of operating at higher frequencies than classical oscillators using bulk wave resonators (see § 9.4).

9.2.3. Acousto-optic Variable Delay Lines

It was pointed out in chapter VIII that diffraction of light by elastic waves at Bragg incidence gives rise to a single deflected beam of frequency F - f where F is the light beam frequency and f the elastic wave frequency. Beats between the deflected beam and a light beam at frequency F will again yield a signal at frequency f, but this latter signal is delayed with respect to the electric input signal by an amount $\tau = \ell/V$ where ℓ is the propagation distance between the input transducer and the interaction zone, and V is the elastic wave propagation velocity. Any variation of ℓ produces a variation of the time delay τ

[19]. Lines with a continuously variable delay have been built following the design of fig. 9.9 [20].

The laser beam is split into two; the diffracted part of one beam is superposed on the transmitted part of the other. Only one resulting beam is detected at a photodiode. The current intensity, proportional to the square of the amplitude, contains a term at the same frequency as the electric signal applied to the transducer.

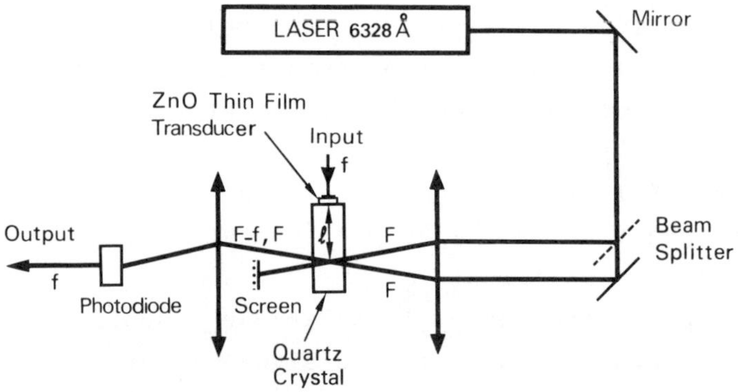

Fig. 9.9. A signal applied to the input is delayed by an amount $\tau = \ell/V$. Since the laser beam is fixed, the delay is varied continuously as the crystal is moved.

There follows an example of the characteristics of delay lines which have been used for calibrating radar altimeters:

- variable delay by manual shift of the crystal in front of the laser: 50 ns to 6 μs;
- centre frequency: 800 MHz;
- bandwidth: 200 MHz;
- loss: 80 dB;
- crystal: quartz ($\vec{X}$): V = 5,720 m/s;
- transducer: thin layer of zinc oxide generating longitudinal waves

Lower loss and shorter delays are obtained with other crystals such as rutile (V_L = 10,670 m/s along [001]).

9.3. PULSE COMPRESSION

Pulse compression is a widely used technique in modern radar and sonar. It has the advantage of increasing the range of the radar (at constant power) without altering the resolution or, conversely, of increasing the resolution at a given range and power. Conventional radars emit constant carrier frequency pulses whose duration is determined by the desired accuracy of detection. Since the pulse amplitude is limited for technological reasons (maximum peak power of the transmitter tubes), once the pulse duration is chosen the range is fixed. This range-accuracy trade-off is only apparent: the range does depend on the signal duration but the resolution capability depends solely on the width of the signal spectrum, and these two quantities (time duration and spectrum width) can be chosen separately. Woodward [21] has shown that the best resolution corresponds to the widest spectrum. For a given pulse duration, the spectrum width can be increased by a suitable modulation of the carrier wave. However, the echo which is reflected by the target is degraded by noise, for modern and conventional radars alike, and removing this noise by means of a bandpass filter is easier when the signal spectrum is narrow (conventional radars). In a modern radar, extraction of a wide spectrum signal from noise requires a more complex detection system, which must be <u>matched</u> to the signal, taking the nature of the noise into account. For white Gaussian noise, extraction of information is best achieved by a filter which yields the <u>autocorrelation</u> of the signal [22, 23]. The resulting time compression accurately localizes the instant of arrival of the echo and improves the resolution.

9.3.1. <u>Filter Matched to a Signal</u>

The impulse response $h(t)$ of the filter which is matched to the real signal $s_1(t)$ should be

$$h(t) = s_1(-t) \qquad (9.2)$$

in order that the output signal $s_2(t)$ given by equation (1.29) be the autocorrelation function of $s_1(t)$ (equation 1.31):

$$s_2(t) = \int_{-\infty}^{+\infty} s_1(\tau)s_1(\tau - t)d\tau \qquad (9.3)$$

Equation 9.2 expresses the fact that the impulse response of the matched filter is the time inverse of the signal (fig. 9.10). Such a filter cannot be implemented in practice: the effect (impulse response) would precede the cause ($\delta(t)$ pulse). Any filter introduces a time delay t_o. The relationship between the signal and the impulse response of the real filter is of the form:

$$h(t) = C\, s_1(t_o - t) \quad C = \text{constant} \qquad (9.4)$$

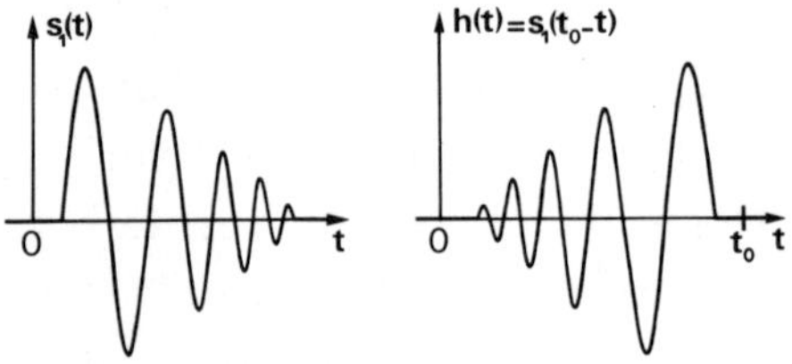

Fig. 9.10: The impulse response h(t) of a real filter matched to a signal $s_1(t)$ is the time inverse $s_1(t)$, apart from a constant delay t_o.

A real matched filter involves a delay line. The transformation of equation (9.4) in the frequency domain is

$$H(\omega) = C\, S_1^*(\omega)e^{-i\omega t_o}. \qquad (9.5)$$

Thus, except for a phase shift due to the delay t_o, the frequency response $H(\omega)$ of the matched filter is the complex conjugate of the signal spectrum $S(\omega)$. Hereafter, the constant t_o, although indispensable physically, will be ignored, and only added at the end of the calculations.

In practice, the signal $s_1(t)$ may have many different forms. Its spectrum can be widened by frequency modulation, as we have already mentioned, or by other methods of coding. The time variation of any of the three characteristic quantities (amplitude, phase, frequency) can be continuous (analogous coding) or discrete (digital coding).

In Figure 9.11 three kinds of signal are compared: a rectangular pulse of time width Θ in which the carrier frequency is frequency modulated, a constant carrier pulse train of total duration Θ divided into N intervals (digits) of duration $\delta = \Theta/N$ occupied or not occupied by elementary pulses and a pulse of duration Θ including phase reversals.

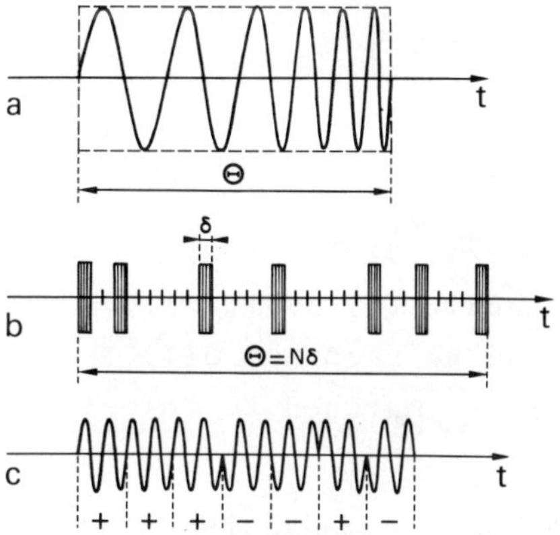

Fig. 9.11. Signals coded by:
(a) frequency modulation;
(b) amplitude modulation;
(c) phase inversion.

Elastic wave delay lines processing these three kinds of signal are described in the following sections. We shall go into more detail about dispersive delay lines matched to a linearly frequency modulated signal. The aim of exercise 9.1 is to determine graphically the response of the filter matched to the pulse sequence of fig. 9.11b.

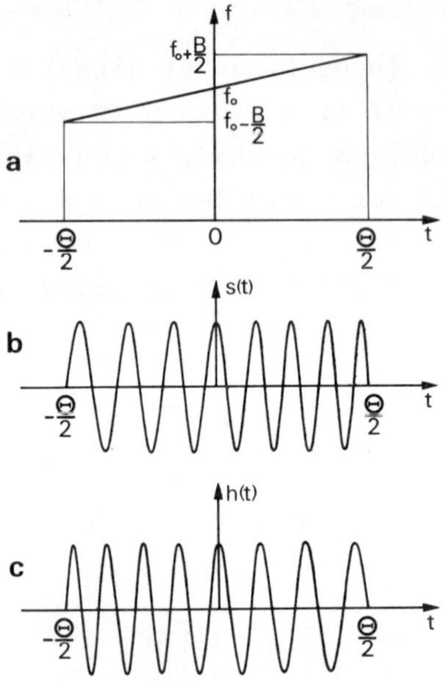

Fig. 9.12. Linear frequency modulation.
(a) instantaneous frequency vs time:
$f = \frac{1}{2\pi} \frac{d\phi_1}{dt}$;
(b) signal $s(t) = \Pi(t/\Theta) \cos \phi_1(t)$;
(c) impulse response $h(t) = s(-t)$ of the filter matched to this signal.

9.3.2. Linearly Frequency Modulated Signal

In the case of a linear frequency modulation as illustrated in fig. 9.12, the time variation of the instantaneous signal angular frequency is

$$\omega = \frac{d\phi_1}{dt} = \omega_0 + \mu t \quad \text{for} \quad -\frac{\Theta}{2} \le t \le \frac{\Theta}{2} \tag{9.6}$$

and the signal $s_1(t)$, with a rectangular envelope, is given by

$$s_1(t) = \Pi(\frac{t}{\Theta}) \cos \phi_1(t) \tag{9.7}$$

where

$$\phi_1(t) = \omega_0 t + \frac{\mu}{2} t^2 . \tag{9.8}$$

The impulse response of the filter matched to this signal is (after 9.2)

$$h(t) = C\ \Pi(\frac{t}{\Theta})\ \cos\ (\omega_0 t - \mu\ \frac{t^2}{2}). \quad (9.9)$$

9.3.2.1. Compressed pulse

The response $s_2(t)$ to the signal $s_1(t)$ applied to the matched filter is the autocorrelation of $s_1(t)$ (equation 9.3). Figure 9.13 shows that:

$$s_2(t) = 0 \qquad \text{for } |t| > \Theta$$

$$s_2(t) = C \int_{\tau_1}^{\tau_2} \cos(\omega_0\tau + \mu\ \frac{\tau^2}{2})\cos[\omega_0(\tau - t) + \mu\ \frac{(\tau-t)^2}{2}]d\tau$$

where

$$\tau_1 = t - \frac{\Theta}{2} \quad \tau_2 = \frac{\Theta}{2} \quad \text{for} \quad 0 < t < \Theta$$

$$\tau_1 = -\frac{\Theta}{2} \quad \tau_2 = t + \frac{\Theta}{2} \quad \text{for} \quad -\Theta < t < 0. \qquad (9.10)$$

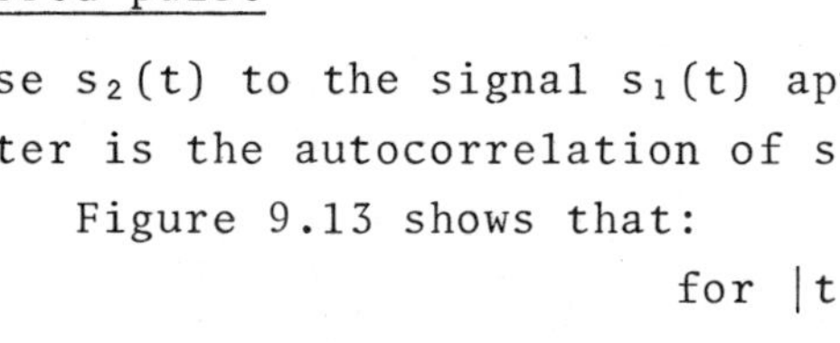

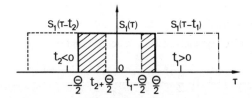

Fig. 9.13. Autocorrelation of a linearly frequency modulated pulse. Depending on the sign of t, the range of integration is $[t - \Theta/2, \Theta/2]$ or $[-\Theta/2, t + \Theta/2]$

The cosine product can be transformed into a sum:

$$s_2(t) = \frac{C}{2} \int_{\tau_1}^{\tau_2} \cos\ [\omega_0 t + \mu t(\tau - \frac{t}{2})]\ d\tau$$

$$+ \frac{C}{2} \int_{\tau_1}^{\tau_2} \cos\ \{\omega_0(2\tau - t) + \frac{\mu}{2}[\tau^2 + (\tau - t)^2]\}\ d\tau$$

The second integral yields a term of frequency $2\omega_0$ which is eliminated in practice and which will be dropped:

$$s_2(t) = \frac{C}{2}\left[\frac{\sin\ [\omega_0 t + \mu t(\tau - t/2)]}{\mu t}\right]_{\tau_1}^{\tau_2}$$

or

$$s_2(t) = \frac{C}{\mu t} \sin\left[\frac{\mu t}{2}(\tau_2-\tau_1)\right] \cos\left[\omega_0 t + \frac{\mu t}{2}(\tau_2+\tau_1-t)\right]$$

According to equations 9.10:

$$\tau_2 + \tau_1 = t \quad \text{and} \quad \tau_2 - \tau_1 = \Theta - |t|$$

$s_2(t)$ becomes:

$$s_2(t) = C \frac{\sin \frac{\mu t}{2} (\Theta - |t|)}{\mu t} \cos \omega_0 t \text{ for } -\Theta < t < \Theta$$

or

$$s_2(t) = C \frac{\sin \frac{\mu t}{2} (\Theta - |t|)}{\frac{\mu t}{2} (\Theta - |t|)} \cdot \frac{(\Theta - |t|)}{2} \cos \omega_0 t \quad (9.11)$$

The output signal, known as the compressed pulse, has as envelope in the interval $[-\Theta, \Theta]$ where it exists, a (sin x)/x curve multiplied by the linear functions $\Theta - |t|$ which limit the autocorrelation triangle of the rectangular envelope of the input pulse. The carrier wave has a fixed frequency equal to the centre frequency $f_0 = \omega_0/2\pi$. Denoting the input signal frequency sweep by $B = \mu\Theta/2\pi$, the envelope $e_2(t)$ of the compressed signal depends only on the time bandwidth product $\Theta B = R$. Expressed as a function of the normalized coordinates t/Θ:

$$e_2(t) = C \frac{\Theta}{2} \frac{\sin\left[\pi R \frac{t}{\Theta}\left(1 - \frac{|t|}{\Theta}\right)\right]}{\pi R \frac{t}{\Theta}} \quad (9.12)$$

Figure 9.14 shows the output signal of a filter for which the product ΘB is equal to 10. At larger values, the compressed pulse envelope is closer to a (sin x)/x curve as far as the first lobes are concerned:

$$e_2(t) \simeq \frac{C\Theta}{2} \frac{\sin\left(\pi R \frac{t}{\Theta}\right)}{\pi R \frac{t}{\Theta}}$$

The 3 dB-width of the central peak of this (sin x)/x curve is

$$\Delta t = \frac{0.885}{R} \Theta = \frac{0.885}{B} \quad (9.13)$$

The signal of duration Θ is compressed by a factor 1.13 R. If the loss of the passive filter is neglected, energy conservation approximately determines the pulse amplitude as $\sqrt{B\Theta} = C\Theta/2$, hence

$$C \simeq \sqrt{2\mu/\pi}.$$

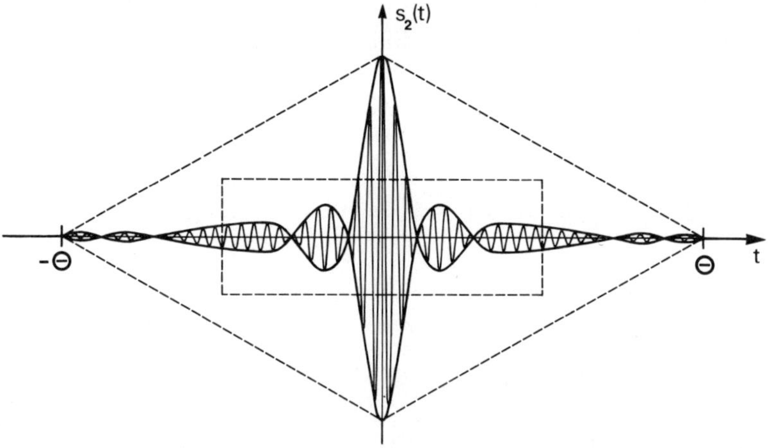

Fig. 9.14. Compressed pulse waveform at the output of a filter for which the product BΘ is equal to 10. The dotted rectangle is the envelope of the input linearly frequency modulated signal.

With respect to the central peak, the first side lobes are at a level -13.3 dB, the second side lobes at -18 dB.

9.3.2.2. Signal spectrum. Frequency response of the matched filter

The spectrum of the linearly frequency modulated signal is the complex conjugate of the frequency response of the matched filter. It is easily derived from equation 1.22:

$$S_1(\omega) = \frac{1}{2} E_1(\omega - \omega_0) + \frac{1}{2} E_1^*(\omega + \omega_0). \qquad (9.14)$$

With $e_1(t) = \Pi(t/\Theta)$ and $\psi(t) = \mu t^2/2$, equation 1.21 yields:

$$E_1(\omega) = \int_{-\Theta/2}^{+\Theta/2} \exp\left[\frac{i\mu}{2}\left(t^2 - 2\frac{\omega}{\mu}t\right)\right]dt.$$

On completing the square:

$$E_1(\omega) = e^{-i(\omega^2/2\mu)} \int_{-\Theta/2}^{+\Theta/2} \exp\left[\frac{i\mu}{2}\left(t - \frac{\omega}{\mu}\right)^2\right]dt$$

the change of variable

$$\sqrt{\mu}\ (t - \frac{\omega}{\mu}) = \sqrt{\pi}x \rightarrow \quad dt = \sqrt{\pi/\mu}\ dx$$

is obvious and the integral becomes

$$E_1(\omega) = \sqrt{\pi/\mu}\ e^{-i(\omega^2/2\mu)} \int_{-X_1}^{X_2} e^{i\pi(x^2/2)}\ dx \quad (9.15)$$

where

$$X_{\frac{1}{2}} = \sqrt{R/2}\ (1 \pm \frac{\omega}{\pi B}) \quad (9.16)$$

For large values of R, the limits of integration are replaced by $-\infty$ and $+\infty$ when $|\omega| < \pi B$. Since

$$\int_{-\infty}^{+\infty} e^{i(\pi x^2/2)}\ dx = \sqrt{2}\ e^{i(\pi/4)}$$

$$\lim_{R\to\infty} E_1(\omega) = \sqrt{2\pi/\mu}\ e^{i(\pi/4-\omega^2/2\mu)} \quad \text{for} \quad -\pi B < \omega < \pi B.$$

Conversely, for $|\omega| > \pi B$, the integral vanishes because the limits of integration are very close to each other. We recover the result of exercise 1.7, established by the method of stationary phase that the amplitude of the spectrum of a linearly frequency modulated signal is a rectangular function centered at f_o, whose width is equal to the overall frequency variation B.

When the product BΘ is finite, the spectrum (9.15) can be expressed in terms of the Fresnel integrals

$$C(X) = \int_o^X \cos \frac{\pi x^2}{2}\ dx \quad \text{and} \quad S(X) = \int_o^X \sin \frac{\pi x^2}{2}\ dx \quad (9.17)$$

by

$$E_1(\omega) = \sqrt{\pi/\mu}\ e^{-i(\omega^2/2\mu)}[C(X_1)+C(X_2)+iS(X_1)+iS(X_2)].(9.18)$$

The curves of figure 9.15 show that the spectrum amplitude is close to a rectangular function when R increases. These curves are reminiscent of those which, in optics [24], account for light diffraction at a finite distance from the source by a slit with parallel straight edges (fig. 9.16). The light intensity incident on a screen at a distance b from a slit of width d, itself at a distance a from the light source, depends on the ratio $R = d^2(a+b)/\lambda ab$ which plays the same rôle as the product BΘ; the position z replaces the frequency and the aperture Δ of the geometric

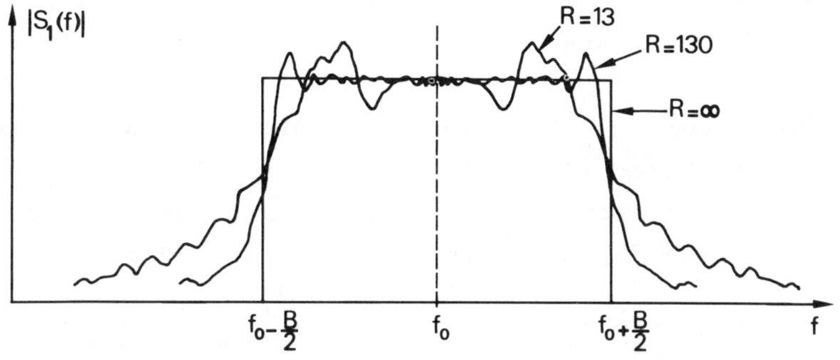

Fig. 9.15. The amplitude of the spectrum of a linearly frequency modulated signal tends to a rectangle as $B\Theta = R$ increases (after fig. 6.6 of ref. 23).

image of the slit replaces the frequency variation B. It is indeed possible to establish the analogy between a matched filter which time compresses a frequency modulated pulse and a lens which focuses in space a parallel light beam [25]. When all points of the slit vibrate in phase ($a = \infty$), the curves in fig. 9.15 (with $R = d^2/\lambda b$ and $B \to d$) yield the radiation pattern of a linear transducer at a finite distance b when the transducer width d is small compared to b.

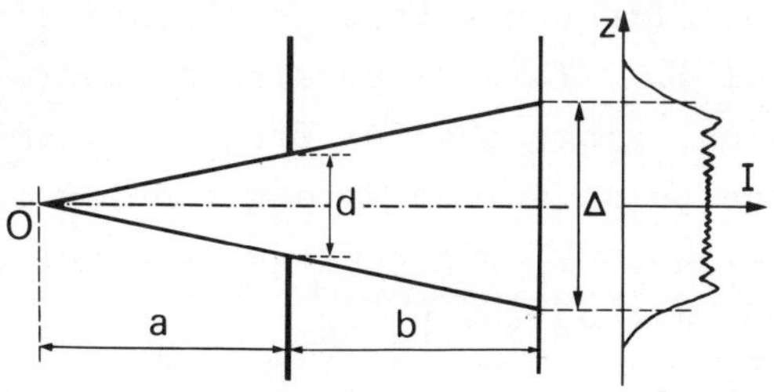

Fig. 9.16. At a finite distance from a slit, the distribution of light intensity on a screen represents the square of the spectral amplitude of a linearly frequency modulated signal.

The frequency response of the matched filter is, from equations 9.5 and 9.14

$$H(\omega) = \sqrt{\frac{2\mu}{\pi}} S_1^*(\omega) e^{-i\omega t_o} = \sqrt{\frac{\mu}{2\pi}} [E_1^*(\omega-\omega_0)+E_1(\omega+\omega_0)] e^{-i\omega t_o}$$

(since $C = \sqrt{2\mu/\pi}$). Assuming (as implicitly we have done so far) that there is nearly no overlap between the two parts of the spectrum centered at $+\omega_0$ and $-\omega_0$, the frequency response for $\omega > 0$ is, from equation 9.18:

$$H(\omega) = \exp i\left[\frac{(\omega-\omega_0)^2}{2\mu} - \omega t_0\right] \frac{(P-iQ)}{\sqrt{2}} \qquad (9.19)$$

where

$$P = C(X_1) + C(X_2) \quad \text{and} \quad Q = S(X_1) + S(X_2) \qquad (9.20)$$

X_1 and X_2 are derived from equation 9.16 by replacing ω by $\omega - \omega_0$. The phase angle of $H(\omega)$ is the sum of two terms:

$$\Phi(\omega) = \Phi_1(\omega) + \Phi_2(\omega). \qquad (9.21)$$

The more important term

$$\Phi_1(\omega) = \frac{(\omega - \omega_0)^2}{2\mu} - \omega t_0 \qquad (9.22)$$

corresponds to a variation of the group delay linearly proportional to the frequency

$$\tau_g = -\frac{d\Phi_1}{d\omega} = t_o - \frac{\omega-\omega_0}{\mu} = t_o - \Theta\,\frac{f-f_0}{B} \quad . \qquad (9.23)$$

The filter matched to a frequency modulated signal is necessarily dispersive. The slope of the delay vs frequency curve is opposite to the slope of the graph of the instantaneous frequency vs time characteristic of the signal (fig. 9.17). This dispersion accounts for the pulse compression, since the low frequency components at the start of the pulse suffer a longer delay in the filter than the high frequency components at the end of the pulse.

The second part of the phase

$$\Phi_2(\omega) = -\tan^{-1}\frac{Q}{P} \qquad (9.24)$$

leads to fluctuations of the group time delay

$$\Delta\tau_g = -\frac{d\Phi_2}{d\omega} = \frac{PQ' - QP'}{P^2 + Q^2}$$

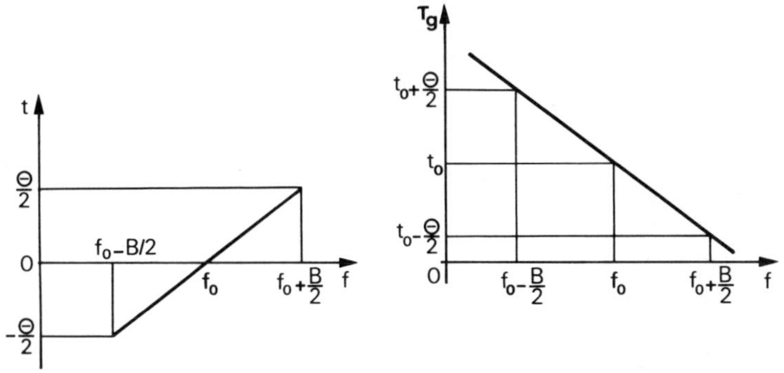

Fig. 9.17. Group delay vs frequency characteristic. The signal frequency increases with time; accordingly, lower frequency components are delayed more.

From 9.20 and 9.17, we derive

$$P' = \frac{dC(X_1)}{d\omega} + \frac{dC(X_2)}{d\omega} = \frac{1}{\pi B}\sqrt{\frac{R}{2}}(\cos\frac{\pi}{2}X_1^2 - \cos\frac{\pi}{2}X_2^2)$$

and

$$Q' = \frac{dS(X_1)}{d\omega} + \frac{dS(X_2)}{d\omega} = \frac{1}{\pi B}\sqrt{\frac{R}{2}}(\sin\frac{\pi}{2}X_1^2 - \sin\frac{\pi}{2}X_2^2)$$

The sums of sines and cosines can be transformed into products; the relative fluctuations of the group delay are then given by the equation:

$$\frac{\Delta\tau_g}{\Theta} = \frac{1}{\pi}\sqrt{\frac{2}{R}}\sin\frac{\pi}{4}(X_1^2-X_2^2)\frac{P\cos\frac{\pi}{4}(X_1^2+X_2^2)+Q\sin\frac{\pi}{4}(X_1^2+X_2^2)}{P^2+Q^2} \qquad (9.25)$$

In the bandwidth $P \simeq Q \simeq 1$, the relative amplitude of the fluctuations, which is of the order of $1/\pi\sqrt{R}$, decreases as R increases. $\Delta\tau_g$ vanishes for frequencies such that

$$X_1^2 - X_2^2 = 4p \qquad p \text{ integer}$$

or, from 9.16:

$$\frac{f - f_o}{B} = \frac{p}{R} \qquad (9.26)$$

The fluctuations are more rapid for large values of R. These results can be checked by the curve in Fig. 9.18, plotted for R = 25.

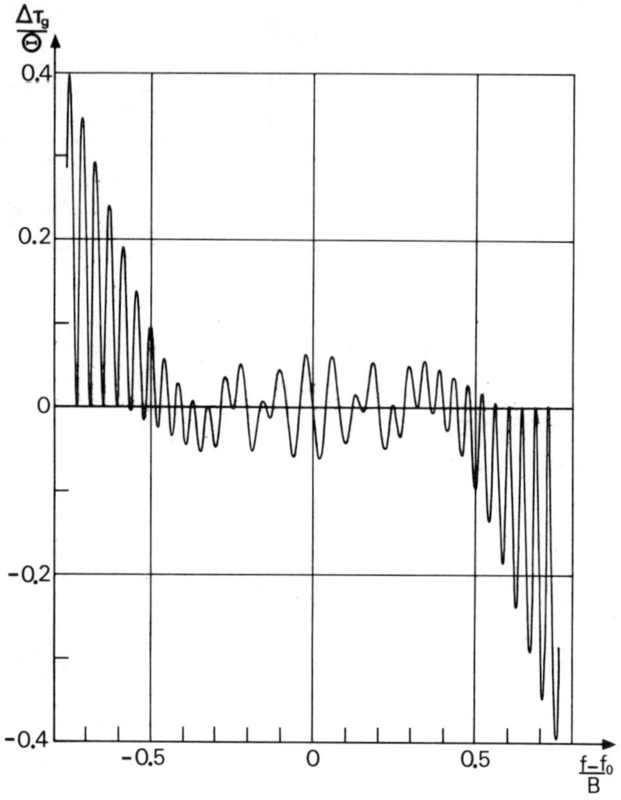

Fig. 9.18. Relative fluctuations of the group delay $\Delta\tau_g/\Theta$ for $B\Theta = 25$. (After P. Hartemann, fig. 14 of reference 26).

The above analysis shows that the time bandwidth product ΘB is the key parameter, in the time domain as in the frequency domain. However, the same value of R is obtained either with a large B and a small Θ (e.g. B = 20 MHz, Θ = 5 μs) or a small B and a large Θ (e.g. B = 100 KHz, Θ = 1 ms). The first class of signals (radar) can be processed only by lines with a high centre frequency; the second class of signals (sonar) requires long delay lines (several meters). Most of the examples we shall describe refer to lines processing signals of relatively short duration, except for the oldest elastic wave delay lines (bulk waveguides), whose characteristics are recalled in the next section. We shall discuss, in turn, three other types of lines, Love waves delay lines, Rayleigh wave delay lines, acousto-optic interaction lines. Of these, Rayleigh wave delay lines, capable of processing a wide variety of signals, are the most commonly employed.

9.3.3. Dispersive Bulk Elastic Wave Delay Lines

The principle of the matched filter has been exploited in several ways. The first filters to have been built were purely electric ones, composed of T-bridge cells. Their drawback was that they required a large number of components and a lot of space. In the frequency range in which they can be used, they are replaced by various compact elastic wave delay lines. Elastic wave propagation in an homogeneous medium is not dispersive, at least when the wavelength is much larger than the atomic spacing (§ 1.2.2). It will be recalled that a 3,000 GHz frequency is needed to obtain $\lambda = 10$ Å. Such a frequency is outside the radio frequency domain, which is why simple elastic bulk wave dispersive delay lines are waveguides. The dispersive effect then arises from the geometry [27] of the propagation medium.

In an isotropic solid, an x_2 polarized shear wave propagating along the x_1 axis, creates only a stress $T_{12} = c_{66}S_{12}$, thus satisfying the boundary condition $T_{i3} = 0$ on any free surface perpendicular to the x_3 axis. Thus, in a thin plate or in a wide strip, there exists a non-dispersive mode of velocity V_S (zero order mode). But this SH wave, which is not altered by reflections on the free surfaces of this waveguide, gives rise to higher order modes whose dispersion curves are similar to those of Fig. 1.8. The variation in group delay, deduced from this diagram, has been exploited, between the cutoff frequencies of the first two modes, to make lines matched to (non linearly) frequency modulated signals [28]. Compression ratios of 100, together with pulses longer than

500 μs at a few MHz are possible. A drawback is the existence of spurious signals, due to the zero order mode, which are hard to remove.

From this point of view, Lamb waves give better results [29]. The dispersion effect comes from the frequency dependence of the proportion of longitudinal (L) and shear (SV) waves. The group delay vs frequency curve is plotted in fig. 9.19 for the first extensional mode. In the vicinity of the point of inflexion, the delay exhibits a linear variation.* Lines have been made [28] of an aluminium alloy for frequencies below 5 MHz, and of steel, which is easier to shape into a thin strip, for frequencies up to 30 MHz. Defects in linearity forbid a relative frequency variation greater than 10 per cent.

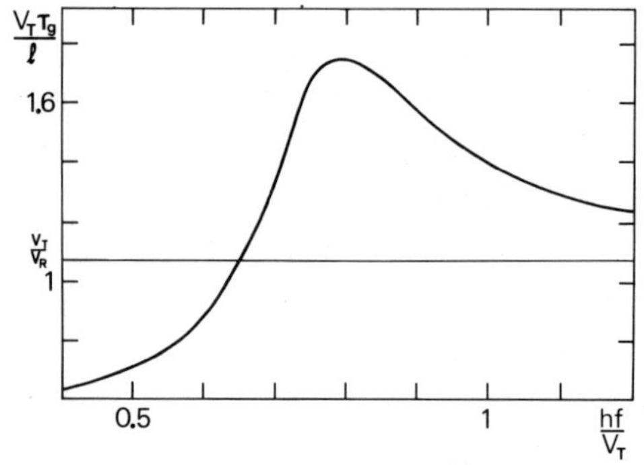

Fig. 9.19. Lamb waves. Group delay τ_g vs frequency f for the first symmetric mode in an isotropic strip, of length 1, h thick, and for $V_T/V_L = .5$. At low frequencies, the proportion of longitudinal waves is greater than that of shear waves. At high frequencies the situation is reversed. For very high frequencies ($\lambda << h$), the velocity tends to the Rayleigh wave velocity V_R.

* This curve is derived from classical dispersion equations [29], known as Rayleigh-Lamb equations. These, like the Pochhammer-Chree equations referring to cylinders, have not been derived in this book, because the general study of guided waves in strips or tubes was excluded.

The compression ratio reaches 100, the pulse duration ranges from 50 μs to several milliseconds (for 1 ms, the aluminium strip length is about 4 meters). Loss is usually below 40 dB and the sidelobe level below -30 dB. In order to reduce the departures from linearity, strips of stepped thickness or of continuously varying thickness have been tried (fig. 9.20) [30]. For the same linearity and compression ratio the bandwidth is increased (40%) and the length and the delay line loss are reduced correspondingly.

Flexural waves, similar to Lamb waves, can be guided by wires or by tubes. Lines have been built with centre frequencies of a few hundreds of KHz and a pulse duration of several milliseconds.

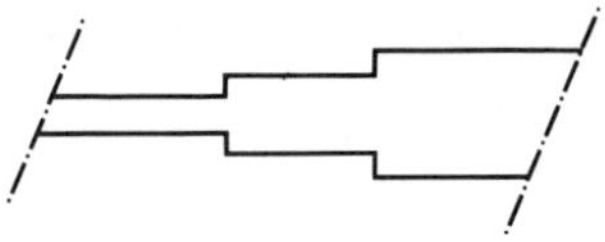

Fig. 9.20. Stepped thickness dispersive delay line.

Techniques other than bulk elastic wave guidance have also been employed, for instance diffraction by a grating [31, 32]. Fig. 9.21 shows a delay line with two perpendicular gratings. Grating E consisting of source elements which are excited in phase, launches non dispersive shear waves. Grating R, whose elements are shunt connected, is the receiving transducer. The dispersion effect is related to the frequency dependent path between the transducers. Since the propagation medium is an insulator (silica), the centre frequency of operation, and therefore the bandwidth of this kind of delay line, are higher, e.g. 45 MHz and 20 MHz [7]. Compression ratios of 400 have been reported.

The wedge line depicted in fig. 9.22 [33, 34] has an analogous geometry but operates on a different principle.

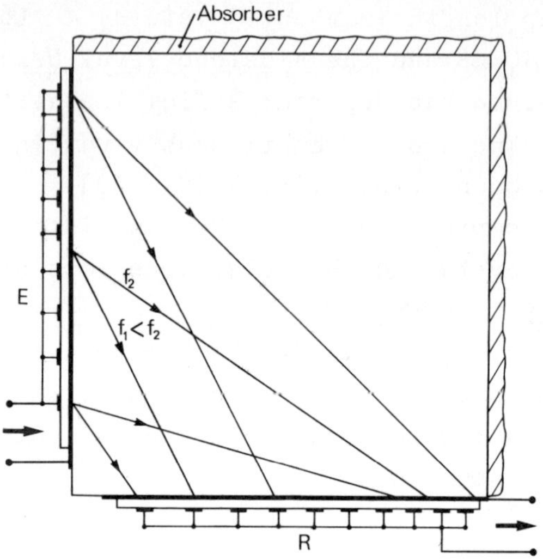

Fig. 9.21. Perpendicular diffraction delay line. The transducer pattern is chosen so that the path is frequency dependent.

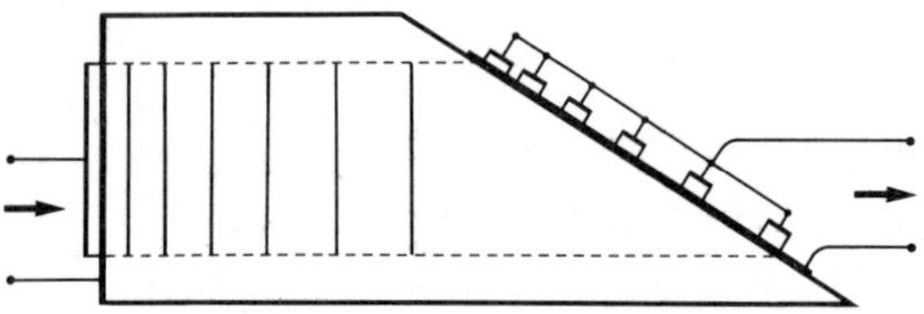

Fig. 9.22. Wedge dispersive delay line. The output transducers have different resonant frequencies.

The elements that constitute the receiving transducer behave as resonators. They are set in such a way that the distances between neighbouring elements and the single emitter transducer differ by one wavelength at the resonant frequency. The desired dispersion law is obtained by adjusting the resonator separations. The slope of delay plotted against frequency characteristic is negative or positive, depending on whether the high frequency resonators are closer to or farther from the emitter transducer.

9.3.4. Dispersive Love Wave Delay Lines

In a medium consisting of a layer on a semi-infinite substrate, the elastic wave propagation velocity is frequency dependent. The case of Love waves has been dealt with in § 5.2.4: at low frequencies the phase velocity is the substrate velocity (V') and at high frequencies the layer velocity (V < V'). This dispersive effect is suited for the design of filters matched to linearly frequency modulated signals, since the curve of group delay against frequency (fig. 5.39) has a point of inflexion in the vicinity of which it may be regarded as a straight line.

Several pairs of materials have been tried [36]: copper on beryllium, tungsten on beryllium, glass on silica, silica on silicon... This last pair offers a technological advantage, since the layer is obtained by thermal oxidization of the single crystal substrate. Polycrystalline metal layers restrict the use of the first two pairs to frequencies below 50 MHz. On the other hand, amorphous layer structures (silica) operate at higher frequencies. The layer thickness, which depends on the frequency and on the materials, is about a few tens of microns at 10 MHz. It is difficult to maintain the uniformity and homogeneity of the layer over the entire substrate, as well as the reproducibility of line characteristics. Fig. 9.23 shows a delay line structure with two transducers soldered to the surface. These are resonators, ceramic or lithium niobate plates, depending on the frequency, creating SH (shear horizontal) waves. The electrode width is about half a wavelength at the centre frequency, so that the radiation is emitted over a wide angle. A fraction of the energy is reflected on the layer-substrate boundary and travels in the layer.

Fig. 9.24 refers to a W/Be line, with centre frequency 32.5 MHz (thickness h = 4 μm) and length 7.5 cm. The linearity is good in a frequency range of nearly 100 per cent. However, significant variation in loss (14 dB) will in practice restrict the frequency sweep bandwidth to less than 10 MHz. The delay vs frequency characteristic

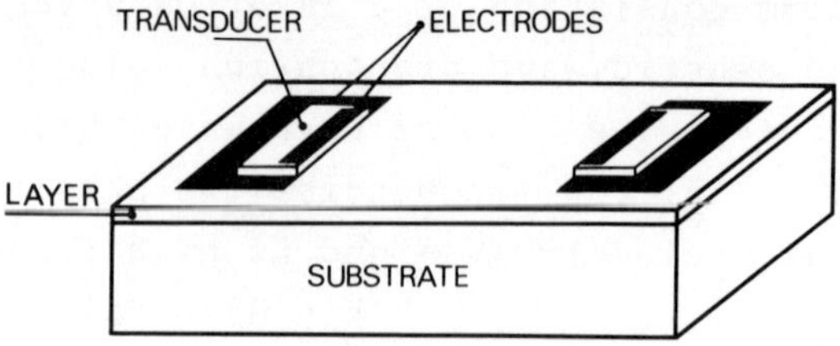

Fig. 9.23. Love wave delay line. The two transducers soldered on the surface have an active width of $\lambda_0/2$ so that the radiation is launched over a wide angle. The resulting SH waves are reflected at the layer-substrate interface and travel in the layer (fig. 5 of reference 36).

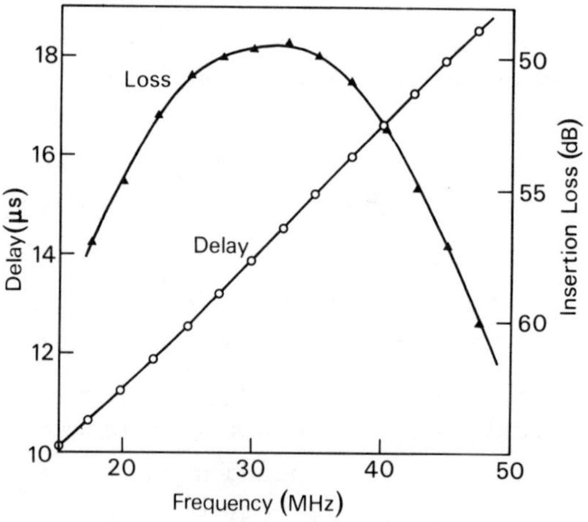

Fig. 9.24. Dispersive Love wave delay line (W/Be). Group delay and loss vs frequency (Fig. 12 and 13 of reference 36).

of a SiO_2/Si line is plotted in fig. 9.25: Θ = .8 μs, B ≃ 80 MHz about a centre frequency of 130 MHz (h = 5 μm). The insertion loss is below 50 dB and has a 5 dB variation over the bandwidth. In the laboratory, a compression ratio of nearly 1,000 has been reached (f_o = 300 MHz, B = 215 MHz, Θ = 4.65 μs), but the variation in insertion loss is more than 10 dB [37].

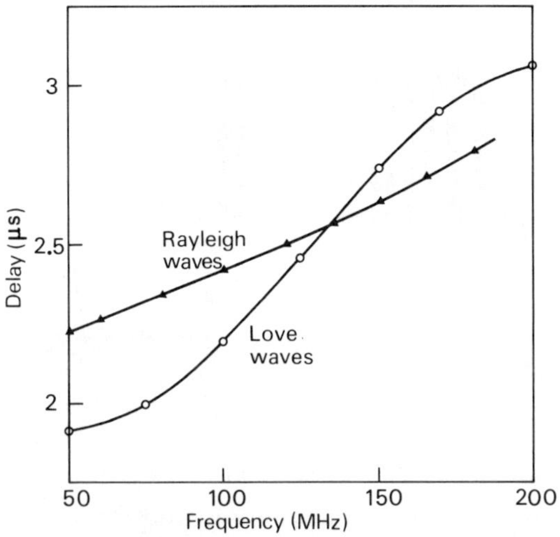

Fig. 9.25. Dispersive delay line (SiO_2/Si). Group delay vs frequency for Love waves and Rayleigh waves (Fig. 15 of reference 36).

This layered structure also permits excitation and propagation of dispersive Rayleigh waves (see fig. 9.25). The transducer is oriented such that the generated wave is polarized in the sagittal plane.

9.3.5. Rayleigh Wave Matched Filters

A major advantage compensates for the complexity of Rayleigh waves. They can be generated or detected by interdigital transducers, whose technology is simple. In addition, these transducers are equivalent to a discrete distribution of ultrasonic sources (or receivers), the relative phase and intensity of which are separately determined by the finger position and active length of the fingers (§ 7.2.4). A great variety of impulse responses can thus be synthesized with appropriate transducer geometries.

9.3.5.1. Dispersive delay lines for frequency modulated signals

Fig. 9.26 illustrates the principle of operation for

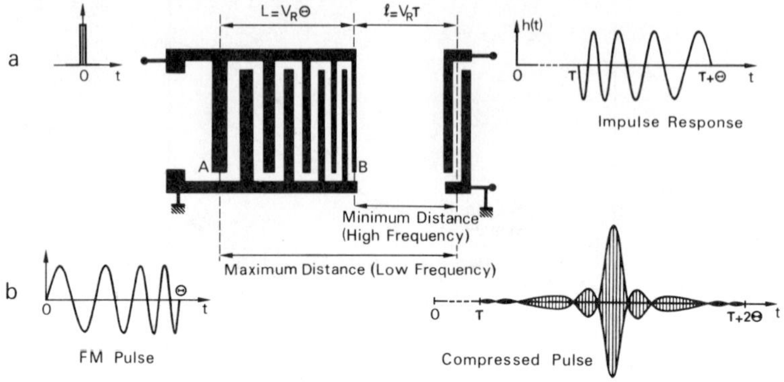

Fig. 9.26. Dispersive Rayleigh wave delay line matched to a linearly frequency modulated signal

a) Impulse response; compare with the signal (it should be its time inverse).

b) Compressed pulse: higher frequency components, from the end of the signal excite extremity B of the transducer just when the low frequency components, generated at the input A, at the start of the signal also arrive at B.

a filter matched to a frequency modulated signal. The dispersive effect is due to the variation in finger spacing along the transducer. Low frequency components are emitted preferentially by zones where the finger spacing is wide, high frequency components by zones of closely spaced fingers. Thus it appears as though the Rayleigh wave source position were frequency dependent. Since the receiving transducer, with a large bandwidth, is well localized, the elastic wave path depends on frequency. The comparison between the signal and the transducer geometry - or its impulse response (fig. 9.26a) - shows that the equation $h(t) = s(t_0 - t)$ is satisfied.

In a filter matched to a linearly frequency modulated signal, the impulse response phase is, from equation 9.9:

$$\frac{\phi(t)}{2\pi} = f_0 t - \frac{B}{\Theta}\frac{t^2}{2} \qquad -\frac{\Theta}{2} < t < +\frac{\Theta}{2}.$$

To conform with the method of discrete sources, mentioned in section 7.2.4, the sources must be distributed at the points $x_n = V_R t_n$, the instants t_n being determined by the condition $\phi(t_n) = n\pi$

$$\frac{B}{\Theta} t_n^2 - 2f_o t_n + n = 0$$

or

$$t_n = \frac{\Theta f_o}{B} [1 - \sqrt{1 - \frac{nR}{(f_o\Theta)^2}}] \qquad (9.27)$$

The duration of the impulse response restricts the allowed values of n:

$$-(f_o + \frac{B}{4}) \Theta \le n \le (f_o - \frac{B}{4})\Theta .$$

For example, when f_o = 50 MHz, Θ = 6 μs, the transducer comprises $N = 2f_o\Theta$ = 600 fingers. Recalling that the envelope is a rectangular function, the overlap between two neighbouring fingers is derived from equation 7.96:

$$\ell_n = \frac{\ell_o f_o}{f(t_n)} = \frac{\ell_o}{\sqrt{1 - \frac{nR}{(f_o\Theta)^2}}} \qquad (9.28)$$

In the present case, where the signal matching is ensured solely by the input transducer, the purpose of the output transducer function is merely to restore an electric signal. Its bandwidth Δf must be greater than the frequency sweep; therefore it has only a few fingers since, from equation 7.37:

$$N - 1 = 1.77 \frac{f_o}{\Delta f} .$$

The above calculation was for the ideal case, ignoring the variation of the Rayleigh wave velocity under the metal fingers which short out the tangential component of the electric field, the mass loading effect of the electrodes (greater with gold, than with aluminium), and the coupling between the sources through inverse piezoelectric effects. However, this analysis is well confirmed by experiment for weakly piezoelectric crystals such as quartz. On the other hand, if needed, it is possible slightly to adjust the transducer shape in order to obtain the desired impulse response. Many lines have been constructed based on this principle, and operate within the following limits:

- Centre frequency		10 MHz < f_o < 400 MHz
- Frequency sweep		B/f_o < 50%
- Signal duration		1 μs < Θ < 50 μs
- Time bandwidth product ΘB = R		10 < R < 1,000
- Insertion loss (α quartz)		30 to 50 dB
- Temperature coefficient $\frac{1}{\tau}\frac{d\tau}{d\theta}$ for α quartz and X propagation [38]	Y cut	$-2.4.10^{-5}$ $(^{o}C)^{-1}$
	ST cut	0

In spite of lower loss, lithium niobate is used less than quartz because of its high temperature coefficient ($8.5.10^{-5}$) and spurious effects due to its high coupling constant.

Figure 9.27 is a photograph of one of the first lines to be built. It is made of quartz (Y cut, X propagation) and operates around 14 MHz (B = 5.5 MHz, Θ = 4 μs) with a compression ratio of 24. The input transducer has been extended on both sides of the active region. Its length (14.7 mm) corresponds to extremal frequencies 10.8 and 17.1 MHz. In addition, all fingers have the same length since it was intended that the impulse response be shaped by a limiter (to remove the effect of convolution with the receiving transducer) and by a clipper (to obtain a constant amplitude). The input impedance is equivalent to a 20 pF capacitance shunted by a 70 kΩ resistance. The receiver transducer has only 6 fingers of width 56 μm. The sidelobes of the compressed pulse (fig. 9.28) are at -16 dB with respect to the central peak. Photograph 9.29 shows the frequency response of this line. In order to check that the frequency response is closer to a square function at higher ΘB, as predicted by the theory of § 9.3.2.2, we have also displayed the frequency response of a dispersive delay line for which the product ΘB is 90.

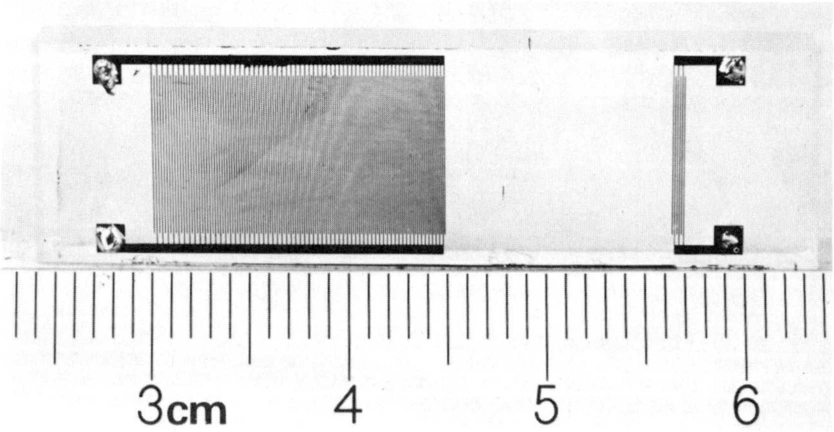

Fig. 9.27. Dispersive Rayleigh wave delay line (quartz). f_o = 14 MHz, B = 5.5 MHz, Θ = 4 μs. The mean finger width is 56 μm.

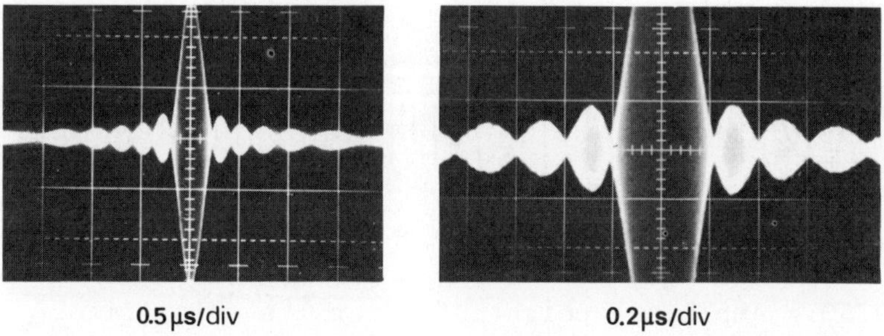

Fig. 9.28. Compressed pulse provided by the line in fig. 9.27. 3 dB width: .17 μs; sidelobe level: -16 dB.

In radar technology, the peak-to-sidelobe ratio must normally be better than the 13.3 dB of a filter matched to a linearly frequency modulated signal with $B\Theta \gg 1$. An improvement can be obtained in two ways:

Firstly, the overlap between the fingers of the input transducer can be varied appropriately, the location of the fingers being fixed [39]. The line in fig. 9.30

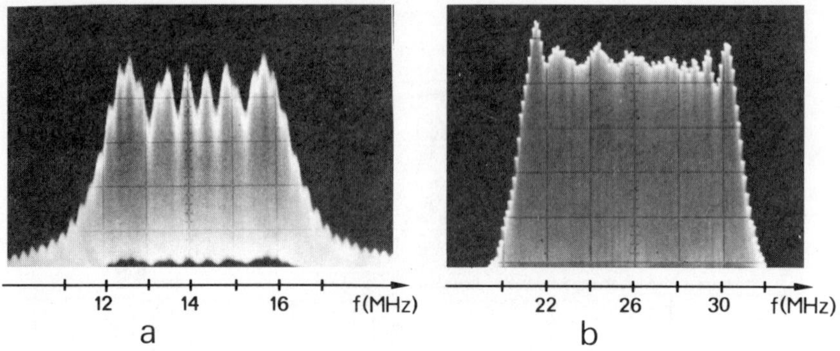

Fig. 9.29. Frequency response.
a) of the line in fig. 9.27 ($\Theta B = 22$);
b) of a line with $\Theta B = 90$ ($f_o = 26$ MHz, $B = 9$ MHz, $\Theta = 10$ μs). Compare to fig. 9.15.

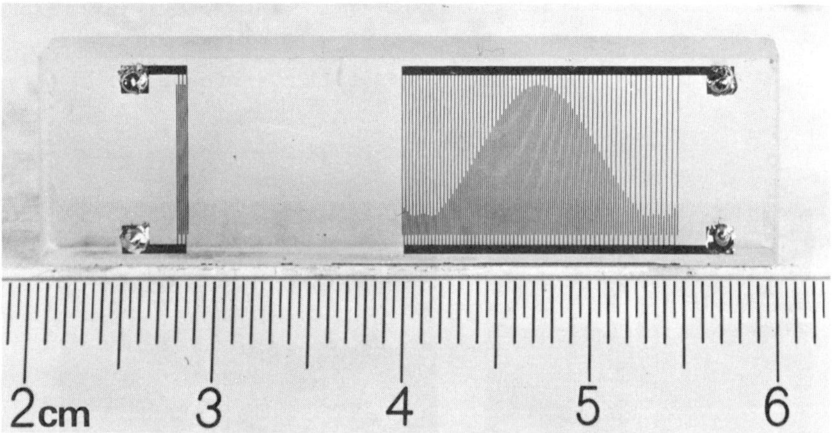

Fig. 9.30. Amplitude weighted dispersive delay line. $f_o = 14$ MHz, $B = 5.5$ MHz, $\Theta = 4$ μs.

has a Dolph-Tchebycheff weighting [40]; it was designed to reach the theoretically attainable sidelobe level of -40 dB for an infinite transducer. In practice the level was only -27 dB. The difference, which is due to the finite number of fingers, is large because the compression ratio is small. However, the improvement must be paid for in the form of a widened compressed pulse (fig. 9.31) and a higher loss. In fact, the filter is no longer matched to the signal since the impulse response no longer obeys equation 9.2.

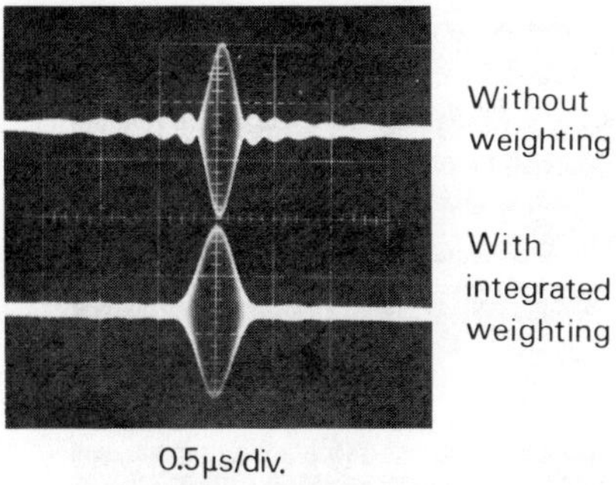

Fig. 9.31. Comparison between the compressed pulses from the lines in fig. 9.27 and 9.30. The weighting reduces the side lobes, but widens the pulse.

The second procedure offers the same advantage, without the related drawback. However it requires a non linear frequency modulation of the signal [41]. The time dependence of the instantaneous frequency may be an S-shaped curve resulting from the combination of a linear relationship and a sine wave (fig. 9.32). The sidelobe level depends on the relative amplitude of the sine function [42].

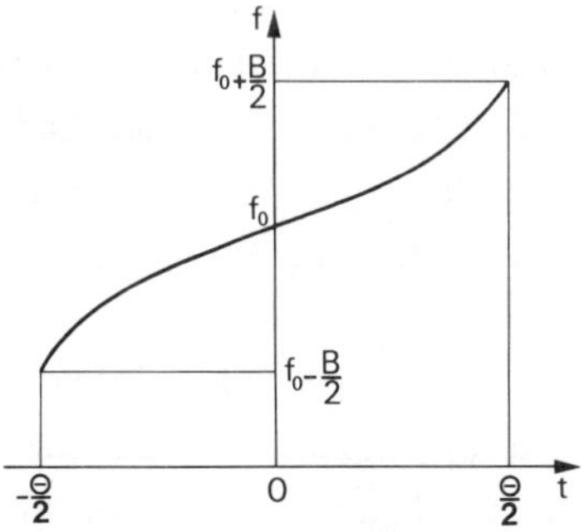

Fig. 9.32. Instantaneous frequency vs time characteristic for a non-linearly frequency modulated signal.

Fig. 9.33 shows the compressed pulse at the output of a line of centre frequency f_o = 14 MHz (B = 5.5 MHz, Θ = 4 μs), designed for a rejection of -27.5 dB. The measured value is -25 dB. The autocorrelation signal is no longer a

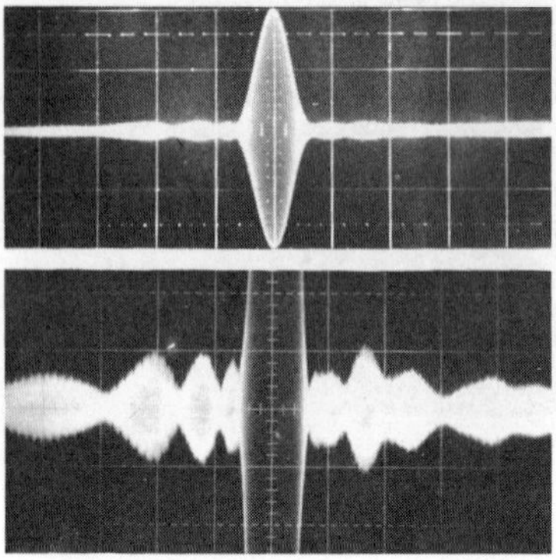

0.5μs/div.

Fig. 9.33. Compressed pulse at the output of a line whose delay curve is S-shaped: f_o = 14 MHz, B = 5.5 MHz, Θ = 4 μs, sidelobes: -25 dB.

(sin x)/x function; its duration Δt at -3 dB is larger than 1/B, so that the ratio of the signal duration to that of the compressed pulse is less than the product $B\Theta$ = 22. A line operating at a slightly higher frequency (f_o = 22 MHz, B = 5 MHz, Θ = 5.6 μs) is depicted in fig. 9.34. The sides of the impulse response (fig. 9.35a) are not steep: this arises from the convolution between the signal emitted by the input transducer and the rectangular impulse response of the receiving transducer (duration .17 μs). The compressed pulse (fig. 9.35b) is .3 μs wide at -3 dB; the sidelobe level is -25 dB. The first lobe is smaller than the others, in agreement with theory (exercise 9.2).

For longer pulses (Θ > 15 μs), as indicated in § 7.2.4, it is possible to reduce the number of fingers by sampling the transducer. The delay line then looks as in fig. 9.36 for a duration Θ = 20 μs and a sampling rate Q = 15. The distribution of sources, consisting of two fingers (λ/4 = 20 μm width, centre frequency f_o = 40 MHz), follows an S-shaped frequency vs time relationship. The distance between neighbouring samples is 15 λ/2. The receiving transducer has a large number of fingers (N = 40) since the relative sweep width is only 1/20 (B = 2 MHz). The compression ratio is 30; the sidelobe level is -24 dB (fig. 9.36b).

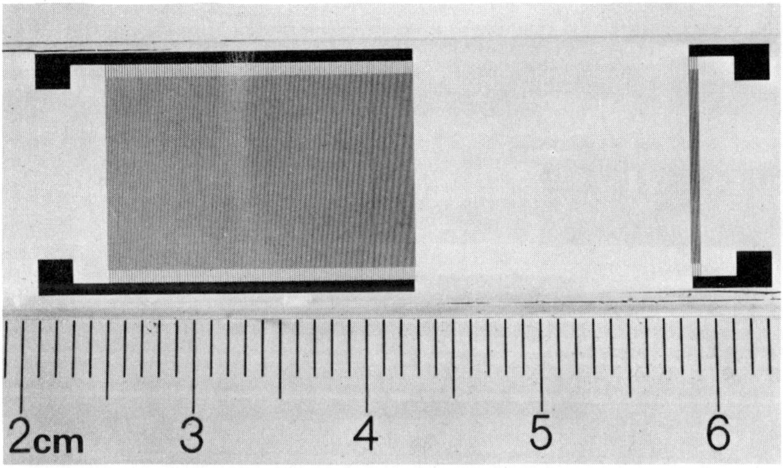

Fig. 9.34. Dispersive delay line having a S-shaped time delay vs frequency characteristic: f_o = 22 MHz, B = 5 MHz, Θ = 5.6 μs.

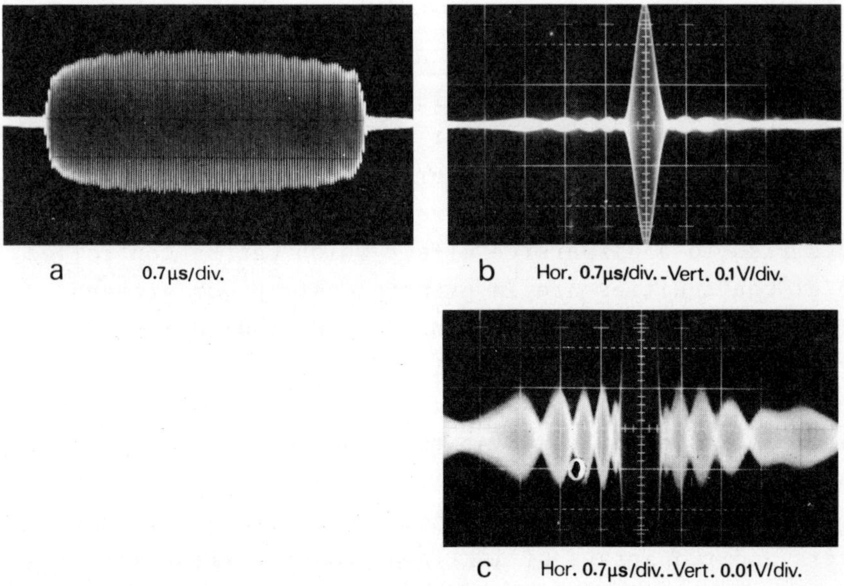

Fig. 9.35. Response of the line in fig. 9.34:

a) to a short pulse;

b) and c) to the signal.

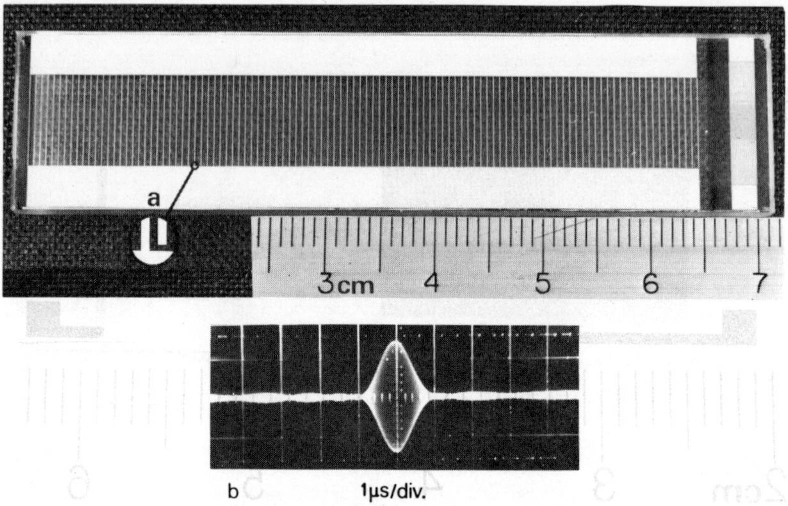

Fig. 9.36. Sampled line with a S-shaped delay curve
$Q = 15$, $f_0 = 40$ MHz, $B = 2$ MHz, $\Theta = 20$ μs.
a) detail of a source;
b) compressed pulse.

Note. Reflective array delay lines. Elastic waves, like all other waves, are reflected at discontinuities. It has been observed [43] that a sequence of discontinuities, with a suitable pattern, either on a ribbon where SH waves propagate or on a surface where Rayleigh waves propagate, gives rise to a dispersive effect which can be controlled. The discontinuities are in principle strips or grooves. Reflexion at grooves has recently been studied in more detail; it has been used in low frequency metal strip lines [44] and also in Rayleigh wave lines [45]. The principle, as established at the Lincoln Laboratory (M.I.T., USA) for Rayleigh waves [46], is depicted in fig. 9.37. The line contains two non dispersive interdigital transducers and two arrays of inclined grooves, with variable spacing. The waves from the input transducer are reflected through 90 degrees by the first array in a zone where the periodicity is equal to the wavelength. Since the spacing increases from the input to the output, high frequency waves travel along a shorter path than low frequency waves. The second array, whose role is identical to that of the first array, directs the waves towards the output transducer.

The reflexion coefficient r is a function of the groove depth h [47]. For $h/\lambda < 10^{-2}$:

$$|r| \simeq \frac{1}{3}\frac{h}{\lambda}$$

and the velocity of the waves which propagate under the grating is almost unchanged. Therefore, when the depth of the grooves is varied (and occasionally also the length), the amplitude curve is modified and the sidelobes of the compressed pulse are reduced.

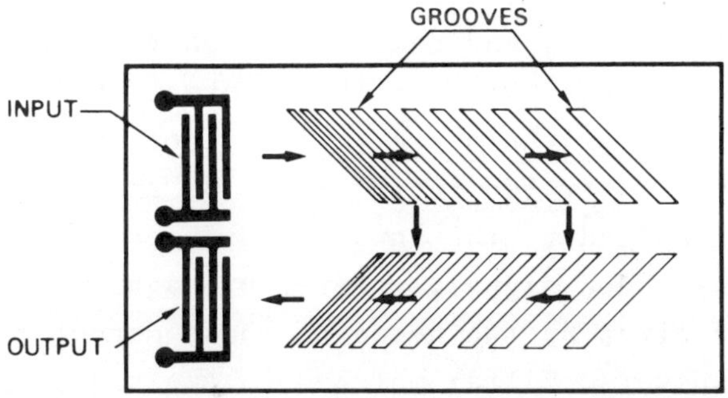

Fig. 9.37. Reflective array dispersive delay line. Rayleigh waves are reflected at 90 degrees to the incident direction in the region where the grating periodicity is synchronous.

This technique gives rise to a new generation of dispersive delay lines with a high compression ratio, for example 1,500 [46] (f_o = 200 MHz, B = 50 MHz, Θ = 30 μs, insertion loss ≃ 40 dB, sidelobe level: -15 dB, substrate: $LiNbO_3$). Figures above 5,000 have been reported [49] (f_o = 1,000 MHz, B = 512 MHz, Θ = 10 μs, loss ≃ 60 dB, sidelobe level: -20 dB with internal weighting, substrate: $LiNbO_3$). The manufacture of many reflecting grooves (12,000 in the latter case) is by means of ion bombardment

[48]. The impedance discontinuity which produces the reflexion is purely mechanical; thus non piezoelectric substrates can be used but then one must solve the problem of the generation of the Rayleigh waves. In principle, this can be achieved by soldering two transducers (one at each end) as with the SH wave strip delay line [44]. In such a case, compression ratios above 1,000 have been reached, but with large pulse durations (f_o = 10 MHz, B = 5 MHz, Θ = 250 μs, loss 30 dB).

9.3.5.2. Transverse filters

As we have already mentioned, frequency modulation is only one among several coding procedures available. Discrete codes are used in telecommunication systems. The simplest code is the binary code: it is a sequence of bits of information, 1, 0 or +1, -1, at regular time intervals. Binary coding and decoding is easy to achieve using Rayleigh waves, whether the information is carried in the amplitude or the phase. In either case, the principle is illustrated in figures 9.38 and 9.39. A short pulse applied to the input transducer of the coder generates the coded signal. After propagation, the signal is received by the decoder, whose output transducer, a transverse filter matched to the signal, provides the autocorrelation function. As an example, fig. 9.40a shows the autocorrelation of a binary phase coded signal, of sequence (+++-++--+-+----). The carrier frequency is 60 MHz. If, instead of the signal s(t), its time reverse s(-t) (i.e. the impulse response of the decoder) is applied, the filter is no longer matched and the output signal looks as in fig. 9.40b.

Among binary phase codes, N-bit-Barker sequences provide an autocorrelation whose central peak is at level N, and whose sidelobes are at level 1(at most). At times equal to an integer multiple of the bit duration, the autocorrelation signal takes values N, 0 and $\pm$ 1. These conditions restrict the code length to 13 bits for odd values of N; no code of this type has been found with more than

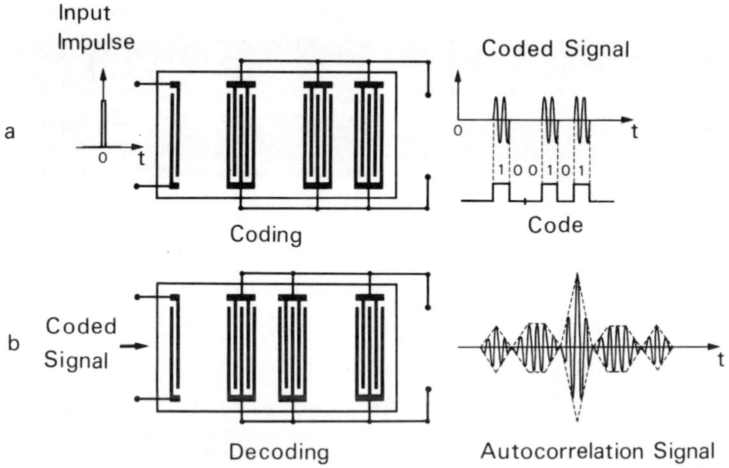

Fig. 9.38. Binary amplitude code.
a) Signal generation by a short pulse applied at the Rayleigh wave line input;
d) Detection by the filter matched to the signal (decoder).

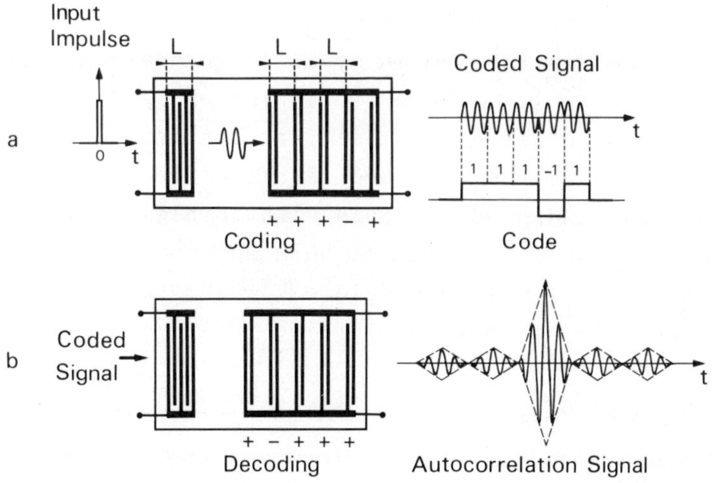

Fig. 9.39. Binary phase code.
a) Signal generation. Phase reversal is obtained by reversing the connections of the fingers at the Rayleigh wave receivers.
b) Detection by the filter matched to the signal (decoder).

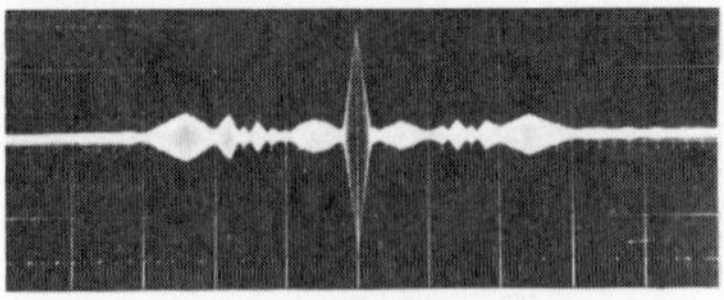

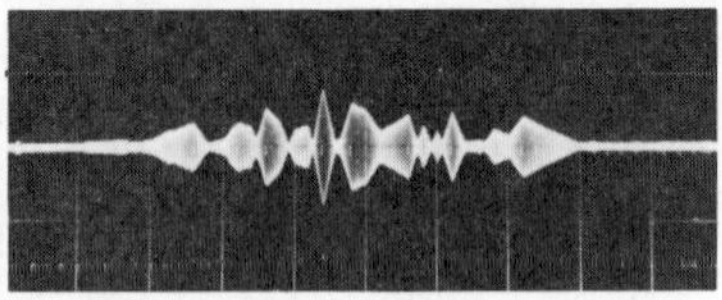

Fig. 9.40. Filter matched to a signal s(t), coded by phase reversal, in the sequence (+++-++--+-+----); the carrier frequency is 60 MHz.
a) autocorrelation of s(t);
b) response to the time reversed signal, which is no longer matched to the filter.

4 bits for even values of N. The code in Fig. 9.39 is a 5-bit Barker sequence.

The linear pattern of the transducer elements can adequately be replaced by an oblique one, especially when spurious signals due to reflexions on a strongly piezo-electric substrate must be avoided. Fig. 9.41 compares the two responses, for a 13-bit Barker sequence and a lithium niobate substrate.

The elements of the output transducer of the coding device can be distributed in order to generate a <u>multi-phase coded signal</u>. The phase is varied by $2\pi/n$ instead of π. On the other hand, modifying the active finger length can reduce the sidelobe level. The same substrate can also bear independent lines with complementary codes (e.g. Golay code): the sum of both autocorrelation signals (from both lines) has no sidelobes. The reader may appreciate this by solving exercise 9.3.

9.3.6. Acousto-optic Matched Filters

Pattern recognition also involves the principle of matched filtering. The reference is inserted into the path of a light beam, focused by a lens. Then, objects to be examined pass in front of the reference like a cine film (fig. 9.42). At the very moment when the object being sought passes in front of the reference the light

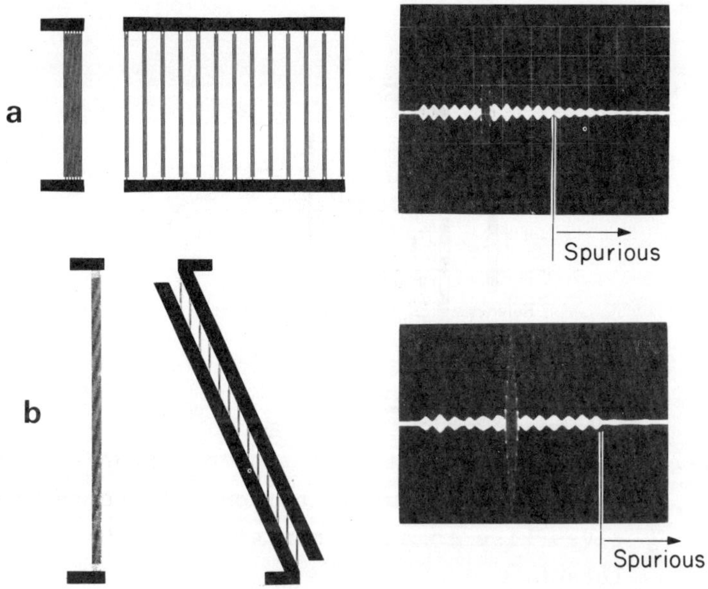

Fig. 9.41. In line (a) and tilted (b) patterns of the summing transducer on a lithium niobate substrate. Comparison between the autocorrelation signals for a 13-bit Barker code, generated by Rayleigh waves (by courtesy of R.H. Tancrel, Raytheon-Waltham, Mass., U.S.A.).

is a maximum at the lens focus. The need to transfer the information on a substrate (e.g. by developing a film) leads to a delay in the output of the analysis. With the aid of light diffraction by elastic waves, real time electrical signal processing is possible, as indicated in fig. 9.43. Consider an amplitude coded signal, with a carrier at a fixed frequency. This signal is applied to the transducer, which immediately launches, for instance, a longitudinal wave packet. The light beam, diffracted at the Bragg angle corresponding to the carrier frequency, is concentrated in the focal plane. The incident light beam is modulated by an amplitude mask which is a spatial replica of the code. The detected light intensity is proportional to the square of the autocorrelation function and takes its maximum value when all elements of the wave packet coincide with the windows of the reference. For a phase coded signal, the amplitude mask is replaced by a phase plate.

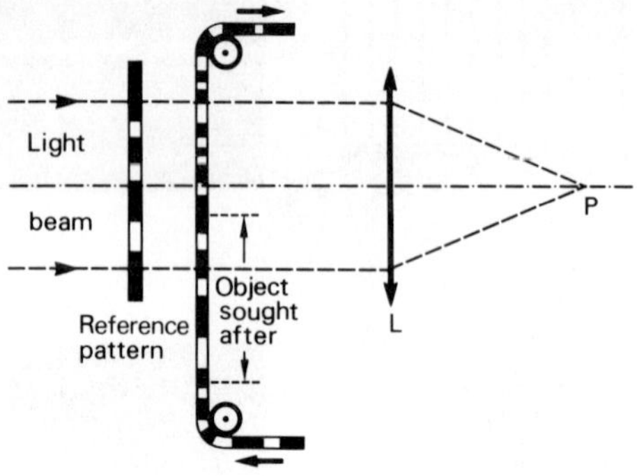

Fig. 9.42. Pattern recognition. The occurrence of the object being sought in front of the reference gives rise to a maximum at the focus P of lens L.

We now list the principal conditions of an experiment performed by R. Torguet and J.M. Bauza with a phase modulated signal carried by longitudinal waves propagating in a lead molybdate crystal, and deflecting a laser beam (6,328 Å):

- code: binary: 255 bits of duration 20 ns;
- carrier frequency: 175 MHz;
- transducer: $LiNbO_3$ plate, soldered with indium; bandwidth 70 MHz;
- line: $PbMoO_4$, [001] propagation (V_L = 3,630 m/s), active length 19.125 mm;
- amount of light deviated: 60 per cent at an electric power level of 1 W:
- phase plate: silica. A π-phase shift is induced bv a variation in thickness of 0.69 um.

The result appears in fig. 9.44. The autocorrelation peak duration (20 ns) is in good accordance with the theoretical compression ratio of 255.

The code itself has been generated from a short pulse applied to the input of a second line similar to the first. bearing the amplitude modulated mask which was previously used to engrave the phase plate (by ion bombardment).

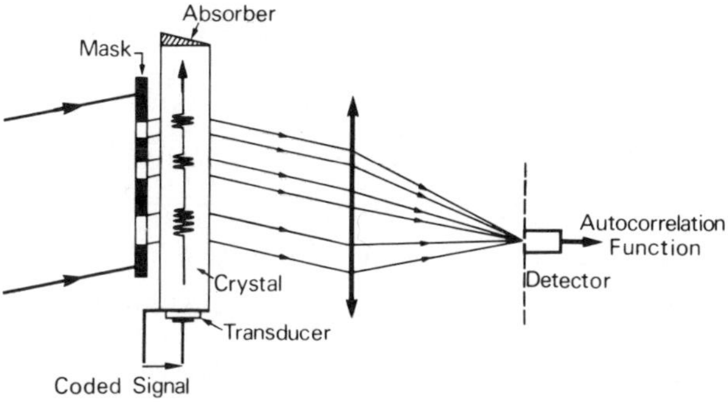

Fig. 9.43. Acousto-optic matched filter. The light beam, diffracted by the signal (in the form of elastic waves), is focused by the lens, and is a maximum when the elements of the wave packet coincide with the windows of the mask (replica of the code).

The elastic pulse propagation in front of the mask generates the code at the output of the photomultiplier which receives the diffracted beam. The signal is obtained by injecting the code and the carrier into a ring modulator which creates the phase inversions.

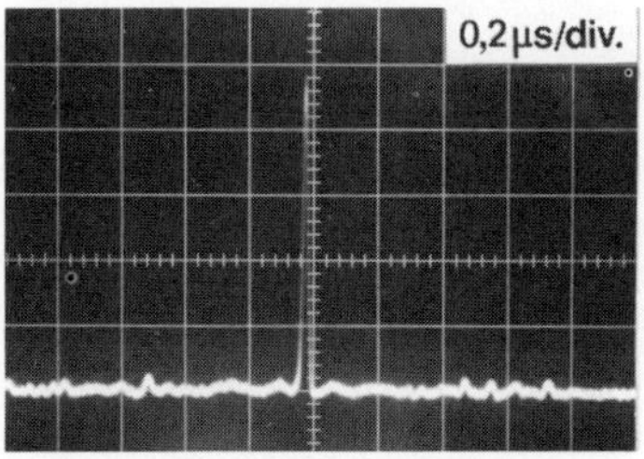

Fig. 9.44. Acousto-optic matched filter. Compressed pulse.

The device is simplified when the signal is linearly frequency modulated, since the mask bearing the code replica is no longer necessary. It was pointed out, in § 8.4.2.1, that the angle of deviation θ for the light beam is proportional to the elastic wave frequency. This angle

$$\theta \simeq \frac{\Lambda_0}{\lambda} = \frac{\Lambda_0}{V} f$$

varies along the line, following the law:

$$\theta \simeq \frac{\Lambda_0}{V} f_o + \frac{\Lambda_0}{V} \frac{B}{\Theta} \left(t - \frac{x_1}{V}\right)$$

which accounts for the frequency modulation (equation 9.6) and for the acoustic signal propagation in the x_1-direction. The diffracted beam is focused along the direction of mean deviation $\theta_0 = \Lambda_0 f_o/V$, at a distance D such that (fig. 9.45):

$$\frac{1}{D} = - \frac{d\theta}{dx_1} = \frac{\Lambda_0 B}{V^2 \Theta} \qquad (9.29)$$

If B = 50 MHz, Θ = 5 μs, Λ_0 = .6328 μm, V = 3,630 m/s ($PbMoO_4$), then the distance D is 2.1 m. This distance can be reduced by a simple optical mounting.

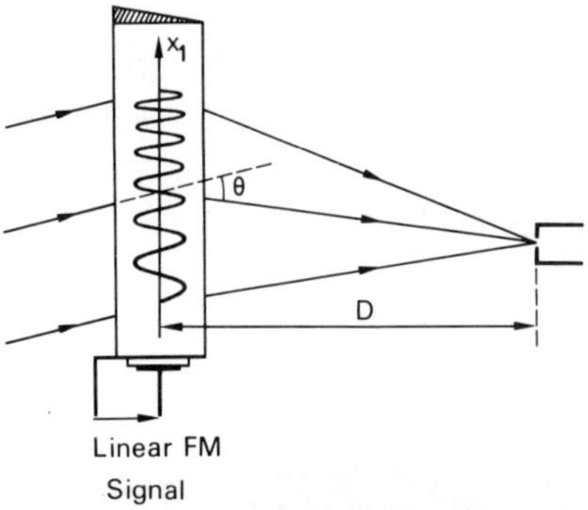

Fig. 9.45. Pulse compression by Bragg diffraction. The deflection angle θ depends upon frequency; the diffracted beam is focused at a distance $D = \Theta V^2/B\Lambda_0$.

If a detector is located at the convergence point, the time variation of the light intensity is proportional to the square of the autocorrelation function of the linearly frequency modulated signal. Compression ratios of several hundred have been reported [50, 51].

9.4. BANDPASS FILTERS

We do not intend to give a detailed description of bulk wave filters. The main features of the piezoelectric resonator, their major component, were discussed in § 7.1.2. Several hundred million conventional quartz filters have already been manufactured. They are employed at frequencies ranging from 50 kHz to 150 MHz; beyond 30 MHz, quartz plates vibrate at an harmonic frequency since their thickness cannot be reduced below 50 μm. Many papers have been published on this subject, since the pioneering work of P. Langevin and the construction of the first quartz-stabilized oscillator by Cady in 1918 [52]. Their use in large numbers is accounted for by the properties of quartz: it is a hard crystal, not very sensitive to its surroundings; a quartz resonator has a high Q-factor ($> 10^4$), an excellent frequency stability with respect to temperature (frequency drift under $10^{-6}\ {}^{o}C^{-1}$) and to time. The bandwidth of such filters is limited by the separation between resonant and antiresonant frequencies, i.e. by the electromechanical coupling constant: $\Delta f/f \lesssim K^2/2$. Quartz filters ($K^2 \simeq 10^{-2}$) are thus very selective. The coefficient K^2 is higher for piezoelectric ceramics ($K^2 \simeq 0.1 \rightarrow \Delta f/f \lesssim 5\%$). However, their poor temperature-stability and loss restrict their applications to low frequencies ($f < 30$ MHz) in the consumer product field. New materials, such as lithium tantalate ($LiTaO_3$) are currently being examined.

In addition to quartz filters, made of discrete elements, <u>monolithic filters</u> are now available. In this integrated structure, the resonators are set on the same plate and are coupled through the mechanical vibration which is evanescent outside the metal-coated zones [53].

The <u>bulk</u> wave filter involves <u>stationary</u> waves. The transfer function is normally a rectangular function about the resonant frequency (itself linked to the plate thickness); the bandwidth is limited by the material; the insertion loss is very low for single crystals with a high Q-factor (< a few dB).

Conversely Rayleigh wave filters, described below, involve progressive waves (except the filters based on surface wave resonators mentioned in note 3). The transfer function is defined by the transducer shape, the finger width being controlled by the frequency, which may reach several hundreds of MHz. Because of matching problems, the losses are greater (> 20 dB) the smaller the coupling constant and the larger the fractional bandwidth. At higher frequencies, propagation loss must be added to those transduction losses; this is now a limitation in the commercial development of bandpass filters.

For a Rayleigh wave delay line, if all the fingers of one transducer are longer then the width of the elastic beam emitted by the other transducer, the transfer function is equal to the product of the transfer functions of the two transducers.* If the receiver has few fingers, and accordingly a large bandwidth, the transfer function and the impulse response are controlled by the design of the input transducer. A $|(\sin x)/x|$ variation of finger length leads to a rectangular transfer function. A rectangular frequency response of width B about the central frequency f_o:

$$H(f) = \Pi\left(\frac{f - f_o}{B}\right) \quad \text{for} \quad f > 0 \qquad (9.30)$$

corresponds to the impulse response (see table 1.18):

$$h(t) = 2B\,\frac{\sin \pi Bt}{\pi Bt}\cos 2\pi f_o t. \qquad (9.31)$$

According to the method of § 7.2.4, the discrete sources should be distributed at times t_n such that

$$\phi(t_n) = 2\pi f_o t_n = n\pi \rightarrow t_n = \frac{n}{2f_o}$$

* This is no longer true if both transducers are apodized. The analysis is then more sophisticated and the synthesis of a filter becomes difficult [54].

i.e. at equidistant points located at

$$x_n = n \frac{V_R}{2f_o}$$

with amplitudes proportional to

$$A_n = \frac{\sin\left(n \frac{\pi B}{2f_o}\right)}{n \frac{\pi}{2} \frac{B}{f_o}} .$$

The filter of figure 9.46a has been implemented on a quartz substrate ($\underline{Y}$, $\underset{\rightarrow}{X}$) with aluminium electrodes [55]. Phase reversal at the zeroes of the (sin x)/x function is produced by a half-wavelength separation between the two adjacent fingers of the same electrode. The envelope of the impulse response is identical to the transducer shape. The frequency response is compared with the theoretical curve in fig. 9.46c. Neither curve is a rectangle, because the transducer has finite size and the (sin x)/x function is determined by a finite number of lobes. The centre frequency is 25 MHz, the bandwidth 2 MHz, the side-lobe level is 21 dB. If the large transducer is matched, losses are about 30 dB.

A <u>triangular</u> transfer function requires, for the transducer, a $((\sin x)/x)^2$ weighting of the active finger length (see table 1.18). Two filters of this type have been linked to make a frequency discriminator [56]. The response of each filter vanishes at the centre frequency of the other (28 MHz, 32 MHz). After the d-c voltages from the output detection diodes have been subtracted, a linear characteristic is obtained in a .7 MHz band around 30 MHz (fig. 9.47); but the losses are high (45 dB) on a quartz substrate.

A <u>narrow</u> fractional bandwidth requires one (or two) many-fingered transducer(s). For example, a 100 kHz bandwidth around 100 MHz implies a line with two identical 1,270 finger-transducers (equation 7.38). One way of reducing this large number is to sample both transducers [57], so that the distance between the samples is an integer multiple (by a factor of n_1 for one transducer, n_2 for the other) of the interdigital period $2d = V_R/f_o$.

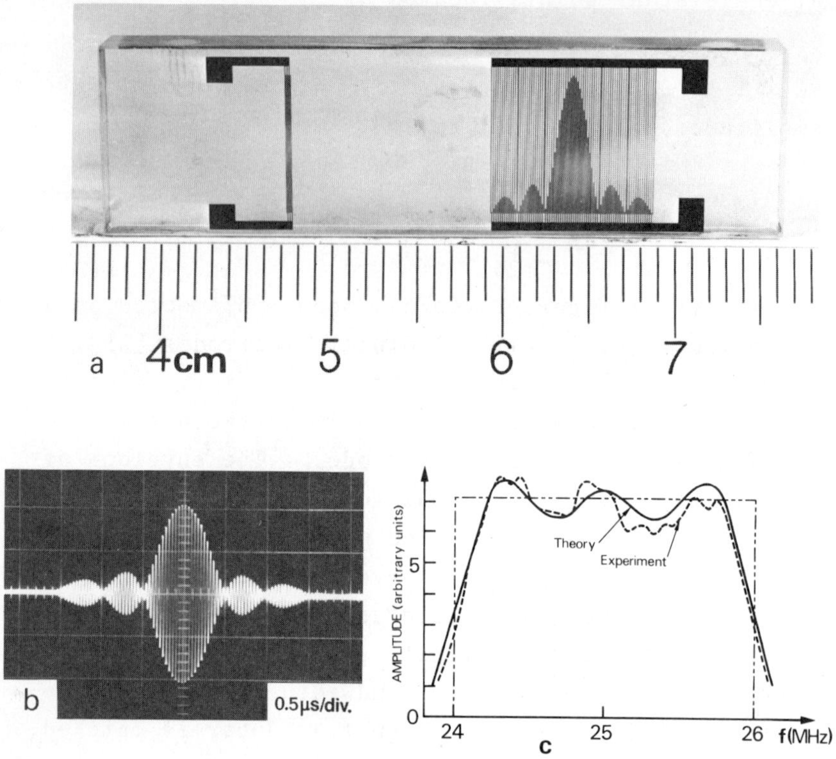

Fig. 9.46. Bandpass filter.
a) |(sin x)/x|-shaped transducer;
b) impulse response;
c) theoretical and experimental frequency responses.

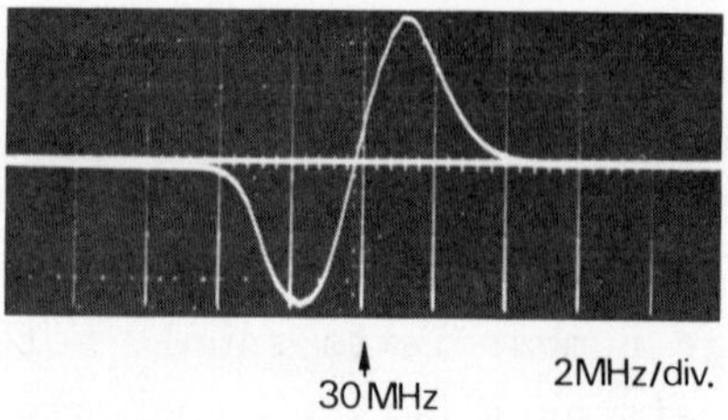

Fig. 9.47. Response of a Rayleigh wave frequency discriminator.

The frequency response of each transducer is a sequence of peaks at frequencies $f_o + pf_o/n$. The frequency interval between neighbouring peaks is the reciprocal of the propagation time n/f_o between two samples (exercise 9.4). The transfer function of the filter is a maximum at the coincidence frequencies $f_o + \delta f$ of the peaks of the two transducers:

$$\delta f = \frac{p_1}{n_1} f_o = \frac{p_2}{n_2} f_o$$

i.e. at the centre frequency f_o and at frequencies that differ from f_o by an integer multiple of f_o/q (q is the highest common factor of n_1 and n_2). Fig. 9.48 shows a filter on quartz ($n_1 = 55$, $n_2 = 45$), with a centre frequency of 102.5 MHz and a bandwidth of 36 kHz. The loss is about 20 dB with matching circuits.

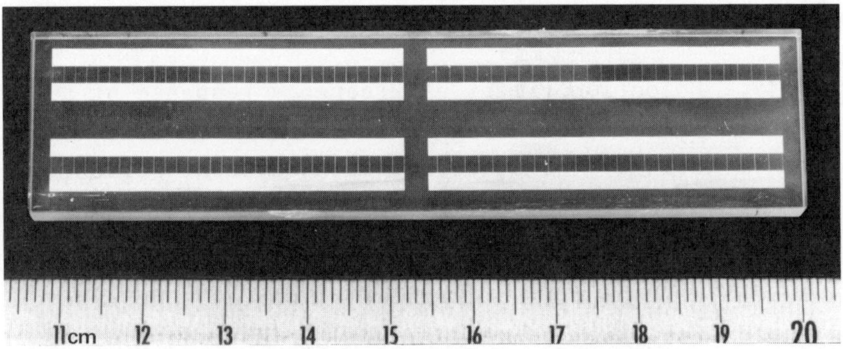

Fig. 9.48. Two identical narrow bandpass filters, on the same quartz single crystal (f_o = 102.5 MHz, Δf = 32 kHz). The distance between samples (each of them has two fingers) of the input (output) transducer is $55\ \lambda_0$ ($45\ \lambda_0$).

If the transfer function is to vanish at selected frequencies, it is possible to associate two lines that are identical, except for the time delays which differ by the amount τ. From the translation theorem (§ 1.3.2), the impulse response of the overall device $h(t) + h(t-\tau)$ corresponds to the frequency response:

$$\mathscr{H}(f) = H(f) + H(f)e^{-i2\pi f\tau} = 2H(f)e^{-i\pi f\tau} \cos \pi f\tau$$

which vanishes at frequencies

$$f_n = (n + \frac{1}{2}) \frac{1}{\tau} .$$

In Fig. 9.49 is an example of response for a filter with centre frequency 100 MHz, the separation between the zeroes being 3.3 MHz.

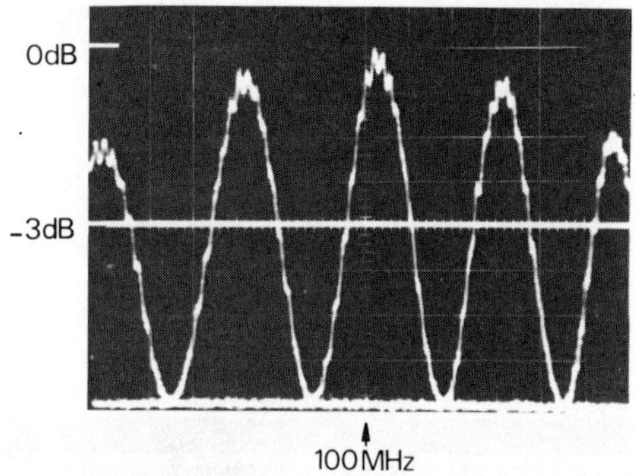

Fig. 9.49. Rejection filter. Frequency response of a filter consisting of two lines with different delays. Distance between adjacent zeroes: 3.3 MHz.

Finally, any frequency response can be synthesized by a suitable shape of the transducer, which is derived from the desired impulse response by the method of discrete sources. This principle is now applied to intermediary frequency television filters (25 to 40 MHz) [58, 59]. The principal problems are the insertion loss, the nature of the substrate and price.

Comments. 1) Reflective array filters. The reflective array technique, described at the end of § 9.3.5.1, has been applied [60] to the construction of Rayleigh wave bandpass filters (e.g.: centre frequency 200 MHz, fractional bandwidth .03, loss due to the reflector 10 dB). The reflective array periodicity and groove size define the filter function. This technique seems to be well suited to narrow band filters (< .5 MHz at 200 MHz).

2) Filters with fingers of constant length. Recalling that

$$s(t) = e(t) \cos \omega_0 t = \cos \psi(t) \cos \omega_0 t \qquad (9.32)$$
$$s(t) = \frac{1}{2} \cos [\omega_0 t + \psi(t)] + \frac{1}{2} \cos [\omega_0 t - \psi(t)]$$

where

$$e(t) = \cos \psi(t) \quad \Rightarrow \quad \psi(t) = \text{Arc}\cos e(t)$$

an amplitude modulated signal can be considered as the sum of two phase modulated signals of constant amplitude. If $e(t)$ is an even function then $\psi(-t) = \psi(t)$. Let $g(t)$ be the first term of the right hand side of (9.32); $s(t)$ reads:

$$s(t) = g(t) + g(-t).$$

The idea of composing $s(t)$ from $g(t)$ and its time reverse function $g(-t)$ has been applied to Rayleigh wave filters [61], especially in order to obtain the impulse response of equation 9.31. The line has a central transducer T_o corresponding to the function g, and two side transducers T_1 and T_2, each with a wide bandwidth. A short pulse applied to T_o launches two waves, propagating in opposite directions, the former representing $g(t)$ and reaching T_1, the latter representing $g(-t)$ and reaching T_2. The addition of the signals from T_1 and T_2 yields $s(t)$. Since the finger length is a constant one advantage of this type of filter lies in the absence of wave front distortion (Note in § 7.2.4).

3) Stationary wave filters. It has been pointed out that reflection at an array of grooves or of metallic strips is used for implementing dispersive delay lines or progressive wave bandpass filters. This procedure is also used for designing resonators as represented in figure 9.50 [62, 63]. Waves launched in both directions by the transducer are reflected by the gratings which act as mirrors. A stationary wave regime is set up at a frequency depending on the period and the spacing of the gratings. With gratings composed of a few hundred strips, Q factors of the order of 20,000 have been reported [64]. Although the Q factors of these cavities are smaller than those provided by bulk waves, these resonators

have the advantage of working at frequencies of a few hundred MHz without being significantly affected by the way they are mechanically mounted. Their planar structure enables more flexible linking cascading of resonators [65], connection by a multistrip coupler (§ 9.7) and weighting of the strip length. In addition, it may be mentioned that reflectors are also fabricated by ion implantation [66] and plasma etching techniques [64].

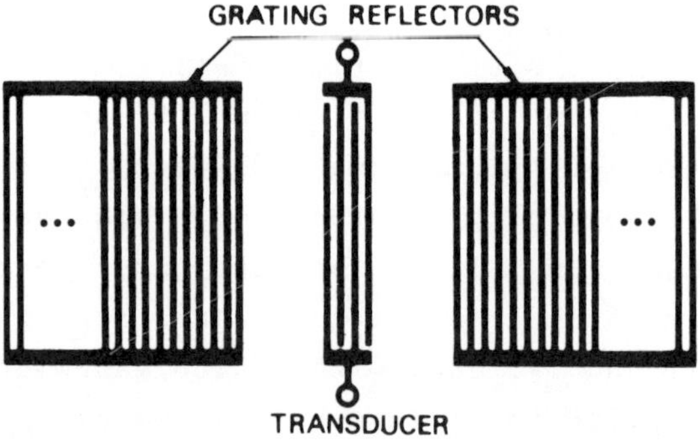

Fig. 9.50. Schematic of a one port surface wave resonator. The waves launched by the transducer are reflected by the grating made of metallic or dielectric strips or of grooves.

9.5. MEMORY FUNCTION

Because of the velocity of elastic wave propagation ($1 < V < 13$ mm/μs) and of the length of the available solid propagation media with acceptable loss, it is possible to store a great amount of information in a delay line. Since it cannot be fixed, at least in a simple way, its circulation around the line is maintained by a loop which ensures loss compensation (by means of an amplifier). The characteristics of these circulating memories are basically those of the delay line which is their major component, thus the repetition frequency f_R is a fraction of the carrier frequency; the longer the line, the greater its capacity and also the access time τ_A. A bulk wave line operating at a repetition frequency of f_R = 100 MHz can keep $f_R\tau_A$ = 1,000 bits in memory, the access time being τ_A = 10 μs.

Surface wave delay lines present two advantages [67]: the information can be coded and tapped by intermediate junctions (access time reduction).

A circulating memory constitutes a time compression device (DELTIC, "DElay Line for TIme Compression") [68] for a signal which has already been sampled. Let τ be the line delay. At the output, the samples, having duration $\delta = \tau/N$ and equal separation $(\tau + \delta)$, are bunched in groups of N, as indicated in fig. 9.51a. The principle is as follows: when sample 1, delayed by a first passage through the line, enters again, sample 2 has then nearly overtaken sample 1 and enters just behind it. After another passage through the line, sample 3 overtakes them, and so on (fig. 9.51b). Where the line is filled, it contains N samples ($\tau = N\delta$) which can be extracted in a single group. Then the time interval between samples is $\tau + \delta = (N + 1)\delta$ before they enter the line and only δ afterwards; the instantaneous signal frequency is multiplied by (N + 1). A very long signal ($\Theta \simeq 1$ s, $B \simeq$ 100 MHz), say a sonar signal, for which the matched filter cannot be implemented, can be processed in this way. The signal is decomposed into N samples ($N \simeq 1{,}000$):

$$\Theta = N(N + 1)\delta = (N + 1)\tau$$

and compressed into a pulse of duration τ ($\simeq$ 1 ms) and of sweep width $(N + 1)B$ ($\simeq$ 100 KHz); this pulse can be processed (after a frequency change) by a dispersive strip delay line (§ 9.3.3).

It is worth mentioning the principle of the static memory, which has been studied in the laboratory [69]. An image of the information is "printed" on a surface with the aid of an electron beam. Fig. 9.52 explains how it works. The signal has been transformed into Rayleigh waves, and its propagation is accompanied by an electric field wave, because the substrate is piezoelectric. The surface is then bombarded by a primary electron beam of

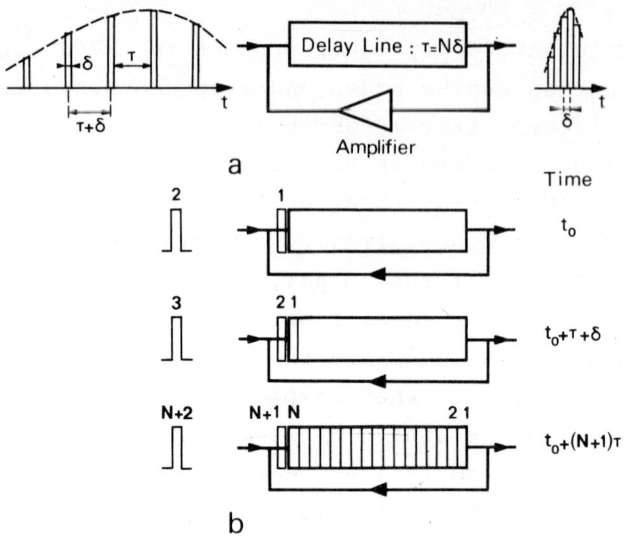

Fig. 9.51. DELTIC.

a) A signal of very long duration, would require a matched filter which is impossible to construct. Its samples are gathered by a circulating memory.

b) Gathering the samples.

a few hundred electron-volts for a period which is short compared to the carrier wave period. These primary electrons extract secondary electrons from the insulating material, which because of their low energy (a few electron-volts) are very sensitive to the elastic wave potential. The electrons which fall back to the surface are most attracted by positive potential domains. Since the substrate is an insulator, these electrons do not move and the elastic wave continues to propagate. The surface pattern of the charges maintains a stress field in the crystal and yields an electrostatic image of the signal; it can be preserved over long periods (several minutes) if the surface is a very good insulator and does not capture any parasitic charge. The information is read by means of a new electron flash on the substrate, which cancels or reduces the electric potential and the stress maintained by the stored charges. This sudden stress release generates two waves, starting in opposite directions: the one which travels towards the output transducer is the replica of the signal, the other, travelling towards the input transducer, is the time reversed signal.

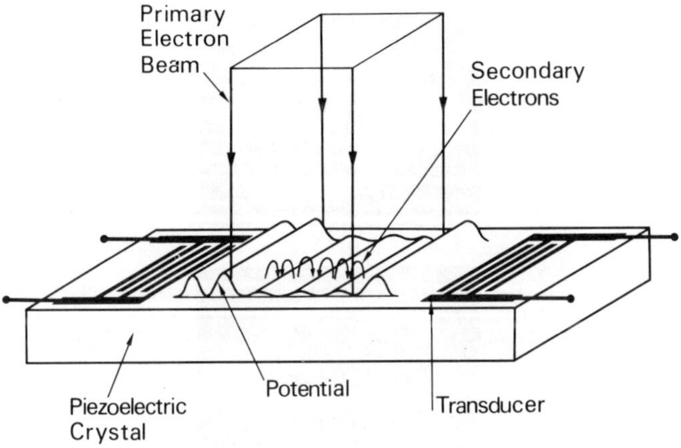

Fig. 9.52. Memory (and time reversal) function. The secondary electrons are distributed following the Rayleigh wave electric potential, and form an electrostatic image of the signal. They remain on the insulating substrate, maintaining a stress. A new electron flash removes these stresses and launches two waves propagating in opposite directions (fig. 1 of reference 69).

Experiments have been performed on quartz at 30 MHz. An example of the results is reproduced in fig. 9.53, showing (a) the signal received at the output transducer, after propagation through the line without having been stored; (b) the signal received at the same transducer after a 160 ms period of storage; (c) the time reversed signal received at the input transducer after 160 ms storage. The storage loss, measured by the power ratio of signals in Fig. 9.53a and 9.53b, is about 60 dB.

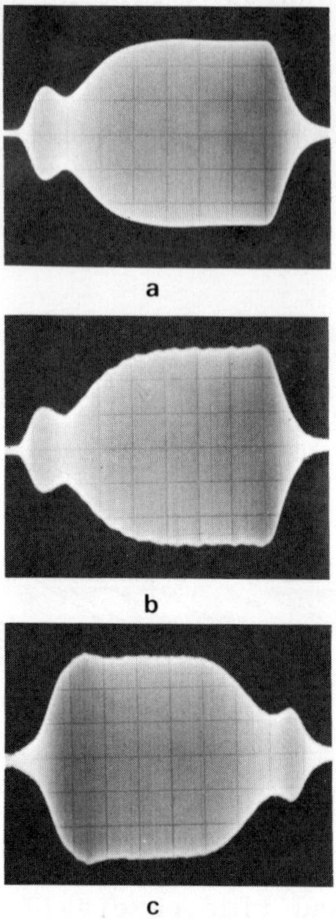

Fig. 9.53. Signals of the device in fig. 9.52 received, at the output transducer
a) without having been stored;
b) after storage for 160 ms;
c) at the input transducer after 160 ms storage.
(Fig. 2 of reference 69, kindly communicated by A.G. Bert).

Since electrical-mechanical energy conversion takes place twice, losses are proportion to K^4. The electronic current density needed to store and read the information is 100 mA/cm^2. The use of electrons travelling in a vacuum is constraining. This is why semiconductor structures are being investigated. One procedure consists in positioning a semiconductor, provided with a matrix of diodes, close to the piezoelectric substrate where Rayleigh waves propagate [70]. The electric field due to the acoustic wave induces a change in the diode states. The modification can be stored if a D.C. voltage pulse whose duration is small compared to the acoustic period, is applied at the instant the wave is under the diodes.

9.6. CONVOLUTION FUNCTION

So far, it has implicitly been assumed that the strain amplitude (i.e. the relative variation of distances in the solid) remains below the threshold ($S < 10^{-4}$) for which the simplified expression for strains (§ 4.1) and Hooke's law (§ 4.3) are no longer valid. When stress and strain are not proportional, the linear theory discussed in the preceding chapters is no longer exact. An analysis of the non linear effects requires additional elastic constants. One possible approach consists of inserting an extra term into the equations of the linear theory, whose degree is immediately above that of the "linear" terms. In equation 4.34, the quadratic form for the elastic potential energy, terms of degree 3 should be added such as $c_{ijk\ell pq}S_{ij}S_{k\ell}S_{pq}$. The strain/stress relationship now reads

$$T_{ij} = c_{ijk\ell}S_{k\ell} + c_{ijk\ell pq}S_{k\ell}S_{pq}.$$

The resulting propagation equation for elastic waves is complicated by the occurrence of these third order elastic constants [71]. However the equation can nevertheless be written down explicitly in a few simple cases (e.g. longitudinal waves in an unbounded medium). The results can be predicted qualitatively as follows: a monochromatic wave, during propagation, generates waves at harmonic frequencies [72]; two waves at frequencies f_1 and f_2, when interacting in the solid, give rise to waves at frequencies $(f_1 + f_2)$ and $(f_1 - f_2)$. Adding boundary conditions to the propagation equation makes the analysis of non linear effects for surface waves extremely complicated [73]. Any discussion of the mathematical aspects of nonlinearity is beyond the scope of the present book; we are interested here in the possible applications of such effects with a view to realizing multiplicative functions.

Consider two amplitude modulated signals

$$s_1(t) = A_1(t) \cos \omega_1 t \quad \text{and} \quad s_2(t) = A_2(t) \cos \omega_2 t$$

which produce two waves, propagating in opposite directions at velocity V:

$$u_1(t,x) = A_1(t - \frac{x}{V}) \cos (\omega_1 t - k_1 x)$$

and

$$u_2(t,x) = A_2(t + \frac{x}{V}) \cos (\omega_2 t + k_2 x).$$

Their interaction generates two waves

$$u_3(t,x) = B\, A_1(t - \frac{x}{V})\, A_2(t + \frac{x}{V}) \cos[(\omega_1 + \omega_2)t - (k_1 - k_2)x]$$

and

$$u_3'(t,x) = B\, A_1(t - \frac{x}{V})\, A_2(t + \frac{x}{V}) \cos[(\omega_1 - \omega_2)t - (k_1 + k_2)x].$$

B is a constant of non-linearity. At the output transducer of length L, of pitch corresponding for example to a wave-vector $(k_1 - k_2)$, one obtains a voltage, of carrier angular frequency $(\omega_1 + \omega_2)$ and of amplitude

$$A_3(t) = \beta \int_{-L/2}^{+L/2} A_1(t - \frac{x}{V}) A_2(t + \frac{x}{V}) dx. \qquad (9.33)$$

If both signals are short (compared to the transit time L/V under the transducer), this transducer can be regarded as having infinite length; a change of variable $t - x/V = \tau$ demonstrates that $A_3(t)$ is the convolution of the envelopes of the two signals.

$$A_3(t) = \beta V \int_{-\infty}^{+\infty} A_1(\tau) A_2(2t - \tau) d\tau. \qquad (9.34)$$

However, by comparing it with the standard convolution product (equation 1.30) we see that $A_3(t)$ has undergone a time compression by a factor 2. In fact, whereas up to now one of the two functions has been in the form of a fixed transducer, both factors of the product are now carried by waves travelling in opposite directions with a relative velocity 2V.

This kind of operation has been performed with bulk waves [74] and surface waves [75, 76]. Experiments are easier with Rayleigh waves since they are accessible on a piezoelectric substrate and they have a high elastic power density. The convolution product is detected by a transducer, whose interdigital spacing $d_3 = \lambda_3/2$ is determined by the wavelength $\lambda_3 = 2\pi/|k_1-k_2|$. When both signals have the same frequency, d_3 is infinite; the transducer is reduced to a plate and the associated electric field, at frequency 2ω, now dependent only on time, is detected between the plate and a ground electrode (fig. 9.54). Figure 9.55 shows the result obtained with a single crystal of $LiNbO_3$ when a 6 μs pulse, of power 1 watt and carrier frequency 150 MHz, is applied to both input transducers. The output signal, at frequency 300 MHz, and below the one microwatt level, is indeed triangular in shape.

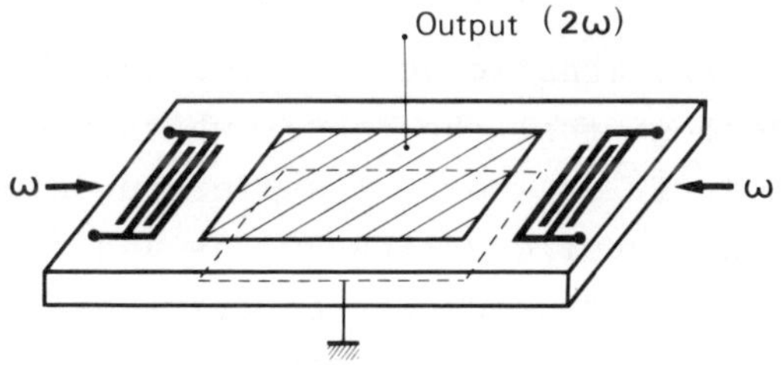

Fig. 9.54. Principle of an experiment aimed at the convolution of two signals (at the same frequency) using the non linear Rayleigh wave effect.

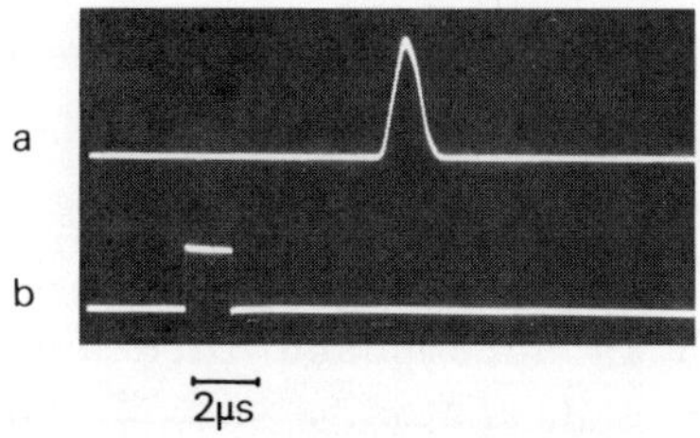

Fig. 9.55. Autoconvolution (a) of a square pulse (b). Carrier frequency 150 MHz. The base of the triangle is equal to the duration of the pulse and not twice as long.

If one of the signals, considered as the reference signal, is the time reverse of the other signal, then the output signal is the autocorrelation function. Compared to the matched filters described in § 9.3.5, the advantage of this device lies in the possibility of varying the reference signal "electronically", so that the autocorrelation of various differently coded signals can be performed rapidly. However, a drawback is that the input level cannot be varied greatly, the dynamic range is currently limited to 30 dB. To our knowledge, this device, which is evolving by the addition of semiconductors [77], has not yet left the laboratory.

9.7. SURFACE WAVE MULTISTRIP COUPLER

In the electromagnetic domain, two wave guides can be coupled together, so that the energy which propagates in the former is partly or wholly transferred into the latter [78]. The practice of coupling guides and arranging apertures in the common side can be transposed to elastic waves: two surface waveguides, consisting of a metal strip on a substrate, are locally joined in order that they are coupled through the substrate. But there is another original and ingenious solution [79, 80]. As depicted in fig. 9.56, the coupling between tracks is via a grating of metal lines on a piezoelectric substrate. A Rayleigh wave propagating in track A generates a voltage between the lines, which also appears in track B thus generating a Rayleigh wave. Complete transfer of the wave from one track to the other occurs after a coupling length L_t which can be derived approximately from the classical theory of modes [81]. Wave A is decomposed into a symmetric mode and an antisymmetric mode (fig. 9.56). The velocity V_S of the symmetric mode is not changed by the presence of the metal lines since no current is allowed to flow in the direction of propagation. On the other hand, the velocity V_A of the antisymmetric mode is equal to the velocity V_o of Rayleigh waves when the surface is shorted by a metal film (§ 6.2.2.4) since the electric potential associated to this

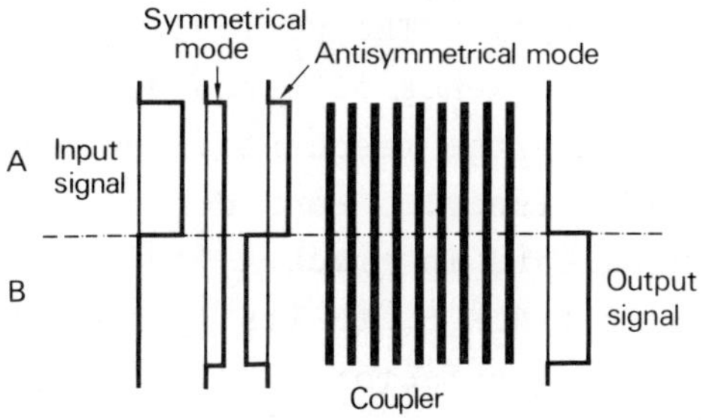

Fig. 9.56. Acoustic surface wave multistrip coupler. Beneath the coupler, both modes propagate, at different velocities; the length is chosen so that the phase shift at the output is π.

mode is cancelled out at the surface by the displacement of the charges along the metal lines. According to equation 6.91:

$$\frac{V_S - V_A}{V_S} = \frac{K^2}{2} \qquad (9.35)$$

The phase shift between the two modes is equal to π if the coupler length is L_t, where

$$2\pi f\left(\frac{1}{V_A} - \frac{1}{V_S}\right)L_t = \pi$$

or

$$L_t = \frac{1}{2f} \cdot \frac{V_S V_A}{V_S - V_A}$$

or, in terms of the coupling constant K and of the Rayleigh wavelength λ

$$L_t = \frac{\lambda}{K^2} .$$

Denoting by d the periodicity of the grating, the number of strips is

$$N_t = \frac{\lambda}{K^2 d} . \qquad (9.36)$$

Therefore the two modes cancel each other out in track A and add constructively in track B: the wave which has been introduced into track A of the coupler of length L_t comes out in track B. Equation 9.36 assumes that the coupler consists of a perfectly insulating layer in the direction of propagation, and a perfectly conducting layer in the orthogonal direction. Such an anisotropy is secured by metal strips. The periodicity of the real device introduces a sequence of frequency cut-offs (the first cutoff occurs at the resonant frequency $f_o = V/2d$) and reduces the efficiency. Considering that the presence of strips is equivalent, in track A, to sampling of a sine wave signal and, in track B, to its synthesis by a periodic structure, the reduction factor [82] is then equal to $[(\sin \theta/2)/\theta/2]^2$ where $\theta = 2\pi\alpha d/\lambda$, where αd is the active width of a metal strip. Equation 9.36 becomes

$$N_t = \frac{\lambda}{K^2 \alpha d (\frac{\sin \theta/2}{\theta/2})^2} = \frac{\pi}{K^2} \frac{\theta}{1 - \cos \theta} . \qquad (9.37)$$

In fact, the parameters α and K^2 are adjusted to fit the experimental results. α is not identical to the fraction a of the coupler surface which is metallized (α = ra) and the value of K^2 must be corrected by a factor F.

$$N_t = \frac{\pi}{FK^2} \frac{\pi raf/f_o}{1 - \cos(\pi raf/f_o)} \quad \text{where} \quad f_o = \frac{V}{2d} . \qquad (9.38)$$

As with the interdigital transducer, the maximum efficiency (maximum value of FK^2) is obtained for a = 1/2. For lithium niobate (YZ) with a = 1/2, r = 1.7 and FK^2 = .043, a working formula [83] is

$$N_t = \frac{195f/f_o}{1 - \cos(153f/f_o)} \qquad (9.39)$$

where the argument of the cosine is in degrees.

Fig. 9.57 plots the number of strips N_t for 100 per cent transfer between tracks, against the normalized frequency f/f_o. The variation in the vicinity of the minimum (f/f_o = .873, N_t = 101) is slow. At a given frequency f, the length of the coupler can be reduced if the spatial period is also decreased i.e. the frequency $f_o = V/2d$ is increased. In figure 9.57 this fact is also revealed by the curve representing the product: length of the coupler x frequency as a function of f/f_o. Compared to the value at the minimum of N_t, the length of the coupler is cut down to 30% of its original size when

$f_o = 2f$. This reduction is appreciable if the overall dimensions of the device are of importance. The characteristics are plotted in dotted lines in the vicinity of the frequency f_o because they do not account for the fact that in this range of frequencies cumulative reflections lead to cut-off.

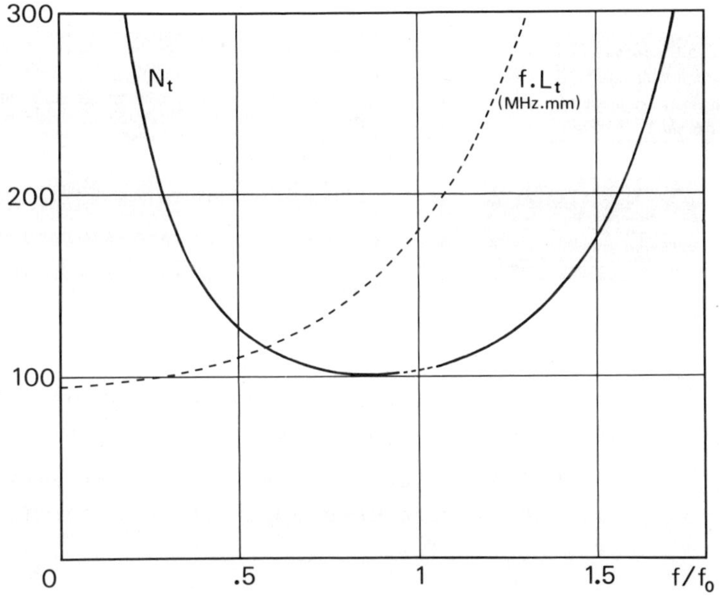

Fig. 9.57. Multistrip coupler characteristics. Number N_t of strips for 100 per cent transfer between tracks versus frequency normalized to the first stopband frequency f_o (solid line). Product length of the coupler x frequency (expressed in mm x MHz) versus normalized frequency (dotted line).

The outstanding performance of this very simple device is illustrated by the photographs of Fig. 9.58, referring to a coupler of 140 lines on $LiNbO_3$. The transfer loss, from one track to the other, is .5 dB at 45 MHz, the rejection between outputs is above 30 dB. Among many applications, already achieved (on YZ $LiNbO_3$) or intended by the inventors of the device [83] are the bulk wave suppressor (the coupler does not transmit the bulk waves generated by the input transducer), the coupler between separate substrates, the beam divider (3 dB coupler, $L = L_t/2$), the magic T and the beam width compressor.

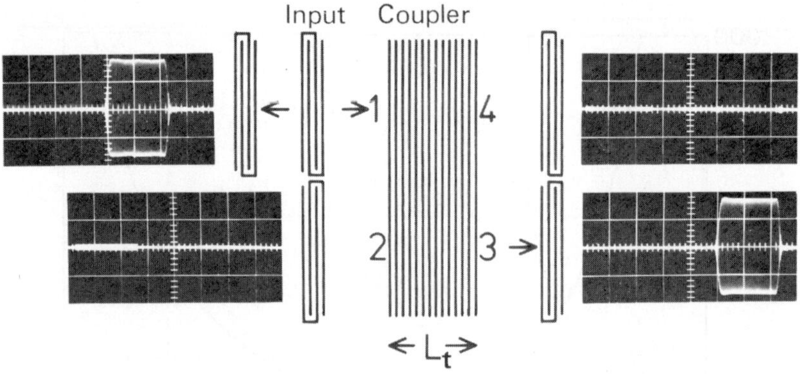

Fig. 9.58. Transmission and reflexion response of the multistrip coupler. For length L_t, the signal at port 1 emerges almost entirely at port 3 (fig. 6 of reference 80, with kind permission of E.G.S. Paige).

9.8. ACOUSTO-OPTIC LIGHT DEFLECTOR AND MODULATOR

At the end of § 8.4.2.5, it was mentioned that colinear interactions between light waves and elastic waves could be applied to make a tunable optical filter and in §§ 9.2.3 and 9.3.6, two applications of this interaction were reported: the variable delay line, and the matched filter. It now seems useful to discuss the structure of an acousto-optic light deflector or modulator. A light beam is an excellent means of transmitting information and elastic waves are a good tool with which to modulate it in space or time.

<u>Structure</u>. Fig. 9.59 shows a deflector operating at Bragg incidence, and at a centre frequency of about 300

Fig. 9.59. Acousto-optic interaction light deflector. Elastic waves propagate in the lead molybdate parallelepiped. The input prisms enlarge the light beam, in order to increase the number of resolvable directions. The output prisms restore a beam of the same size as the incident beam.

MHz. The interaction medium is a lead molybdate crystal. One of the ends bears a lithium niobate transducer, vacuum soldered with indium. The other extremity is coated with an acoustic absorber. The electric matching of the transducer to the signal generator is achieved by means of a strip line. The role of the prisms, on each side of the crystal [84] is to enlarge the incident beam along the $PbMoO_4$ bar axis - so as to increase the elastic wave transit time (§ 8.4.2.4) - and to reduce the output beam width. Thus the deflector restores an output beam identical to the input beam. Prisms are used rather than lenses because they keep waves planar and lead to larger angles of deflection. If the electric power applied to the transducer is of the order of 1 watt, about 50 per cent

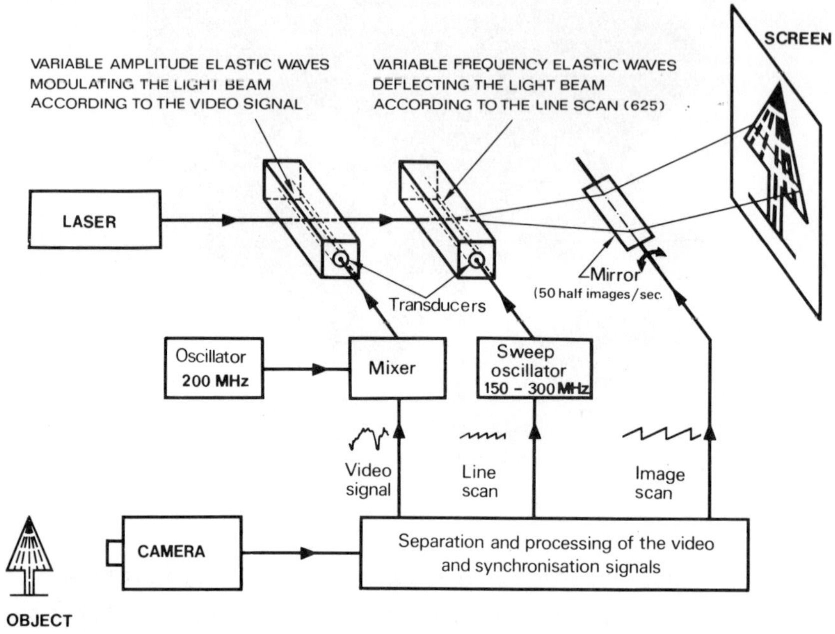

Fig. 9.60. Layout of a television experiment. The light beam is intensity modulated by elastic waves of variable intensity and horizontally deflected by elastic waves of variable frequency.

of the incident light beam is deviated and the number of distinct directions is close to 500 [85]. An acousto-optic light modulator looks very much like a deflector but has no beam enlarging prisms; the crystal is also shorter. Some typical values for a modulator are:

- fraction of light deflected: 70 per cent;
- pulse rise time: 30 ns;
- controlling power: 1 W.

Television experiment. In order to illustrate the possibilities of the acousto-optic light deflector or modulator several television experiments have been designed in laboratories [86]. Such an experiment (fig. 9.60) has

Fig. 9.61. Photograph of the image projected onto the screen (width 50 cm).

been carried out by R. Torguet in our laboratory. The video frequency signal is applied to the modulator which directly modifies an argon laser beam. The modulated light beam is deflected horizontally by a deflector whose control frequency corresponds to the scanning rate. Since an image of duration 20 ms is defined by 625 lines, each line is scanned in 32 μs. The slower vertical deflection is ensured by a galvanometric mirror which receives a 50 Hz saw-tooth shaped signal. The image may be displayed onto a one metre square screen with a 2 W laser [87].

It is obviously possible to combine an X-deflector and a Y-deflector if fast deflections in both directions are needed.

In conclusion, a device occupying a volume of a few cubic centimetres, and consisting of a crystal bearing a transducer, which transforms electric control signals into elastic waves can strongly, almost entirely, modulate an optical signal, and deflect the light beam along several hundreds of resolvable directions in a very short time (10 μs). This component, now developing beyond the laboratory stage, should be applied in such areas as telereprography, display screens and the fast printing of graphic symbols.

REFERENCES

[1] P.V. Hunt. "Electroacoustics. The analysis of transduction, and its historical background", Chap. 1. New York: John Wiley & Sons (1954).
[2] D.L. Arenberg. Ref. 5 of Chap. 5.
[3] J.R. Klauder, A.C. Price, S. Darlington and W.J. Aldersheim, "The theory and design of chirp radars", Bell Syst. Tech. J., 39, 745 (1960).
[4] J.H. McFee. "Transmission and amplification of acoustic waves in piezoelectric semiconductors". In "Physical Acoustics" (W.P. Mason and R.N. Thurston Eds.), Vol. 4A, Chap. 1. New York: Academic Press (1966).
[5] J.H. Collins, K.M. Lakin and P.J. Hagon, Proc. IEEE, 57, 740 (1969).
[6] H.C. Huang, J.D. Knox, Z. Turski, R. Wargo and J.J. Hanak. Appl. Phys. Lett., 24, 109 (1974).
[7] J.H. Eveleth. Proc. IEEE, 53, 1406 (1965).
[8] Ph. Defranould. European Microwave Conference, Stockholm (1971).
[9] Reference 25 of Chap. 7.
[10] P. Hartemann and C. Arnodo. Electron. Lett., 8, 265 (1972).
[11] B. Rulf. J. Acoust. Soc. Am., 45, 493 (1969).
[12] Ref. 8 of Chap. 5, pp. 29-42.
[13] W.L. Bond, T. Reeder, H.J. Shaw. Electron. Letters, 7, 79 (1971).
[14] M.F. Lewis and E. Patterson. Appl. Phys. Letters, 18, 143 (1971).
[15] L.R. Adkins and A.J. Hughes. J. Appl. Phys., 42, 1819 (1971).
[16] G. Kino and T.M. Reeder. IEEE Trans. Electron. Devices, ED.18, 909 (1971).
[17] L.A. Coldren. IEEE Trans. Son. Ultrason., SU.20, 17 (1973).
[18] M.F. Lewis. Ultrasonics, 12, 115 (1974).
[19] M.J. Brienza and A.J. Demaria. Appl. Phys. Letters, 9, 312 (1966).

[20] O. Cahen, E. Dieulesaint and R. Torguet. C.R. Acad. Sc. Paris, 266, 1009 (1968).

[21] P.M. Woodward. "Probability and information theory, with application to radar", Oxford: Pergamon Press (1953).

[22] M. Carpentier. "Radars. Concepts nouveaux", chap. 2, Paris: Dunod (1966).

[23] C.E. Cook and M. Bernfeld. "Radar Signals. An Introduction to Theory and Application", Chap. 2. New York: Academic Press (1967).

[24] G. Bruhat and A. Kastler. "Cours de physique générale - Optique", pp. 202-209. Paris: Masson (1965).

[25] P. Tournois. Ann. de Radioélectricité, 78, 267 (1964).

[26] P. Hartemann. "Filtres dispersifs à ondes élastiques de Rayleigh", Thèse, université de Paris-Sud (Orsay) (1973).

[27] T.R. Meeker and A.H. Meitzler. "Guided wave propagation in elongated cylinders and Plates". In "Physical Acoustics" (W.P. Mason Ed.), Vol. 1A, pp. 111-167. New York: Academic Press (1964).

[28] J.E. May, Jr. "Guided wave ultrasonic delay lines". In "Physical Acoustics" (W.P. Mason Ed.), Vol. 1A, pp. 417-483. New York: Academic Press (1964).

[29] Reference 8 of chap. 5, chap. 2.

[30] A.H. Fitch. J. Acoust. Soc. Am., 35, 709 (1963).

[31] R.S. Duncan and M.R. Parker, Jr. Proc. IEEE, 53, 413 (1965).

[32] G.A. Coquin and R. Tsu. Proc. IEEE, 53, 581 (1965).

[33] D.E. Miller and M.R. Parker. Proc. IEEE, 54, 891 (1966).

[34] W.S. Mortley. Marconi Rev., 28, 273 (1965).

[35] P. Tournois and C. Lardat. IEEE Trans. Son. Ultrason., SU.16, 107 (1969).

[36] C. Lardat, C. Maerfeld and P. Tournois. Proc. IEEE, 59, 355 (1971).

[37] C. Lardat and P. Tournois. IEEE Ultrasonic Symposium Proc., p. 280 (1972).

[38] M.B. Schulz, B.J. Matsinger and M.G. Holland. J. Appl. Phys., 41, 2755 (1970).

[39] P. Hartemann and E. Dieulesaint. Electron. Letters, 5, 219 (1969).

[40] Reference 23, p. 178.

[41] E.L. Key, E.N. Fowle and R.D. Haggarty. Lincoln Lab., M.I.T., Lexington Massachusetts, Tech. Rept., 207 (Sept. 1969).

[42] P. Hartemann and E. Dieulesant. Onde électrique, 51, 523 (1971).

[43] E.K. Sittig and G.A. Coquin. IEEE Trans. Son. Ultrason., SU.15, 111 (1968).

[44] T.A. Martin. IEEE Trans. Microwave Theory Tech., MTT-21, 186 (1973).

[45] R.C. Williamson and H.I. Smith. Electron. Lett., 8, 401 (1972).

[46] R.C. Williamson and H.I. Smith. IEEE Trans. Microwave Theory Tech. MTT-21, 195 (1973).

[47] R.C.M. Li. IEEE Ultrasonics Symposium Proc., p. 263 (1972).

[48] H.I. Smith, R.C. Williamson and W.T. Brogan. IEEE Ultrasonics Symposium Proc., p. 198 (1972).

[49] R.C. Williamson, V.S. Dolat and H.I. Smith. IEEE Ultrasonics Symposium Proc., p. 490 (1973).

[50] M.B. Schulz, M.G. Holland and L. Davis, Jr. Appl. Phys. Lett., 11, 237 (1967).

[51] J.H. Collins, E.G.H. Lean and H.J. Shaw. Appl. Phys. Lett., 11, 240 (1967).

[52] Reference 1, pp. 53-57.

[53] W.J. Spencer. "Monolithic Crystal Filters". In "Physical Acoustics" (W.P. Mason and R.N. Thurston, Eds.), Vol. 9, pp. 167-220. New York: Academic Press (1972).

[54] Reference 16 of chap. 7.

[55] P. Hartemann and E. Dieulesant. Electron. Letters, 5, 657 (1969).

[56] P. Hartemann and O. Menager. Electron. Letters, 8, 214 (1972).

[57] P. Hartemann. Electron. Letters, 7, 674 (1971).

[58] A.J. De Vries, R. Adler, J.F. Dias and T.J. Wojcik. paper G-5, IEEE Ultrasonics Symp, (St. Louis U.S.A. 1969).

[59] D. Chauvin, G. Coussot and E. Dieulesaint. Electron. Letters, 7, 491 (1971).

[60] J. Melngailis, J.M. Smith and J.H. Cafarella. 1972 IEEE Ultrasonics Symposium Proc., p. 221 (1972).

[61] C. Atzeni, G. Marnes and L. Masotti. 1973 IEEE Ultrasonics Symposium Proc., p. 414 (1973).

[62] E.J. Staples, J.S. Schoenwald, R.C. Rosenfeld and C.S. Hartmann. 1974 IEEE Ultrasonics Symposium Proc., p. 245 (1974).

[63] R.C.M. Li, R.C. Williamson, D.C. Flanders and J.A. Alusow. 1974 IEEE Ultrasonics Symposium Proc., p. 257 (1974).

[64] S.P. Miller, R.E. Stigall and W.R. Shreve. 1975 IEEE Ultrasonics Symposium Proc., p.474 (1975).

[65] P.S. Cross, R.S. Smith and W.H. Haydl. Electron. Lett., 11, 244 (1975).

[66] P. Hartemann. 1975 IEEE Ultrasonics Symposium Proc., p. 303 (1975).

[67] H. Van De Vaart and L.R. Schissler. IEEE Trans. Microwave Theory Tech., MTT-21, 236 (1973).

[68] V.C. Anderson. Harvard Acoust. Lab., Cambridge, Mass., Tech. Mem., No. 37 (1956).

[69] A.G. Bert, B. Epsztein, G. Kantorowicz. Appl. Phys. Lett., 21, 50 (1972).

[70] K.A. Ingebrigtsen, R.A. Cohen and R.W. Mountain. Appl. Phys. Lett., 26, 596 (1975).

[71] R.N. Thurston. "Wave propagation in fluids and normal solids". In "Physical Acoustics" (W.P. Mason, Ed.), Vol. 1A, pp. 91-109. New York: Academic Press (1964).

[72] E. Adler, E. Bridoux, G. Coussot and E. Dieulesaint. IEEE Trans. Son. Ultrason., SU-20, 13 (1973).

[73] V.E. Ljamov, Tzu-Hwa Hsu and R.M. White. J. Appl. Phys., 43, 800 (1972).

[74] C.F. Quate and R.B. Thompson. Appl. Phys. Lett., 16, 494 (1970).

[75] L.O. Svaasand. Appl. Phys. Letters, 15, 300 (1969).

[76] M. Luukkala and G.S. Kino. Appl. Phys. Lett., 18, 393 (1971).

[77] P. Das, M.N. Araghi and W.C. Wang. Appl. Phys. Lett., 21, 152 (1972).

[78] P. Grivet. "Physique des lignes HF et UHF", tome III, chap. 10. Paris: Masson et Cie (1974).

[79] F.G. Marshall and E.G.S. Paige. Electron. Letters, 7, 460 (1971).

[80] F.G. Marshall, C.O. Newton and E.G.S. Paige. IEEE Trans. Microwave Theory Tech., MTT-21, 206 (1973).

[81] W.H. Louisell. "Coupled mode and parametric electronics", chap. 1. New York: Wiley (1960).

[82] J. Max. "Méthodes et techniques de traitement du signal et applications aux mesures physiques", tome I, chap. 7. Paris: Masson (1972).

[83] F.G. Marshall, C.O. Newton and E.G.S. Paige. IEEE Trans. Microwave Theory Tech., MTT-21, 216 (1973).

[84] F. Gires. Revue Phys. Appliquée, 4, 505 (1969).

[85] R. Torguet and E. Dieulesaint. Electron. Letters, 5, 632 (1969).

[86] A. Korpel, R. Adler, P. Desmares and W. Watson. Appl. Optics, 5, 1667 (1966).

[87] R. Torguet. "Etude théorique et expérimentale de l'interaction de la lumière avec des ondes acoustiques de forte puissance". Thèse, université de Paris VI (1973).

[88] R.W. Damon. IEEE Spectrum, p. 87 (June 1967).

BIBLIOGRAPHY

The reference list includes books and review papers referring to the various topics dealt with in this chapter, for instance references 22, 23, 27, 28, 53. We may add a few other papers of general interest.

R.M. White. "Surface elastic waves", Proc. IEEE, 58, 1238 (1970).

E. Dieulesaint. "Lignes à ondes élastiques: application au traitement des signaux électriques", l'onde électrique, 50, 899 (1970).

J. Burnsweig, E.H. Gregory and R.J. Wagner. "Surface-wave device applications and component developments", IEEE J. Solid-state circuits, SC-5, 310 (1970).

H. Sabine and P.H. Cole. "Surface acoustic waves in communications engineering", Ultrasonics, 9, 103 (1971).

G.S. Kino and H. Matthews. "Signal processing in acoustic surface wave devices", IEEE Spectrum, 18, 22 (August 1971).

D.A. Gandolfo, C.L. Grasse and E.J. Schmitt. "Surface acoustic waves: new processing tools for ew", Microwaves, 10, Nb 10, 44 (1971).

E. Dieulesaint. "Traitement de signaux électriques par ondes élastiques de Rayleigh", J. Physique, 33, C6-176 (1972).

J. de Klerk. "Past, present and future of surface elastic waves", J. Physique, 33, C6-182 (1972).

E. Dieulesaint and P. Hartemann. "Acoustic surface wave filters", Ultrasonics, 11, 24 (1973).

G.S. Kino and J. Shaw. "Acoustic surface waves", Scientific American, Oct. 1972, p. 51.

R.H. Tancrell, M.B. Schulz, H.H. Barret, L. Davis Jr. and M.G. Holland. "Dispersive delay lines using ultrasonic surface waves", Proc. IEEE (letters), 57, 1211 (1969).

C. Atzeni. "Sensor number minimization in acoustic surface-wave matched filters", IEEE Trans. Son. Ultrason., SU-18, 193 (1971).

M.E. Pedinoff, H.M. Gerard, G.W. Judd. "Electrical probe measurements of wavefront phase distortion from apodized wide band surface wave transducers", IEEE Trans. Son. Ultrason., SU-19, 395 (1972).

W.D. Squire, H.J. Whitehouse and J.M. Alsup. "Linear signal processing and ultrasonic transversal filter", IEEE Trans. Microwave Theory and Tech., MTT-17, 1020 (1969).

W.S. Jones, C.S. Hartmann and L.T. Claiborne. "Evaluation of digitally coded acoustic surface-wave matched filter", IEEE Trans. Son. Ultrason., SU-18, 21 (1971).

C.C. Tseng. "Signal multiplexing in surface-wave delay lines using orthogonal pairs of Golay's complementary sequences", IEEE Trans. Son. Ultrason., SU-18, 103 (1971).

C. Atzeni and L. Masotti. "Bandpass filtering by acoustic planar processing techniques", Alta Frequenza, 52, 84 (1973).

J.D. Maines and E.G.S. Paige. "Surface-acoustic-wave components, devices and applications", IEEE Reviews, 120, 1078 (1973).

M.G. Holland and L.T. Claiborne. "Practical Surface Acoustic Wave Devices", Proc. IEEE, 62, 582 (1974).

E. Dieulesaint and D. Royer. "Acoustic Surface Waves". In "Handbook of surfaces and interfaces" (L. Dobrzynski Ed.), Vol. 2, Chap. 2, New York: Garland Press (1978).

"Surface acoustic wave devices and applications"

1. "Introductory review", by D.P. Morgan, Ultrasonics, 11, 121 (1973).
2. "Pulse compression filters", by J.D. Maines and J.N. Johnston, Ultrasonics, 11, 211 (1973).
3. "Spread spectrum processors", by B.J. Hunsinger, Ultrasonics, 11, 254 (1973).
4. "Bandpass filters", by R.F. Mitchell, Ultrasonics, 12, 29 (1974).
5. "Signal processing using programmable non-linear convolvers", by D.P. Morgan, Ultrasonics, 12, 74 (1974).
6. "Oscillators - the next successful surface acoustic wave device?", by M.F. Lewis, Ultrasonics, 12, 115, (1974).

EXERCISES

9.1. Find the response of the filter matched to the signal of fig. 9.11b.

Solution. The autocorrelation of the signal can be determined by graphical means, as described in § 1.3.4, and it is depicted in fig. 9.62.

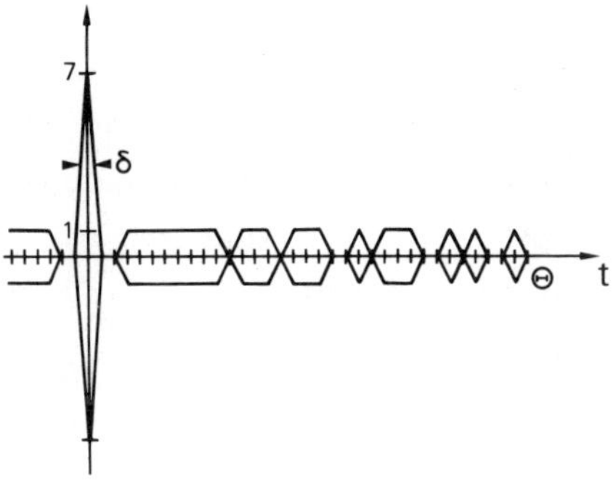

Fig. 9.62.

9.2. By applying the principle of stationary phase (as explained in § 1.3.2), calculate the compressed pulse $s_2(t)$ at the output of a filter matched to a signal

$$s_1(t) = \Pi(t/\Theta) \cos [2\pi f_o t + \psi(t)]$$

whose S shaped frequency modulation follows a law of the form

$$t = \Theta \left[\frac{f - f_o}{B} + \frac{a}{2\pi} \sin \left(2\pi \frac{f - f_o}{B}\right)\right] \qquad a = \text{constant}$$

for $-\frac{\Theta}{2} < t < \frac{\Theta}{2}$, or $f_o - \frac{B}{2} < f < f_o + \frac{B}{2}$.

Solution. The signal $s_2(t)$ is derived from its spectrum $S_2(f) = H(f)S_1(f) = |S_1(f)|^2$ through

$$s_2(t) = \int_{-\infty}^{+\infty} |S_1(f)|^2 e^{i2\pi ft} df$$

Equation 1.26 together with $e_1[t(f)] = \Pi(\frac{f-f_o}{B})$ provides

$$|E_1(f - f_o)|^2 \simeq \Pi\left(\frac{f-f_o}{B}\right) \frac{\Theta}{B} \left[1 + a \cos \left(2\pi \frac{f-f_o}{B}\right)\right].$$

According to 1.22:

$$|S_1(f)|^2 = \frac{1}{4}|E_1(f - f_o)|^2 + \frac{1}{4}|E_1(f + f_o)|^2$$

and the output signal $s_2(t) = e_2(t) \cos (2\pi f_o t)$ has an envelope:

$$e_2(t) = \frac{\Theta}{2B} \int_{-B/2}^{+B/2} (1 + a \cos \frac{2\pi f}{B}) e^{i2\pi ft} \, df$$

or

$$e_2(t) = \frac{\Theta}{2} \frac{\sin \pi Bt}{\pi Bt} [1 + a \frac{(Bt)^2}{1 - (Bt)^2}]$$

Figure 9.63 shows that the function in brackets, which vanishes for $Bt = 1/\sqrt{1-a}$, widens the central peak of $(\sin \pi Bt)/\pi Bt$ and weakens the first side lobe much more than the others, which are reduced by a factor of about (1-a).

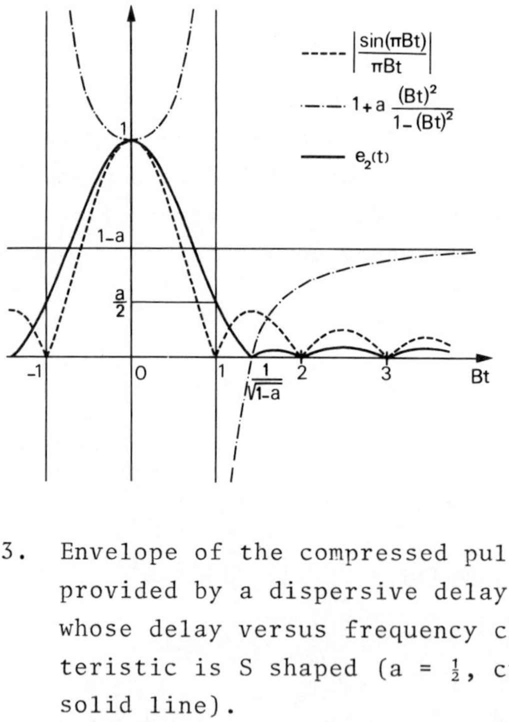

Fig. 9.63. Envelope of the compressed pulse provided by a dispersive delay line whose delay versus frequency characteristic is S shaped ($a = \frac{1}{2}$, curve in solid line).

9.3. Prove that the sum of the autocorrelation signals for each pair of signals in Fig. 9.64a has no sidelobe.

<u>Solution</u>. See Fig. 9.64b.

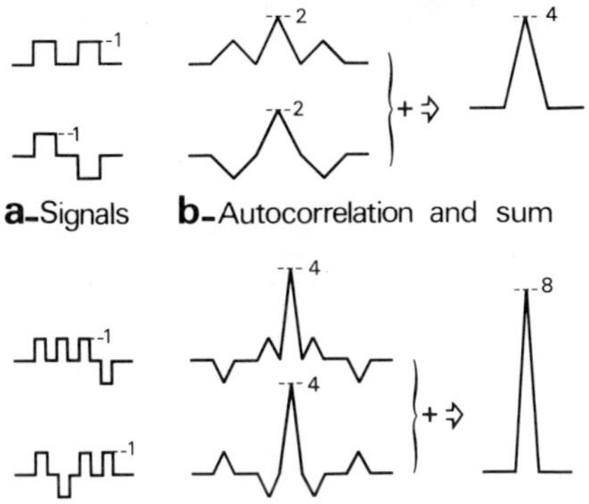

Fig. 9.64. Golay complementary codes.

9.4. Derive the frequency response of a transducer consisting of an infinity of samples, of separation $n\lambda_o$, each sample having five fingers.

Solution. The impulse response is a sequence of pulses of width $2/f_o$, and spacing $\tau = n/f_o$

$$h(t) = \left[\sum_{m=-\infty}^{+\infty} \Pi\left[\frac{f_o(t - m\tau)}{2}\right]\right] \cos 2\pi f_o t.$$

From the results of § 1.3.2, the spectrum of the envelope is

$$E(f) = \frac{\sin \frac{2\pi f}{f_o}}{\pi f} \sum_{m=-\infty}^{+\infty} e^{-i2\pi m f\tau}$$

or, making use of equation 1.40:

$$E(f) = \frac{\sin \frac{2\pi f}{f_o}}{\pi f} \sum_{p=-\infty}^{+\infty} \delta(f\tau - p).$$

The frequency response $H(f) = \frac{1}{2}E(f - f_o) + \frac{1}{2}E(f + f_o)$ is a sequence of peaks at frequencies $f = \pm f_o + p\, f_o/n$, whose intensity is modulated by a $(\sin x)/x$ curve.

9.5. How should the sampling period T at the input of a DELTIC be chosen to obtain the time inverse of the signal?

Solution. $T = (N - 1)\delta$.

General Bibliography

The reader who wishes to be informed of advances in the topics dealt with in this book should know that symposia and colloquia are regularly held, where recent researches and applications in the field of elastic waves are discussed. The most suited to the field of this work is the IEEE Ultrasonics Symposium;

IEEE Ultrasonics Symposium Proceedings, Publication: 72 CHO 708-8SU (1972), 73 CHO 807-8SU (1973), 74 CHO 896-1SU (1974), 75 CHO 994-4SU, 76 CH 1120-5SU, 77 CH 1264-1SU.

Reading specialized periodicals is also a good means of keeping up to date. Amongst these should be mentioned:

IEEE Transactions on Sonics and Ultrasonics,

Ultrasonics,

The Journal of the Acoustical Society of America.

From time to time, a special issue of a periodical is devoted to the high frequency elastic wave domain. For example, see:

IEEE Transactions on Microwave Theory and Techniques, MTT-17, November 1969, special issue on microwave acoustics.

IEEE Transactions on Microwave Theory and Techniques, MTT-21, April 1973, special joint issue on microwave acoustic signal processing.

Proceedings IEEE, 64, May 1976, special issue on surface acoustic wave devices and applications.

Microwave Journal 13, March 1970, special issue on microwave acoustics.

There is also the series:

Physical Acoustics, vol. 1 to 11, W.P. Mason and R.N. Thurston eds., Academic Press, New York, 1964-75, the main aspects of elastic waves are presented by separate authors.

Finally there should be mentioned a recent book:

B.A. Auld. Acoustic fields and waves in solids, vol. I and II. New York: Wiley - Interscience (1973).

Index

Symbols

c : elastic stiffness
e : piezoelectric constants
V : bulk wave velocity
K : bulk wave electromechanical coupling coefficient
V_s: surface wave velocity
K_s: surface wave electromechanical coupling coefficient
ρ : mass density
M : acoustooptic figure of merit
ε : dielectric constants
n : optical index
p : acoustooptic constants

S

T